17th Aceh International Workshop and Expo on Sustainable Disaster Recovery (AIWEST-DR 2025)

E3S Web of Conferences Volume 651 (2025)

London, United Kingdom
20-22 August 2025

ISBN: 979-8-3313-2723-1

Printed from e-media with permission by:

Curran Associates, Inc.
57 Morehouse Lane
Red Hook, NY 12571

Some format issues inherent in the e-media version may also appear in this print version.

For additional information, please contact EDP Sciences – Web of Conferences at the address below.

EDP Sciences – Web of Conferences
17, Avenue du Hoggar
Parc d'Activité de Courtabœuf
BP 112
F-91944 Les Ulis Cedex A
France

Phone: +33 (0) 1 69 18 75 75
Fax: +33 (0) 1 69 28 84 91

contact-edps@webofconferences.org

Additional copies of this publication are available from:

Curran Associates, Inc.
57 Morehouse Lane
Red Hook, NY 12571 USA
Phone: 845-758-0400
Fax: 845-758-2633
Email: curran@proceedings.com
Web: www.proceedings.com

TABLE OF CONTENTS

Statement of Peer review

In submitting conference proceedings to E3S Web of Conferences, I certify to the Publisher that I adhere to the **Policy on Publishing Integrity** of the journal in order to safeguard good scientific practice in publishing.

1. All articles have been subjected to peer review administered by the proceedings editors.
2. Reviews have been conducted by expert referees, who have been requested to provide unbiased and constructive comments aimed, whenever possible, at improving the work.
3. Proceedings editors have taken all reasonable steps to ensure the quality of the materials they publish and their decision to accept or reject a paper for publication has been based only on the merits of the work and the relevance to the journal.

The 17th Aceh International Workshop and Expo on Sustainable Disaster Recovery (AIWEST-DR 2025), August 20-22, 2025, University College London-United Kingdom

Proceedings editor(s):

Chief Editor: Dr. Harapan
Members of Editor:
Dr. Alfi Rahman
Dr. Katerina Stavrianaki
Dr. Elizabeth Maly
Dr. Rina Suryani Oktari
Dr. Rizanna Rosemary
Dr. Syafruddin Chan
Prof. Dr. Syamsidik
Dr. Yunita Idris

Date and editor's signature. October 4th, 2025

Enhancing Dam Infrastructure Safety through CFD-Based Spillway Assessment

Muhammad Hafiz Aslami[1], *Eldina* Fatimah[2,*], *Ikhlasul* Amal[3], *Davikia* Nastyakshan[4], *Cut Dwi* Refika[2], *Nina* Shaskia[2]

[1]Civil Engineering Department, Faculty of Engineering, Binus University, 11480, Indonesia
[2]Civil Engineering Department, Faculty of Engineering, Syiah Kuala University, 23111, Indonesia
[3]Master Program, Civil Engineering Department, Faculty of Engineering, Syiah Kuala University, 23111, Indonesia
[4] Bachelor Program, Civil Engineering Department, Faculty of Engineering, Syiah Kuala University, 23111, Indonesia

Abstract. Spillways play a crucial role in dam safety by controlling the release of excess floodwater. However, high discharge rates through these structures may lead to adverse hydraulic effects, such as cavitation and slug flow. This study examines these risks in the side spillway of the *Krueng Kluet* Dam, South Aceh, using a Computational Fluid Dynamics (CFD) approach. The model evaluates the flow conditions under operational discharges through velocity and pressure field simulations. The results indicate cavitation risk levels of 1 and 2—corresponding to no cavitation and potential cavitation—across most of the spillway. However, localized cavitation damage was detected near the downstream end of the chute at approximately 553 m and 580 m. These areas exhibited cavitation index values of 0.24 and 0.20, respectively. No signs of slug flow were observed in any section of the flow path. As a preventive measure, installing aerators or redesigning the chute with a stilling basin is proposed to reduce downstream flow velocity. This research provides design recommendations to improve spillway systems' hydraulic performance and safety and demonstrates the utility of CFD modelling for early identification of failure-prone zones.

1 Introduction

Flood events pose substantial hydrological and socio-economic challenges, particularly in regions with limited floodplain management and infrastructure capacity. In response to the floods, the Indonesian government initiated the development of the *Krueng Kluet* Dam, a multipurpose infrastructure in South Aceh Regency. The dam integrates several functions, including irrigation support, hydropower generation, aquaculture, tourism development, and flood regulation [1, 2]. Its presence is central to the region's water resource allocation and hydraulic safety strategies.

*Corresponding author: eldina.fatimah@usk.ac.id

The dam includes a side spillway comprising a crest, side channel, transition zone, and chute. This configuration is designed to safely convey excess discharge during peak hydrological conditions, ensuring the structural and operational reliability of the dam [3]. A physical hydraulic model of the structure was constructed at a scale of 1:60 to simulate the behavior of flow throughout the spillway system under various design discharges [4]. The results confirmed the overall functionality of the structure but indicated localized flow anomalies, particularly cavitation and transient surging, in the chute channel. To address these risks, design modifications were introduced, including changes in sill location, channel slope adjustment, and transition floor extension [4].

While physical modeling provides valuable insights into overall spillway performance, it remains limited in resolving localized pressure fluctuations and transient hydraulic responses. Accordingly, numerical methods such as Computational Fluid Dynamics (CFD) have been integrated into spillway assessments. Due to its mesh-free, Lagrangian framework, the Smoothed Particle Hydrodynamics (SPH) approach has demonstrated significant capability in simulating free-surface flows with complex geometries [5,6]. SPH enables the detailed prediction of velocity and pressure fields, making it suitable for assessing critical hydraulic phenomena that may compromise structural safety.

Cavitation remains a significant concern in high-velocity spillway flows. It arises when local static pressures fall below vapor pressure, forming vapor bubble and implosive collapse. This process can erode concrete surfaces and reduce structural durability [7]. Slug flow, which occurs in long chute channels, represents another risk as it creates unsteady surging regimes that impose uneven hydrodynamic forces and potential vibration effects [8,9]. Both phenomena underscore the need for predictive modeling and design interventions that minimize the likelihood of hydraulic instabilities.

In this context, the present study applies SPH-based CFD modeling to evaluate the hydraulic response of the *Krueng Kluet* Dam spillway under multiple discharge scenarios. The analysis emphasizes pressure distribution, velocity profiles, cavitation index, and flow regime stability. The findings are expected to provide refined insights into vulnerable sections of the spillway and support design improvements that strengthen dam safety under operational and extreme hydrological conditions.

2 Materials and methods

2.1 Smoothed Particle Hydrodynamics

Smoothed Particle Hydrodynamics (SPH) is one of the methods in numerical models, which is a particle-based numerical modeling method. SPH is a particle-type Lagrangian method with information in the form of position, velocity, pressure, and others, and then uses the particles to discretize the Navier-Stokes equations. After being discretized, the particles are considered weakly compressible (WCSPH), and particle updates are based on the interpolation of all particles in the characteristic length h [5]. 2D simulation forms particles in circles, while 3D is in the form of balls with characteristic lengths in the form of h according to simulation needs.

The law of conservation of continuum fluids has been changed from a partial differential form to one that is in accordance with particle-based simulations using integral equations based on interpolation functions that provide estimates of values at specific points. If a function *F(r)* in the domain *Ω* can be approached by an integral approach, such as Equation (1), (2), and (3), where *W* is the kernel function and *r* is the position vector [5].

$$F_{(r)} = \int_{\Omega} F_{(r')} W_{(r - r',\, h)}\, dr' \quad (1)$$

$$F(r_a) \approx \Sigma_b F(r_b)\, W(r_a - r_b, h)\, \Delta v_b \tag{2}$$

$$F(r_a) \approx \Sigma_b F(r_b)\, W(r_a - r_b, h)\, (\frac{m_b}{\rho_b}) \tag{3}$$

$$\frac{dv_a}{dt} = -\Sigma_b m_b (\frac{P_b + P_a}{\rho_b \cdot \rho_a} + \prod_{ab}) \nabla_a W_{ab} + g \tag{4}$$

The Equation above is the discrete procedure of the SPH method, and Figure 1 shows the particle approximation process of the SPH method. The calculation of particle i can be interpolated according to the surrounding particles in the domain. Therefore, Equation (3) can discretize the momentum conservation equation and obtain the discrete acceleration form in Equation (4).

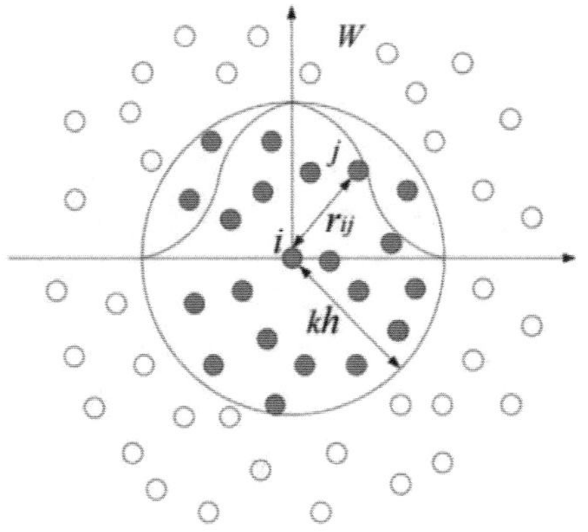

Fig. 1. The interpolation principle of the SPH method includes the kernel and particle approaches.

P is the pressure, v_a is particle a's velocity, ρ is the density, *m* is the particle's mass, *g* is the acceleration due to gravity (g = 9.8 m/s^2), ∏ and *ab* is the given artificial viscosity. Because the particle mass is the same, the fluid density calculation can only be obtained by discretizing the continuity equation through Equation (5).

$$\frac{d\rho_a}{dt} = \Sigma_b m_b v_{ab} \nabla_a W_{ab} \tag{5}$$

In the SPH method, which uses discrete solutions, fluid particles are generally assumed to have weak compressibility. Equation (4) and the Equation of state (5) determine the calculation of pressure particles. Therefore, Equation (6) can explain the calculation of pressure particles.

$$P = B\, [(\frac{\rho}{\rho_0})^{\gamma} - 1] \tag{6}$$

Where $\gamma = 7$ for water, ρ_o = 1000 kg/m^3, $B = c^2_o \rho_o / \gamma$, and $c_o = c(\rho_o) = \sqrt{(\partial P/\partial \rho)}$ when $\rho = \rho_o$, c_o represents the speed of sound.

2.2 CFD Model Setup and Simulation Process

The numerical simulation of the spillway flow was carried out using the Smoothed Particle Hydrodynamics (SPH) method, implemented via the DualSPHysics platform. The kernel function applied in the simulation was the cubic spline or quintic Wendland, known for stability and interpolation accuracy [6].

The 3D spillway geometry was modelled with particles discretized using a particle spacing (Δp) determined based on the smallest geometric feature, as shown in Figure 2 [4,6]. Boundary conditions were defined with fixed-wall particles for structural interfaces, and inlet/outlet zones were adjusted for discharge control.

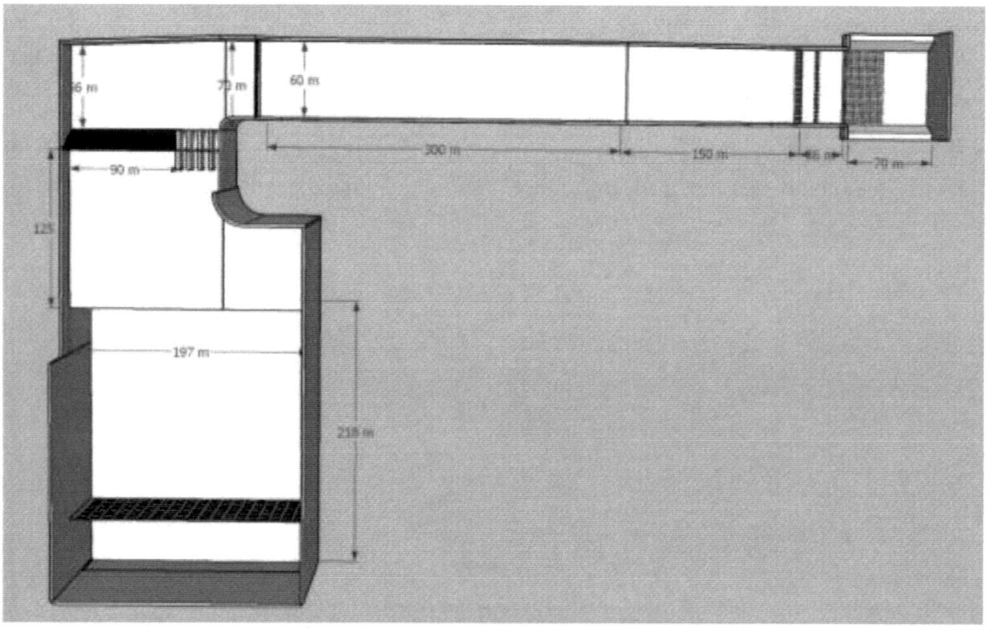

Fig. 2 3D Model Results of the Krueng Kluet Dam [6].

The simulation employed an artificial viscosity formulation based on Monaghan's model to stabilize high-gradient flows and suppress numerical noise. The inflow was based on the design discharge Q from dam planning documents, and the analysis focused on pressure and velocity distributions throughout the spillway, particularly to detect cavitation-prone zones [4,6].

2.3 Cavitation

Cavitation is a process related to surface damage to channel walls, pipes, and blades characterized by noise in a flow system [7]. Cavitation occurs when cavities or 'bubbles' of water vapor and other liberated gases form in flowing water. The flow may carry these to regions of higher pressure where they implode [7]. The occurrence of cavitation damage to a spillway can be predicted using the cavitation index. In this method, to prevent cavitation damage to a spillway, it is required to keep the flow cavitation index more than the cavitation inception index ($\sigma > \sigma_i$) everywhere on the structure.

$$\sigma = \frac{P - P_v}{\rho \frac{v^2}{2}} \tag{7}$$

$$P = P_a + P_g \tag{8}$$

Where σ is the cavitation inception index, P is reference flow pressure, P_a is atmospheric pressure, P_g is pressure on the surface of a structure, P_v is water vapor pressure, ρ is water density, and v is reference flow velocity. The value of the cavitation inception index is determined according to surface irregularities or surface roughness, and various equations have been presented based on research and experiments.

The cavitation damage level for a spillway has been predicted by two parameters: the flow velocity and cavitation index [8]. Five levels have been considered for cavitation damage risk to a spillway. The levels are presented based on the flow velocity and cavitation index (Table 1).

Table 1. Cavitation damage level.

Level	Cavitation damage risk	Intervals of velocity (m/s)	Intervals of cavitation index
1	No cavitation damage	$v \leq 5$	$\sigma > 1$
2	Possible cavitation damage	$5 < v \leq 16$	$0.45 < \sigma \leq 1$
3	Cavitation damage	$16 < v \leq 25$	$0.25 < \sigma \leq 0.45$
4	Serious damage	$25 < v \leq 40$	$0.17 < \sigma \leq 0.25$
5	Major damage	$v > 40$	$\sigma \leq 0.17$

2.4 Slug Flow

Slug flow is a hydraulic phenomenon that must be considered in constructing a chute channel structure. If this phenomenon occurs, it can cause uneven flow so that the resulting hydrodynamic force can endanger the stability of the construction. In addition, due to uneven flow, the flow velocity at the bottom of the chute channel becomes uneven, reducing damping effectiveness in the stilling basin [9].

If the design of the chute channel is more than 30.0 meters long, there is a danger of instability in the flow called slug flow. To control this risk, calculations need to be made with the Vendernikov number (V) and the Montuori number (M) in equations (8) and (9) [10].

$$\underline{V} = \frac{2bv}{3P\sqrt{gd\cos\theta}} \tag{8}$$

$$\underline{M}^2 = \frac{v^2}{gI\cos\theta} \tag{9}$$

Where b is the width of the channel base, *v* is the velocity, *P* is the wetted circumference (m), *g* is the acceleration of gravity (9.8 m/s^2), d is the average water depth, θ is the energy gradient angle, *I* is the average slope of the energy gradient, and *L* is the intended length.

After obtaining the calculated values of the Vendernikov and Montuori numbers, the next step is plotting these values on a graph representing the slug flow criteria, as shown in Figure 3. The graph provides a two-zone classification: the slug flow zone and the no slug flow zone. The diagonal line dividing these two regions serves as a boundary that indicates the critical condition at which the flow regime changes from stable to unstable.

This plotting method allows researchers to determine the flow regime by visually locating the calculated parameters on the graph. If the point falls within the zone of slug flow, it indicates the potential presence of unstable surging conditions, whereas a point located in the zone of no slug flow corresponds to a stable regime. The process transforms dimensionless numbers into a practical diagnostic tool that simplifies the evaluation of flow stability in hydraulic channels.

To support design analysis, several alternatives are commonly discussed in the literature for reducing the likelihood of slug flow [11]. These include adjustments to the channel cross-section, such as narrowing the width to regulate discharge distribution, or reducing the channel bed slope to decrease acceleration. Such considerations are integrated into the methodological framework to guide the assessment and control of slug flow phenomena in hydraulic structures.

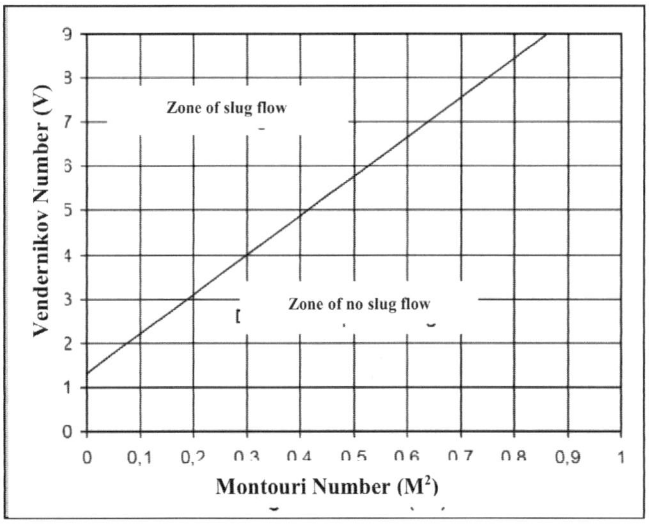

Fig. 3. Slug flow criteria graph [11].

3 Results

The numerical modeling of the *Krueng Kluet* Dam spillway was conducted and subsequently evaluated against analytical calculations for design return periods of Q_{50}, Q_{100}, and Q_{1000}, corresponding to discharge values of 1,210.78 m³/s, 1,495.66 m³/s, and 1,702.32 m³/s, respectively. The comparison focused on the root-mean-square (RMS) values derived from numerical simulations and analytical methods, as presented in Figure 4. To facilitate a consistent comparison, the RMS curves were normalized by the maximum values of the analytical data. The graphical representation illustrates that the RMS values from the numerical approach closely followed the analytical trends for all discharge scenarios, with discrepancies consistently below 10%.

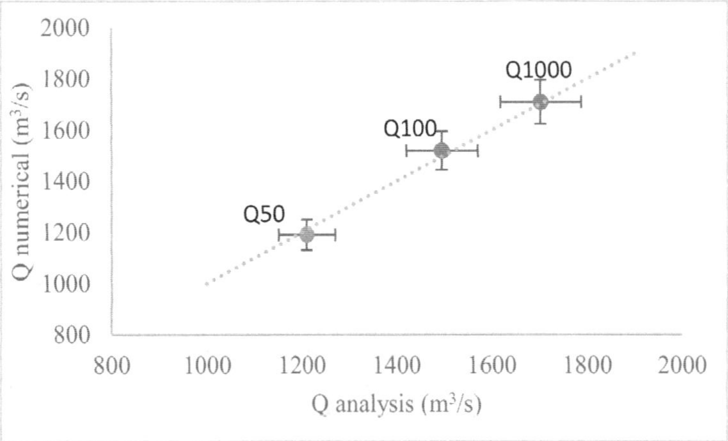

Fig. 4. RMS numerical and analysis results for discharge with 50, 100, and 1000 years return periods.

These results confirm that the numerical model provides a reliable representation of the hydraulic performance of the *Krueng Kluet* Dam spillway. The limited deviation indicates that the applied modeling framework successfully reproduces key flow characteristics and discharge capacities across flood scenarios. In particular, the close alignment with analytical predictions supports the robustness of the numerical approach in capturing flow dynamics under both moderate and extreme flood conditions. This agreement validates the computational method and provides confidence in extending the analysis toward more complex hydraulic phenomena, such as cavitation risk and flow regime stability, which are difficult to capture through purely analytical formulations.

Figure 5 presents the velocity field along the spillway components under the Q_{100} discharge condition (1,495.66 m^3/s). At the crest (Figure 5a), flow velocities ranged between 3.0 and 6.0 m/s and remained uniformly distributed across the spillway width, demonstrating effective crest regulation. In the side channel (Figure 5b), the velocity increased progressively from less than 5.0 m/s near the inlet to 12.0–15.0 m/s downstream, consistent with geometric convergence and gravitational acceleration.

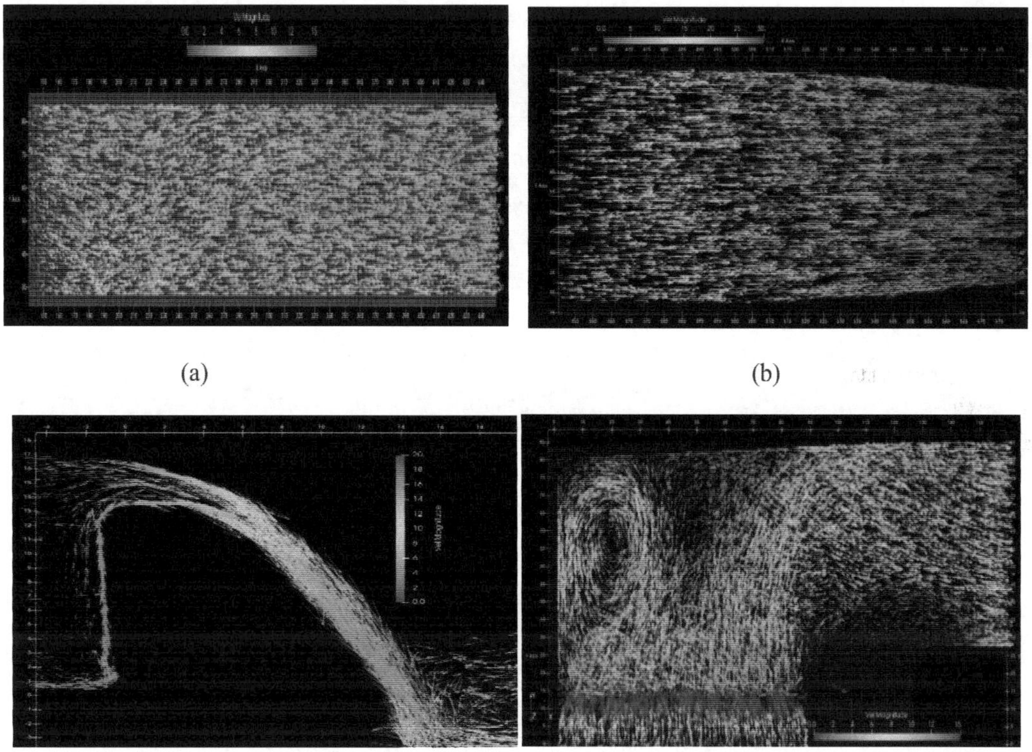

(a) (b) (c) (d)

Fig. 5. Flow pattern through (a) crest of spillway, (b) side channel, (c) transition channel, and (d) chute channel for Q_{100}.

The transition channel (Figure 5c) exhibited the highest velocity gradients, with peak velocities exceeding 15.0 m/s. These rapid accelerations and slope variations correspond to localized pressure drops, in line with cavitation indices of 0.24 and 0.20 observed at 553 m and 580 m. This confirms that the transition–chute section is the most critical zone concerning cavitation risk. Finally, the chute and stilling basin (Figure 5d) displayed jet velocities of 15.0–20.0 m/s entering the basin, followed by large vortices and recirculation zones with velocities below 5 m/s. This significant energy dissipation indicates that the stilling basin effectively stabilizes downstream conditions and reduces erosion risk.

Cavitation index analysis requires pressure and flow velocity data. The recapitulation of the cavitation index calculation can be seen in Table 2. Based on the results of the analysis that has been carried out, it was found that the cavitation index value is at risk level 1 and 2, where the building does not or is likely to experience cavitation damage. A solution for cavitation hazards at risk level 2, requiring surface treatment, such as applying cement to the surface, and smoothing the channel surface [12].

Table 2. Recapitulation of index cavitation on the spillway.

Distance (m)	Structure	Q_{50}	Q_{100}	Q_{1000}	Distance (m)	Structure	Q_{50}	Q_{100}	Q_{1000}
0,52	Crest of spillway	51,5	11,8	8,8	296,80	Transition channel	15,4	2,7	1,9
1,56		50,4	12,5	8,9	328,30		17,6	4,2	2,0
3,06		9,9	9,7	6,6	354,10		8,4	2,0	1,7
4,26		3,7	9,2	7,9	383,80		15,0	3,7	1,8
5,80		2,3	1,6	2,0	409,60		10,2	6,2	1,9
7,15		4,7	1,3	2,4	421,30		9,8	7,5	1,7
8,40		2,4	1,9	2,1	433,30		19,5	3,4	1,9
9,80		1,8	1,7	1,3	439,00		16,8	3,4	1,7
11,15		1,3	1,6	2,6	441,40		9,9	2,9	1,7
12,35		1,9	2,4	3,8	444,70		15,2	3,4	1,7
13,45		12,1	9,1	11,8	454,20	Chute channel	6,8	6,5	1,4
10,92	Side channel	39,6	32,0	28,3	463,90		6,9	6,5	1,2
44,88		71,4	68,5	16,7	475,00		5,9	3,0	0,8
78,60		92,5	9,6	6,9	486,40		2,0	2,2	0,7
107,25		83,7	59,7	24,9	503,40		1,4	1,1	0,5
149,50	Transition channel	57,3	7,9	3,2	519,80		1,2	0,7	0,3
174,10		19,6	3,4	2,5	536,50		0,6	0,3	0,3
193,60		16,6	4,3	2,4	553,10		1,0	0,4	0,2
220,90		16,6	2,7	2,3	565,30		1,1	0,5	0,3
245,50		18,3	3,2	2,2	579,70		1,4	0,9	0,2
276,40		23,3	3,0	2,5					

Based on the results presented in Table 2, cavitation was observed at the downstream end of the chute channel under the Q_{1000} discharge condition, specifically at distances of 553m and 580 m. The cavitation index values at these locations were 0.24 and 0.20, respectively, indicating a Level 3 cavitation risk. Mitigation measures are necessary to minimize structural damage when cavitation is detected in a spillway channel [12]. Recommended interventions include installing aerators or modifying the stilling basin to reduce velocity at the chute toe by generating controlled backflow.

The occurrence of cavitation at these locations is closely associated with abrupt changes in channel slope and rapid flow acceleration, which produce localized pressure drops below the vapor pressure of water. These hydraulic and geometric discontinuities create favorable conditions for vapor bubble formation and subsequent collapse. The implosive collapse of these bubbles can result in localized erosion and surface deterioration, posing long-term risks to the structural integrity of the spillway.

In addition to cavitation, the potential for slug flow was assessed using the Vendernikov and Montuori numbers. This evaluation required data on mean flow depth and velocity in the side, transition, and chute channels. The energy loss (H_f) and flow angle (Θ) were calculated from these values to characterize flow conditions. The computed Vendernikov and Montuori numbers were plotted, as shown in Figure 6, to determine whether the results fell within the slug flow or no-slug-flow zones. The findings demonstrated that all side, transition, and chute channel points fell within the no-slug-flow zone, confirming hydraulic stability against surging flow conditions.

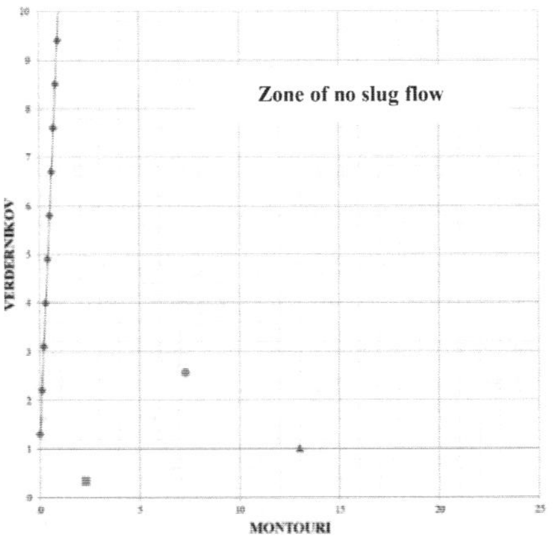

(a)

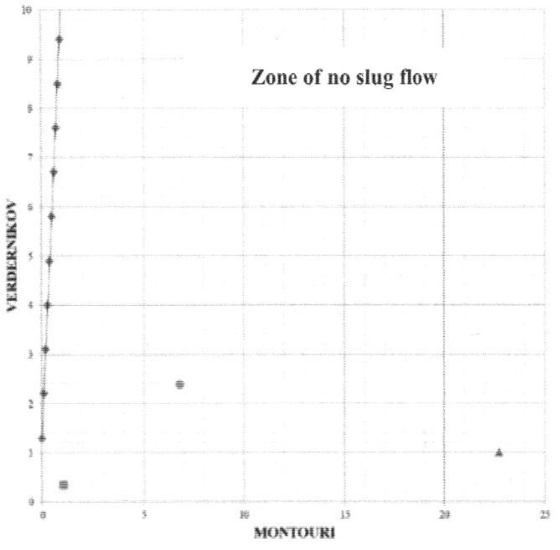

(b)

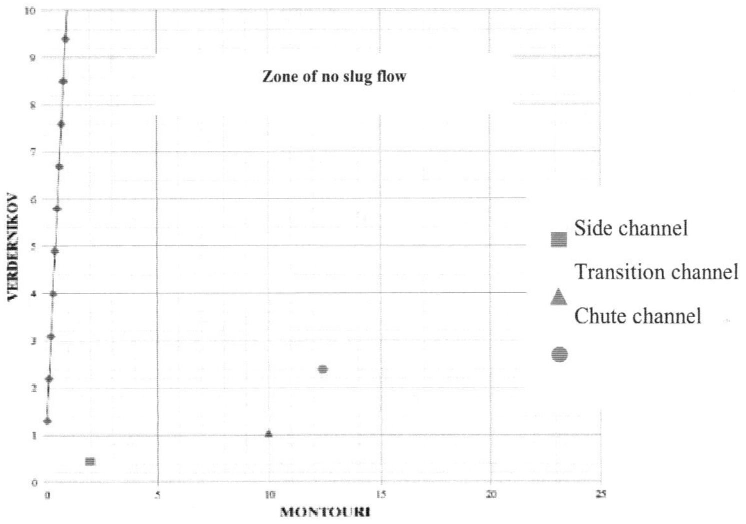

(c)

Fig. 6. Slug flow analysis graph for (a) Q_{50}, (b) Q_{100}, and (c) Q_{1000}.

The side channel recorded Montuori and Vendernikov numbers of approximately 0.35 and 11.5, respectively, placing it firmly in the no-slug-flow zone and reflecting deep, uniform flow. The Montuori number in the transition channel increased to about 4.8, while the Vendernikov number decreased to around 3.2, indicating accelerated but stable flow. The chute channel exhibited the highest Montuori value of about 12.5 with a corresponding Vendernikov number near 2.3, characteristic of high-velocity jet flow. Despite these variations, all calculated points remain within the no-slug-flow zone.

This progression—from high V and low M^2 in the side channel to low V and high M^2 in the chute—corresponds to the expected hydraulic behavior of a side spillway system. Previous studies have documented similar trends: Rajaratnam [13] and Khatsuria [14]. They emphasized that properly designed transition and chute geometries prevent slug flow formation. At the same time, Wibowo and Dermawan [15] demonstrated through numerical modeling of a side spillway that CFD analysis provides reliable predictions of flow regime stability. The alignment of the present results with these validated studies underscores the accuracy of the numerical modeling approach in characterizing flow regime transitions in the Krueng Kluet Dam spillway.

4 Conclusion

This study employed CFD-based numerical modeling to assess cavitation and slug flow risks in the Krueng Kluet Dam side spillway. The results show that most spillway sections remain at cavitation risk levels 1–2. However, at the downstream chute under Q_{1000}, cavitation risk level 3 was identified at 553 m and 580 m with index values of 0.24 and 0.20, requiring mitigation such as aerators or a sinking-basin stilling basin to reduce flow velocity. In contrast, slug flow analysis using Vendernikov and Montuori numbers confirmed that the side channel, transition, and chute lie within the no-slug-flow zone, demonstrating hydraulic stability. These findings validate the effectiveness of CFD modeling in identifying localized risks and provide a basis for targeted design improvements to enhance spillway safety and dam reliability.

References

1. D. C. Froehlich and D. G. Diaz, Dam Safety—Living with the Risk of Failure. A Syst. Eng. Approach to Disaster Resilience. Lect. Notes Civ. Eng. **205** (2022). https://doi.org/10.1007/978-981-16-7397-9_30
2. V. Variadi, K. Iqbal, F. Mufti, and R. Adriman, Keureuto Dam Operation for Flood Control System. E3S Web Conf. **476**, 01049 (2024). https://doi.org/10.1051/e3sconf/202447601049
3. Q. Aini, E. Fatimah, and A. Azmeri, Kajian Panjang Loncatan Hidrolik Pada Model Fisik Kolam Olak Bendungan Krueng Kluet Kabupaten Aceh Selatan. Teras J. **12**, 1, pp. 81–92 (2022). http://dx.doi.org/10.29103/tj.v12i1.619
4. S. Kamal, Cavity and Slug Flow on The Physical Model Chutes Channel of Krueng Kluet Dam, South Aceh Regency, Bachelor's Thesis, Syiah Kuala Univ, Banda Aceh, Faculty of Engineering (2021).
5. A. Trimulyono, H. Hashimoto, and A. Matsuda, Experimental Validation of Single- and Two-Phase Smoothed Particle Hydrodynamics on Sloshing in a Prismatic Tank. J. Mar. Sci. Eng. 7, 8, 247 (2019). https://doi.org/10.3390/jmse7080247.
6. Amal, Hydraulic Modeling of Krueng Kluet Dam, South Aceh Regency Using Computational Fluid Dynamic, Master Thesis, Syiah Kuala Univ, Banda Aceh, Faculty of Engineering (2024).
7. Y. Sutopo, Kavitasi di Dasar Saluran Curam, (Andi, Yogyakarta 2014).
8. E. F. Kermani, G. A. Barani, and M. Ghaeini-Hessaroeyeh, Investigation of Cavitation Damage Levels on Spillways. World Appl. Sci. J. **12**, 1, pp. 73–78 (2013). https://doi.org/10.5829/idosi.wasj.2013.21.1.2630
9. C. Wibowo, V. Dermawan, and P. T. Juwono, Pemodelan Numerik Pelimpah Samping Waduk Telagawaja Bali Kabupaten Karangasem Dengan Analisa Komputasi Fluida Dinamis. J. Tek. Pengairan. Univ. Brawijaya, Malang. **7**, 2, pp. 184-192 (2016). Available at:

https://jurnalpengairan.ub.ac.id/index.php/jtp/article/view/294

10. Departemen Pekerjaan Umum, Standar Perencanaan Bangunan Utama (KP-02), (Galang Persada, Jakarta, 2013).
11. C. A. Sulistiono, Kajian Hidraulika Pelimpah Samping Bendungan Raknamo Kupang NTT dengan Uji Model Fisik 1:50, Bachelor's Thesis, Universitas Brawijaya, Malang, Faculty of Engineering (2016).
12. J. A. Kells and C. D. Smith, Reduction of cavitation on spillways by induced air entrainment. Can. J. Civ. Eng. **18**, 3, pp. 358–377 (1991). https://doi.org/10.1139/l91-047
13. Rajaratnam, N. (1990). *Skimming flow in side-weir channels. Journal of Hydraulic Research*, 28(4), 477–493. https://doi.org/10.1080/00221686900949904
14. Khatsuria, R. M. (2005). *Hydraulics of Spillways and Energy Dissipators*. CRC Press, Taylor & Francis Group. https://doi.org/10.1201/9780203020593
15. Wibowo, A. C., & Dermawan, V. (2016). *Pemodelan numerik pelimpah samping Waduk Telagawaja Bali Kabupaten Karangasem dengan analisa komputasi fluida dinamis. Jurnal Teknik Pengairan (JTP)*, 7(2), 126–136. https://jurnalpengairan.ub.ac.id/index.php/jtp/article/view/294

Potential Economic Losses due to Earthquakes Induced Building Damages in Ambon-Indonesia: An Event Based Probabilistic Seismic Hazards Assessment Approach

Aulia Khalqillah[1], *Syamsidik*[1,2,*], *Ferad* Puturuhu[3], *Teuku* Andri Renaldi[1], *Muhammad* Daffa Al Farizi[4], *Anawat* Suppasri[4], and Al Mukarramah[5]

[1]Tsunami and Disaster Mitigation Research Center (TDMRC), Universitas Syiah Kuala, Banda Aceh, 23111, Indonesia
[2]Civil Engineering Department, Faculty of Engineering, Universitas Syiah Kuala, Banda Aceh, 23111, Indonesia
[3]Faculty of Agriculture, Universitas Pattimura, Ambon, 97234, Indonesia
[4]International Research Institute of Disaster Science (IRIDeS), Tohoku University, Sendai, Miyagi Prefecture, 9800845, Japan
[5]Civil Engineering Department, Faculty of Engineering, Universitas Teuku Umar, Meulaboh-Aceh, 23651, Indonesia.

Abstract. Ambon, in the eastern part of Indonesia, has historical records of damaging earthquake events. In the last ten years, at least there were two times of earthquakes that caused severe damages on the city. This research was aimed at estimating potential economic losses due to earthquake induced building damages in Ambon. There were over 80,000 buildings included in the analysis based on a series of surveys conducted in 2023 and in 2024. The earthquake hazards were calculated using Event based Probabilistic Seismic Hazards Analysis (PSHA). The damage ratios were computed using Global Earthquake Model (GEM)'s vulnerability curves. A stochastic event set method which later produce Event Loss Table was introduced to calculate Probable Maximum Losses (PML) and Average Annual Losses (AAL) of buildings in Ambon. The sources of the seismic hazards were taken from two segments, namely North Banda Segment and South Banda Segment. There were 16,910 earthquake events ranging from 6.55 Mw until 9.05 Mw included in the analysis. The intensity values for the buildings were computed based on Peak Ground Acceleration (PGA) and Spectral Acceleration (SA). This study found that the average annual losses (AAL) for the city is 249,964 USD per year due to the earthquake or about 1.56% comparing to the AAL of Maluku Province provided by GEM for latest version in 2023. The highest loss would be from residential type buildings that could reach probable maximum loss (PML) of around 87 million USD.

*Corresponding author: syamsidik@usk.ac.id

1 Introduction

Earthquakes is one of important geological hazards in Indonesia. Some major earthquakes also caused tsunamis to the coastal areas as in the case of the 2004 Aceh Tsunami that hit around 21 countries in the Indian Ocean. Yet, comparing to western part of Indonesia, the studies on the two types of hazards are not well conducted for the eastern part of Indonesia including in Ambon of Maluku Province [1]. This limited research in eastern Indonesia has led to gaps in understanding earthquake and tsunami risks, even though disaster trends mirror those in the west. For instance, in 2018, earthquake, tsunami, and liquefaction devastated Palu Bay, Central Sulawesi. In 2019, a series of earthquakes hit Ambon Island, damaging thousands of buildings and injuring hundreds. The complex seismic setting in eastern Indonesia makes risk assessment especially challenging, with some earthquake sources potentially causing tsunamis that threaten Ambon [2].

The building's qualities contribute to the potential damage caused by earthquakes. Previous study has shown that the majority of buildings in Ambon are confined-masonry types, predominantly 1-storey buildings [3]. They are mostly for residential purposes. In the latest case of earthquakes in Cianjur, West Java, in 2022, most of the damaged buildings were attributed to the use of weak materials and a lack of structural strengthening in residential-type buildings. However, there is a limited number of studies investigating the buildings' vulnerability in the eastern part of Indonesia due to the potential impacts of earthquakes.

Annually, Indonesia loses about 4.5 billion rupiah due to the impacts of disasters. The most frequent impacts were reported due to the effects of earthquakes. Despite this, there remains a limited number of methods developed to assess potential losses from damaged buildings resulting from earthquake impacts. This research aimed to provide earthquake-loss estimations based on the characteristics of buildings in Ambon, one of the important cities in the eastern part of Indonesia. We started with building classifications and later combined them with the probabilistic seismic hazard analysis. The results are expected to fill gaps in building vulnerability assessments in eastern Indonesia, combined with seismic hazard analysis.

2 The Study Area

Ambon is a city situated in Ambon Island of Maluku Province, and the city has a long bay dividing the city into north and south bay (see **Fig. 1**). The bay has also two parts, i.e., outer bay and inner bay, where the outer bay has a deeper bathymetry compared to the inner bay. The north and the south bays are connected by a low-lying area at the eastern part named Passo and a bridge named Merah Putih Bridge. The South Bay has a more dense population compared to the north. Meanwhile, both the south and north bays are also characterized by steep topographic slopes. **Fig. 2** shows photos taken from the southern part of the Ambon Bay.

In the last decade, there were a number of major earthquakes that caused severe damage on the city. The latest one was a 6.5 Mw earthquake that occurred on September 26, 2019. The epicentre is located approximately 42 km northeast of Ambon Bay. The earthquake caused more than 2,600 unit of building damaged [4]. Around 440 unit of the damaged buildings are in Ambon. Forty-one people were killed due to impacts of the earthquakes. It took more than a week to convince the affected community to return to their homes due to frequent aftershocks.

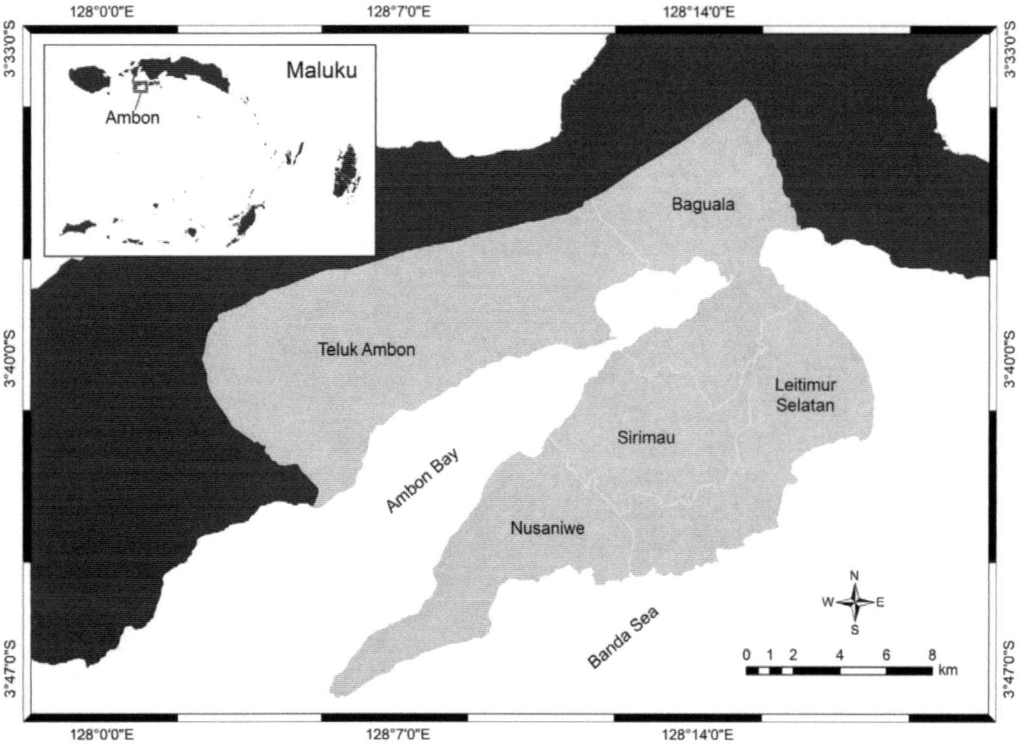

Fig 1. The Ambon Bay: the study area.

Fig 2. Examples of buildings in Sirimau sub-district of Ambon (left). Buildings located in Teluk Ambon sub-district (right). The photos were taken in September, 2024.

3 Methodology

3.1 Buildings Survey and Classifications

Buildings survey in Ambon was performed between July 2023 until June 2024. The survey involved four trained surveyors to identify types, number of storeys, coordinates, and functions of the buildings. There were equipped with electronic devices that enabled them to locate the targeted buildings and online forms for the building classifications. The structural classifications of the buildings followed Global Earthquake Model that has been defined for exposure's Indonesia [5]. **Fig. 3** shows the classification of the buildings and occupancies. Calculations of the building values were based on building unit price per square meter published by the Government of Ambon in 2023 and in 2024.

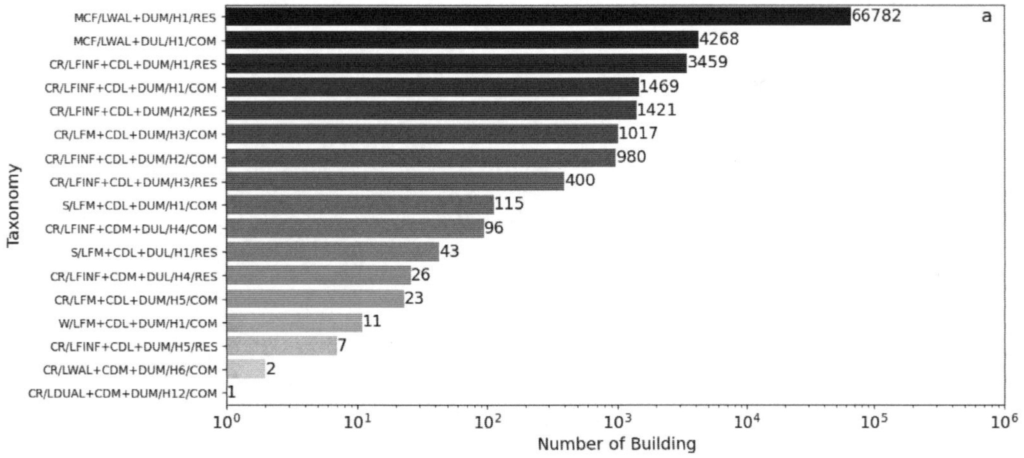

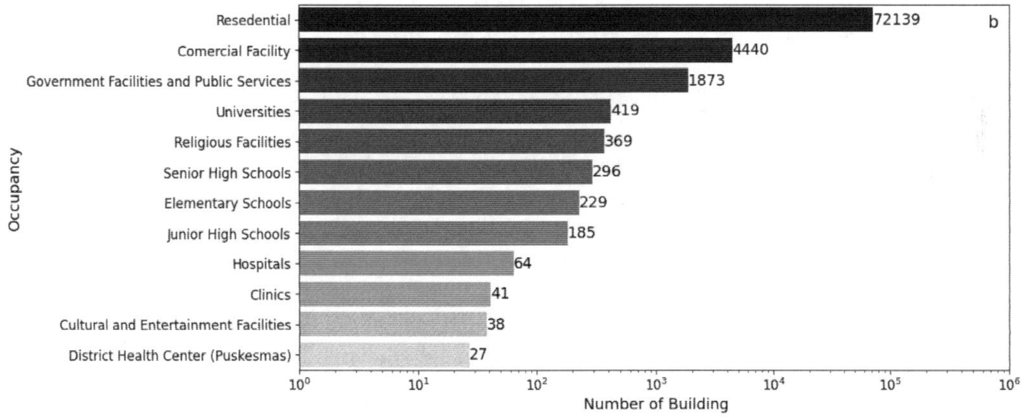

Fig 3. (a) Type of taxonomies used for exposure's taxonomy based on Global Earthquake Model (GEM) where CR=Reinforced-Concrete, LDUAL = Lateral-Dua Frame, LFINF = Lateral-infilled frame, LFM = Lateral-Moment Frame, LWAL= Lateral Load Resist System Wall, CDM = Moderate Code, CDL = Low Code, DUM = Moderate Ductility, Hi = i- number of storey, COM =Commercial/Public, RES = Residential, and (b) number of building for each occupancy category in Ambon City.

3.2 Earthquake-Losses Estimations

We conducted the analysis by grouping into three stages: hazard, exposure, and risk. Each stage has its own steps. All the processes are illustrated in **Fig. 4**. The first stage of hazard preparation done by passing several steps; retrieved earthquake catalogue data, declustered catalogue data, estimated the occurrence rates (a-value and b-value), and estimated the ground motion value. The earthquake catalog data have been downloaded from United State Geological Survey (USGS) catalog during 1985 – 2021 (https://earthquake.usgs.gov/earthquakes/search/). There are 13.058 earthquakes recorded that have been reviewed in this catalog. The earthquake magnitude types are non-uniform, then it should be uniformed to the moment magnitude scale (M_w) by using empirical relationship from [6]. Since the earthquakes distribution are composed by three earthquake types; foreshock, mainshock, and aftershock, we used mainshock only for the ground motion analysis due to the independent consideration effect.

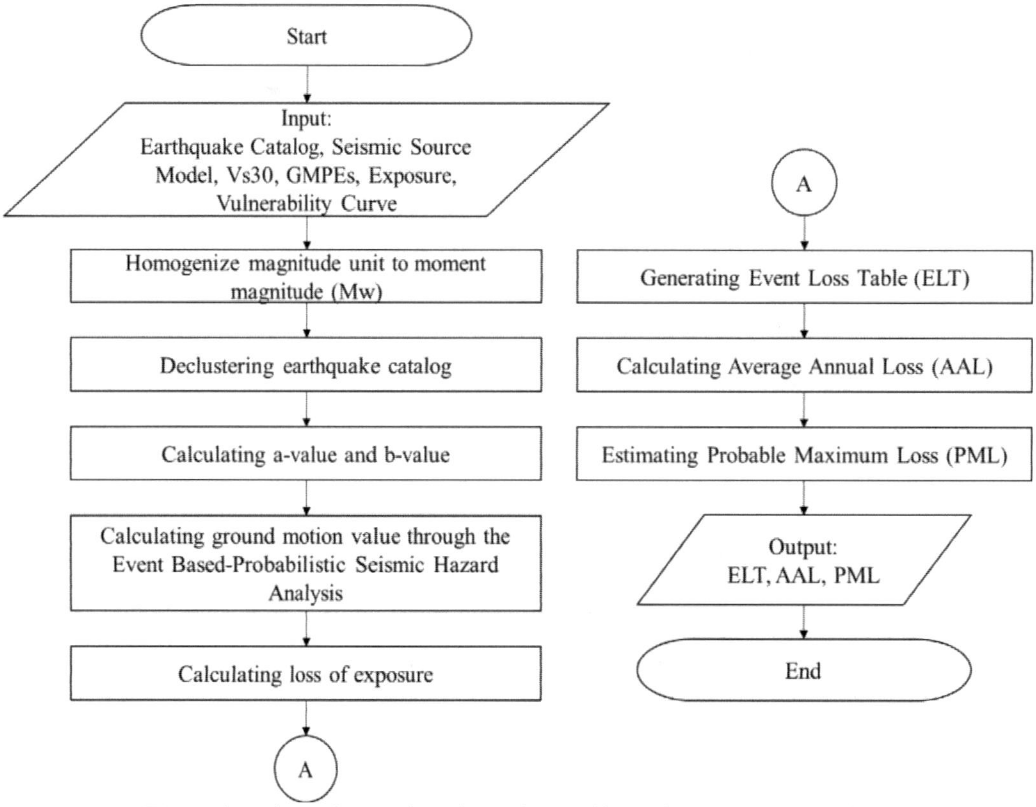

Fig 4. Flowchart of ground motion value and loss of exposure calculation.

We utilized delustering method of Reseanberg's algorithm [7] through the ZMAP MATLAB® package software [8] to obtain the mainshock. The mainshock statistically tested by using Kolmogorov-Smirnov test to evaluate the earthquake distribution that should fit to the Poisson distribution. Furthermore, this mainshock catalog was used to calculate the occurrence rates in terms of a-value and b-value for two megathrust segments as seismic source models of subduction interface that have earthquake distribution greater than 40 Km depth: North Banda Sea and South Banda Sea. The b-value was calculated by using the Maximum-Likelihood [9] as written in Equation 1, where the $\log e$ is equal to 0.4343, $\overline{M}$ is the average of magnitude and M_o is the minimum magnitude from the earthquake distribution for each segment. The M_o was estimated through the highest number of magnitudes of the non-cumulative distribution. Then, the a-value was calculated by using the Equation 2 which is another term of Gutenberg-Richter's relationship [10],

$$b = \frac{\log e}{\overline{M} - M_o} \quad (1)$$

$$a = \log_{10} N(M \geq m) + b \quad (2)$$

where the $\log_{10} N(M \geq m)$ is the cumulative number of earthquakes when the magnitude greater than and equal to minimum magnitude. These parameters then were fitted into the Gutenberg-Richter's relationship as written in Equation 3,

$$\log_{10} N(M \geq m) = a - bM \quad (3).$$

Moreover, we utilized four Ground Motion Prediction Equations (GMPEs) that have been applied in the national hazard maps of Indonesia [11] to estimate the ground motion value in terms of Peak Ground Acceleration (PGA) and Spectral Acceleration (SA) for the seismic source model of subduction interface that were arranged to the logic tree as shown in **Fig 5**. It was considered to the full realization. The calculation was stochastically sampled by using Monte Carlo algorithm. It created stochastic event sets (SES) through the relationship as written in Equation 4 for given time span T. Then, the number of occurrences given a T was simulated by the sampling process of probability distribution of $P_{rup}(k|T)$ for each rupture generated by seismic source models from the logic tree that have been configured, where the k is the number of occurrences and rup is the simulated rupture. In this analysis, the T was configured to 1 year and 1000 for stochastic event set per logic tree.

$$SES\ (T) = \{k\ x\ rup, k \sim P_{rup}(k|T)\ \forall\ rup\ in\ Src\ \forall\ in\ SSM\} \qquad (4)$$

We also considered the site condition effects by using Vs30 slope from USGS [12] sampled for 50 districts site in Ambon City. Finally, all these processes were simulated by using Event-Based Probabilistic Seismic Hazard Analysis (EB-PSHA) through the OpenQuake engine [13]. Finally, we simulated the ground motion value for magnitude 6.55 Mw to 9.05 Mw.

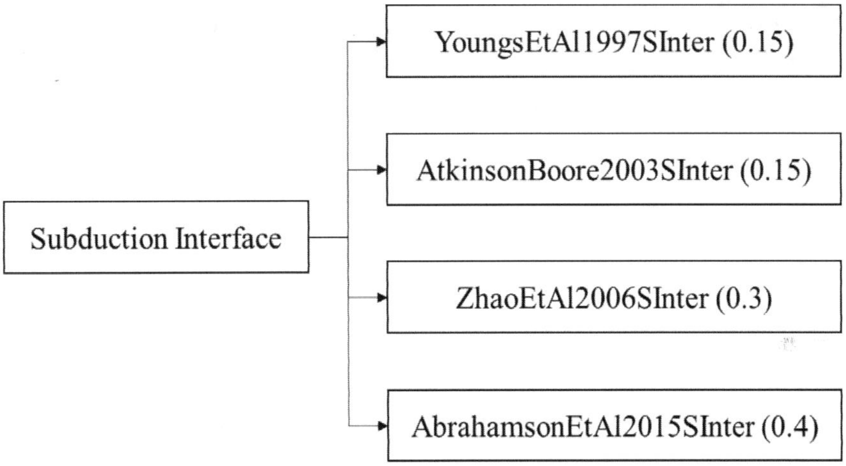

Fig 5. Logic tree of the Ground Motion Prediction Equations (GMPEs) for the subduction interface as the seismic source model.

The second stage is the exposure preparation. We digitized about 80.120 exposures in Ambon City that consist of 12 occupancies as shown in the spatial distributions in **Fig 6**. The taxonomy and vulnerability curve for each exposure was configured to 17 taxonomies that follow by Global Earthquake Model's (GEM) term [5] as shown in **Fig 3(a)**. The exposure's price or exposure valuation was estimated by using Highest Standard Unit Price (*SHST*) per meter square from the regional government of Ambon City multiplied by the buildings' footprint area and number of storeys of the exposure. The exposure valuation is then called as reconstruction value.

The third stage is the loss (risk) calculation. The losses of each exposure were calculated by multiplication between the vulnerability as a function of corresponding ground motion and reconstruction value for each event generated from the previous stage. The aggregate potential loss of exposures was estimated in terms of Average Annual Loss (AAL) and Probable Maximum Losses (PML) through the Event Loss Table (ELT). The AAL and PML were calculated by using Equation 5-7, respectively

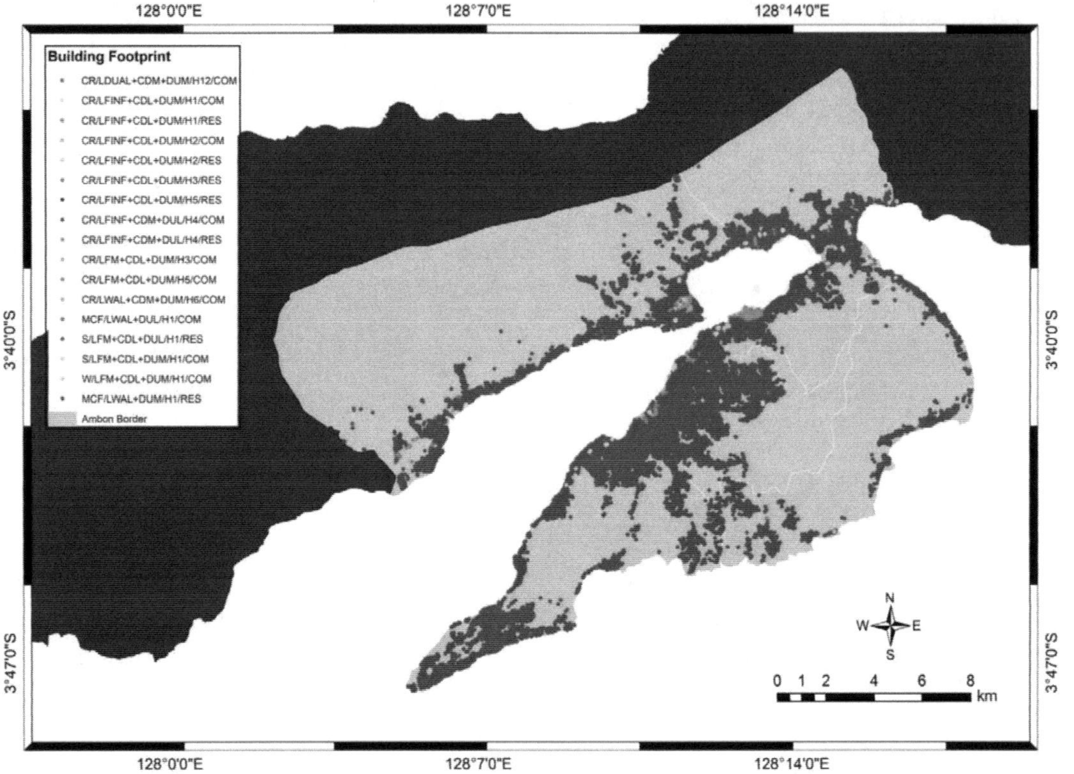

Fig 6. Distribution of type of buildings in Ambon as surveyed between 2023-2024. The classifications follow Global Earthquake Model (GEM).

$$AAL = \sum_i p_i \, L_i \tag{5}$$

$$PML\ (L_i|T_i) = 1/OEP \tag{6}$$

where the OEP is the occurrence exceedance probability,

$$OEP = 1 - \prod_{j=1}^{i-1}(1 - p_i) \tag{7}$$

the p_i is the annual probability of occurrence and L_i is the associated loss for each event i. The AAL and PML are [14].

4 Result

The potential losses of exposure in Ambon City due to the earthquake induced by two subduction interfaces; North Banda Sea and South Banda Sea have been estimated models through the EB-PSHA approach. More than 10.000 unique events are generated from the stochastic process, with each event corresponding to a specific loss. As shown in **Fig. 7**, the simulation shows the highest loss (PML) for all exposures at around Rp. 29,725 billion, which is equivalent to 1,813 million USD. This loss is corresponding to the more than 1.000.000-year of return period. Furthermore, the potential AAL of all exposures is about Rp. 4,097 million or 249,964 USD. The AAL of Ambon in this calculation has ratio about 0.0156 or equivalent to 1.56% comparing to the AAL of Maluku Province which is about 15.9 million USD, that has been provided by GEM [15]. Additionally, the PML of 250-year return period has been estimated, as indicated by a red dot in **Fig. 7**. The PML for this return period is approximately Rp. 10 billion, equivalent to 615,156 USD. The 250-year of return period was considered due to the minimum reinsurance standard for the natural risks that should be retained, as stipulated by Indonesia Financial Services Authority Regulation (*OJK*), number of 14/POJK.05/2015. These results partially show the earthquake risks in Ambon only based on the two subduction segments. It is important to highlight that there are also numerous active earthquake faults lie on land of the island. Some of them have not been fully investigated.

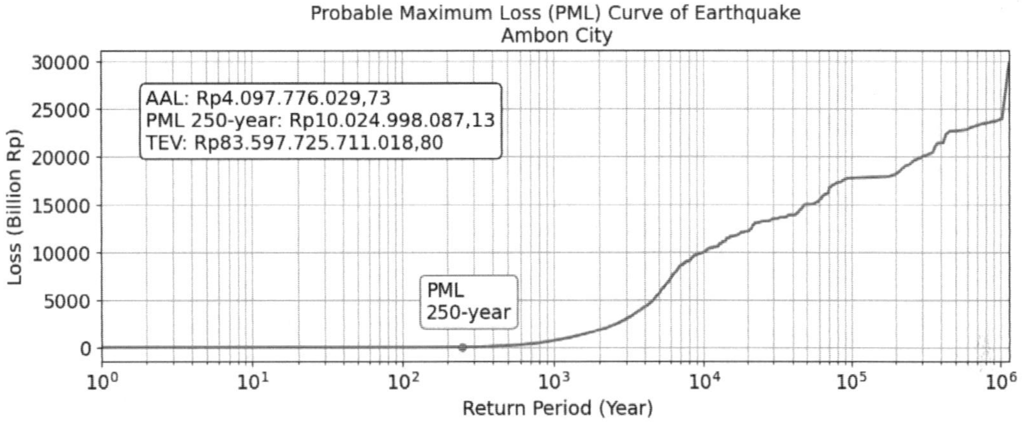

Fig 7. The blue line is the Probable Maximum Losses (PML) of earthquake for the exposure of Ambon City in the range of return period. The red dot corresponds to the PML of 250-year.

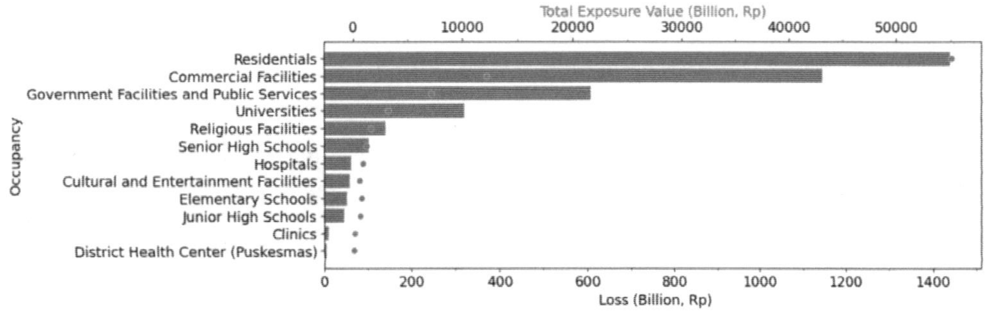

Fig 8. The estimated potential losses of exposure *(gray bar) and total exposure value (red dot)* for each building's occupancy in Ambon City due to the subduction interface model of earthquakes.

The simulation then found that the residential is the highest loss at around Rp. 1.4 trillion, similar to USD 87 million. Furthermore, the district health center (*Puskesmas*) is the type of building with the lowest loss. The Puskesmas-type buildings have potential losses of about Rp. 5.2 billion or about USD 319 thousand (see **Fig. 8** and **Fig. 3 (b)**).

Mitigation of economic losses due to the earthquake's impacts should be undertaken by the Indonesian government through several scenarios. First, it could be done by increasing the government's annual On-Call Budget (OCB). In reality, the OCB not only functions as financial protection due to the impacts of disasters but also for other emergency purposes, such as pandemic crises, as seen between 2019 and 2022. Nevertheless, the OCB could serve as at least a financial retention limit to mitigate the impacts of the earth. Secondly, financial mitigation can be achieved by introducing disaster insurance policies, especially for large public buildings that are vital to the community's needs, such as hospitals, clinics, and educational facilities. Promoting disaster insurance policies remains a challenge in Indonesia. To date, only one city has adopted these policies, namely Padang in West Sumatra Province. Nevertheless, as a disaster-prone country, financial strategies to mitigate the impacts of disasters on properties are urgent. This study could be a milestone in estimating the costs of applying disaster insurance policies in Ambon.

5 Conclusions

Residential-type buildings are the most concerning type of building in terms of the potential impacts of megathrust earthquakes in Ambon. With a 250-year probabilistic earthquake, the city could lose approximately 10 trillion rupiahs (approximately $ 606 million US dollars). Should the government of Ambon City decide to provide full financial protection for all buildings, it would need to allocate approximately 4.1 billion rupiahs per year. The amount could be allocated across various financing schemes, including the government's on-call budget and disaster insurance scheme. The urgency of applying effective and efficient Disaster Risk Financing and Insurance (DRFI) in Indonesia is imminent. This could be part of financial mitigation for large investments on public and private buildings in earthquake-prone areas, such as in Ambon, one of the most earthquake-prone cities in the eastern part of Indonesia.

Acknowledgements. Authors are grateful for research grant under *Penelitian Fundamental Regular* (PFR) funded by Ministry of Higher Education, Science and Technology (*Kemendiktisaintek*) Indonesia Year 2025 Contract No. 113/C3/DT.05.00/PL/2025 titled "Development of A Dashboard for Earthquake and Tsunami Risks for Ambon". Collaborations with Prof. Anawat Suppasri and Muhammad Daffa Al Farizi from Tohoku University was made under SATREPS Building Sustainable System for Resilience and Innovation in Coastal Community (BRICC) Program.

References

[1] F. Løvholt, D. Kühn, H. Bungum, C. B. Harbitz, and S. Glimsdal, "Historical tsunamis and present tsunami hazard in eastern Indonesia and the southern Philippines," *J Geophys Res Solid Earth*, vol. 117, no. B9, Sep. 2012, doi: 10.1029/2012JB009425.

[2] T. A. Renaldi, Syamsidik, A. Suppasri, E. Fatimah, K. Pakoksung, and F. Puturuhu, "Estimating Impacts of Tsunami on Buildings around Ambon Bay of Indonesia," *E3S Web of Conferences*, vol. 447, p. 01006, Nov. 2023, doi: 10.1051/e3sconf/202344701006.

[3] E. Lahcene *et al.*, "Characteristics of building fragility curves for seismic and non-seismic tsunamis: case studies of the 2018 Sunda Strait, 2018 Sulawesi–Palu, and 2004 Indian Ocean tsunamis," *Natural Hazards and Earth System Sciences*, vol. 21, no. 8, pp. 2313–2344, Aug. 2021, doi: 10.5194/nhess-21-2313-2021.

[4] BNPB (Indonesia Disaster Management Agency), "Badan Nasional Penanggulangan Bencana." Accessed: Jul. 22, 2025. [Online]. Available: https://bnpb.go.id/berita/2675-rumah-rusak-teridentifikasi-akibat-gempa-maluku-m-65

[5] L. Martins and V. Silva, "Development of a fragility and vulnerability model for global seismic risk analyses," *Bulletin of Earthquake Engineering*, vol. 19, no. 15, pp. 6719–6745, Dec. 2021, doi: 10.1007/s10518-020-00885-1.

[6] Tim Pusat Studi Gempa Nasional, *PETA SUMBER DAN BAHAYA GMEPA INDONESIA TAHUN 2017*. Kabupaten Bandung: Pusat Penelitian dan Pengembangan Perumahan dan Permukiman, 2017.

[7] P. Reasenberg, "Second-Order Moment of Central California Seismicity, 1969-1982," *J Geophys Res*, vol. 90, pp. 5479–5495, 1985.
[8] S. Wiemer, "A Software Package to Analyze Seismicity ZMAP," *Seismological Research Letters*, vol. 72, no. 2, p. 373, 2001.
[9] K. Aki, "Maximum likelihood estimated of b in the formula log N = A – b*M and its confidence limits," 1965.
[10] B. Gutenberg and C. F. Richter, "Frequency of earthquakes in California," *Nature*, p. 185, 1944, doi: 10.1038/156371a0.
[11] M. Irsyam *et al.*, "Development of the 2017 national seismic hazard maps of Indonesia," *Earthquake Spectra*, vol. 36, no. 1_suppl, pp. 112–136, 2020, doi: 10.1177/8755293020951206.
[12] D. C. Heath, D. J. Wald, C. B. Worden, E. M. Thompson, and G. M. Smoczyk, "A global hybrid VS30 map with a topographic slope–based default and regional map insets," *Earthquake Spectra*, vol. 36, no. 3, pp. 1570–1584, Aug. 2020, doi: 10.1177/8755293020911137.
[13] M. Pagani *et al.*, "Openquake engine: An open hazard (and risk) software for the global earthquake model," *Seismological Research Letters*, vol. 85, no. 3, pp. 692–702, 2014, doi: 10.1785/0220130087.
[14] P. Grossi and H. Kunreuther, "Clarification and Errata to Catastrophe Modeling: A New Approach to Managing Risk," 2021.
[15] V. Silva *et al.*, "Country/Territory Seismic Risk Profiles," Oct. 2023, *Zenodo*. doi: 10.5281/zenodo.10355885.

Probabilistic Tsunami Hazard Assessment Using Supervised Learning with Nonstationary Sea-Level Rise: Bali Case Study

Abdul Kadir Alhamid[1,2,*], *Muhamad Farhan* Permana[3], *Mohammad* Farid[1,2], *Mohammad Bagus* Adityawan[1,4], *Alamsyah* Kurniawan[2,5]

[1]Water Resources Engineering Research Group, Faculty of Civil and Environmental Engineering, Bandung Institute of Technology, Jalan Ganesha No. 10, Bandung 40132, Indonesia
[2]Center for Coastal and Marine Development, Directorate of Research and Innovation, Bandung Institute of Technology, Jalan Ganesha No. 10, Bandung 40132, Indonesia
[3]Master of Water Resources Management Study Program, Faculty of Civil and Environmental Engineering, Bandung Institute of Technology, Jalan Ganesha No. 10, Bandung 40132, Indonesia
[4]Center for Water Resources Development, Directorate of Research and Innovation, Bandung Institute of Technology, Jalan Ganesha No. 10, Bandung 40132, Indonesia
[5]Coastal Engineering Research Group, Faculty of Civil and Environmental Engineering, Bandung Institute of Technology, Jalan Ganesha No. 10, Bandung 40132, Indonesia

Abstract. Indonesia has been considered one of the most tsunami-prone countries due to its proximity to three major tectonic plates. Particularly, Bali is at a greater risk because it is a major tourist destination, since it is located between the Sunda Megathrust and the Flores Back-Arc Thrust. However, future tsunamis are expected to become more destructive due to the effects of sea-level rise. Therefore, this study aims to provide a framework for Probabilistic Tsunami Hazard Assessment (PTHA) in Bali considering nonstationary sea-level rise effects based on a supervised learning approach. Earthquake magnitude and sea-level rise are statistically analyzed, and earthquake parameters, including rupture width, length, and slip, are estimated using empirical scaling. Tsunami propagation analyses are performed considering several cases of earthquake magnitude and sea-level rise. Furthermore, Gaussian Process Regression (GPR) response surface models are developed to correlate the tsunami inundation depth to magnitude and sea-level rise to accelerate computation time. Finally, hazard curves are evaluated using a Monte Carlo simulation based on the uncertainty associated with earthquake magnitudes and sea-level rise. A case study in Denpasar is provided in the illustrative example, and the effects of sea-level rise and different earthquake rupture locations on tsunami hazard are evaluated.

*Corresponding author: abdul@itb.ac.id

1 Introduction

Indonesia is considered one of the most tsunami-prone countries in the world due to the convergence of the Indo-Australian, Eurasian, and Pacific plates. Such a geodynamic setting makes the region prone to megathrust earthquakes and submarine landslides that cause destructive tsunamis. Indonesia alone has documented at least 24 events of tsunamis throughout the past century, and more than 231,900 casualties have been recorded, with the devastating 2004 Indian Ocean tsunami resulting in approximately 226,900 casualties by itself [1]. Additionally, recent studies showed that around 5.5 million people are subjected to tsunami risks of 500-year return periods [2], necessitating the onset of better risk mitigation, specifically in attractive, heavily populated, and economically important islands such as Bali.

The exposure of Bali to tsunami risk is attributed to its location between two active subduction zones: the Sunda Megathrust and the Flores Back-Arc Thrust. Based on historical and geological evidence, it is known that the run-up heights of Indonesian tsunamis have been more than 10 meters [3], which is a significant threat to coastal communities and infrastructure. However, most of the tsunami hazard studies in Indonesia predominantly rely on a deterministic approach, which only considers the worst-case scenarios and does not consider the uncertainties associated with source parameters, including magnitude, location, and slip distribution. The Probabilistic Tsunami Hazard Assessment (PTHA) approach can overcome these shortcomings by considering a set of possible earthquake scenarios and their probabilities [4,5]. Such a probabilistic approach can be used to make risk-based decision-making in vulnerable regions like Bali.

In addition to the uncertainties of tsunami sources, climate-induced sea-level rise is expected to exacerbate tsunami risks further. A study indicates that a 30 cm increase in sea level may amplify inundation depths by a factor of two to four [6], especially critical in low-lying, vulnerable areas such as Bali. However, full-scale PTHA with sea-level rise integration is computationally intensive due to the large number of required simulations. Therefore, it is necessary to develop a PTHA framework, which can accommodate both nonstationary sea level rise and tsunami source uncertainties to support effective tsunami risk management for vulnerable coastal areas such as Bali.

This study provides a framework for PTHA in Bali that considers the uncertainties associated with nonstationary sea-level rise and the tsunami source based on supervised learning using Gaussian Process Regression (GPR). First, necessary inputs associated with digital elevation and bathymetric data are collected, combined with seismic source parameters obtained from the United States Geological Survey (USGS) earthquake catalogue. The tsunami fault dimensions are obtained via empirical scaling laws based on the earthquake magnitude. Subsequently, tsunami propagation and inundation analyses are performed based on the nonlinear shallow-water equations. Since these simulations are computationally expensive, a GPR-based surrogate model is developed to correlate the earthquake magnitude, sea-level rise, and the inundation depth at a particular location. Finally, a Monte Carlo simulation is employed on the response surface to evaluate the tsunami hazard curve considering sea-level rise effects. The results can provide a reliable decision-making tool in coastal risk management in Bali.

2 Framework for PTHA considering sea-level rise effects based on GPR

To provide a computationally efficient PTHA in Bali, Fig. 1 shows the procedure to perform PTHA considering sea-level rise effects using GPR-based surrogate models. The procedure comprises seven steps: (1) topography and earthquake data collection, (2) probabilistic modelling of earthquake magnitude, (3) probabilistic modelling of sea-level rise, (4) tsunami model initial setup, (5) tsunami propagation analyses, (6) GPR-based surrogate modelling of tsunami inundation depth, and (7) tsunami hazard assessment considering sea-level rise. The details are described in the following subsections.

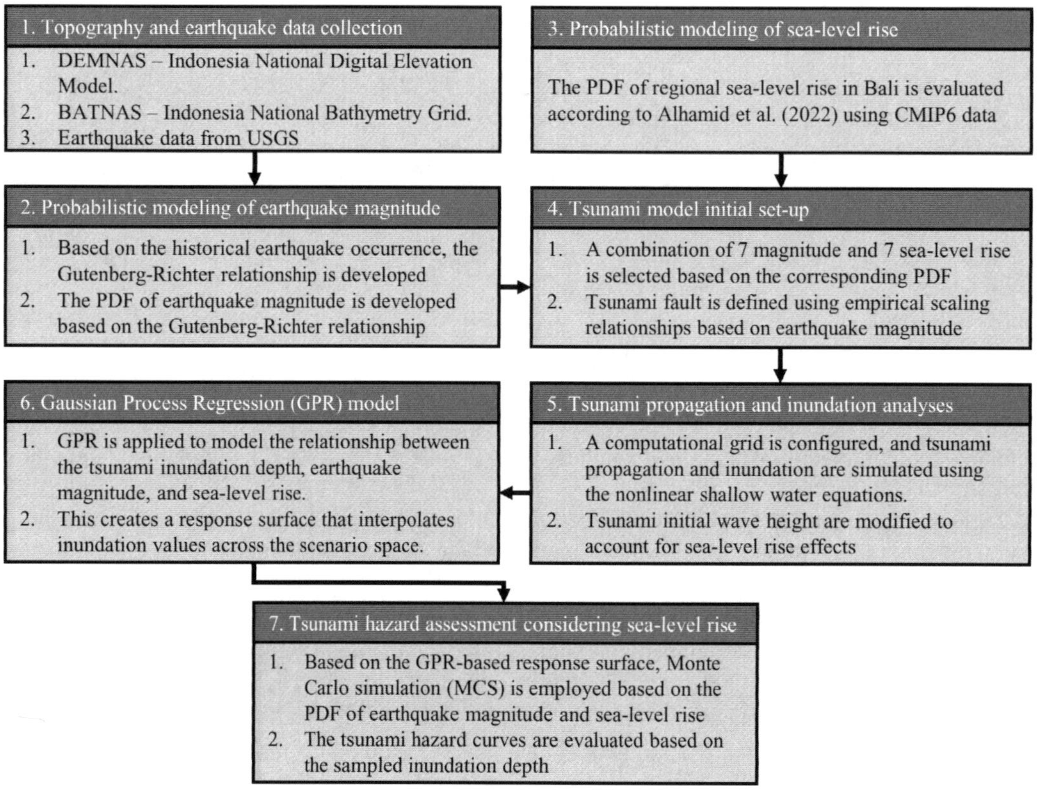

Fig. 1. Procedure to perform PTHA considering sea-level rise effects based on GPR.

2.1 Probabilistic modelling of earthquake magnitude

The probabilistic modeling of earthquake magnitudes is conducted using the earthquake catalog compiled by the USGS. The frequency and magnitude of the recorded earthquake events are analyzed and fitted to the Gutenberg–Richter law as follows:

$$\log\left[\lambda_f\left(M_w \geq m\right)\right] = a - bM_w \tag{1}$$

where λ_f represents the average annual rate of earthquakes generated from fault f with magnitude M_w equal to or greater than a value of m, and a and b are the empirical constants derived from linear fitting of the sorted catalog data. Assuming that the earthquake occurrences follow a Poisson process, the probability of at least one earthquake of magnitude M or greater occurring within a time window of duration t can be evaluated as:

$$P\left(M_w \geq m\right) = 1 - \exp\left(-\lambda_f\left(M_w \geq m\right)\cdot t\right) \tag{2}$$

In particular, when t = 1, the expression simplifies to represent the annual probability of exceeding magnitude M_w as follows:

$$P\left(M_w \geq m\right) = 1 - \exp\left(-\lambda_f\left(M_w \geq m\right)\right) \tag{3}$$

The cumulative density function (CDF) associated with moment magnitude Mw can be obtained as the complementary of Equation (3). This distribution is utilized further in the tsunami hazard assessment during the MCS process, which is described in greater detail in the following subsections.

2.2 Probabilistic sea-level rise hazard assessment

The methodology to conduct sea-level rise hazard assessment is adopted from Alhamid et al. [10] and briefly summarized herein. The exceedance probability of sea-level rise at a particular location (i.e., regional sea-level rise hazard) is evaluated by considering three primary components: regional sterodynamic, glacier, and ice-sheet sea-level rise. Each component is treated as a random variable with its own corresponding PDF. The PDF of regional sterodynamic sea-level rise is derived from 19 CMIP5 General Circulation Models (GCMs) [7], while the regional glacier and ice-sheet sea-level rise components are derived from 11 GlacierMIP and 6 SeaRISE experiment models, respectively. Assuming a perfect correlation between the sea-level rise components, the CDF of total regional sea-level rise can be evaluated as follows:

$$F_\Psi\left(\psi\left(\mathbf{x}_\mathbf{r},t\right)\right) = \int_0^\psi \left\{ \sum_{RCP=1}^{4} \left[\iint_{\Omega_{G,I}} f_{SGI}\left(\omega\left(\mathbf{x}_\mathbf{r},t\right), \xi\left(\mathbf{x}_\mathbf{r},t\right), \eta\left(\mathbf{x}_\mathbf{r},t\right) \mid \Pi_{RCP}\right) d\xi\, d\eta \right] \cdot P\left(\Pi_{RCP}\right) \right\} d\psi \tag{4}$$

where $F_\Psi(\psi(\mathbf{x_r},t))$ represents the CDF of the total regional sea-level rise, indicating the probability that sea-level rise intensity Ψ at a given location vector $\mathbf{x_r}$ (specified by latitude and longitude) and time t is less than a specified value $\psi(\mathbf{x_r},t)$, $f_{SGI}\,(\omega(\mathbf{x_r},t), \xi(\mathbf{x_r},t), \eta(\mathbf{x_r},t) \mid \Pi_{RCP})$ is the multivariate probability distribution of regional sterodynamic (ω), glacier (ξ), and ice-sheet (η) sea-level rise components, conditioned on a specific climate change scenario Π_{RCP}. The scenarios Π_1, Π_2, Π_3, Π_4 correspond to Representative Concentration Pathways (RCPs) of 2.6, 4.5, 6.0, and 8.5, respectively. The integration is carried out over the joint domain $\Omega_{G,I}$, which spans the possible values of glacier and ice-sheet contributions. Each RCP scenario is assumed to have an equal probability of occurrence, $P(\Pi_{RCP}) = 1/4$.

2.3 Tsunami propagation and inundation analyses

Tsunami propagation and inundation analyses in this study were conducted using Delft Dashboard and Delft3D with a five-level nested grid system. Grid-1 (11 km) and Grid-2 (2.2 km) cover the open ocean and offshore-coastal transition, respectively, both using the General Bathymetric Chart of the Oceans (GEBCO) bathymetry data with a spatial resolution of 15 arc-seconds (450 m). Grid-3 (440 m) integrates the National Digital Elevation Model (DEMNAS) topography (~8.3 m) and the National Bathymetric Map (BATNAS) bathymetry (180 m) for coastal zone modeling. Grid-4 (146.7 m) refines the land-sea interface, and Grid-5 (29.3 m) resolves run-up and inland inundation using high-resolution DEMNAS data. Sea-level rise scenarios were incorporated by adjusting the bathymetry maps. It is acknowledged that the accuracy of tsunami inundation simulations is strongly influenced by the resolution of DEMS and bathymetric data. Comparative studies have shown that bathymetric grids with resolutions of 10–50 m can reproduce tsunami heights and arrival times within 5 to 10% of those obtained from very high-resolution (≤5 m) reference models, while significantly coarser grids may underestimate wave heights by up to 70% [8–10]. Since a DEM with a finest spatial resolution of approximately 30 m was employed herein, the resolution is considered sufficient for regional-scale tsunami simulations.

Furthermore, numerical tsunami simulations were performed for a range of earthquakes along the Java megathrust (M_w 6.0–8.7) with a 0.5 increment. In addition, each scenario was evaluated under various sea-level rise conditions ranging from 0 to 1.2 meters with a 0.2 m increment. This combination yielded a total of 49 scenarios for Java megathrust fault rupture. The simulations are performed by numerically solving the nonlinear shallow water theory. The tsunami fault rupture dimensions associated with the length L and width W are estimated using scaling laws from Thingbaijam et al. [11], which can be generally expressed as follows:

$$\log(L) = a_L + b_L \cdot M_w \tag{5}$$

$$\log(W) = a_W + b_W \cdot M_w \tag{6}$$

where a_L and b_L are the scaling parameters associated with tsunami rupture length L, while a_W and b_W correspond to the tsunami rupture width W. The scaling parameters associated with the Java megathrust are given in Table 1.

Table 1. Scaling parameters associated with the fault rupture [11].

Fault rupture scenario	Scaling parameters for length L		Scaling parameters for width W	
	a_L	b_L	a_W	b_W
Java megathrust	-2.412	0.583	-0.880	0.366

In this study, sea-level rise was incorporated into the tsunami simulations through dynamic boundary condition adjustments within Delft3D, allowing the nonlinear interactions between elevated baseline sea levels and tsunami wave dynamics to be represented. Moreover, Manning's roughness coefficients were assigned to account for different land-cover types, following the classification developed in the tsunami reconstruction modelling for Banda Aceh by Tursina and Syamsidik [12]. These coefficients, summarized in Table 2, were applied to represent urban areas, vegetation, and bare land. The land-cover information required for this classification was obtained from the *Rupa Bumi Indonesia* topographic dataset provided by the Geospatial Information Agency of Indonesia (*Badan Informasi Geospasial*).

Table 2 Manning's roughness coefficient from reconstruction tsunami in Banda Aceh by Tursina and Syamsidik [12].

Land Use	**Manning's Roughness Coefficient (*n*)**
Coastal Vegetation	0.035
Fish Ponds	0.017
Building	0.04
Sea	0.013
Soil	0.02

2.4 Gaussian Process Regression (GPR) surrogate model

Full-scale PTHA is computationally expensive since it requires numerous high-resolution tsunami models to capture the vast number of possible earthquake scenarios, each with different magnitudes, locations, rupture geometries, and sea-level rise conditions. Most recent studies have explored machine learning approaches, especially surrogate models, to reduce computation time with a sufficient level of accuracy [7–9].

GPR is a probabilistic and non-parametric machine learning approach that has been widely used for regression. Within the context of this study, suppose that the combination of magnitudes, sea-level rise, and inundation depth associated with training data for the GPR has a total of n data points. Particularly, the training inputs $\mathbf{X}$ associated with magnitude m and sea-level rise ψ are represented as an $(n\times 2)$ matrix, while the training outputs $\mathbf{z}$ associated with tsunami inundation depth are a vector of $(n\times 1)$, such that:

$$\mathbf{X}=\begin{bmatrix} m_1 & \psi_1 \\ m_2 & \psi_2 \\ \vdots & \vdots \\ m_n & \psi_n \end{bmatrix}; \quad \mathbf{z}=\begin{bmatrix} h_1 \\ h_2 \\ \vdots \\ h_n \end{bmatrix} \tag{7}$$

The first step of GPR is to define a kernel (covariance) function. In this study, an exponential kernel function is adopted since it is more suitable for data with abrupt transitions or non-smooth transitions [13]. For a particular input data $\mathbf{X}_i = [m_i, \psi_i]^T$ and $\mathbf{X}_j = [m_j, \psi_j]^T$, the exponential kernel function can be mathematically expressed as:

$$k(\mathbf{X}_i,\mathbf{X}_j)=\sigma_f^2\cdot\exp\left(-\frac{|m_i-m_j|}{l_m}-\frac{|\psi_i-\psi_j|}{l_\psi}\right) \tag{8}$$

where σ_f^2 is the signal variance that maintains the amplitude of functions, l_m and l_ψ are the length-scales for earthquake magnitude and sea-level rise, respectively. The covariance matrix associated with the multivariate Gaussian distribution can then be derived based on the kernel function as follows:

$$\mathbf{K}=\begin{bmatrix} k(\mathbf{X}_1,\mathbf{X}_1) & k(\mathbf{X}_1,\mathbf{X}_2) & \cdots & k(\mathbf{X}_1,\mathbf{X}_n) \\ k(\mathbf{X}_2,\mathbf{X}_1) & k(\mathbf{X}_2,\mathbf{X}_2) & \cdots & k(\mathbf{X}_2,\mathbf{X}_n) \\ \vdots & \vdots & \ddots & \vdots \\ k(\mathbf{X}_n,\mathbf{x}_1) & k(\mathbf{X}_n,\mathbf{X}_2) & \cdots & k(\mathbf{X}_n,\mathbf{X}_n) \end{bmatrix} \tag{9}$$

Furthermore, the predicted distribution associated with inundation depth z_* given an input data of $\mathbf{X}_* = [m_*, \psi_*]^T$ can be evaluated according to the posterior distribution of the multivariate Gaussian distribution with mean and variance as:

$$\mu_* = \mathbf{k}_*^T \left(\mathbf{K} + \sigma_n^2 \mathbf{I}\right)^{-1} \mathbf{z} \tag{10}$$

$$\sigma_*^2 = k\left(\mathbf{X}_*, \mathbf{X}_*\right) - \mathbf{k}_*^T \left(\mathbf{K} + \sigma_n^2 \mathbf{I}\right)^{-1} \mathbf{k}_* \tag{11}$$

$$\mathbf{k}_* = \begin{bmatrix} k\left(\mathbf{X}_*, \mathbf{X}_1\right) \\ k\left(\mathbf{X}_*, \mathbf{X}_2\right) \\ \vdots \\ k\left(\mathbf{X}_*, \mathbf{X}_n\right) \end{bmatrix} \tag{12}$$

Other necessary parameters, such as σ_f, l_m, l_ψ, and σ_n, can be estimated based on hyperparameter optimization algorithms.

2.5 Probabilistic tsunami hazard assessment (PTHA) considering sea-level rise effects

The PTHA to evaluate the tsunami hazard curve is carried out by employing MCS to the GPR surrogate model based on the distributions of earthquake magnitude M_w and sea-level rise Ψ, which are described in Equations (3) and (4), respectively. For each trial i, a pair of input data $\mathbf{X}_*^{(i)} = (m_*^{(i)}, \psi_*^{(i)})$ is sampled and are fed into the GPR model to obtain the predictive mean $\mu_*(\mathbf{X}_*^{(i)})$ and variance $\sigma_*^2(\mathbf{X}_*^{(i)})$. To consider the bias uncertainties associated with the GPR model, the tsunami inundation depth at MCS trial i is sampled by assuming a standard-normal distribution, such that:

$$H_f^{(i)} = \mu_*\left(\mathbf{X}_*^{(i)}\right) + \sigma_*\left(\mathbf{X}_*^{(i)}\right) \cdot U \tag{13}$$

where H_f denotes the random variable of inundation depth stemming from fault rupture f, and U is a uniform random variable between 0 and 1. $P(H_f > h)$ is estimated as the fraction of simulations where $H_f^{(i)} > h$ and the total number of simulations.

The earthquake magnitude distribution in Equation (3) is stationary, which can be treated as independent Bernoulli trials across years. In contrast, sea-level rise is nonstationary and is a function of time t. Without considering the effects of sea-level rise, the distribution of maximum inundation depth from time $t = 1$ until an investigated time of interest t_{inv} can be developed as follows:

$$P\left[H_{max,f}\left(t_{inv}\right) > h\right] = 1 - \left[1 - P\left(H_f > h\right)\right]^{t_{inv}} \quad \forall t_{inv} \in \square \tag{14}$$

On the other hand, when considering sea-level rise effects, the annual tsunami hazard in Equation (16) becomes nonstationary due to the time-dependent nature of sea-level rise distribution in Equation (4). Therefore, the distribution of maximum tsunami inundation depth considering sea-level rise effects can be described as:

$$P\left[H_{max,f}\left(t_{inv}\right)>h\right]=1-\prod_{t=1}^{t_{inv}}\left[1-P\left(H_f\left(t\right)>h\right)\right] \qquad \forall\{t,t_{inv}\}\in\square \tag{15}$$

where $P(H_f(t) > h)$ is the tsunami hazard curve at time t, evaluated based on the aforementioned MCS process using Equation (4) at time t. To demonstrate the effectiveness of this framework, an illustrative example for the island of Bali, Indonesia, is provided in the following section.

3 Illustrative example

3.1 Analyzed region and the distribution of earthquake magnitude and sea-level rise

Fig. 2 presents the geographic setting of Denpasar, the capital of Bali, which is located on the southern coast of the island in a low-lying area. With an area of 127.78 km^2 and an average elevation of 4 meters, Denpasar is located in a flat deltaic plain washed by the Indian Ocean, with the Badung River flowing within the city towards the Gulf of Benoa. This coastal morphology, together with the fact that the city is located between two active tectonic sources, namely the Java Megathrust in the south and the Flores Back-Arc Thrust in the north, makes southern Denpasar and locations such as Sanur and Semawang Beach extremely prone to tsunamis. Flat topography, riverine systems, reclaimed areas, and dense coastal urbanization also enhance vulnerability by exposing coastal communities and infrastructure to more tsunami risks.

Fig. 2. Analyzed region for illustrative example: Denpasar of Bali.

The distribution of earthquake magnitude Mw is evaluated according to the procedure described in Subsection 2.1 based on the USGS earthquake catalogue, while the sea-level rise CDF is adopted from Alhamid et al. [14]. Figure 3 presents their CDF, which is used as input in the probabilistic tsunami hazard assessment. The CDF of earthquake magnitude for the Sumba Megathrust indicates a steep rise around 6.0 to 7.5, suggesting a relatively high probability of moderate-to-large tsunami events in Bali to occur. On the other hand, the CDFs of projected sea-level rise for the southern coast of Bali under different future time horizons (i.e., 2050, 2075, and 2100). The distribution shifts rightward over time, indicating the nonstationary nature of sea-level rise due to climate change.

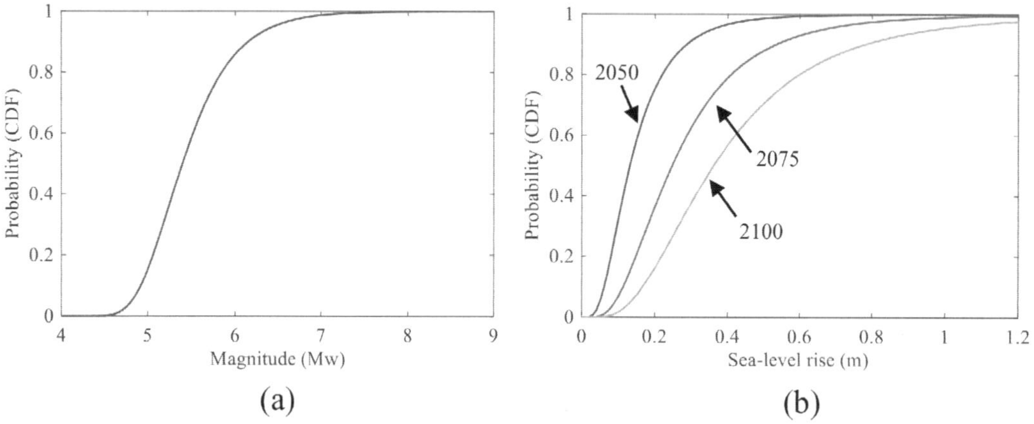

Fig. 3. (a) CDF of earthquake magnitude Mw in Sumba Megathrust, and (b) Sea-level rise for the southern part of Bali from Alhamid et al. [14].

3.2 Tsunami propagation and inundation analyses

Fig. 4 illustrates the initial sea surface displacement generated by a hypothetical megathrust earthquake off the southern coast of Bali with a magnitude M_w of 8.7. The source rupture is assumed to occur along the subduction interface of the Sumba Megathrust, as indicated by the concentration of vertical displacement aligned roughly parallel to the trench axis. The tsunami source model is constructed using the finite fault deformation approach based on Okada's [15] analytical solution for elastic dislocation in a half-space. The approach allows the spatially distributed slip on the fault plane to be converted into vertical sea surface displacement, which is the initial condition for the tsunami propagation modeling.

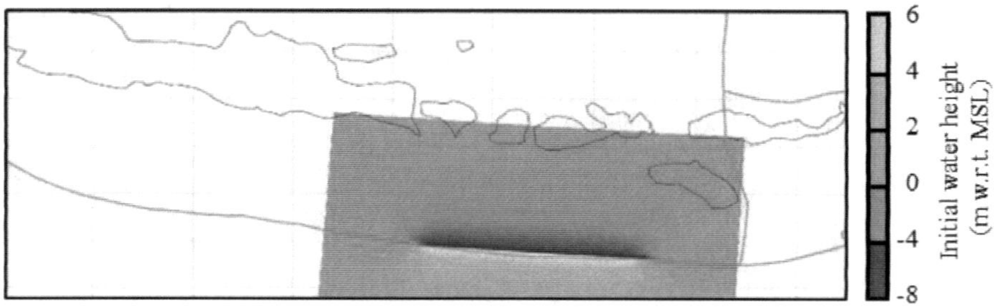

Fig. 4. Initial tsunami height generated near the Sumba Megathrust under a magnitude M_w 8.7 based on Okada [15].

Furthermore, based on the tsunami's initial height in Fig. 4, tsunami propagation and inundation analyses were conducted under various scenarios of earthquake magnitude and sea-level rise, as described in Subsection 2.3. Fig. 5 presents an example of the inundation analysis result for a scenario with an earthquake magnitude of M_w 8.7, comparing two sea-level rise conditions: 0 m and 1.2 m. In the northern part of Denpasar, the inundation depth of the tsunami becomes substantially greater with the rise of the sea level. On the other hand, the extension of the inundation area due to the tsunami is observed in the southern region. This intensification is due to the rise in the baseline sea level, which enables the tsunami waves to reach the mainland further. Such conditions could increase the risk of loss and casualties, particularly in low-lying, densely populated areas like southern Denpasar.

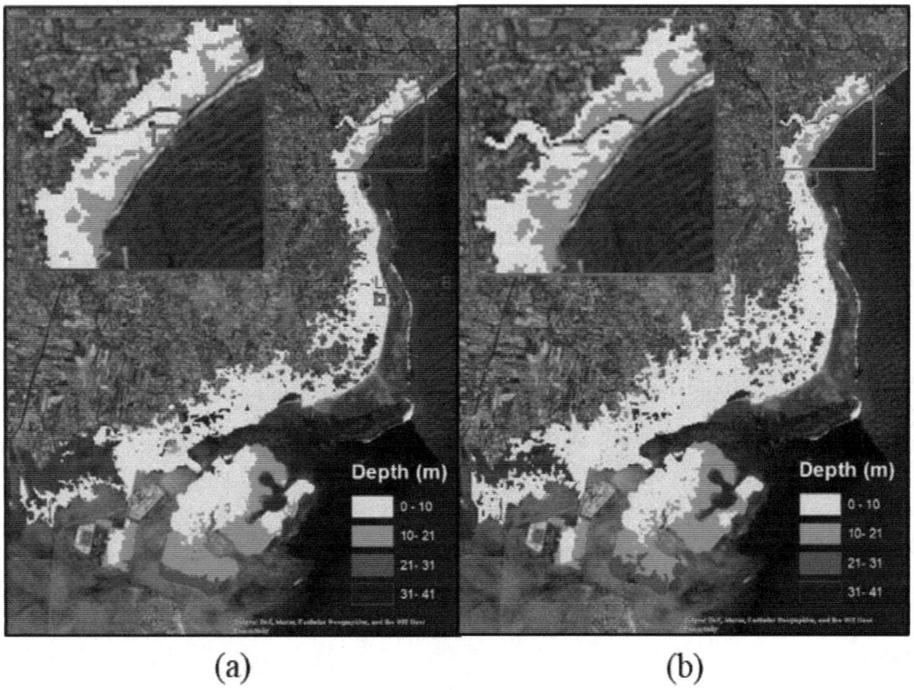

Fig. 5. Example of tsunami inundation analysis results for an earthquake scenario with magnitude M_w 8.7 under two sea-level rise conditions: (a) 0 meters and (b) 1.2 meters.

The validation of tsunami inundation simulations for Bali is constrained by the absence of detailed observational data. The most recent tsunami that affected the island was the 3 June 1994 Java event, yet its impact on Bali was minimal and no systematic measurements of inundation depth were reported. More recent tsunamis, such as the 2018 Sunda Strait and Palu Bay events, did not affect Bali, and therefore no field survey data exist for this region. In the absence of direct validation data, the results of this study were compared with the tsunami hazard map developed under the German–Indonesian Tsunami Early Warning System (GITEWS) collaboration, as shown in Fig. 6. This map has been widely recognized as a national reference, produced through extensive modeling process. The comparison indicates that the simulated inundation extents in this study are in good agreement with the GITEWS hazard zones, thereby providing confidence in the reliability and applicability of the hazard estimates for Bali.

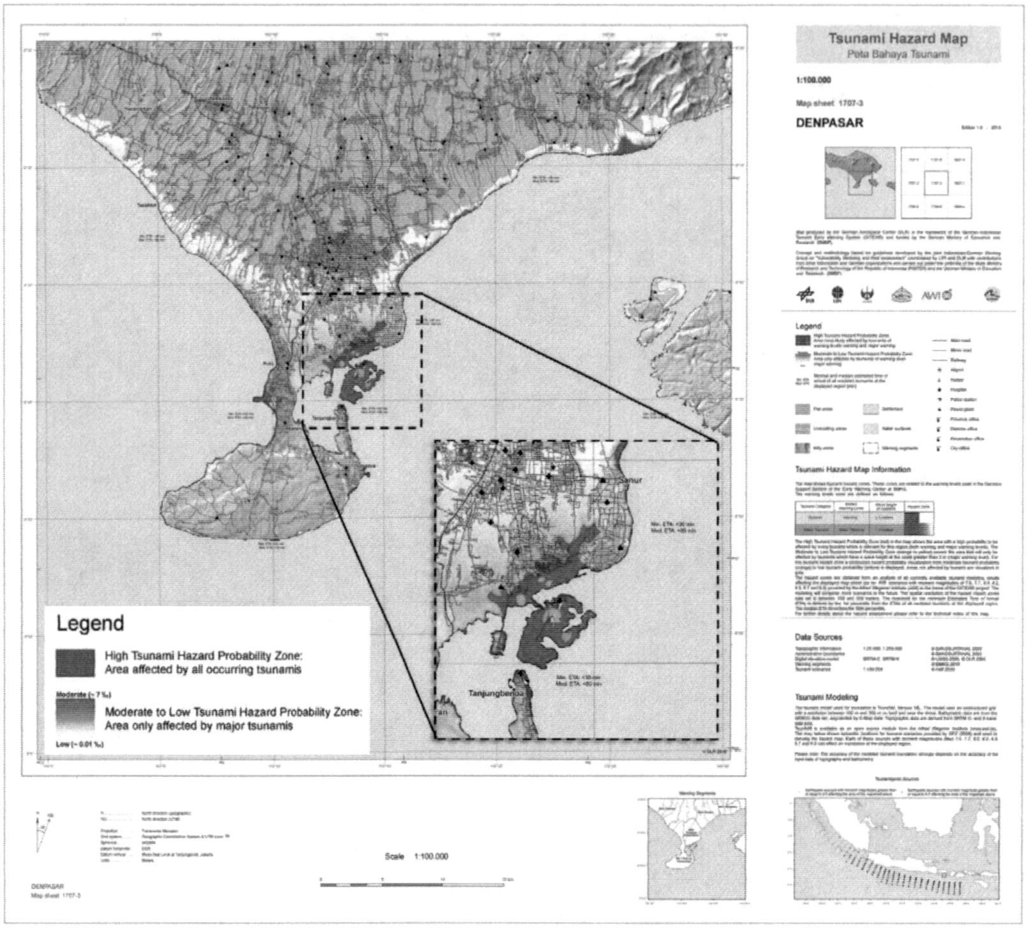

Fig. 6. Tsunami hazard map for Denpasar City of Bali, developed by GITEWS.

3.3 GPR surrogate model of tsunami inundations

To evaluate the spatial variation of tsunami inundation depth in greater details, Fig. 7 shows the GPR surrogate model at two representative locations along the Denpasar coastline, referred to as Location A and Location B in Fig. 5. The distance of Location A and Location B from the coastline is 100 m and 300 m, respectively. The GPR surfaces constructed from training data points represent the joint effect of the earthquake magnitude and rise of sea level on tsunami inundation depth. In general, the inundation depth attenuate as the analyzed site is located farther from the coast, as indicated by the lower values of Location B. In addition, sea-level rise does not exhibit a linear impact on the inundation depth. Moderate sea-level rise increases inundation depth dramatically at Location A since it is located closer to the coastline, and the increase is less dramatic but still substantial at higher sea-level conditions at Location B. These results highlight the geographic variability of tsunami damage and the nonlinear effect of sea-level rise.

 https://doi.org/10.1051/e3sconf/202565101003

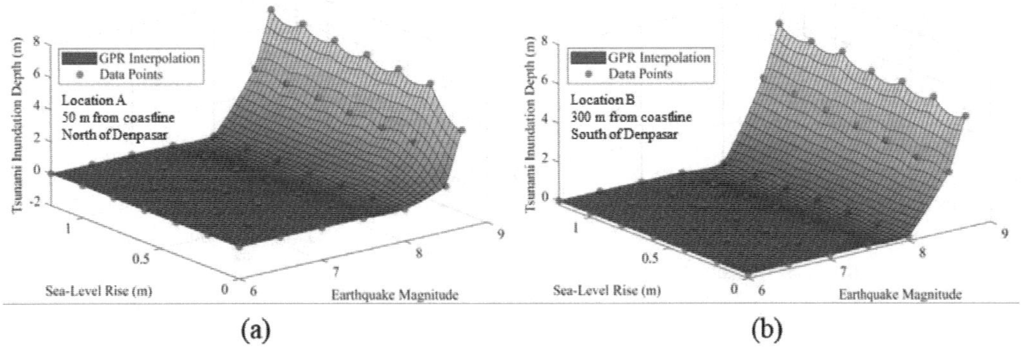

Fig. 7. Predicted tsunami inundation depth surfaces using Gaussian Process Regression (GPR) surrogate models at two locations in Denpasar: (a) Location A, 50 m from the coastline (north), and (b) Location B, 300 m from the coastline (south).

3.4 PTHA considering nonstationary sea-level rise effects

The tsunami hazard curve is estimated by employing the uncertainties associated with earthquake magnitude and sea-level rise, as shown in Fig. 3, to the GPR surrogate model in Fig. 7 based on MCS. Fig. 8 shows the tsunami hazard curves computed according to the procedure described in Subsection 2.5, which represent the exceedance probability of maximum inundation depth until the year 2100 at the two analyzed locations. In both sites, an initial steep decline in the curves shows a high likelihood that no tsunami events will occur during the assessment period, which is represented by the high number of zero inundation depths in Fig. 7. However, the impact of sea-level rise becomes more evident at the nearer coastal site (Location A), in which the effects of sea-level rise consistently increase the exceedance probability across all inundation depths compared to that of without sea-level rise effects. Nevertheless, the impact is less significant at Location B, which indicates that the closer to the coastline, the greater the impact of sea-level rise on the tsunami risk. These results highlight the critical need to consider future sea-level effects in tsunami risk assessments, particularly for low-lying areas close to the shore.

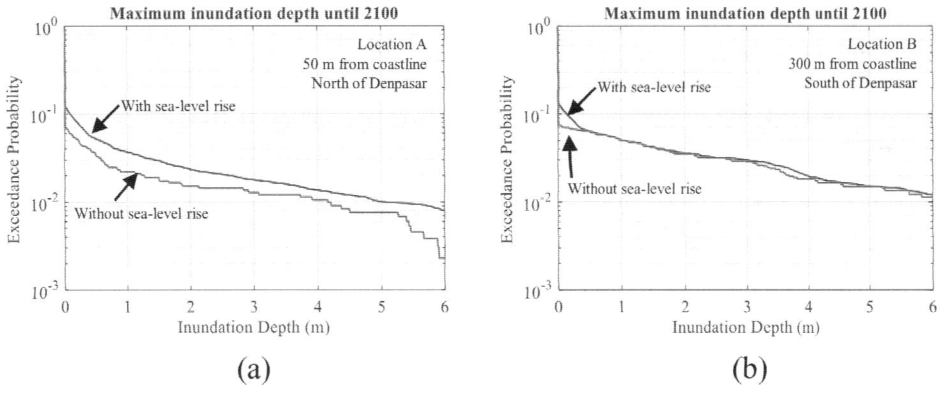

Fig. 8. Tsunami hazard curves associated with maximum inundation depth until 2100 at (a) Location A (50 m from coastline, north Denpasar) and (b) Location B (300 m from coastline, south Denpasar), under scenarios with and without sea-level rise effects.

4 Conclusions

The following conclusions are drawn from this study:

1. A PTHA framework for Bali is developed considering sea-level rise effects through a supervised learning approach using a GPR surrogate model. MCS is employed based on the probabilistic uncertainties of earthquake magnitude and sea-level rise in the GPR surrogate model.
2. Sea-level rise significantly amplifies tsunami inundation depth and extent, especially in low-lying coastal regions of Denpasar. It is found that a 1.2 m sea-level rise leads to deeper and more widespread inland flooding.
3. The GPR surrogate model accurately captures the nonlinear relationship between earthquake magnitude, sea-level rise, and inundation depth, providing a reliable response surface for probabilistic analysis. In addition, by comparing different locations, it is found that the tsunami inundation depth may intensify nonlinearly depending on the relative distance of the analyzed location to the shore.
4. The use of MCS on the GPR model allows the computationally efficient tsunami hazard assessment under the uncertainties of earthquake magnitude and sea-level rise. This integration minimises the dependency on large-scale numerical simulations and maintains acceptable hazard information.

This PTHA framework offers a computationally efficient and climate-informed tool to support tsunami risk management and coastal resilience planning. In this study, tsunami propagation and inundation analyses were carried out under the assumption of uniform slip distributions within the fault rupture plane and disregarding the uncertainties associated with fault rupture location. Therefore, future studies should consider these aspects to improve the accuracy of PTHA. In addition, the proposed framework provides a practical basis for updated hazard maps, land-use planning, and building regulations in Bali. Its computational efficiency allows potential integration with the Indonesian Tsunami Early Warning System (InaTEWS) to support agencies such as Indonesian Agency for Meteorology, Climatology, and Geophysics (BMKG) and the Regional Disaster Management Agencies (BPBD) in evacuation planning and decision-making. Future work should extend this framework toward full risk assessment by incorporating fragility curves and exposure data, enabling evaluation of both structural and non-structural mitigation strategies.

Acknowledgements. The authors would like to express gratitude for the support of this research from the SATREPS Building Sustainable System for Resilience and Innovation in Coastal Community (BRICC) Project and the NRFI Community and Infrastructure Resilience to Climate-geological Long-term Effects (CIRCLE) Project.

References

1. NOAA, NOAA Natl. Centers Environ. Information. (2023)
2. UNISDR, Int. J. Disaster Resil. Built Environ. **4**, (2013)
3. L. Hamzah, N. T. Puspito, and F. Imamura, J. Nat. Disaster Sci. **22**, 25 (2000)
4. E. L. Geist and T. Parsons, Nat. Hazards **37**, 277 (2006)
5. H. K. Thio, P. Somerville, and G. Ichinose, J. Earthq. Tsunami **1**, 119 (2007)
6. L. Li, A. D. Switzer, Y. Wang, C. H. Chan, Q. Qiu, and R. Weiss, Sci. Adv. **4**, (2018)
7. K. E. Taylor, R. J. Stouffer, and G. A. Meehl, Bull. Am. Meteorol. Soc. **93**, 485 (2012)
8. R. Felix, J. Hubbard, K. Wilson, and A. Switzer, Geosci. Lett. **11**, (2024)
9. J. Griffin, H. Latief, W. Kongko, S. Harig, N. Horspool, R. Hanung, A. Rojali, N. Maher, A. Fuchs, J. Hossen, S. Upi, S. Edi Dewanto, N. Rakowsky, and P. Cummins,

Front. Earth Sci. **3**, (2015)
10. J. Song and K. Goda, Front. Earth Sci. **7**, (2019)
11. K. K. S. Thingbaijam, P. M. Mai, and K. Goda, Bull. Seismol. Soc. Am. **107**, 2225 (2017)
12. Tursina and Syamsidik, in *IOP Conf. Ser. Earth Environ. Sci.* (2019)
13. Y. Jiang, J. Jia, Y. Li, Y. Kou, and S. Sun, Flow Meas. Instrum. **81**, (2021)
14. A. K. Alhamid, M. Akiyama, H. Ishibashi, K. Aoki, S. Koshimura, and D. M. Frangopol, Struct. Saf. **94**, (2022)
15. Y. Okada, Bull. - Seismol. Soc. Am. **82**, 1018 (1992)

Development of Rigid Landslide Tsunami Physical Modelling: A Simplified Model for The Palu Bay, Indonesia

Syamsidik[1,*], *Arifullah* [2], *Haekal* Azief Haridhi[1,3], *Muhammad* Zawil Kiram[1], *Dimas* Salomo Sianipar[4], *Muzli*[4], *Munawar*[4], *Marzuki* Sinambela[4], *Mujibur Rahman*[2], Azrul Rahmat Fadhillah[2], and *Indah* Fajerianti[4]

[1]Tsunami and Disaster Mitigation Research Center (TDMRC), Universitas Syiah Kuala, Banda Aceh, 23111, Indonesia
[2]Civil Engineering Department, Faculty of Engineering, Universitas Syiah Kuala, 23111, Indonesia
[3]Marine Science Department, Faculty of Marine and Fisheries, Universitas Syiah Kuala, 23111, Indonesia
[4]Sekolah Tinggi Meteorologi, Klimatologi dan Geofisika (STMKG), South Tangerang, Banten, 15221, Indonesia

Abstract. The Palu Bay Tsunami on September 28, 2018 still left some unresolved hydraulic phenomena. One of them was the arrival time of the tsunami to certain locations as the event mechanisms are still not fully known. This research was aimed at estimating tsunami arrival times due to landslide tsunami in the Palu Bay through a physical modelling. A simplified shape of the Palu Bay was constructed at Tsunami-Wave Flume facility at Tsunami and Disaster Mitigation Research Center (TDMRC). The simplified model was 9.5 m long, 2.5 m wide and 1.0 m deep. The slopes of the bay were made 42° for the east side of the modelled bay and 54° at the west side of the bay following average slopes of both sides of the bay. It has one open boundary at the north direction, while Palu city was hypothetically located at the closed boundary at the south part of the bay. The physical model was made in distorted scales of 1:3,200 for horizontal and 1:550 for vertical. The landslide material was made as a half-spherical shape to approach the volume of submarine landslide at *Buluri* of Palu at the west bay side as published by Pakoksung et al. (2019). In the experiment, the location was marked as S5 which is the largest landslide volume during the 2018 Palu Bay tsunami. A series of wave gauges were installed at some observation points, namely at *Pantoloan* Port, North *Taipa*, *Tondo-Tadulako*, *Lere* of West Palu, Tipo, *Watusampu* of West Palu, *Banawa* of Donggala, and *Donggala* Port. No capillary wave effects were detected during the experiments proving that the surface tension effects could be excluded. The experiments results showed that there is no significant difference of tsunami arrival times in related to the position of the rigid landslide material relative to the water surface. A significant trend of influence was observed in terms of tsunami heights monitored at the wave gauges. The results will be beneficial for further development model of landslide tsunamis for the Palu Bay

*Corresponding author: syamsidik@usk.ac.id

 https://doi.org/10.1051/e3sconf/202565101004

1 Introduction

The 2018 Palu Bay tsunami remains a puzzling event as the source of the tsunami was difficult to confirm [1], [2]. One of the most convincing arguments was that the tsunami was triggered by multiple submarine landslides inside the Palu Bay [3]. The tsunami was preceded by a-7.5 Mw earthquake with epicenter was located 0.20°S and 119.89°E, which was about 80 km to northwest of the Palu city. The frequency of landslides induced tsunami is rather longer than tectonogenic tsunami. Only 8% of the tsunami that occurred around Indonesia are classified as landslide tsunamis [4]. This adds to another complexity to understand the mechanism of the landslide tsunamis. Challenges are on waveforms created after the landslide and the propagation of the waves to inundation. In recent years, researchers have been conducting a number of studies to investigate the mechanism. Those include a series of physical modelling to simulate the tsunami. However, physical modelling-based research that was aimed to investigate landslide tsunami inside a bay is still rare. The Lituya Bay in Alaska of the United States of America was one good example of the landslide tsunami event [5]. Landslide tsunamis are relatively less impactful compared to tectono-genic tsunamis. The landslide tsunamis tend to create larger wave amplitudes near the sources and wider inundation area. Nonetheless, landslide tsunamis are more difficult to estimate the initiation of the process and it creates a more complicated hydrodynamic condition when it happens inside constricted areas such as a bay.

Challenges in the physical model for landslide tsunami events have been revealed by previous researchers [6], [7]. Among the most anticipated problems are the different results when the landslide materials are made from granular or rigid. Furthermore, the complexity will be added up when it involves a combination between aerial and submarine landslides. In the case of the 2018 Palu Tsunami, some of the identified landslides were submarine landslides where the depths of the sources could be deeper than 300 m inside the bay. However, there were some sources where they are situated near the coastal area of the bay (coastal landslides). The sources were between 0-300 meter deep. With different slopes of both sides of the bay, this created another uncertainty to the tsunami mechanisms.

This research was motivated by the urgency to develop a tsunami database for the Palu Bay area due to non-seismic sources, namely landslides-tsunami. This research was aimed at estimating tsunami wave heights and arrival times at several observation points inside a hypothetical shape of the Palu Bay based on tsunami physical models. The modelling was performed using a hypothetical bathymetry of the Palu Bay. Therefore, the interpretation of the results was made as three-dimensional experiments. This research is considered the first step to develop an integrated non-seismic tsunamis database for the Palu Bay. Further use of the database is expected to support the initiation of a non-seismic tsunami early warning system in Indonesia.

2 Methods

2.1 A Hypothetical Bathymetry of the Bay Model

A real bathymetry of the Palu Bay is still difficult to model at this stage. As a preliminary study, a hypothetical bathymetry of the bay was developed in this study based on the average slopes and depths of the bay. The bathymetric profile of the Palu Bay was constructed using GeoTIFF raster data sourced from the BATNAS (Geospatial Information Agency) at 6 arc-second resolutions. The 30 km length, 9 km wide and 550 m depth of raster data was extracted by generating ten evenly spaced cross-sections along the Palu Bay at approximately ~3 km intervals (see **Fig. 1(a)**). Each cross-section provided seafloor elevation data relative to horizontal distance, producing a set of local bathymetric profiles. These profiles were averaged to produce a representative bathymetric profile that reflects the general morphology of the bay's seafloor. The average slopes of the east and west sides of the Palu Bay can be seen in **Fig. 1(b).**

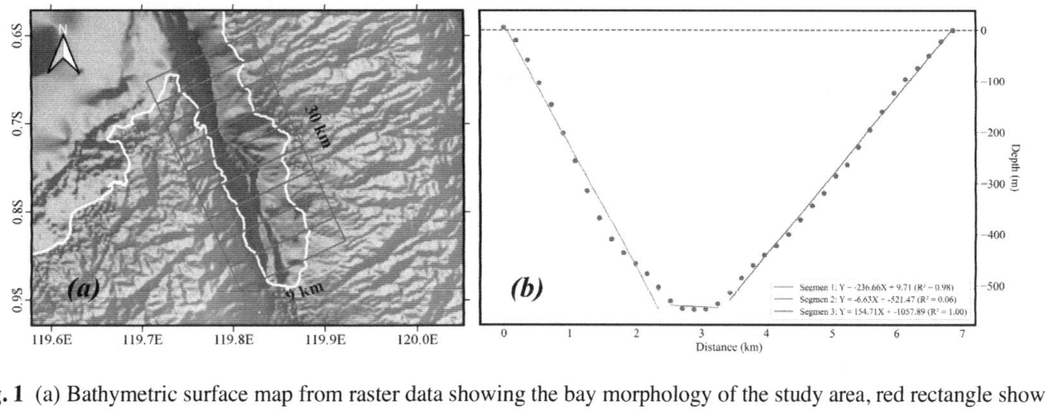

Fig. 1 (a) Bathymetric surface map from raster data showing the bay morphology of the study area, red rectangle show the clipped profile to extract the slope, (b) final slope geometry from average cross section profile.

The average profile was plotted as a depth–distance graph and simplified through a linearization process, in which the bathymetric curve was approximated as a series of straight-line segments with fixed angles. **Fig. 1** shows the bathymetry map of the Palu bay and cross sections considered in calculating average slopes of both sides of the bay. The elevation at 0 m above mean sea level (MSL) was excluded during the process of averaging the bathymetric profile, to ensure a clearer representation of the submerged terrain. The resulting profile was then segmented into linear sections to represent key features of the bathymetry. For physical modelling purposes, the two primary slope angles were calculated based on the height difference and horizontal distance of the seafloor on each side of the domain.

2.2 Physical Model and Similitude

The experiments were designed to investigate the generation and propagation of tsunami waves by submarine landslides. The experiment was conducted at Tsunami Wave Flume (TWF) Facility at the Tsunami and Disaster Mitigation Research Center of Universitas Syiah Kuala [8]. Considering the scale of the wave flume facility, the physical model was set in distorted scales of 1:3200 for horizontal and 1:550 for vertical. The physical model of the Palu Bay has a total length of 9.5 meters, a width of 2.5 meters, and a height of 1.0 meter measured from the base of the flume. The western sliding slope was designed with a length of 1.24 meters and an inclination angle of 54 degrees, while the eastern slope had a length of 1.49 meters with a gentler angle of 42 degrees. The base width of the bay model was 0.31 meters, representing the narrowest section of the actual bay geometry. Details of the geometry of the bay model can be seen in **Fig. 2**.

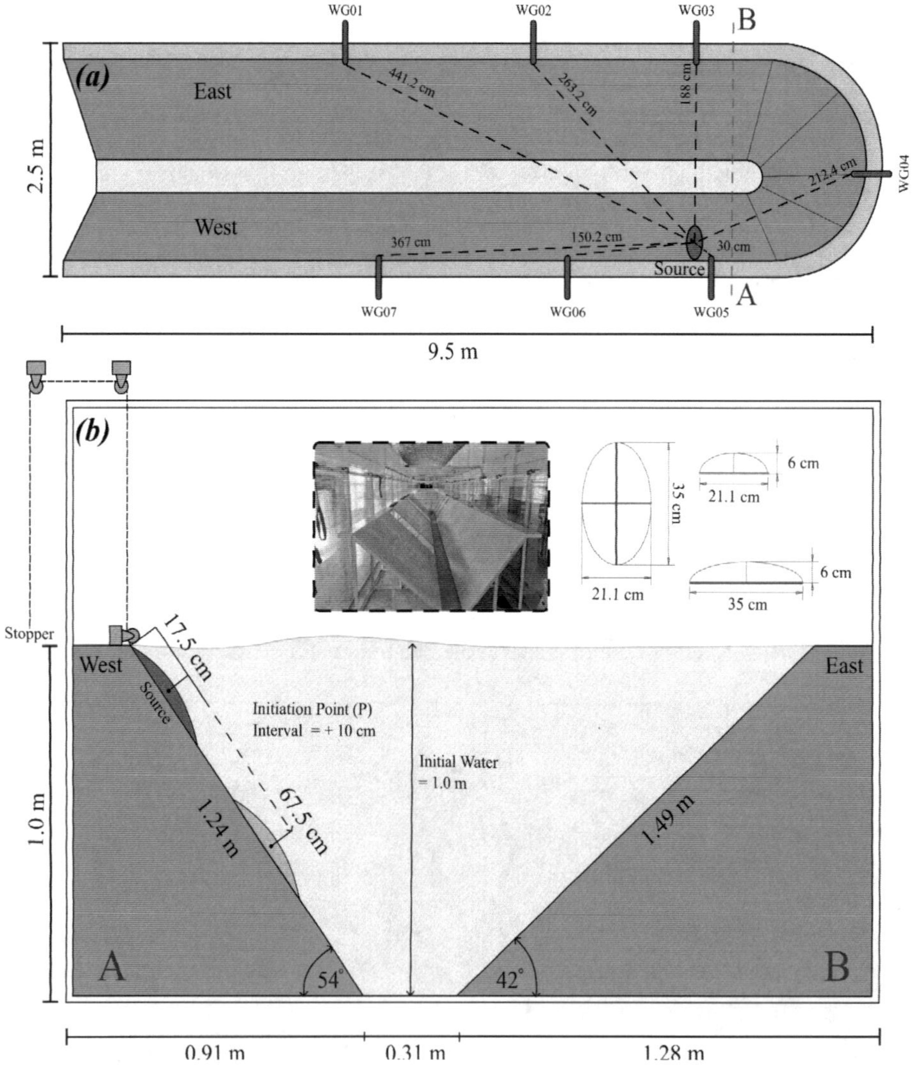

Fig. 2 Experimental setup of physical simulation (a) top view of source location and gauges; dash line is a distance of source to gauges (b) Cross-sectional view of the bathymetry model and the initiation point of source; dark red and light red shapes indicating the initial and final positions of the source initiation point, released at 10 cm intervals. The photos on the top show the installed bathymetry model in the Tsunami Wave Flume Facility.

Resonance of the model was calculated based on Merian's formula as can be seen in Equation (1). Here, f_n is a resonance frequency at n-mode (Hz), L is the length of the model, h is the average bathymetry of the model (m), and g is the gravitational acceleration (m^2/s). The average depth of the model was calculated by divided the cross-section area of the model with the top width of the model. Based on Fig. 1(b\), the average depth of this model was 58.68 cm. Using Eq. (1), therefore, f_1, f_2, f_3, f_4 and f_5 for the model are 0.108 Hz, 0.323 Hz, 0.538 Hz, 0.753 Hz, and 0.968 Hz, respectively. The values were later be compared with dominant frequencies recorded by the wave gauges to assess the similarity between the hydrodynamic results generated by the physical model and the natural frequency responses of the model [9].

$$f_n = \frac{(2n-1)}{4L}\sqrt{\frac{g}{h}} \qquad (1)$$

The landslide material was made as a half-spherical shape to approach the volume of submarine landslides at *Tipo* of Palu at the west bay (S5) side as published by Pakoksung et al. (2019). The landslide model has a 35 cm in length, 21.1 cm in width, and 6 cm of thickness at the center. The total volume of the sliding mass was 2320.07 cm^3, and the material weighed 5.39 kg, resulting in an average density of approximately 2.32 g/cm^3. The semi-elliptical geometry was selected to represent the typical three-dimensional shape of subaqueous landslides observed in natural settings. This configuration also allows for a realistic distribution of mass and initial momentum, essential for reproducing the impulsive wave generation mechanism in the flume-based tsunami simulation.

The flume was constructed using metal framing with a transparent acrylic section at the sliding point to allow visual monitoring of the landslide movement. An initial water depth (Ho) of 1.0 m was maintained throughout all experiments. The experiment was designed to simulate realistic landslide dynamics and tsunami wave generation in a controlled laboratory environment, employing Froude similarity scaling to ensure dynamic consistency between the physical model and its prototype. The landslide material was placed on an inclined acrylic plane, representing a submarine slope, at selected positions considered as initiation points for submarine landslides. Initially, the material was held in place using a restraining string. The release was controlled by timing: at the designated moment, the string was manually detached to initiate downslope movement of the material under gravity. This approach ensured consistent initial conditions across trials. The material was released in a single direction—downslope only. Seven measurement stations were installed at specific locations in the flume to represent the relative positions of cities within the Palu Bay, enabling analysis of wave propagation characteristics across different locations (**Table 1**).

Table 1. Observation Point and gauge locations in Palu Bay for wave evaluation

STA	Latitude	Longitude	Observation Point
WG01	-0.7104703722998741	119.85577580865373	*Pantoloan Port*
WG02	-0.7783855185511501	119.85919309192072	North *Taipa*
WG03	-0.8289567606726337	119.88175744868256	*Tondo-Tadulako*
WG04	-0.8853081447244794	119.85514422128446	*Lere* of West Palu
WG05	-0.8545556397618889	119.82639750224531	Tipo
WG06	-0.8095550042276154	119.80937598520609	*Watusampu* of West Palu
WG07	-0.7519654275113956	119.78354292593136	*Banawa* of Donggala

The wave captured on gauges was set 65 seconds, which was sufficient to capture the complete tsunami wave period along with its reflections within the flume. This duration ensured that both the initial wave generation and the subsequent wave behavior were adequately recorded for further analysis. The release time was recorded and used as the origin time for the landslide event. Each initiation point scenario was repeated three times to ensure repeatability and reduce experimental uncertainty.

2.3 Numerical model

In this study, a numerical simulation approach was also employed to complement the physical modelling results. The numerical simulation solver of the landslide tsunami was conducted using the open-source code smoothed particle hydrodynamics (SPH) [10]. This numerical setup (**Fig. 3**) was designed to replicate the landslide generated tsunami scenario under laboratory-scale conditions to provide wave propagation.

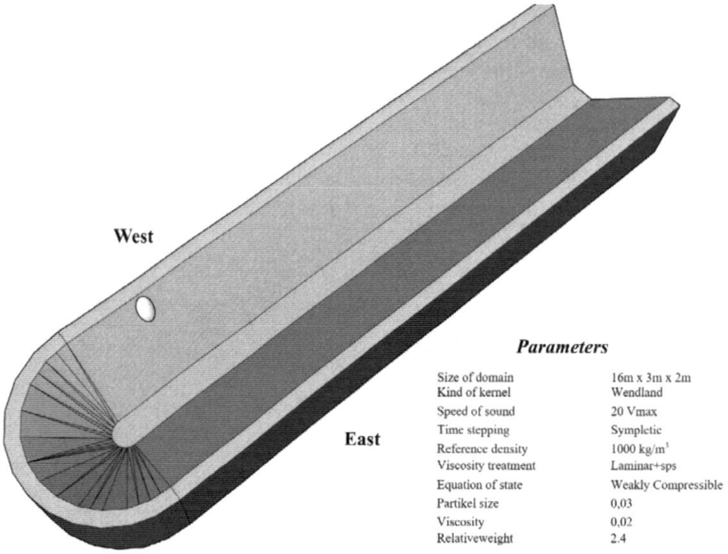

Fig 3. The 3D model of landslide geometry and parameters used in DualSPHysics.

2.4 Power Spectral Density Analysis

The wave data recorded at the wave gauges along the model were analysed using Power Spectral Density Analysis (PSD). The time series data of the wave fluctuations were converted into frequency domain using Fourier Transform as in Equation (2) [11], [12].

$$X(f) = \int_{-\infty}^{\infty} x(t) \, . \, e^{-j2\pi f t} dt \qquad (2)$$

Where *x(t)* is water level at time *t* second. The equation was transformed into Discrete Fourier Transform using the Equation (3).

$$X[k] = \sum_{n=0}^{N-1} x[n] \, . \, e^{-j2\pi kn/N} \quad (3)$$

Here, N is the number of data.

The PSD illustrates how the signal power distributed in frequency domain as calculated using Eqs. (4-6).

$$P_{xx}(f) = \frac{1}{T} |X_T(f)|^2 \quad (4)$$

The periodogram was calculated as in Eq. (5).

$$\widehat{P_{xx}}(f_k) = \frac{1}{N} |X[k]|^2, k = 0, 1, 2, 3, ..., N\text{-}1 \qquad (5)$$

For averaging the periodogram as in Eq. (5), this research used Welch's Method as it can be found in Eq. (6).

$$\hat{P}_{xx}^{Welch}(f) = \frac{1}{L} \sum_{l=1}^{L} \hat{P}_{xx}^{l}(f), \qquad (6)$$

3 Results and Discussion

A total of 18 tsunami events were successfully recorded from 6 different test cases. The submarine tsunami wave propagation observed immediately after the landslide release exhibited a consistent pattern of water level fluctuations across all gauges. A sudden drawdown of the water surface occurred near the source, forming a localized trough of measurable diameter as the sliding mass was released. Simultaneously, a wave crest propagated outward from the source. The initial drawdown was followed by an upward reflection, generating a second wave that traveled to multi-directions from the source (see Fig. 2). The wave gauges installed at each specific location captured the coming wave, starting approximately 10 seconds (release time set-up) before and continuing until about 55 seconds after the landslide release. The recorded waveforms at P=17.5 cm from all gauges are presented in **Fig. 4**, illustrating the time-varying water surface response during each scenario.

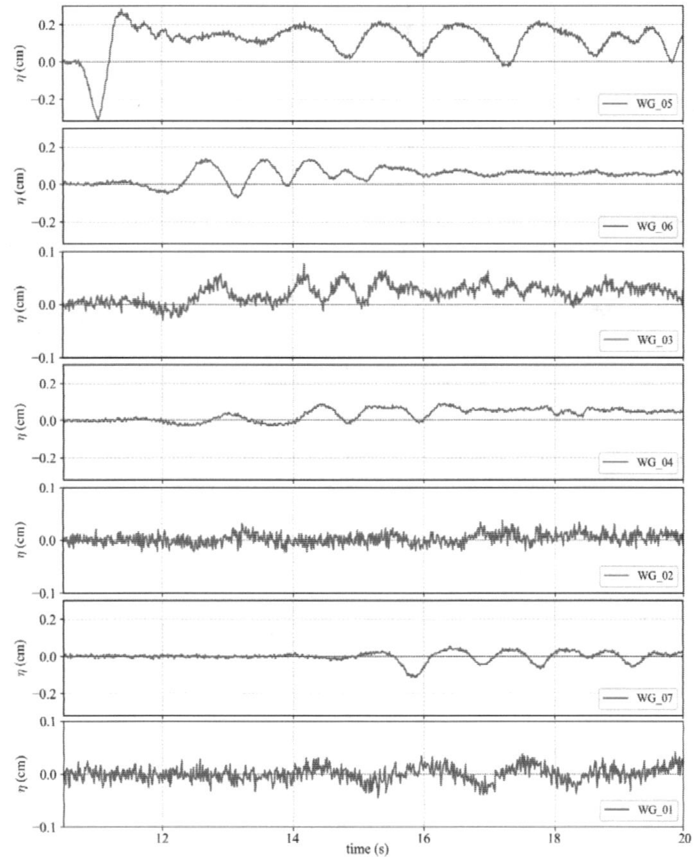

Fig. 4. Tsunami wave profile sort by distance of gauges to the source at P= 17.5 cm, material release at t = 10.50 s

Fig. 4 shows the time series of free surface elevation (η) recorded at seven wave gauge stations (WG_05 to WG_01). The gauges are arranged in order of increasing distance from the landslide source. The station closest to the source, WG_05, recorded the highest amplitude and earliest arrival of the wave. The initial waveform exhibited a pronounced leading depression, an abrupt drop in free surface elevation followed by two to three major oscillations, indicating a typical tsunami generated by landslides. The first-arriving wave has a similar amplitude to the second wave. However, especially of the source (WG05), the amplitude of the initial wave is very large compared to subsequent amplitude. The wave amplitude decreases linearly as it moves away from the source.

When the tsunami waves propagated farther from the source, a systematic reduction in wave amplitude was observed, particularly at stations located opposite the source region (WG01, WG02, and WG03). At WG04, the recorded waveforms displayed a more complex structure: the initial wave exhibited smooth crest and trough shapes, followed by sharper and more symmetric oscillations. This pattern suggests localized frequency modulation or interactions with the bay's geometric features. In contrast, WG06 (peak $\eta < 0.2$ cm) and WG07 (peak $\eta < 0.1$ cm), located on the same side as the landslide source (west), retained the overall waveform profile, although the amplitude dropped significantly after t = 16 s. These observations highlight the damping effects along the propagation path, likely due to bottom friction, geometric spreading, and attenuation of the impulsive energy over distance.

The estimated arrival time was determined based on the first significant change in the free surface elevation (η) detected at each gauge. The picking was performed visually by identifying by the first positive or negative phase of the wave relative to the mean water level during the first 8 seconds of the record. Station WG05, located closest to the initiation point P, recorded an initial wave with negative polarity, reflecting the typical characteristics of landslide generated tsunamis. The estimate of arrival time (AT) and η at each of initiation P show in **Table 2**.

Table 2. Arrival time, physical simulated wave height and velocity, and scaled prototype wave height based on froude similarity

[WG01]
Distance to source 441.2 cm

P (cm)	Arrival Time (s)			Height (cm)			Velocity (cm/s)			Prototype Wave Height (cm)		
	a	*b*	*c*	*a*	*b*	*c*	*a*	*b*	*c*	*a*	*b*	*c*
17.5	3.27	2.25	3.38	0.034	0.056	0.053	134.92	196.09	130.53	21.71	30.57	29.09
27.5	0.56	2.3	2.32	0.033	0.088	0.045	787.86	191.83	190.17	18.34	48.46	24.67
37.5	4.95	3.36	3.28	0.035	0.056	0.089	89.13	131.31	134.51	18.97	30.75	48.46
47.5	3.09	3.1	4.95	0.067	0.071	0.033	142.78	142.32	89.13	36.91	39.00	18.34
57.5	4.97	5.18	3.13	0.027	0.026	0.040	88.77	85.17	140.96	14.76	14.55	22.33
67.5	4.95	2.93	3.51	0.031	0.033	0.056	89.13	150.58	125.70	17.08	18.37	30.58

[WG03]
Distance to source 188 cm

P (cm)	Arrival Time (s)			Height (cm)			Velocity (cm/s)			Prototype Wave Height (cm)		
	a	*b*	*c*	*a*	*b*	*c*	*a*	*b*	*c*	*a*	*b*	*c*
17.5	1.01	1.2	1.24	0.0718	0.0675	0.0628	186.14	156.67	151.61	39.47	37.13	34.57
27.5	1.06	0.68	1.03	0.0516	0.2172	0.0407	177.36	276.47	182.52	28.38	119.46	22.40
37.5	0.53	1.08	0.97	0.0365	0.0807	0.1145	354.72	174.07	193.81	20.06	44.39	62.98
47.5	0.88	0.99	1.32	0.0795	0.0738	0.0287	213.64	189.90	142.42	43.73	40.59	15.79
57.5	1.57	1.02	0.87	0.0237	0.0209	0.0322	119.75	184.31	216.09	13.02	11.52	17.71
67.5	0.72	0.16	0.5	0.0186	0.0283	0.0369	261.11	1175.00	376.00	10.24	15.57	20.30

[WG04]
Distance to source 212.4 cm

P (cm)	Arrival Time (s)			Height (cm)			Velocity (cm/s)			Prototype Wave Height (cm)		
	a	*b*	*c*	*a*	*b*	*c*	*a*	*b*	*c*	*a*	*b*	*c*
17.5	0.66	1.31	1.25	0.0629	0.0562	0.0372	321.82	162.14	169.92	34.59	30.89	20.45
27.5	1.21	1.51	1.15	0.0605	0.1166	0.0281	175.54	140.66	184.70	33.28	64.13	15.44
37.5	1.32	1.06	1.14	0.0467	0.0810	0.0910	160.91	200.38	186.32	25.67	44.55	50.05
47.5	1.18	1.23	1.43	0.0573	0.0574	0.0202	180.00	172.68	148.53	31.52	31.57	11.09
57.5	1.47	2.03	1.29	0.0285	0.0574	0.0336	144.49	104.63	164.65	15.66	31.57	18.48
67.5	0.35	1.09	1.09	0.0206	0.0253	0.0376	606.86	194.86	194.86	11.31	13.92	20.68

[WG07]
Distance to source cm

P (cm)	Arrival Time (s)			Height (cm)			Velocity (cm/s)			Prototype Wave Height (cm)		
	a	*b*	*c*	*a*	*b*	*c*	*a*	*b*	*c*	*a*	*b*	*c*
17.5	3.53	1.96	2.94	0.1562	0.1148	0.0891	103.97	187.24	124.83	85.92	63.15	49.02
27.5	2.85	2.82	3.63	0.0841	0.0934	0.0911	128.77	130.14	101.10	46.26	51.37	50.12
37.5	3.58	3.55	3.56	0.1195	0.0441	0.0601	102.51	103.38	103.09	65.72	24.26	33.06
47.5	2.62	3.5	3.17	0.0584	0.0384	0.0237	140.08	104.86	115.77	32.12	21.12	13.03
57.5	3.85	2.73	3.4	0.0334	0.0347	0.0304	95.32	134.43	107.94	18.36	19.09	16.72
67.5	3.47	3.4	3.53	0.0267	0.0285	0.0334	105.76	107.94	103.97	14.69	15.68	18.37

The prototype wave heights were also examined based on Froude scaling using a vertical scale ratio of 1:550, with a particular focus on the wave gauges located across and alongside the source region (**Table 2**). The gauges analysed include WG01 and WG03, positioned across to the landslide source, and WG04 and WG07 located laterally aligned with the source. The highest prototype wave height recorded at WG01 occurred when the landslide mass was released at positions P = 27.5 cm and P = 37.5 cm, resulting in a wave height of 48.46 cm. At WG03, the maximum wave height was observed at P = 27.5 cm, reaching 119.46 cm. On the source side, WG04 recorded its maximum wave height of 64.13 cm at P = 27.5 cm, while WG07 showed the highest response at P = 17.5 cm, with a wave height of 85.92 cm. The results indicate that wave heights were generally higher along the source side (WG04 and WG07) compared to the opposite side (WG01 and WG03). This suggests stronger wave energy concentration and near-field amplification effects in the immediate vicinity of the landslide source.

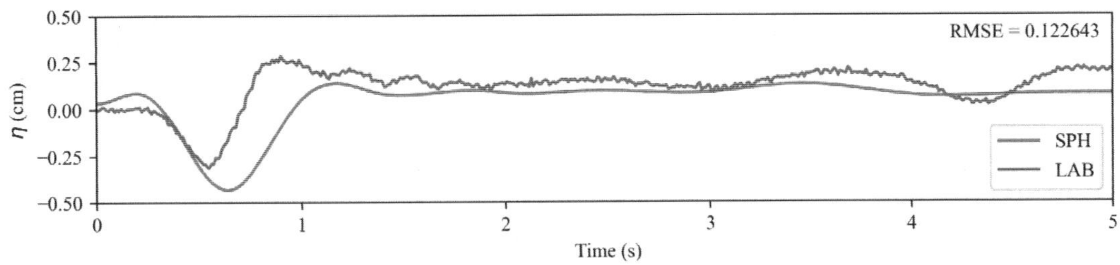

Fig. 5. A comparison of WG05 between the numerical simulation results obtained using the Smoothed Particle Hydrodynamics (SPH) method (red line) and the experimental wave data from the physical laboratory model (blue line). The horizontal axis represents time in seconds, while the vertical axis shows the normalized wave amplitude.

The physical experiment (LAB) displayed a rapid initial drawdown followed by a strong wave crest, indicating the immediate response of the water surface to the submarine landslide release (**Fig. 5**). The SPH model captures the trend of wave propagation, including the initial drawdown and subsequent crest, although with a smoother curve and less variability compared to the laboratory measurements. Differences in amplitude are observed at first peak (0-1 s), indicating the influence of physical effects such as viscosity or friction of slope. The Root Mean Square Error (RMSE) between the two datasets is 0.122643 cm, indicating a relatively good agreement between the SPH simulation and the experimental data within the first 5 seconds of the event.

Power Spectral Density (PSD) from six wave gauges along the bay are displayed in **Figs. 6** dan **7**. Wave Gauge 05 is the closest wave gauge to the source of the submarine landslide at Tipo, at the south bay of the physical model of the bay. The first peak of the graph is at 0.78 Hz with 0.002 cm^2/Hz of PSD. The second peak was seen at 11.5 Hz with much lower PSD, which was at 2.10^{-5} cm^2/Hz. The first peak was clearly identified as a tsunami wave caused by the landslide material. The frequency is significantly close to the natural frequency of the bay model, which is at 0.75Hz as stated in Section Methods. This indicates that the tsunami wave created by the model was amplified by the resonance frequency of the bay model. This could add another severe impacts and larger tsunami waves as it involved nonlinear process of the tsunami wave propagation. The first peak also implies that the submarine landslide caused an immediate primary wave (a tsunami) around the source.

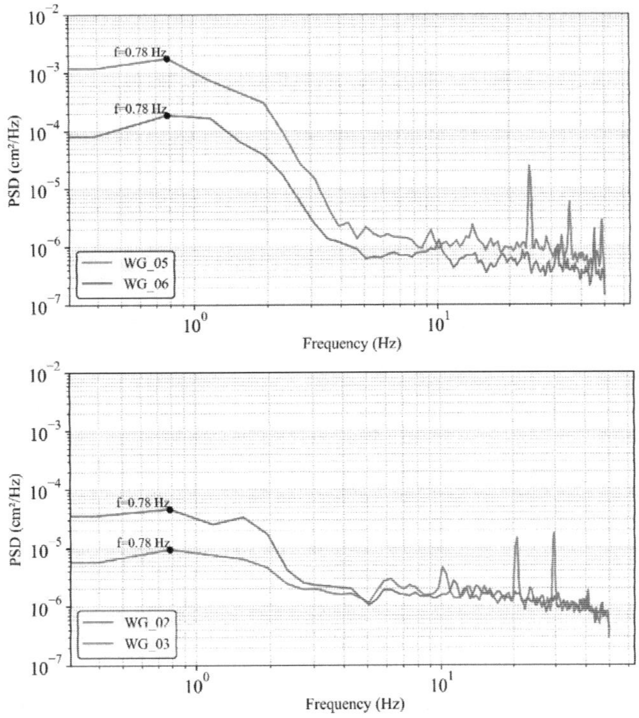

Fig. 6. Power Spectral Densities (PSD) for Wave Gauges 05 (WG_05) and 06 (WG_06) on the left and for Wave Gauges 02 (WG_02) and Wave Gauge 03 (WG_03) on the right.

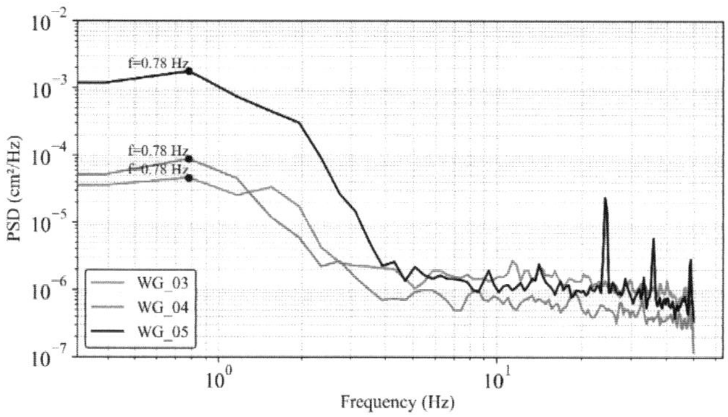

Fig. 7. Power Spectral Densities (PSD) for Wave Gauges 03 (WG_03), 04 (WG_04), and Wave Gauge 05 (WG_05).

The second peaks at some PSD graphs that are located around 11.5 Hz frequency could be attributed to gauge vibrations. They were certainly not part of the tsunami waves caused by the submarine landslide. Meanwhile, a declining trend of the PSD graphs between 0.8 Hz and 10 Hz at all wave gauges imply that the tsunami wave energies generated around the WG05 were distributed to all parts of the bay harmonically. This was due to non-linier interactions of the waves, diffraction process, reflections from the bay, and side-wall effects of the model. At frequencies higher than 15Hz, there were noise signals. The noises could be related to the sensitivity of the sensors used in the experiments. The tsunami wave energy at WG04 located at a closed end of the Bay (Palu city) is higher than the energy recorded at WG03, a location that is opposite to the WG05 as the source of the landslide tsunami.

4 Conclusions

Based on the physical modelling and numerical results, we confirm that the estimated wave height at the observation points correlated to the initiation points of the landslides relative to the surface of the water. The deeper of the landslide initiation point relative to the surface, the lower tsunami wave height measured at the observation points. Accumulation of water volume at the closed boundary of the bay could be observed. The location of the slides was set up at Tipo on the west side of the Palu Bay (WG05). The accumulation of water volume was observed at the western part of the landslide location, the same side as the source, suggesting that the displaced mass directly pushed the water toward that direction. This occurs because the momentum and geometry of the sliding material transfer energy toward the adjacent basin, causing localized water build-up before the waves propagate further. However, this modelling could not solidly confirm any relationship between the landslide initiation points to the tsunami arrival times. The sloping bathymetry of the bay alters the propagation paths of waves from different initiation depths to the sensors. Deeper landslides tend to produce longer or deflected pathways due to seabed contours, while shallower events result in shorter propagation paths. Seiche effects during the tsunami wave propagation could be detected at all observation points as the frequency of the first peak of the tsunami wave was significantly close to the natural frequency of the bay model. This implies that during the submarine landslide tsunami, the effects of the tsunami were exacerbated by the resonance frequency occurred around the same time. The results of this study also confirm the nonlinearity process of the landslide tsunami inside the hypothetical model of the Palu Bay.

5 Acknowledgements

Authors are grateful to *Sekolah Tinggi Meteorologi, Klimatologi, dan Geofisika* (STMKG) of Indonesia for a collaborative research project under Indonesia Disaster Resilience Initiative Project (IDRIP) 2024 funded by World Bank Contract No. 595/PL.101/PPK-IDRIP/XII/STMKG/2024 (Physical Modelling of Non-Seismic Tsunamis, Development of Formulation of Modelling Parameters for Multi-Source Tsunami for Palu Region). The grant has enabled this research to be published. This research is also funded by Ministry of Higher Education, Science and Technology of Indonesia (Kemdiktisaintek) under Grant No. 166/UN11.L1/PG.01.03/DPPM/2025 (*Penelitian Terapan*) 2025 title "SIMuSTi 1.0: Sistem Evaluasi Tsunami Non-Seismik Multi Sumber".

References

1. Syamsidik, Benazir, M. Umar, G. Margaglio, and A. Fitrayansyah, "Post-tsunami survey of the 28 September 2018 tsunami near Palu Bay in Central Sulawesi, Indonesia: Impacts and challenges to coastal communities," *International Journal of Disaster Risk Reduction*, vol. 38, p. 101229, Aug. 2019, doi: 10.1016/j.ijdrr.2019.101229.

2. E. Lahcene *et al.*, "Characteristics of building fragility curves for seismic and non-seismic tsunamis: case studies of the 2018 Sunda Strait, 2018 Sulawesi–Palu, and 2004 Indian Ocean tsunamis," *Natural Hazards and Earth System Sciences*, vol. 21, no. 8, pp. 2313–2344, Aug. 2021, doi: 10.5194/nhess-21-2313-2021.
3. K. Pakoksung, A. Suppasri, F. Imamura, C. Athanasius, A. Omang, and A. Muhari, "Simulation of the Submarine Landslide Tsunami on 28 September 2018 in Palu Bay, Sulawesi Island, Indonesia, Using a Two-Layer Model," *Pure Appl Geophys*, vol. 176, no. 8, pp. 3323–3350, Aug. 2019, doi: 10.1007/s00024-019-02235-y.
4. J. A. Reid and W. D. Mooney, "Tsunami Occurrence 1900–2020: A Global Review, with Examples from Indonesia," *Pure Appl Geophys*, vol. 180, no. 5, pp. 1549–1571, May 2023, doi: 10.1007/s00024-022-03057-1.
5. J. M. González-Vida *et al.*, "The Lituya Bay landslide-generated mega-tsunami – numerical simulation and sensitivity analysis," *Natural Hazards and Earth System Sciences*, vol. 19, no. 2, pp. 369–388, Feb. 2019, doi: 10.5194/nhess-19-369-2019.
6. S. Yavari-Ramshe and B. Ataie-Ashtiani, "Numerical modeling of subaerial and submarine landslide-generated tsunami waves—recent advances and future challenges," *Landslides*, vol. 13, no. 6, pp. 1325–1368, Dec. 2016, doi: 10.1007/s10346-016-0734-2.
7. G.-B. Kim *et al.*, "Three Dimensional Landslide Generated Tsunamis: Numerical and Physical Model Comparisons," *Landslides*, vol. 17, no. 5, pp. 1145–1161, May 2020, doi: 10.1007/s10346-019-01308-2.
8. S. Syamsidik *et al.*, "A New Multi-Purposes Flume Experiments Facility: Challenges and Opportunity for Tsunami and Coastal Engineering in Indonesia," *International Journal of Disaster Management*, vol. 6, no. 3, pp. 345–354, Mar. 2024, doi: 10.24815/ijdm.v6i3.34568.
9. S. Monserrat, I. Vilibić, and A. B. Rabinovich, "Meteotsunamis: atmospherically induced destructive ocean waves in the tsunami frequency band," *Natural Hazards and Earth System Sciences*, vol. 6, no. 6, pp. 1035–1051, Dec. 2006, doi: 10.5194/nhess-6-1035-2006.
10. J. M. Domínguez *et al.*, "DualSPHysics: from fluid dynamics to multiphysics problems," *Comput Part Mech*, vol. 9, no. 5, pp. 867–895, Sep. 2022, doi: 10.1007/s40571-021-00404-2.
11. A. V. . Oppenheim, R. W. . Schafer, and J. R. . Buck, *Discrete-time signal processing*. Prentice Hall, 1999.
12. R. E. . Thomson and W. J. . Emery, *Data analysis methods in physical oceanography*. Elsevier Science, 2014.

Seismic Vulnerabilities Disaggregation Chart in the 2nd District of Cagayan de Oro City, Philippines

Baguhin Israel A[1,*]

[1]University of Science and Technology of Southern Philippines, Cagayan de Oro City, Philippines

Abstract. This study demonstrates the development of a site-specific response spectrum for the 2nd District of Cagayan de Oro City, Philippines, utilizing the R-CRISIS seismic hazard analysis software. The response spectrum was generated using a short vibration period of 0.01 second, a design life of 50 years, and an epsilon value of 5, representing an uncommon seismic event with high ground motion levels. The analysis incorporated local geological and seismotectonic parameters to reflect the unique seismic hazard conditions of the area. Through the use of R-CRISIS, exceedance rates of ground motion were computed and disaggregated by earthquake magnitude, source-to-site distance, and epsilon. The disaggregation charts revealed the dominant earthquake scenarios contributing to the hazard at the designated period and location. The results show that Lapasan, Nazareth, Consolacion, Macabalan, Puntod, and Barangay 1-40 are most vulnerable to earthquake hazard; Cugman, Indahag, Camaman-an, Gusa, and Macasandig are highly vulnerable to earthquake hazard; Balubal, Bugo, Puerto, and FS Catanico are moderately vulnerable to earthquake hazard; and Agusan and Tablon are least vulnerable to earthquake hazard. Additionally, vulnerability charts were generated to visualize ground tremor intensity and hazard levels across the study area. These charts serve as critical instruments for identifying high-risk zones and guiding local land use planning, infrastructure design, and disaster preparedness strategies. This work contributes to the ongoing efforts of seismic risk reduction in the region by providing localized, science-based hazard information tailored for engineering and planning applications.

1 Introduction

In support of the site-specific seismic assessment, the Seismic Vulnerability Disaggregation Chart for the 2nd District of Cagayan de Oro City was developed as part of this study. Seismic hazard disaggregation is a crucial component of Probabilistic Seismic Hazard Analysis (PSHA), as it decomposes the total hazard into the most probable combinations of earthquake magnitude, source-to-site distance, and ground motion levels that contribute to the hazard at a specific site [9]. This process enables engineers and planners to identify the earthquake scenarios that most critically influence the seismic design parameters for structures.

* Corresponding author: israel.baguhin@ustp.edu.ph

Using the R-CRISIS 2015 software [7,8], and leveraging ground motion data from USGS [12] and DOST-PHIVOLCS [4], the study generated disaggregation results under the 10% probability of exceedance in 50 years, corresponding to a 475-year return period. This level of assessment aligns with the seismic performance objectives established by building codes like the National Structural Code of the Philippines (NSCP 2015) and its American counterparts such as ASCE/SEI 7-10 (2010), UBC-1997, and the International Building Code [1,2]. By identifying the predominant earthquake magnitude-distance pairs that contribute to the seismic risk in the district, the disaggregation chart aids in characterizing the Maximum Considered Earthquake (MCE) scenarios. This is pivotal in adopting performance-based seismic design strategies [5] that not only ensure life safety but also enhance structural resilience against collapse and major damage.

Furthermore, this approach responds to the limitations of the generalized design spectra in existing codes [6,10], which may not accurately capture the site-specific seismic threats of Cagayan de Oro City. By utilizing disaggregation outputs, local engineers can develop tailored design spectra that better reflect the actual seismic hazard of the area, contributing to safer and more sustainable infrastructure development. For urban planners and policymakers, these insights provide a scientific basis for risk-informed decision-making, allowing for the integration of seismic risk reduction strategies in land use planning, disaster preparedness, and infrastructure investment [4]. Ultimately, the seismic vulnerability disaggregation enhances the understanding of the earthquake risks that Cagayan de Oro City faces, particularly in the context of rapid urbanization and infrastructure expansion.

2 Material and methods

2.1 Study area

The focus area of this study is the 2nd District of Cagayan de Oro City (figure 1), an urban region experiencing rapid development and infrastructure growth. The area is geographically located in Northern Mindanao, Philippines, within proximity to various active fault systems and tectonic sources, making it vulnerable to seismic hazards.

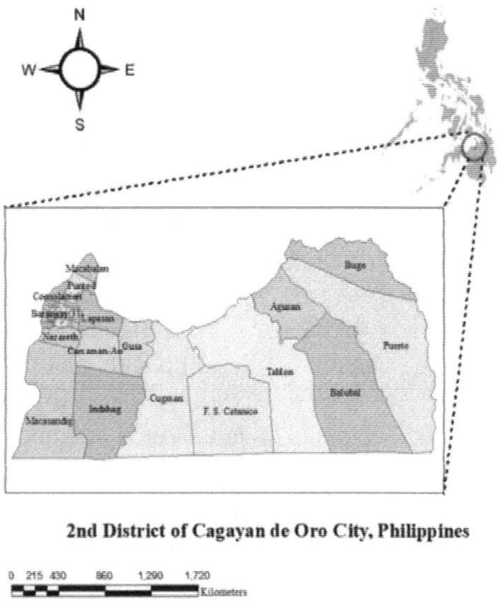

Fig. 1. Study Area.

2.2 Data acquisition

Historical earthquake records from 1907 to 2018 were acquired from the United States Geological Survey (USGS) online earthquake catalog and Department of Science and Technology – Philippine Institute of Volcanology and Seismology (DOST-PHIVOLCS) earthquake database. The collected data comprised of magnitude (M_w) of historical seismic events, epicentral coordinates (latitude and longitude), focal depth, and source-to-site distances. These data served as inputs for defining regional seismic sources and characterizing the recurrence behavior of earthquakes relevant to the study area.

2.3 Probabilistic seismic hazard analysis (PSHA) process

The R-CRISIS 2015 software was utilized to perform the Probabilistic Seismic Hazard Analysis (PSHA). This software is designed to integrate seismicity, seismic source characterization (geometry), ground motion prediction equation (attenuation), site effects, and combination of seismicity, geometric, attenuation, site effects (probabilistic calculations) to estimate seismic hazard levels and conduct disaggregation analysis as shown in figure 2.

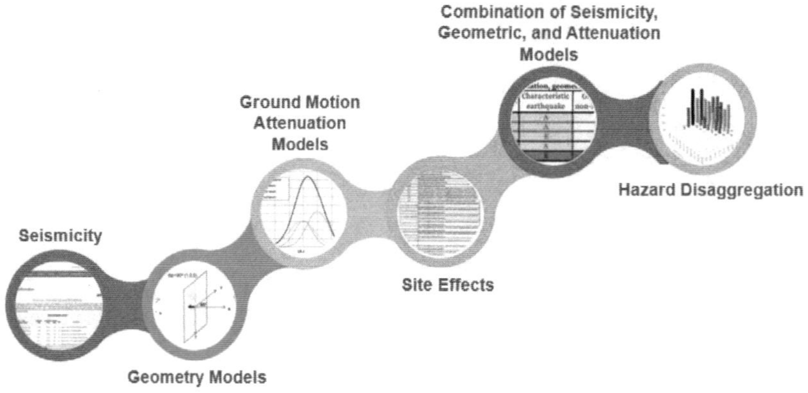

Fig. 2. PSHA Process.

2.3.1 Seismicity and seismic sources

The historical earthquake catalog was gathered from the United States Geological Survey (USGS) website and Department of Science and Technology – Philippine Institute of Volcanology and Seismology (DOST-PHIVOLCS), exported to the ZMAP v7.0 software for de-clustering, to check for identical earthquake events and removal of earthquake foreshocks and aftershock. Then seismic sources have been added to the R-CRISIS software that correspond to a polyline with varying coordinates and depths, consisting of Cabanglasan fault, Cebu Lineament, Central Mindanao fault, Central Negros fault, Cotabato trench, Davao River fault, East Bohol fault, Lanao fault, Mindanao Fault (MF) Daguma Extension, Mindanao Fault (MF) Western Mindanao, Philippine Fault Zone (PFZ) Central Leyte fault, Philippine Fault Zone (PFZ) Eastern Mindanao fault, Philippine Fault Zone (PFZ) Lianga fault, Philippine Fault Zone (PFZ) Mati fault, Philippine Trench, Sulu Trench, Tagoloan River fault, Tangbulan fault, and Zamboanga fault. Figure 3 shows a sample of the Cabanglasn fault added to the R-CRISIS software with indicated coordinates and depth. The ZMAP software was also used to determine earthquake seismic parameters "a" and "b" in every seismic source using the average value and maximum sampling radius of 60 km. Values of"a" and "b" indicate the seismicity rate and the ratio of small to large earthquakes. Equations

1 and 2 were used to calculate the corresponding values of α and β, and subsequently, the mean annual exceedance using equation 3. Values of M_O and M_{max} are determined from the ZMAP software.

$$\alpha = a\ x\ \ln 10 \qquad (1)$$
$$\beta = b\ x\ \ln 10 \qquad (2)$$

$$\lambda_{\alpha,\beta,M_O,\,M_u} = e^{(\alpha-\beta M_o)}\left[e^{(-\beta(M-M_o))} - e^{(-\beta(M_u-M_o))}\right] * \left[\frac{1}{1-e^{(-\beta(M_u-M_o))}}\right] \qquad (3)$$

where,

a = seismicity rate of earthquakes
b = ratio of small to large earthquakes $\alpha = a\ x\ \ln(10)$
$\beta = b\ x\ \ln(10)$
M = earthquake magnitude (usually $M = M_O$)
M_O = threshold magnitude = $M_c - 0.05$
M_c = magnitude of completeness determined from the ZMAP software
M_u = ultimate or maximum magnitude
$\lambda_{\alpha,\beta,M_O,M_u}$ = mean annual rate of exceedance

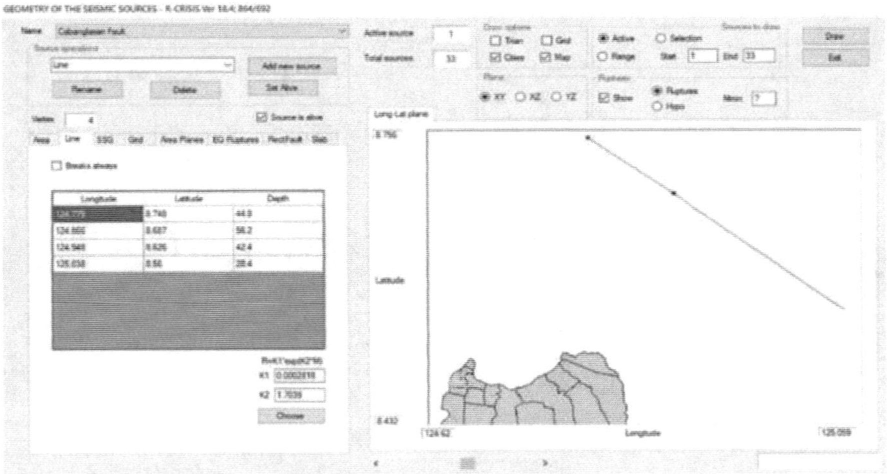

Fig. 3. Line source geometry of Cabanglasan fault in R-CRISIS

2.3.2 Ground motion prediction equation and site effects

The Stewart et al. [11] Ground Motion Prediction Equation (see figure 4) for soil sites was used in this study because of the geologic history of the study area which is an alluvial soil deposit (https://www.cagayandeoro.gov.ph) to predict spectral accelerations as a function of magnitude, distance to the seismic source, and site, where the average upper shear-wave velocity (Vs_{30}) are explicitly used.

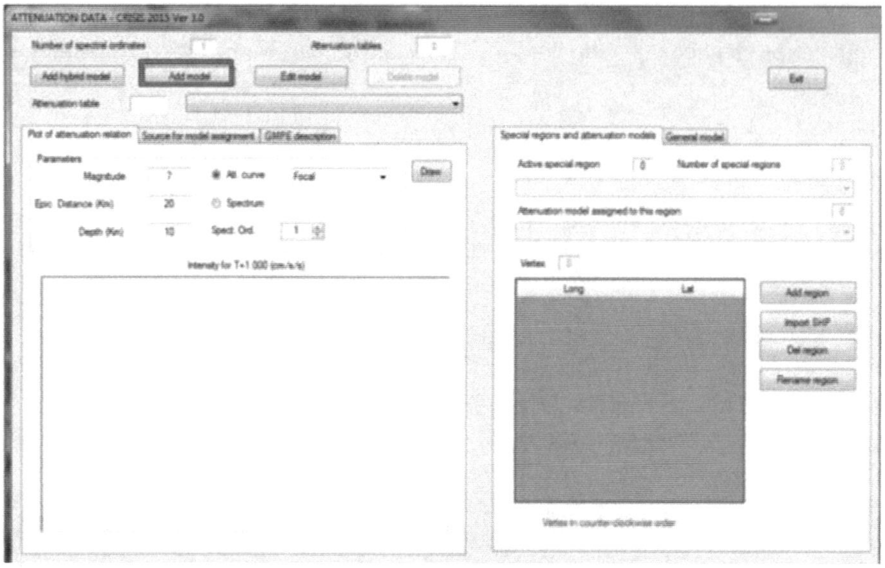

(a)

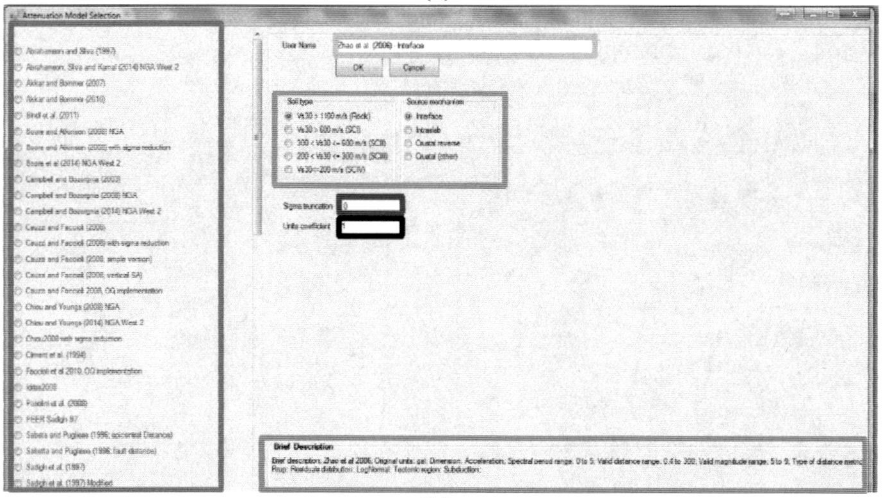

(b)

Fig. 4. Adding a built-in GMPE in R-CRISIS

From geotechnical soil test reports, SPT-N values of ninety-five (95) boreholes are correlated [3] to predict the average upper 30-m shear-wave velocities of soil in fifty-six (56) sites in the 2nd District of Cagayan de Oro City using the equation below.

$$V_{sc} = 59.44N^{0.109}z^{0.426} \tag{4}$$

$$\log V_{s30} = c_o + c_1 \log V_{sz} + c_2(\log V_{sz})^2 \tag{5}$$

The coefficients and standard deviation of the residuals to the fit (σ_{res}) are shown in Table 1.

 https://doi.org/10.1051/e3sconf/202565101005

Table 1. Coefficients relating $\log V_{S30}$ and $\log V_{SZ}$

Depth, z	c_0	c_1	c_2	σ_{RES}
5	1.146	0.5810	0.02573	0.114
6	0.8962	0.7366	0.001817	0.107
7	0.6788	0.8675	-0.01782	0.100
8	0.5684	0.9224	-0.02416	0.094
9	0.4219	1.002	-0.03468	0.087
10	0.3462	1.033	-0.03680	0.082
11	0.2453	1.081	-0.04223	0.076
12	0.1932	1.097	-0.04211	0.071
13	0.08882	1.151	-0.04915	0.066
14	0.02964	1.174	-0.05075	0.061
15	-0.02178	1.191	-0.05150	0.056
16	-0.05916	1.201	-0.05115	0.051
17	-0.1287	1.235	-0.05555	0.047
18	-0.1725	1.252	-0.05679	0.043
19	-0.1992	1.256	-0.05610	0.038
20	-0.2088	1.250	-0.05346	0.034
21	-0.2353	1.255	-0.05317	0.030
22	-0.2462	1.251	-0.05127	0.027
23	-0.2409	1.236	-0.04735	0.023
24	-0.2231	1.212	-0.04213	0.019
25	-0.1929	1.180	-0.03543	0.016
26	-0.1564	1.144	-0.02814	0.013
27	-0.1270	1.114	-0.02205	0.009
28	-0.08924	1.079	-0.01512	0.006
29	-0.04862	1.042	-0.007949	0.003

Further, site-specific conditions (site effects) for the study area were categorized based on soil profiles and geotechnical studies, which influence ground motion amplification (see figure 5). A site class was assigned in accordance with NSCP 2015 and ASCE 7-10 standard.

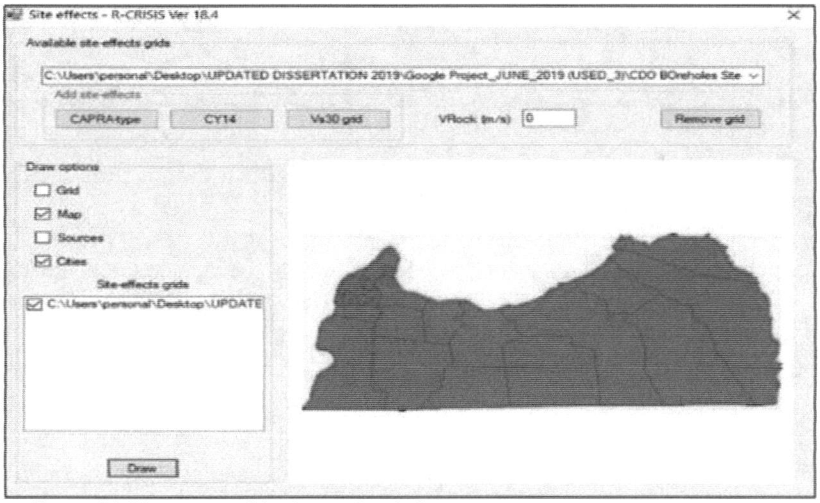

Fig. 5. Site effect grid file exported in the R-CRISIS.

The hazard analysis was calibrated for a 10% probability of exceedance in 50 years, equivalent to a 475-year return period, which corresponds to the Design Basis Earthquake

(DBE) per NSCP 2015 provisions. Using R-CRISIS 2015, the site-specific response spectrum was developed in the study area. The response spectrum represents the maximum expected structural response (acceleration) at varying natural periods of vibration, critical for structural design and performance assessment. The generated site-specific response spectrum was compared against the generalized design response spectrum from UBC-1997, referenced in NSCP 2015 Section 208.

2.3.3 Disaggregation chart

Disaggregation analysis was then conducted at the spectral acceleration corresponding to the first-mode period of typical building structures in the study area using a 0.01s period with a design period of 50 years and an epsilon of 5. The procedure involved identifying the contributing magnitude-distance pairs that dominate the seismic hazard at the site, quantifying the relative contribution of each scenario to the total hazard, and generating disaggregation charts and plots that visualize the dominant seismic sources and their respective impacts (see figure 6). The results of the disaggregation chart were interpreted to inform engineers and designers about the most critical earthquake scenarios to consider in structural design, and guide urban planners and policymakers in seismic risk reduction planning, and recommend the potential for updating local seismic design parameters based on site-specific findings.

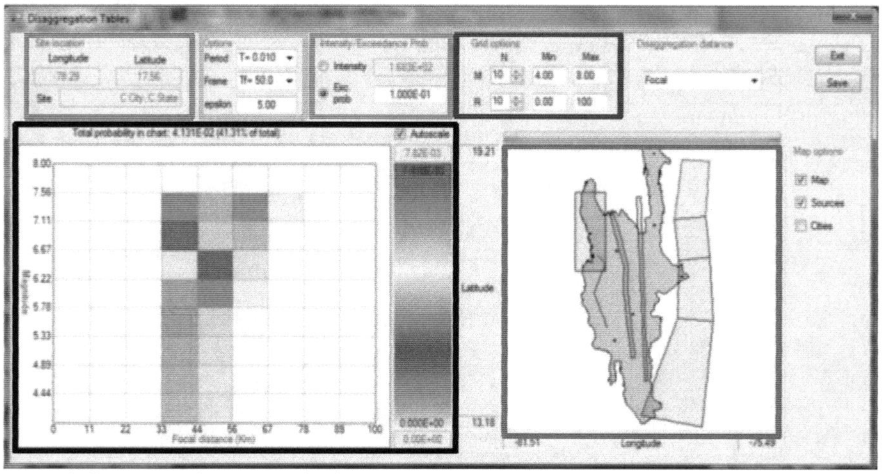

Fig. 6. Hazard disaggregation screen in the R-CRISIS.

3 Results and discussions

Before de-clustering, the earthquake catalog consists of 8,957 magnitude events and after de-clustering, it was reduced to 6,804 magnitude events using a 300-km radius. The de-clustering of the earthquake catalog was executed to check for identical earthquake events and removal of earthquake foreshocks and aftershock, i.e., only earthquake mainshocks are considered in the analysis as shown in figure 7a (before de-clustering) and 7b (after de-clustering).

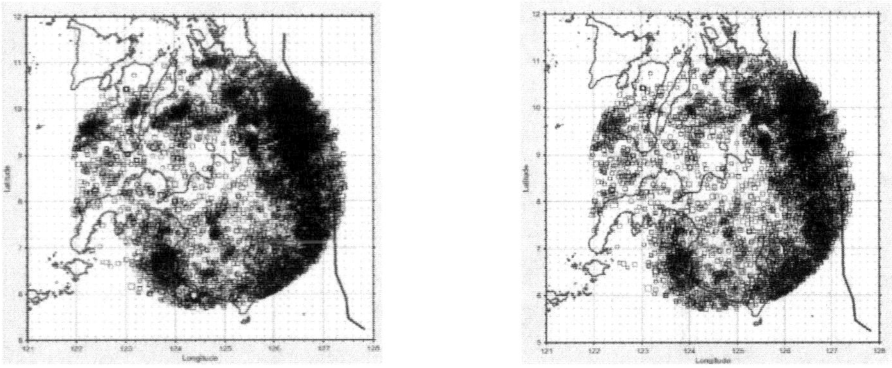

(a) Before de-clustering (b) After de-clustering

Fig. 7. Seismicity of the Study Area.

Seismic parameters for seismic sources were then computed using equation 1-3 as shown in Table 2.

Table 2. Seismic Parameters of Seismic Sources

Seismic Source	a-value	b-value	A	β	M_o	M_u	$\lambda_{\alpha,\beta,M_o,M_u}$
Tagoloan River	7.36	1.17	16.947026	2.694025	5.20	6.30	18.88
Cabanglasan	8.23	1.37	18.950275	3.154542	5.10	6.30	17.50
Central Mindanao	7.08	1.08	16.302302	2.486792	5.00	7.20	47.86
Davao River	6.86	1.04	15.795734	2.394688	4.90	7.00	58.08
PFZ Mati	7.02	0.97	16.164147	2.233508	5.00	8.00	147.91
PFZ Lianga	8.56	1.28	19.710128	2.947309	5.10	7.60	107.65
PFZ Eastern Mindanao	7.49	1.07	17.246362	2.463766	5.00	8.00	138.04
Tangbulan	6.80	0.97	15.657579	2.233508	4.90	7.30	111.43
MF Daguma	6.11	0.83	14.068795	1.911146	4.90	7.60	110.41
Lanao	6.16	0.89	14.183924	2.049301	5.30	7.30	27.73
MF Western Mindanao Ext.	5.91	0.86	13.608278	1.980223	5.10	7.40	33.42
Zamboanga	6.25	0.99	14.391157	2.279559	5.30	7.20	10.07
East Bohol	5.41	0.73	12.456985	1.680887	4.60	7.10	112.72
Central Negros	6.00	0.89	13.815511	2.049301	4.80	7.00	53.46
Cebu Lineament	5.69	0.85	13.101709	1.957197	4.80	7.10	40.74
PFZ Central Leyte	5.94	0.85	13.677355	1.957197	4.80	7.00	72.44
Philippine Trench	7.55	1.07	17.384517	2.463766	4.90	7.60	202.77
Sulu Trench	7.09	1.12	16.325328	2.578895	5.00	6.30	30.90
Cotabato Trench	6.59	0.96	15.174036	2.210482	4.90	7.60	76.91

These seismic parameters were then inputted in the R-CRISIS software as shown in figure 8a-8s.

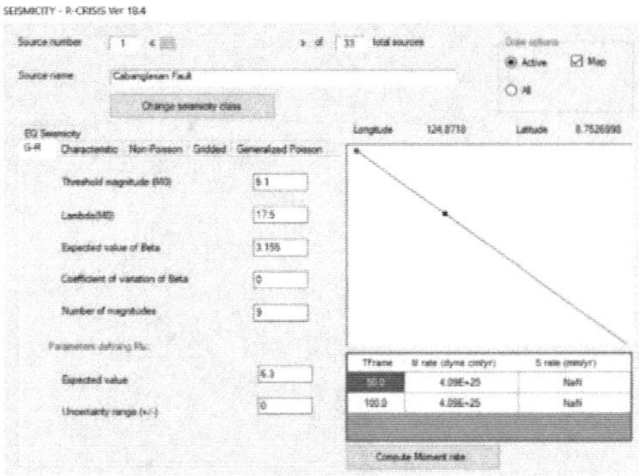

a) Cabanglasan Fault

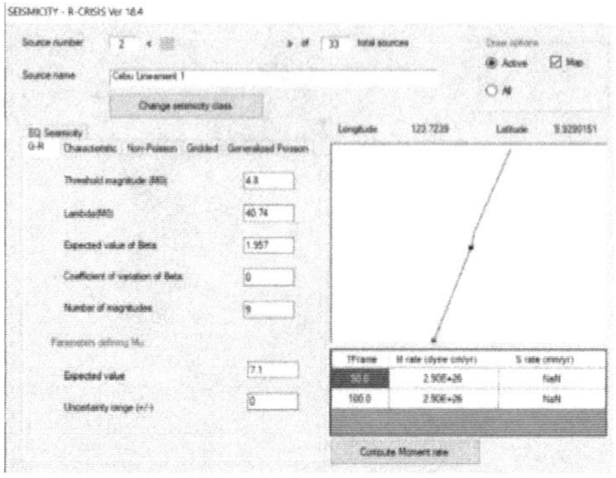

b) Cebu Lineament

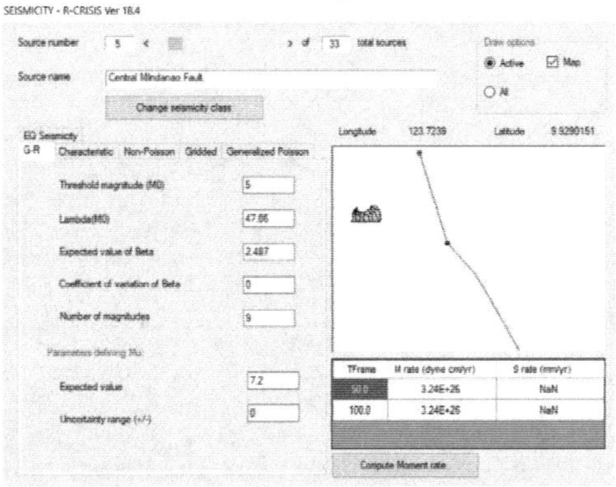

c) Central Mindanao Fault

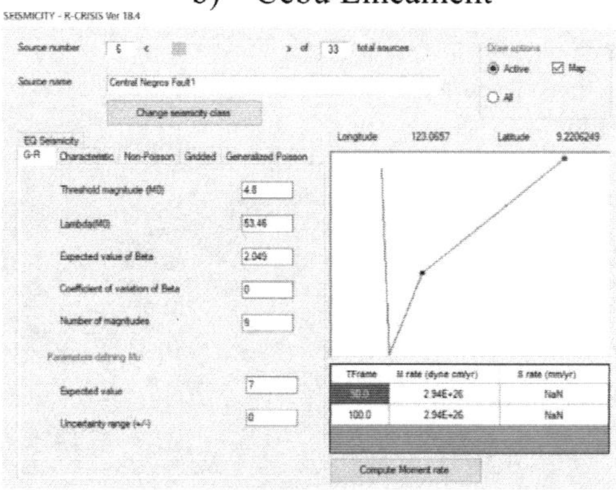

d) Central Negros Fault

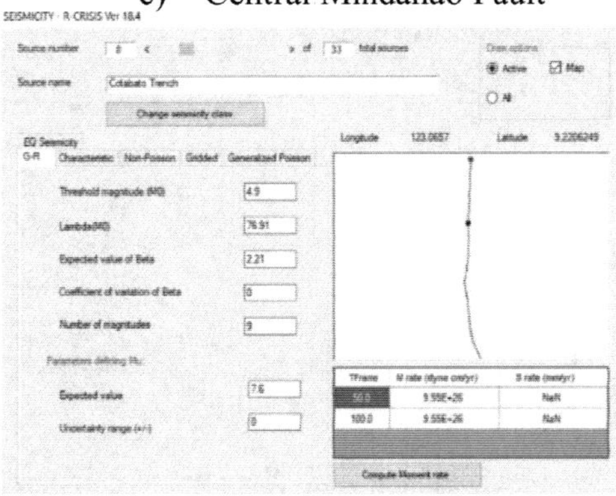

e) Cotabato Trench

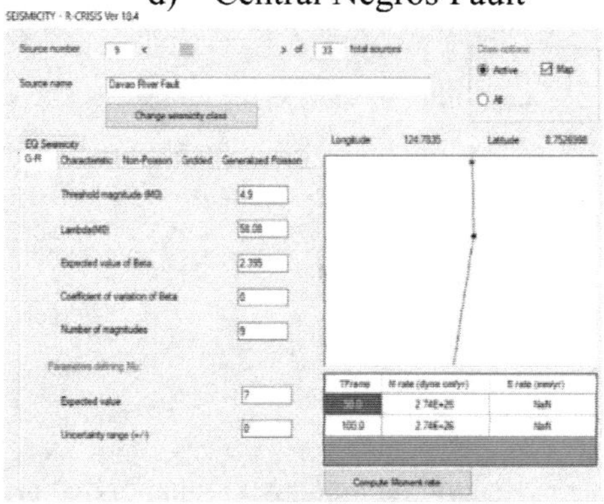

f) Davao River Fault

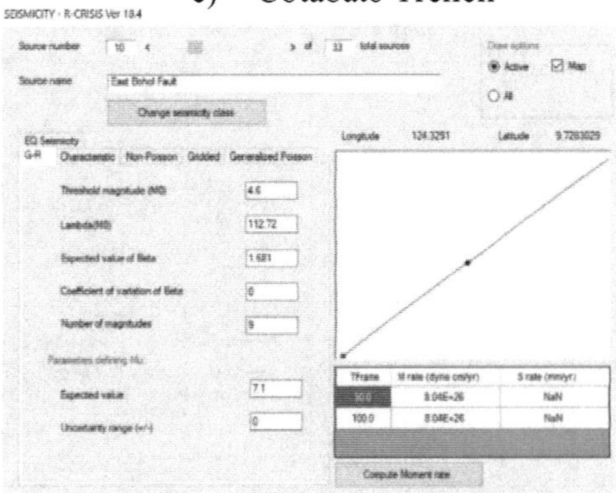

g) East Bohol Fault

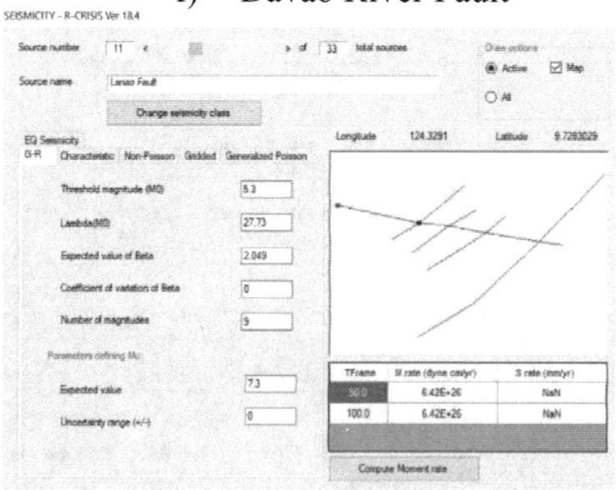

h) Lanao Fault

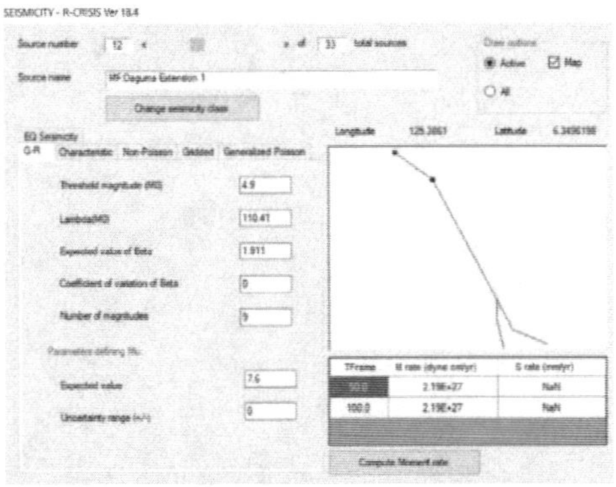

i) MF Daguma Extension

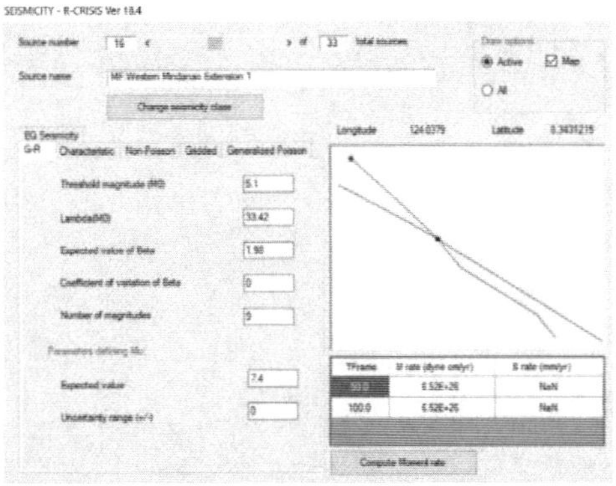

j) MF West Mindanao Extension

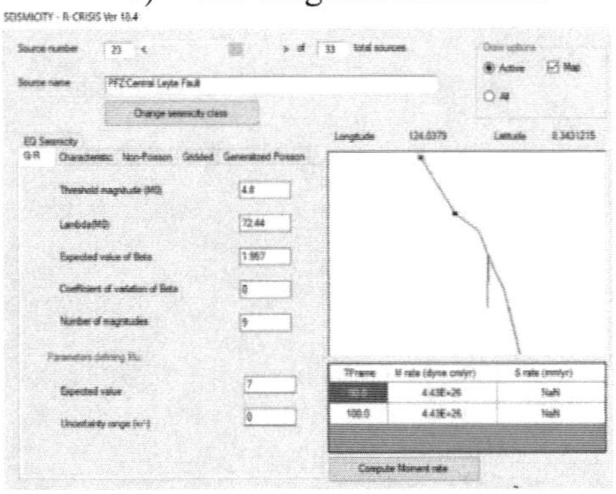

k) PFZ Central Leyte Fault

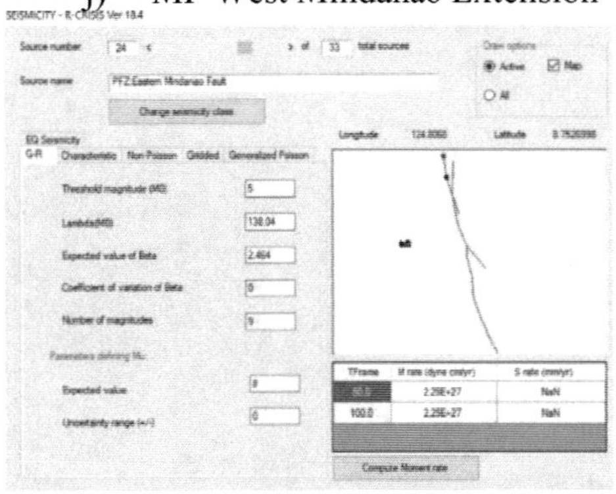

l) PFZ Eastern Mindanao Fault

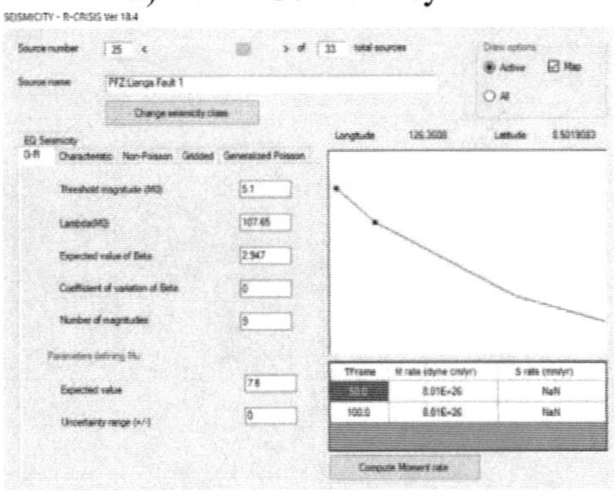

m) PFZ Lianga Fault

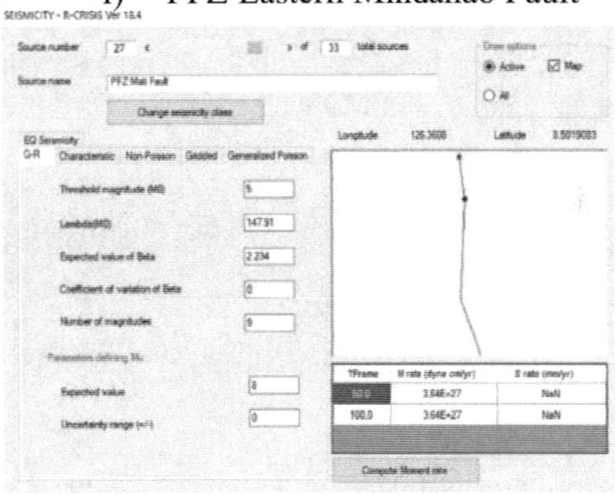

n) PFZ Mati Fault

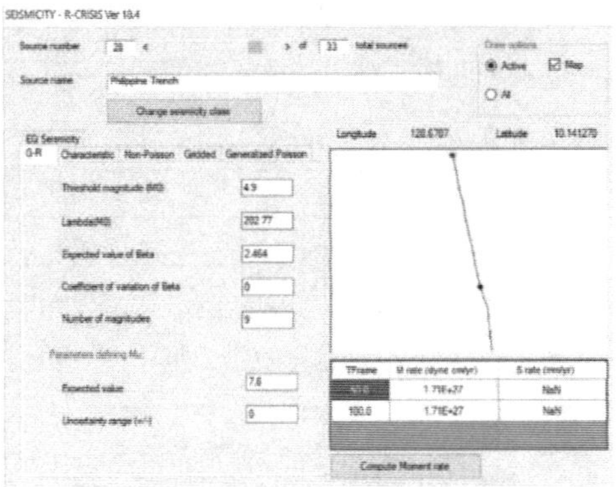

o) Philippine Trench

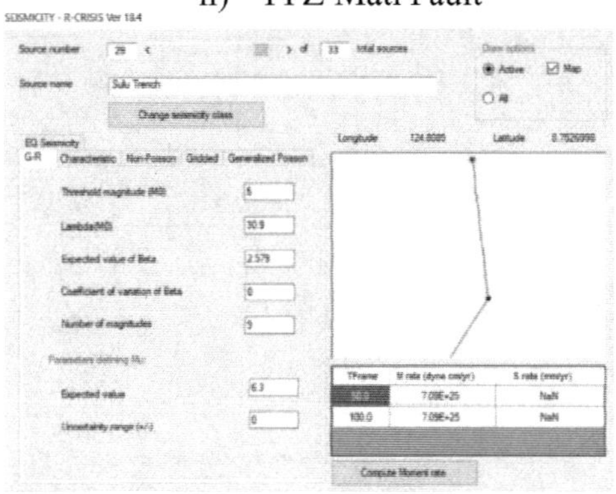

p) Sulu Trench

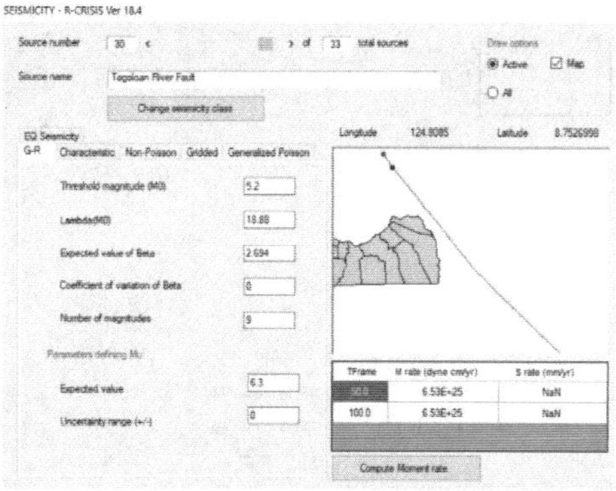

q) Tagoloan River Fault

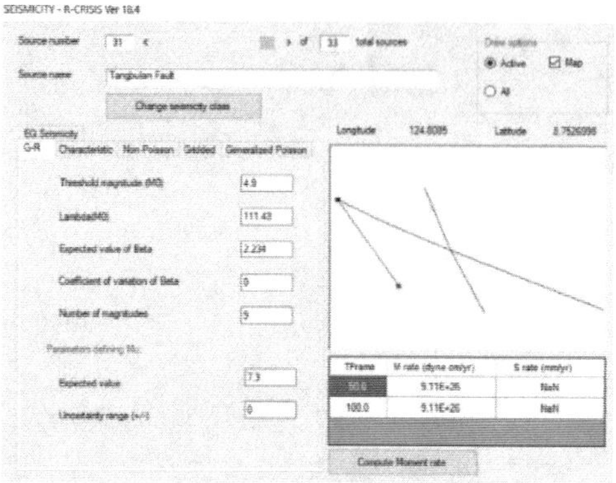

r) Tangbulan Fault

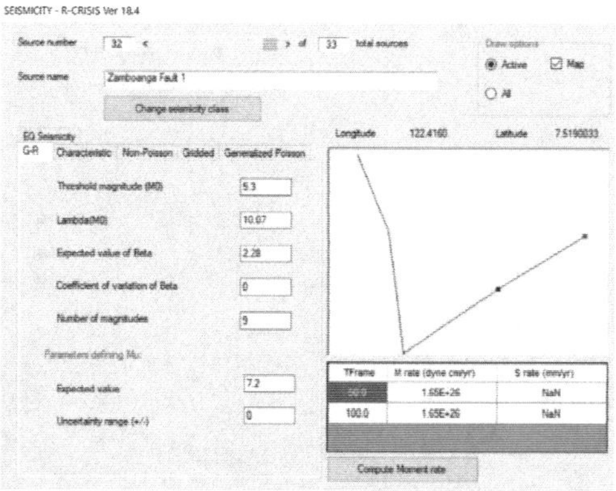

s) Zamboanga Fault

Fig. 8. Input of Seismic Parameters to Seismic Source in the Study Area.

In addition, using equation 4-5, and Table 1, the average upper 30-m shear-wave velocities of soil in fifty-six (56) sites in the 2nd District of Cagayan de Oro City are shown in Table 3.

Table 3. Vs_{30} Values in the 2nd District of Cagayan de Oro City, Philippines

Sites	**Vs_{30}**	**Sites**	**Vs_{30}**
Agusan	383	Barangay 13	366
Balubal	401	Barangay 14	364
Bugo	394	Barangay 15	367
Camaman-an	389	Barangay 16	363
Consolacion	360	Barangay 17	367
Cugman	404	Barangay 18	364
FS Catanico	403	Barangay 19	362
Gusa	411	Barangay 20	360
Indahag	389	Barangay 21	353
Lapasan	372	Barangay 22	341
Macabalan	357	Barangay 23	354
Macasandig	396	Barangay 24	344
Nazareth	371	Barangay 25	355
Puerto	403	Barangay 26	343
Puntod	354	Barangay 27	357
Tablon	381	Barangay 28	350

Barangay 1	368	Barangay 29	363
Barangay 2	368	Barangay 30	358
Barangay 3	364	Barangay 31	351
Barangay 4	366	Barangay 32	359
Barangay 5	366	Barangay 33	358
Barangay 6	368	Barangay 34	357
Barangay 7	367	Barangay 35	359
Barangay 8	365	Barangay 36	362
Barangay 9	362	Barangay 37	362
Barangay 10	367	Barangay 38	363
Barangay 11	364	Barangay 39	361
Barangay 12	365	Barangay 40	364

Using R-CRISIS 2015, the site-specific response spectrum was developed in the fifty-six (56) sites (see figure 9) and compared to the UBC-1997 spectrum. It is noticeable that the UBC-1997 is much conservative than the site-specific response spectrum of the study area which entails higher cost.

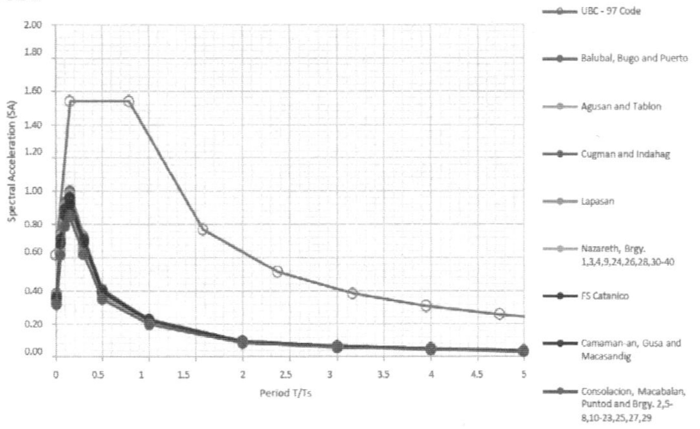

Fig. 9. Site-specific response spectra of the study area.

Fig. 10. Five-Story Commercial Building

A sample of an idealized Five-Story commercial building using the site-specific response spectrum and UBC-1997 design response spectrum was initiated for comparison (see figure 10).

Table 4 shows the comparison of dynamic base shears, lateral forces, displacements, inter-story drift, and Inelastic drift, showing significant differences between the two spectra.

Table 4. Comparison of Site-Specific Spectrum vs UBC-1997

Item/Description of the Five-Story Commercial Building	**Response Spectrum**		**Difference**	**Difference (%)**
	Site-Specific (This Study)	**Uniform Building Code (UBC-1997)**		
1. Dynamic Base Shear	4,857 kN	5,163 kN	306 kN	6.30
2. Maximum Lateral Forces				
5th Floor	581 kN	553 kN	-28 kN	-4.82
4th Floor	892 kN	927 kN	35 kN	3.92
3rd Floor	1,059 kN	1,135 kN	76 kN	7.18
2nd Floor	1,138 kN	1,177 kN	39 kN	3.43
1st Floor	1,187 kN	1,372 kN	185 kN	15.59
3. Maximum Displacement				
5th Floor	0.723 mm	1.075 mm	0.352 mm	48.69
4th Floor	1.377 mm	2.048 mm	0.671 mm	48.73
3rd Floor	1.903 mm	2.830 mm	0.927 mm	48.71
2nd Floor	2.264 mm	3.370 mm	1.106 mm	48.85
1st Floor	2.611 mm	4.182 mm	1.571 mm	60.17
4. Inter-story Drift				
5th Floor	0.72 mm	1.08 mm	0.36 mm	50.00
4th Floor	0.65 mm	0.97 mm	0.32 mm	49.23
3rd Floor	0.53 mm	0.78 mm	0.25 mm	47.17
2nd Floor	0.36 mm	0.54 mm	0.18 mm	50.00
1st Floor	0.35 mm	0.81 mm	0.46 mm	131.43
5. Inelastic Drift				
5th Floor	4.303 mm	6.399 mm	2.096 mm	48.71
4th Floor	3.888 mm	5.786 mm	1.898 mm	48.82
3rd Floor	3.129 mm	4.652 mm	1.523 mm	48.67
2nd Floor	2.149 mm	3.213 mm	1.064 mm	49.51
1st Floor	2.064 mm	4.831 mm	2.767 mm	134.06

One noticeable disparity is the dynamic base shear of the site-specific response spectrum of 4857 kN which is 306 kN less than the 5163 kN of the UBC-1997, a significant difference of 6.3%. Generally, site-specific response spectrum is preferably had the advantage over the UBC-1997 in terms of the dynamic base shear, maximum lateral forces, maximum displacement, inter-story drift and inelastic drift. Further, a simple cost comparison of the five-story building in terms of beams, slabs, columns, walls, and footings is shown in Table 5.

Table 5. Cost Comparison of Site-Specific Spectrum vs UBC-1997

Structural Members	Material Cost (Php)		Difference (Php)	Difference (%)
	Site-Specific Response Spectrum	UBC-1997		
Beams @ 3.0M	₱345,718.08	₱347,564.48	₱1,846.40	0.53
Beams @ 6.0M	₱491,122.13	₱492,427.18	₱1,305.05	0.27
Beams @ 9.0M	₱297,681.72	₱304,259.18	₱6,577.46	2.21
Beams @ 12.0M	₱300,038.79	₱265,671.05	-₱34,367.74	-11.45
Beams @ 15.0M	₱193,928.03	₱201,374.94	₱7,446.91	3.84
Slab @ 3.0M	₱2,636,860.64	₱2,537,005.64	-₱99,855.00	-3.79
Slab @ 6.0M	₱2,298,094.18	₱2,537,005.64	₱238,911.46	10.40
Slab @ 9.0M	₱2,574,740.26	₱2,617,315.88	₱42,575.62	1.65
Slab @ 12.0M	₱2,574,595.77	₱2,777,488.60	₱202,892.83	7.88
Slab @ 15.0M	₱2,770,656.71	₱2,828,732.53	₱58,075.82	2.10
Columns/Walls	₱2,380,852.13	₱2,565,720.57	₱184,868.44	7.76
Footings	₱12,845,789.39	₱13,967,137.68	₱1,121,348.29	8.73
MATERIAL COST	**₱29,710,077.83**	**₱31,441,703.37**	**₱1,731,625.54**	**5.83**

In Table 5, an over-all savings of Php 1,731,625.54 or 5.83% of the building material was achieved when the site-specific response spectrum was used instead of the UBC-1997.

After considering seismicity, seismic source characterization (geometry), ground motion prediction equation (attenuation), and site effects in the R-CRISIS software to estimate seismic hazard levels and disaggregation analysis, exceedance rates disaggregated by magnitude, distance and epsilon value were generated, using a period of 0.01s, a design period of 50 years and an epsilon of 5. For Balubal, Bugo, and Puerto, most of the hazard comes from the earthquake moment magnitude M_w of 6.11 where the probability that the probability of exceedance (P_E) is exceeded, using a 10% probability of exceedance and design period of 50 years, is 0.00934 located 33 kilometers of focal distance from the site (see figure 11). This implies that there is a 37% probability of exceedance that the M_w of 6.11 at a disaggregation focal distance of 33 kilometers from the site is exceeded in a 50-year exposure period (see Table 6). The calculation is shown below;

Annual rate of exceedance, $\lambda_D = 0.00934$
Exposure period, $T_e = 50$ years
Return period, $T_R = 1/\lambda_D$
Probability of exceedance, $P_E = 1 - e^{-\lambda_D * T_e}$
Probability of exceedance, $P_E = 1 - e^{-0.00934*50} = 37\%$

Table 6. P_E for Balubal, Bugo, and Puerto

Sites/Locations	Moment Magnitude (Mw)	Disaggregated Distances (km)							
		33	67	100	133	167	200	233	267
Balubal, Bugo and Puerto	4.98	-	-	-	-	-	-	-	-
	5.36	3%	-	-	-	-	-	-	-
	5.73	22%	-	-	-	-	-	-	-
	6.11	37%	2%	2%	-	-	-	-	-
	6.49	27%	7%	13%	6%	3%	1%	-	-
	6.87	34%	27%	27%	18%	13%	7%	1%	-
	7.24	-	19%	29%	22%	25%	17%	4%	1%
	7.62	-	-	28%	25%	20%	27%	1%	-
	8.00	-	-	-	-	-	-	-	-

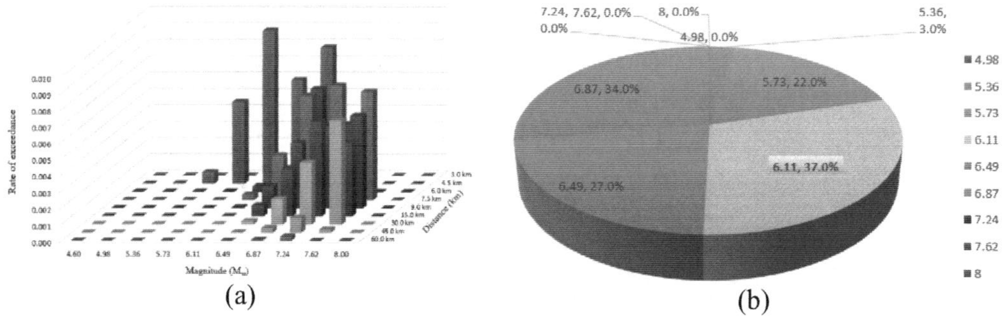

(a) (b)

Fig. 11. a. Disaggregation chart, b. Pie chart for Balubal, Bugo, and Puerto.

For Agusan and Tablon, most of the hazard comes from the earthquake moment magnitude M_w of 6.87 where the probability that the P_E is exceeded is 0.00895 located 67 kilometers of focal distance from the site (see figure 12). This implies that there is a 36% probability of exceedance that the M_w of 6.87 at a disaggregation focal distance of 67 kilometers from the site is exceeded in a 50-year exposure period (see Table 7).

Table 7. P_E for Agusan and Tablon

Sites/Locations	Moment Magnitude (Mw)	Disaggregated Distances (km)							
		33	67	100	133	167	200	233	267
Agusan and Tablon	4.98	-	-	-	-	-	-	-	-
	5.36	2%	-	-	-	-	-	-	-
	5.73	13%	-	-	-	-	-	-	-
	6.11	23%	2%	2%	-	-	-	-	-
	6.49	19%	10%	16%	10%	4%	1%	-	-
	6.87	20%	36%	32%	26%	14%	9%	2%	-
	7.24	-	22%	31%	29%	28%	20%	6%	1%
	7.62	-	-	28%	32%	16%	32%	1%	-
	8.00	-	-	-	-	-	-	-	-

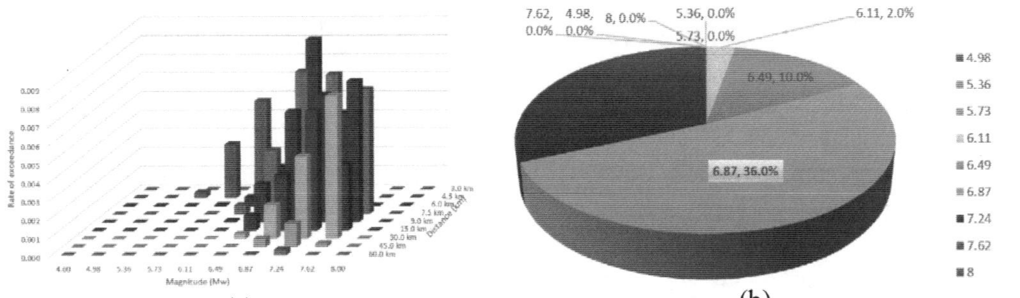

(a) (b)

Fig. 12. a. Disaggregation chart, b. Pie chart for Agusan and Tablon.

For FS Catanico, most of the hazard comes from the earthquake moment magnitude M_w of 6.87 where the probability that the P_E is exceeded is 0.00928 located 67 kilometers of focal distance from the site (see figure 13). This is an indication that there is a 37% probability of exceedance that the M_w of 6.87 at a disaggregation focal distance of 67 kilometers from the site is exceeded in a 50-year exposure period (see Table 8).

Table 8. P_E for FS Catanico

Sites/Locations	Moment Magnitude (Mw)	Disaggregated Distances (km)							
		33	67	100	133	167	200	233	267
FS Catanico	4.98	-	-	-	-	-	-	-	-
	5.36	2%	-	-	-	-	-	-	-
	5.73	15%	-	-	-	-	-	-	-
	6.11	27%	2%	2%	-	-	-	-	-
	6.49	17%	12%	13%	10%	4%	1%	-	-
	6.87	18%	37%	32%	24%	15%	9%	2%	-
	7.24	-	21%	29%	31%	30%	20%	5%	2%
	7.62	-	-	28%	33%	21%	28%	1%	-
	8.00	-	-	25%	-	-	-	-	-

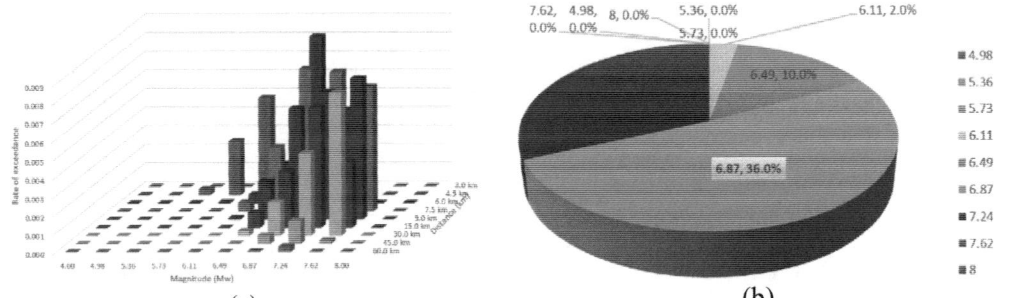

(a) (b)

Fig. 13. a. Disaggregation chart, b. Pie chart for FS Catanico.

For Cugman, Indahag, Camaman-an, Gusa, and Macasandig, most of the hazard comes from the earthquake moment magnitude M_W of 7.24 where the probability that the P_E is exceeded is 0.01298 located 133 kilometers of focal distance from the site (see figure 14). This is an indication that there is a 48% probability of exceedance that the M_w of 7.24 at a disaggregation focal distance of 133 kilometers from the site is exceeded in a 50-year exposure period (see Table 9).

Table 9. P_E for Cugman, Indahag, Camaman-an, Gusa, and Macasandig

Sites/Locations	Moment Magnitude (Mw)	Disaggregated Distances (km)							
		33	67	100	133	167	200	233	267
Cugman, Indahag, Camaman-an, Gusa and Macasandig	4.98	-	-	-	-	-	-	-	-
	5.36	-	-	-	-	-	-	-	-
	5.73	5%	-	-	-	-	-	-	-
	6.11	11%	3%	2%	-	-	-	-	-
	6.49	5%	11%	16%	2%	-	-	-	-
	6.87	5%	32%	32%	16%	5%	1%	4%	-
	7.24	-	20%	31%	39%	18%	9%	10%	2%
	7.62	-	-	28%	48%	31%	19%	13%	-
	8.00	-	-	-	29%	24%	22%	-	-

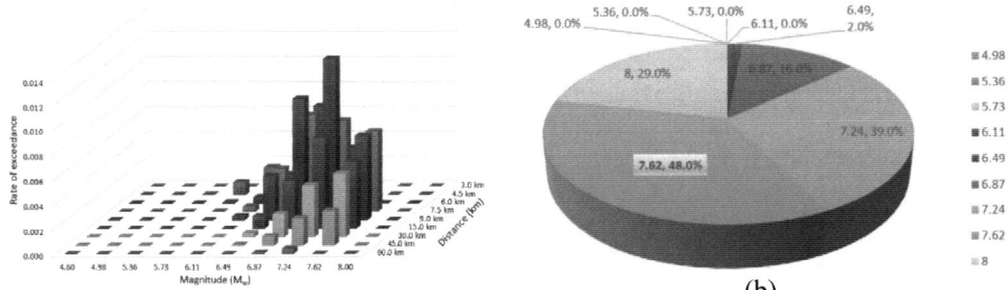

(a) (b)

Fig. 14. a. Disaggregation chart, b. Pie chart for Cugman, Indahag, Camaman-an, Gusa, and Macasandig.

For Lapasan, Nazareth, Consolacion, Macabalan, Puntod, and Barangay 1-40, most of the hazard comes from the earthquake moment magnitude M_w of 7.24 where the probability that the P_E is exceeded is 0.01298 located 133 kilometers of focal distance from the site (see figure 15). This is an indication that there is a 48% probability of exceedance that the M_w of 7.24 at a disaggregation focal distance of 133 kilometers from the site is exceeded in a 50-year exposure period (see Table 10).

Table 10. P_E for Lapasan, Nazareth, Consolacion, Macabalan, Puntod, and Barangay 1-40

Sites/Locations	Moment Magnitude (Mw)	Disaggregated Distances (km)							
		33	67	100	133	167	200	233	267
Lapasan, Nazareth, Consolacion, Macabalan, Puntod and Barangays 1-40	4.98	-	-	-	-	-	-	-	-
	5.36	-	-	-	-	-	-	-	-
	5.73	5%	1%	-	-	-	-	-	-
	6.11	9%	3%	2%	2%	-	-	-	-
	6.49	5%	13%	12%	20%	4%	1%	-	-
	6.87	6%	38%	24%	45%	14%	8%	4%	-
	7.24	-	25%	23%	53%	26%	16%	12%	2%
	7.62	-	-	23%	39%	17%	22%	12%	-
	8.00	-	-	-	-	-	-	-	-

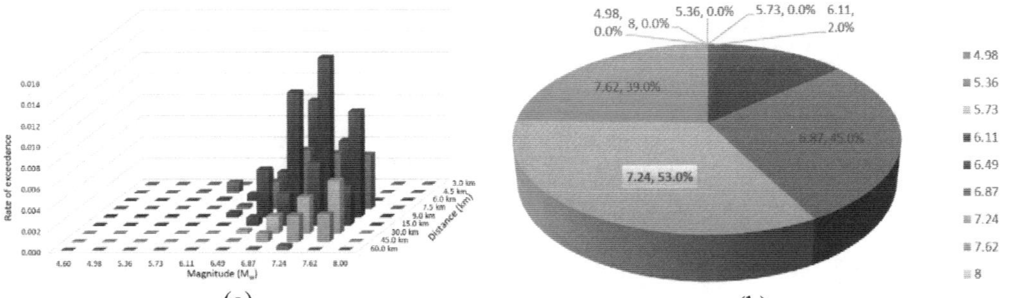

(a) (b)

Fig. 15. a. Disaggregation chart, b. Pie chart for Lapasan, Nazareth, Consolacion, Macabalan, Puntod, and Barangay 1-40.

4 Conclusion

Based on the results, it was concluded that:

- Lapasan, Nazareth, Consolacion, Macabalan, Puntod, and Barangay 1-40 are **most vulnerable** to earthquake hazard with 53% probability of exceeding a moment magnitude M_W of 7.24 at a disaggregation focal distance of 133 kilometers from the site in a 50-year exposure period.
- Cugman, Indahag, Camaman-an, Gusa, and Macasandig are **highly vulnerable** to earthquake hazard with 48% probability of exceeding a moment magnitude M_W of 7.24 at a disaggregation focal distance of 133 kilometers from the site in a 50-year exposure period.
- Balubal, Bugo, Puerto are **moderately vulnerable** to earthquake hazard with 37% probability of exceeding a moment magnitude M_W of 6.11 at a disaggregation focal distance of 33 kilometers from the site in a 50-year exposure period.
- FS Catanico are **moderately vulnerable** to earthquake hazard with 37% probability of exceeding a moment magnitude M_W of 6.87 at a disaggregation focal distance of 67 kilometers from the site in a 50-year exposure period.
- Agusan and Tablon are **least vulnerable** to earthquake hazard with 36% probability of exceeding a moment magnitude M_W of 6.87 at a disaggregation focal distance of 67 kilometers from the site in a 50-year exposure period.

Acknowledgement. The author would like to thanks the Commission on Higher Education K-12 (CHED K-12) Scholarship, Department of Public Works and Highways (DPWH) Region 10, and the Department of Science and Technology (DOST)-PHIVOLCS for the assistance in providing the data and documents needed for the completion of the study.

References

1. American Society of Civil Engineers. (2022). *Minimum design loads and associated criteria for buildings and other structures (ASCE/SEI 7-22).* American Society of Civil Engineers. https://doi.org/10.1061/9780784415789.
2. American Society of Civil Engineers / International Code Council. *2024 International Building Code (IBC)*. 2024.

3. Dehghanian, K., Çaltılı, E., Dinç, B. K., & Soysal, H. M. (2022). Implementation of Shear Wave Velocity and Standard Penetration Test Correlation for Edirne District, Turkey. *Journal of Sustainable Construction Materials and Technologies, 7*(1), 24-29. https://doi.org/10.14744/jscmt.2022.10.
4. Department of Science and Technology – Philippine Institute of Volcanology and Seismology (DOST-PHIVOLCS): Earthquake Information and Philippine Seismicity (2023). Available at: https://www.phivolcs.dost.gov.ph/index.php/earthquake/earthquake-information3
5. Federal Emergency Management Agency. (2021). *NEHRP recommended seismic provisions for new buildings and other structures (FEMA P-2082).* National Institute of Standards and Technology / Building Seismic Safety Council. https://www.fema.gov/emergency-managers/risk-management/building-science/2020-nehrp-provisions.
6. National Structural Code of the Philippines (NSCP): Volume 1: Buildings, Towers, and Other Vertical Structures (Association of Structural Engineers of the Philippines, 2015).
7. Ordaz, M., Martinelli, F., Aguilar, A., Arboleda, J., Meletti, C., D'Amico, V.: CRISIS 2015: Program for computing seismic hazard. Available at https://www.researchgate.net/deref/https%3A%2F%2Fsites.google.com%2Fsite%2Fcodecrisis2015%2Fhelp?_tp=eyJjb250ZXh0Ijp7ImZpcnN0UGFnZSI6InB1YmxpY2F0aW9uIiwicGFnZSI6InB1YmxpY2F0aW9uIiwicG9zaXRpb24iOiJwYWdlQ29udGVudCJ9fQ.
8. Ordaz, M., Salgado-Gálvez, M.A.: R-CRISIS Validation and Verification Document, ERN Technical Report (Mexico City, Mexico, 2018).
9. Sawires, R. (2024). *Deaggregation of probabilistic seismic hazard results for 15 cities in Western Mexico*. Natural Hazards. https://doi.org/10.1080/17499518.2023.2251125.
10. Shakeel, S., Khan, S. H., Saqib, S. A., Awais Khan, M., & Abdul Moiz, M. (2024). *Seismic performance and cost comparison of RC moment resisting and dual frames using UBC-97 and IBC 2021. Scientific Reports, 14*, Article 16349. https://doi.org/10.1038/s41598-024-67373-3.
11. Stewart, J. P., Parker, G. A., Atkinson, G. M., Boore, D. M., Hashash, W. J., & Silva, W. J. "Ergodic site amplification model for central and eastern North America." *Earthquake Spectra* 36(1), 42–68 (2020).
12. United States Geological Survey (USGS): Earthquake Catalog and Data (2024). Available at: https://earthquake.usgs.gov/

E3S Web of Conferences **651**, 01006 (2025) https://doi.org/10.1051/e3sconf/202565101006
AIWEST-DR 2025

Generation of Seismic Time Histories for Disaster Risk Reduction: A Case Study in the 2nd District of Cagayan de Oro City, Philippines

Baguhin Israel A[1,*]

[1]University of Science and Technology of Southern Philippines, Cagayan de Oro City, Philippines

Abstract. In the 2nd District of Cagayan de Oro City, Philippines, this work describes the construction of a target site-specific response spectrum and the subsequent synthesis of compatible ground motion time-histories for seismic analysis and structural design. The site-specific response spectrum was developed utilizing the seismic hazard parameters specified in the 2015 National Structural Code of the Philippines (NSCP 2015) and soil reports from geotechnical investigations as part of a performance-based earthquake engineering strategy. This target spectrum captures the expected seismic demand specific to the site's tectonic and geologic circumstances. Finding relevant ground motion data required submitting the target spectrum to the Pacific Earthquake Engineering Research (PEER) Ground Motion Database. Thousands of unscaled ground motion data that closely matched the target's spectral properties within a given period range were found after an extensive search. After that, the response spectrum for each site was sent to the SeismoMatch program for spectral matching. A matching procedure was iteratively used to scale each ground motion to match the desired spectrum over the designated period range. A collection of spectrum-compatible, site-specific, and scaled ground motion time histories that may be applied to nonlinear time-history analysis was the end result. When assessing a structure's structural performance under seismic loads, these temporal histories are crucial. The method takes into consideration soil conditions, spectral compatibility, and local seismicity to improve the dependability of earthquake-resistant designs.

1 Introduction

Earthquakes are among the most catastrophic natural phenomena that cause substantial damage to infrastructures and threaten lives, particularly in seismically active regions like the Philippines. To mitigate these risks, structural designs have evolved with strict compliance to building codes that incorporate seismic provisions to ensure life safety and structural integrity [10]. The National Structural Code of the Philippines [11] has adapted provisions from globally recognized standards, yet it continues to rely on generalized design response spectra such as those from UBC-97, which are not site-specific and may not represent the true seismic demand of a particular location. Generalized design response

* Corresponding author : israel.baguhin@ustp.edu.ph

spectra, such as those from UBC-97 or the default provisions of NSCP, may misrepresent seismic demand because they are based on broad regional averages that neglect local soil conditions, site effects, and source-to-site variations [7]. This limitation often leads to either underestimation or overestimation of seismic forces. For instance, Bazzurro and Cornell [4] demonstrated that generalized spectra can deviate by more than 30% from site-specific demands, potentially resulting in unsafe or overly conservative designs. Likewise, [5] showed that the Uniform Hazard Spectrum (UHS) frequently mischaracterizes the frequency content of expected ground motions, especially for structures with fundamental periods greater than 1.0 s.

One significant limitation of generalized response spectra is their inability to capture the local site effects, soil conditions, and the unique seismic characteristics of a specific region [7]. This often leads to either overestimation or underestimation of design parameters, potentially resulting in inefficient or unsafe designs [4]. To address this, Probabilistic Seismic Hazard Analysis (PSHA) is employed to develop site-specific response spectra that account for the actual seismicity, attenuation relationships, and geotechnical conditions of a locality [9]. PSHA provides a more robust estimation of peak ground acceleration (PGA) and associated hazard levels by considering uncertainties in earthquake magnitude, location, and frequency of occurrence.

A site-specific response spectrum is vital as it serves as the foundation for generating ground motion time histories, which include acceleration, velocity, and displacement records, thereby ensuring that design outcomes reflect actual seismic demands. These time histories are essential for dynamic analysis, enabling engineers to simulate the actual behavior of structures during seismic events [8]. By converting the response spectrum into time-history records, structural engineers can perform nonlinear dynamic analyses that provide insights into structural performance, energy dissipation, and potential damage mechanisms [12].

In the Philippine context, particularly in the 2nd District of Cagayan de Oro City, there is a growing need to develop localized seismic models given the city's proximity to active fault systems and historical seismic activities recorded by DOST-PHIVOLCS and USGS databases [14]. Utilizing these data, the site-specific response spectrum can be derived through software tools like R-CRISIS 2015, employing a 10% probability of exceedance in a 50-year period (return period of 475 years).

Following the development of the site-specific response spectrum, PEER Ground Motion Database serves as a resource for selecting appropriate seed ground motions that match the target spectrum [13]. These records can then be spectrally matched using Seismosoft software, allowing the generation of synthetic yet realistic acceleration time histories. From these, corresponding velocity and displacement time histories are derived, which are critical for assessing dynamic structural response [15].

By generating acceleration, velocity, and displacement time histories anchored on a site-specific response spectrum, engineers can more accurately predict structural demands and potential failure mechanisms. The objective of this research is to develop such site-specific ground motion records for the 2nd District of Cagayan de Oro City to support reliable seismic analysis and resilient structural design. This localized approach not only aligns with international best practices but also supports the Philippine government's initiatives for disaster risk reduction and resilient infrastructure development.

2 Material and methods

2.1 Study area

Using a quantitative research design, this study generated acceleration, velocity, and displacement time histories by developing a site-specific response spectrum using probabilistic seismic hazard analysis (PSHA). To guarantee accuracy and applicability to the seismic features of Cagayan de Oro City's 2nd District, the procedure included worldwide seismic data, specialist software tools, and defined techniques (see figure 1).

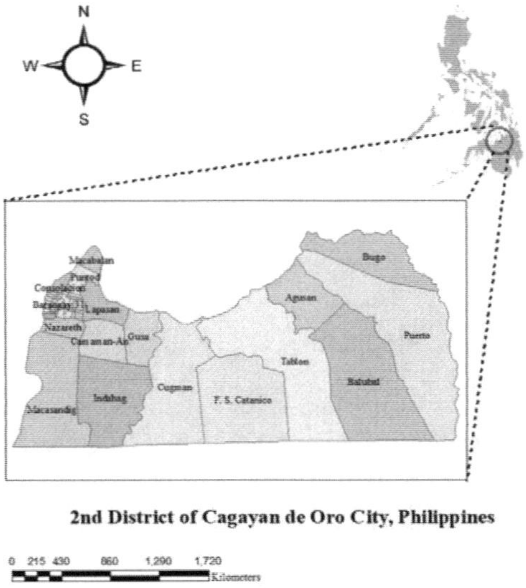

Fig. 1. Study Area.

2.2 Seismic data collection

Historical earthquake ground motion records were collected from the Department of Science and Technology – Philippine Institute of Volcanology and seismology (DOST-PHIVOLCS) and United States Geological Survey (USGS), covering events from 1901-2018. Using a 300-km radius, with coordinates of 8.235° (latitude) and 124.024° (longitude), as the center point of the study area, the catalog is exported to ZMAP v7.0 software for de-clustering using

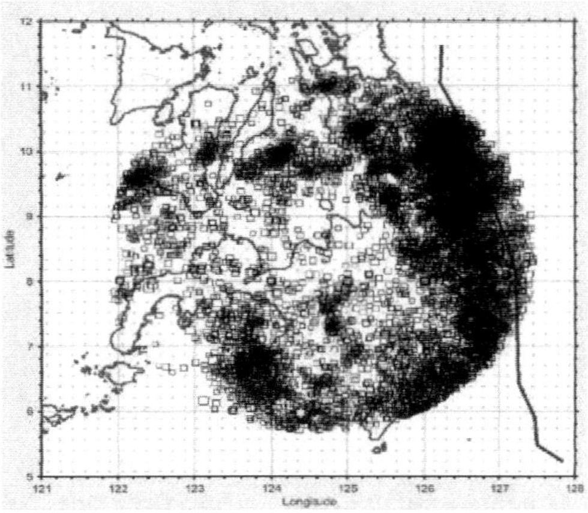

Fig. 2. Using a 300-km radius for earthquake events from 1901-2018.

Reasenberg method (figure 2). Records were observed based on magnitude, epicentral distance, fault mechanism, and data completeness to ensure that only high-quality event data with sufficient duration and frequency content were included for hazard analysis. These databases provided comprehensive earthquake event data for hazard analysis.

The completeness analysis of the earthquake moment magnitude was analyzed using the STEPP method and the number of moment magnitude in every magnitude range is shown in Table 1.

Table 1. Earthquake events in every magnitude range

Magnitude Range	Number of Moment Magnitude Earthquakes
4.0 – 5.0	3985
5.1 – 6.0	2612
6.1 – 7.0	175
> 7.1	32
Total	6804

2.3 Probabilistic seismic hazard analysis (PSHA)

R-CRISIS 2015, a program for seismic hazard calculation based on probabilistic techniques, was used to conduct PSHA. Seismicity parameters (magnitude-frequency distributions), ground motion prediction equations (GMPEs) customized for soil site conditions, local geotechnical data, and soil classification, also seismic source characterization (active fault lines and seismic sources surrounding the study area) were among the input parameters for the PSHA. In accordance with NSCP 2015 rules, the hazard estimation is considered for a 10% possibility of exceedance in 50 years, or a 475-year of return period. PSHA follows the process as indicated in figure 3 below.

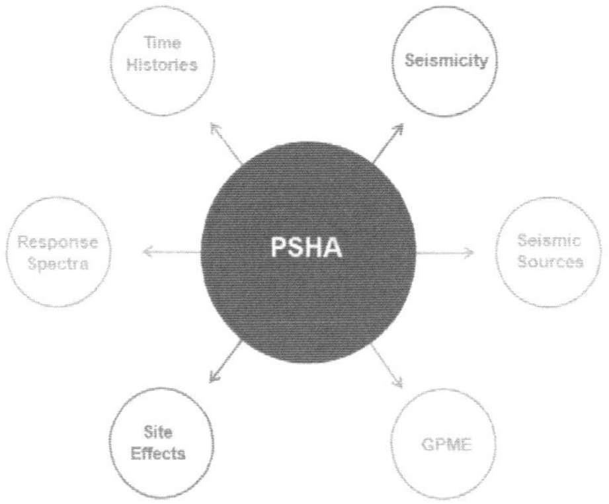

Fig. 3. PSHA Process.

2.3.1 Seismicity and seismic sources

The historical earthquake catalog was gathered from the United States Geological Survey (USGS) website and Department of Science and Technology – Philippine Institute of Volcanology and Seismology (DOST-PHIVOLCS), exported to the ZMAP v7.0 software for de-clustering, to check for identical earthquake events and removal of earthquake foreshocks and aftershock. Then seismic sources have been added to the R-CRISIS software that correspond to a polyline with varying coordinates and depths, consisting of Cabanglasan fault, Cebu Lineament, Central Mindanao fault, Central Negros fault, Cotabato trench, Davao River fault, East Bohol fault, Lanao fault, Mindanao Fault (MF) Daguma Extension,

Mindanao Fault (MF) Western Mindanao, Philippine Fault Zone (PFZ) Central Leyte fault, Philippine Fault Zone (PFZ) Eastern Mindanao fault, Philippine Fault Zone (PFZ) Lianga fault, Philippine Fault Zone (PFZ) Mati fault, Philippine Trench, Sulu Trench, Tagoloan River fault, Tangbulan fault, and Zamboanga fault. Figure 4 shows a sample of the Cabanglasn fault added to the R-CRISIS software with indicated coordinates and depth. The ZMAP software was also used to determine earthquake seismic parameters "a" and "b" in every seismic source using the average value and maximum sampling radius of 60 km. Values of"a" and "b" indicate the seismicity rate and the ratio of small to large earthquakes. Equations 1 and 2 were used to calculate the corresponding values of α and β, and subsequently, the mean annual exceedance using equation 3. Values of M_O and M_{max} are determined from the ZMAP software.

$$\alpha = a\ x\ \ln 10 \tag{1}$$

$$\beta = b\ x\ \ln 10 \tag{2}$$

$$\lambda_{\alpha,\beta,M_o,\ M_u} = e^{(\alpha-\beta M_o)}\left[e^{(-\beta(M-M_o))} - e^{(-\beta(M_u-M_o))}\right] * \left[\frac{1}{1-e^{(-\beta(M_u-M_o))}}\right] \tag{3}$$

where,

a = seismicity rate of earthquakes
b = ratio of small to large earthquakes α = a x ln (10)
β = b x ln (10)
M = earthquake magnitude (usually M = M_O)
M_O = threshold magnitude = M_c – 0.05
M_c = magnitude of completeness determined from the ZMAP software
M_u = ultimate or maximum magnitude
$\lambda_{\alpha,\beta,M_o,M_u}$ = mean annual rate of exceedance

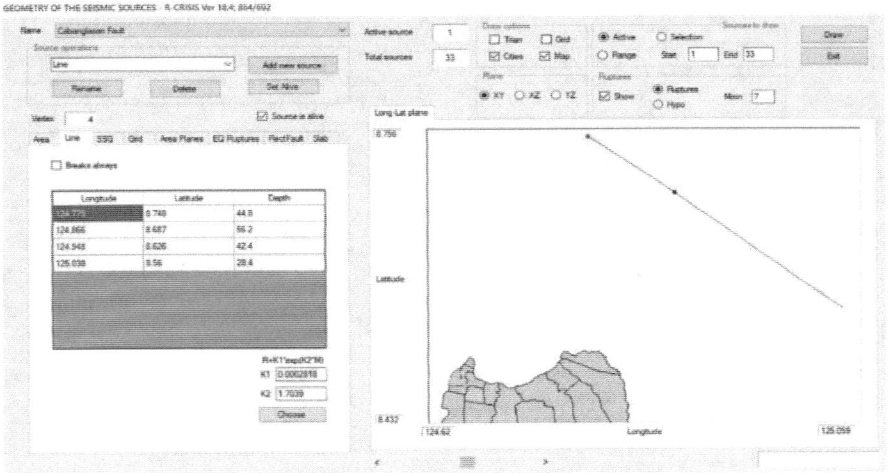

Fig. 4. Line source geometry of Cabanglasan fault in R-CRISIS

2.3.2 Ground motion prediction equation (GMPE) and site effects

The Atkinson and Boore [2] Ground Motion Prediction Equation (figure 5) for soil sites was used in this study because of the geologic history of the study area which is an alluvial soil deposit (https://www.cagayandeoro.gov.ph) to predict spectral accelerations as a function of magnitude, distance to the seismic source, and site, where the average upper shear-wave velocity (Vs_{30}) are explicitly used.

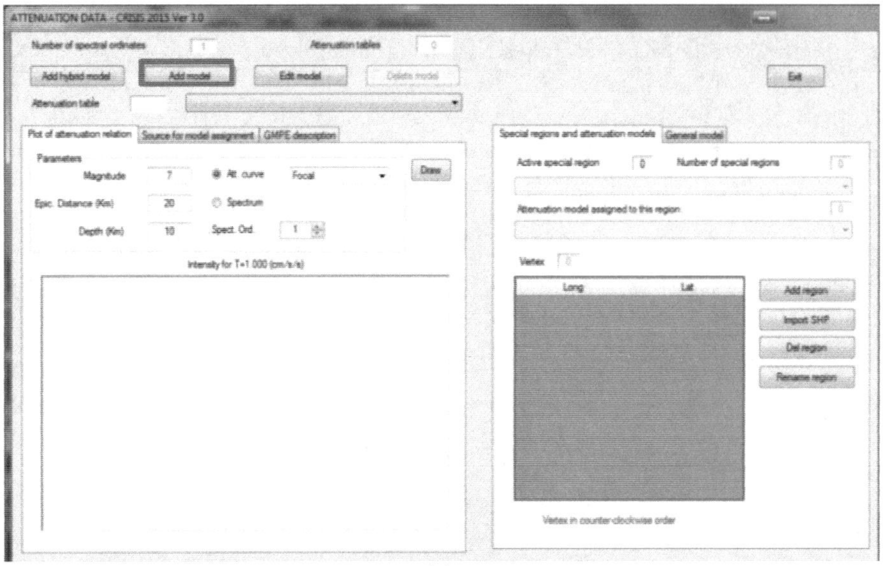

(a)

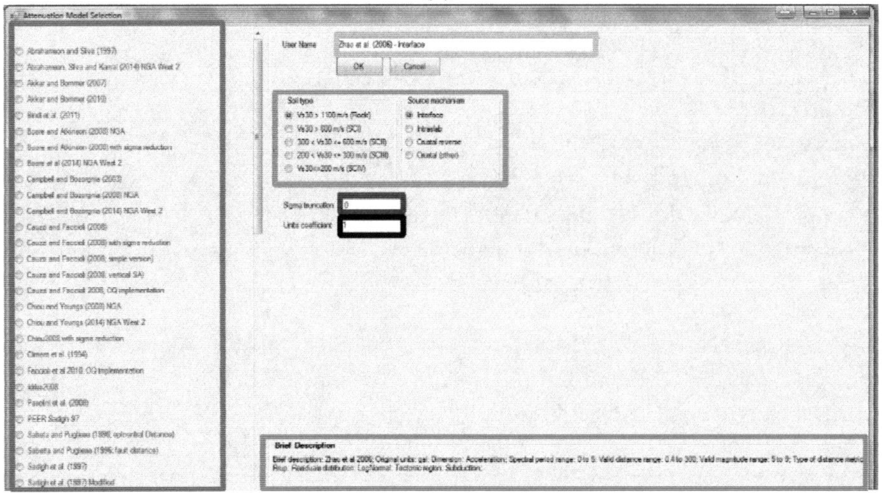

(b)

Fig. 5. Adding a built-in GMPE in R-CRISIS

From geotechnical soil test reports, SPT-N values of ninety-five (95) boreholes are correlated [1] to predict the average upper 30-m shear-wave velocities of soil in fifty-six (56) sites in the 2nd District of Cagayan de Oro City using the equation below.

$$V_{sc} = 59.44N^{0.109}z^{0.426} \tag{4}$$

$$\log V_{s30} = c_o + c_1 \log V_{sz} + c_2(\log V_{sz})^2 \tag{5}$$

The coefficients and standard deviation of the residuals to the fit (σ_{res}) are shown in Table 2.

Table 2. Coefficients relating $\log Vs_{30}$ and $\log Vs_Z$

Depth, z	c_0	c_1	c_2	σ_{RES}
5	1.146	0.5810	0.02573	0.114
6	0.8962	0.7366	0.001817	0.107
7	0.6788	0.8675	-0.01782	0.100
8	0.5684	0.9224	-0.02416	0.094

https://doi.org/10.1051/e3sconf/202565101006

9	0.4219	1.002	-0.03468	0.087
10	0.3462	1.033	-0.03680	0.082
11	0.2453	1.081	-0.04223	0.076
12	0.1932	1.097	-0.04211	0.071
13	0.08882	1.151	-0.04915	0.066
14	0.02964	1.174	-0.05075	0.061
15	-0.02178	1.191	-0.05150	0.056
16	-0.05916	1.201	-0.05115	0.051
17	-0.1287	1.235	-0.05555	0.047
18	-0.1725	1.252	-0.05679	0.043
19	-0.1992	1.256	-0.05610	0.038
20	-0.2088	1.250	-0.05346	0.034
21	-0.2353	1.255	-0.05317	0.030
22	-0.2462	1.251	-0.05127	0.027
23	-0.2409	1.236	-0.04735	0.023
24	-0.2231	1.212	-0.04213	0.019
25	-0.1929	1.180	-0.03543	0.016
26	-0.1564	1.144	-0.02814	0.013
27	-0.1270	1.114	-0.02205	0.009
28	-0.08924	1.079	-0.01512	0.006
29	-0.04862	1.042	-0.007949	0.003

Furthermore, site-specific conditions (site effects) for the study area were categorized based on soil profiles and geotechnical studies, which influence ground motion amplification (figure 6). A site class was assigned in accordance with NSCP 2015 and ASCE 7-10 standard.

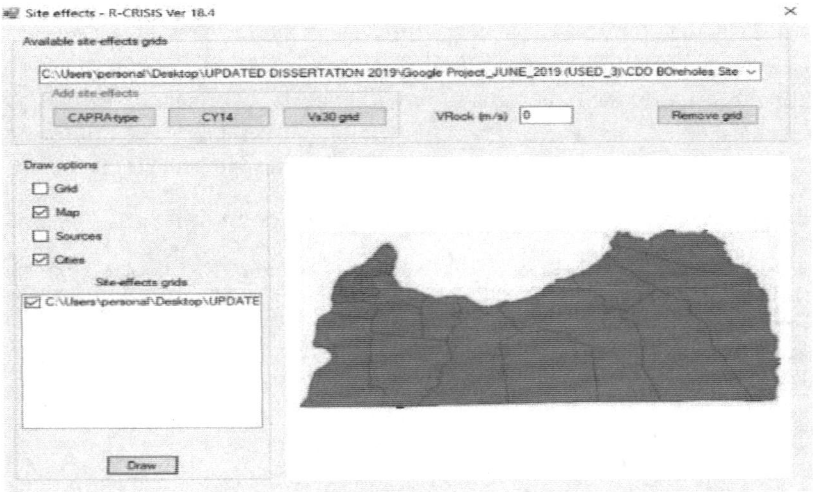

Fig. 6. Site effect grid file exported in the R-CRISIS.

The hazard analysis was calibrated for a 10% probability of exceedance in 50 years, equivalent to a 475-year return period, which corresponds to the Design Basis Earthquake (DBE) per NSCP 2015 provisions. Using R-CRISIS 2015, the site-specific response spectrum was developed in the study area. The response spectrum represents the maximum expected structural response (acceleration) at varying natural periods of vibration, critical for structural design and performance assessment. The generated site-specific response spectrum was compared against the generalized design response spectrum from UBC-1997, referenced in NSCP 2015 Section 208.

2.3.3 Development of site-specific response spectra

The PSHA generated the site-specific response spectrum (figure 7), which is a representation of the seismic demand unique to Cagayan de Oro City's 2nd District, while accounting for uncertainties in earthquake magnitude, location, and recurrence; however, it remains limited by the accuracy of available seismic models and ground motion prediction equations.

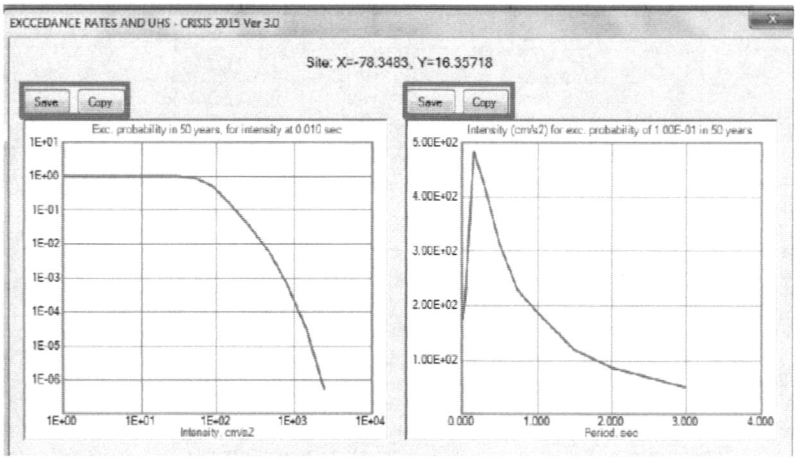

Fig.7. Hazard Curve and Response Spectrum in R-CRISIS

2.3.4 Selection and scaling of ground motion records

To match the properties of the target spectrum, suitable ground motion records were selected from the PEER NGA-West2 Ground Motion Database, fault mechanism similarity, and source-to-site distance (figure 8). These records were then processed using SeismoMatch software, where a wavelet-based algorithm iteratively adjusted their frequency content to achieve compatibility with the target spectrum within a ±10% tolerance over the period range of 0.1–5.0 seconds. The final matched records exhibited average misfit values ranging from 3.3% to 4.4%, indicating a high level of spectral agreement with the target spectrum.

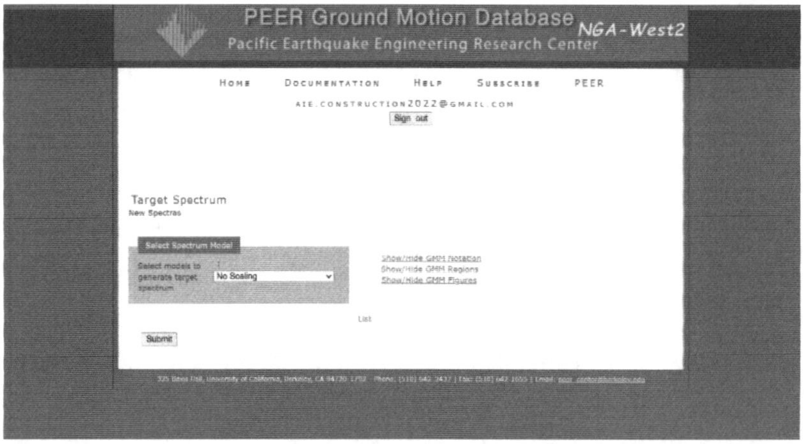

Fig. 8. Selection and Scaling of Ground Motion in PEER NGA-West2

2.3.5 Spectral matching and time history generation

Utilizing spectral matching techniques, the chosen ground motion records were modified using the Seismosoft SeismoMatch software (figure 9). The wavelet-based matching algorithm in SeismoMatch modified the frequency content of the seed records to match the site-specific response spectrum, and the outputs were acceleration time histories that were compatible with the target spectrum. The acceleration and displacement time histories were then acquired by integrating the acceleration and velocity data, respectively, with baseline correction applied using standard procedures such as those outlined by Boore [6] to avoid long-period drifts and numerical inaccuracies.

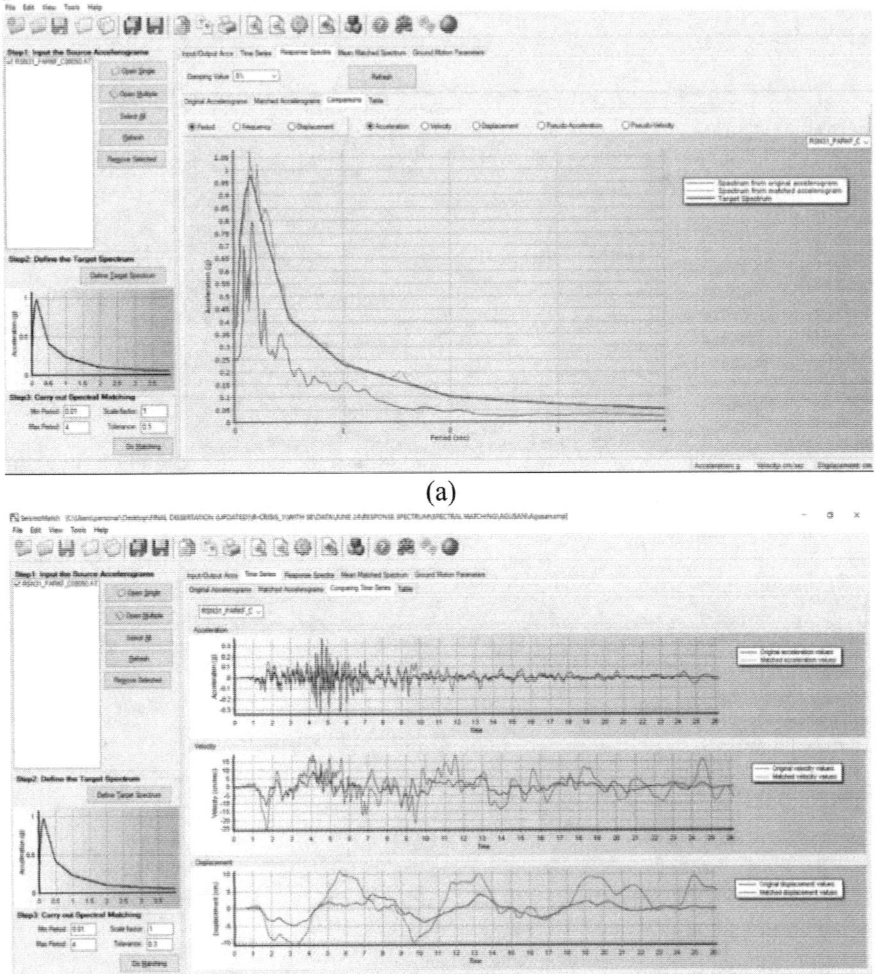

(a)

(b)

Fig. 9. Spectral Matching and Time History Generation using the SeismoMatch

2.4 Data Validation and Analysis

The generated acceleration, velocity, and displacement time histories were validated by comparing the response spectra of the modified records against the target site-specific spectrum, and ensuring that the time histories reflected realistic ground motion characteristics without numerical artifacts.

3 Results

Utilizing the results of earthquake magnitude de-clustering, seismic sources parameter and shear-wave velocity determination [3], site-specific response spectra of the study area were generated as shown in figure 10.

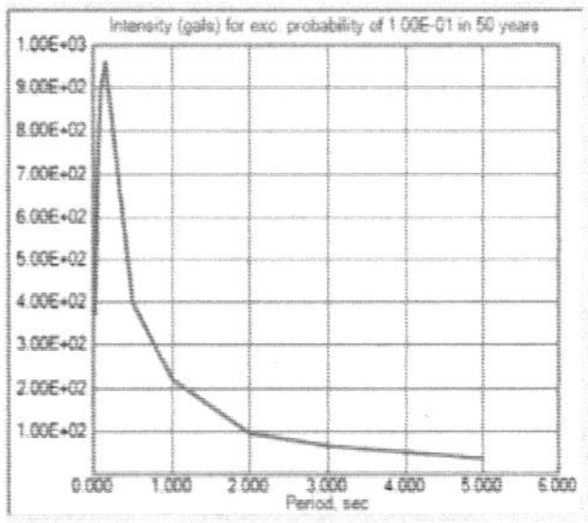

(a) Agusan

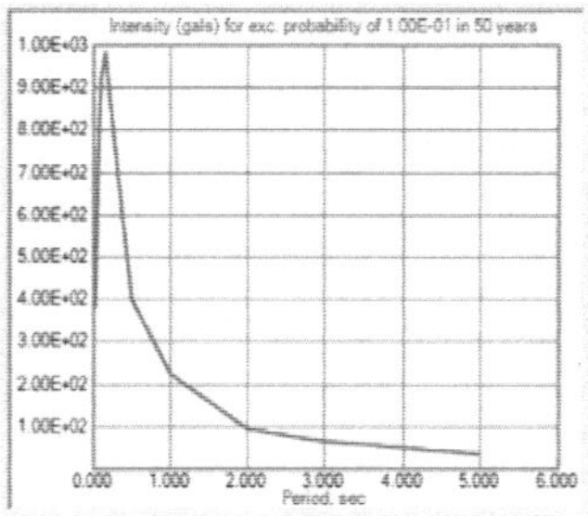

(b) Balubal

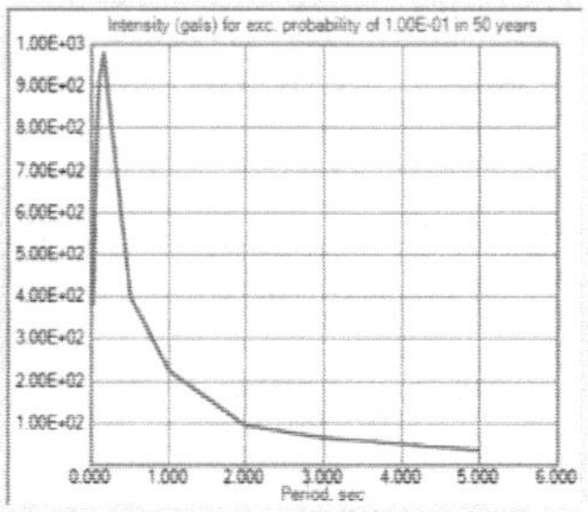

(c) Bugo

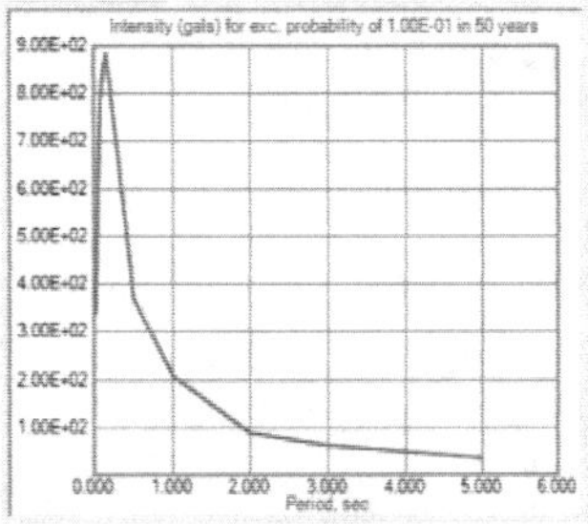

(d) Camaman-an

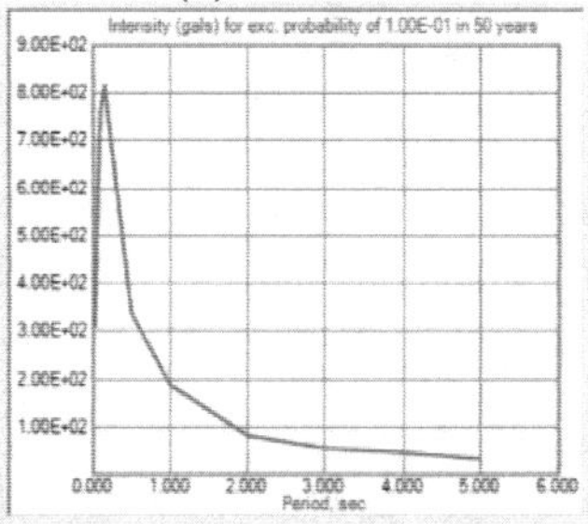

(e) Consolacion

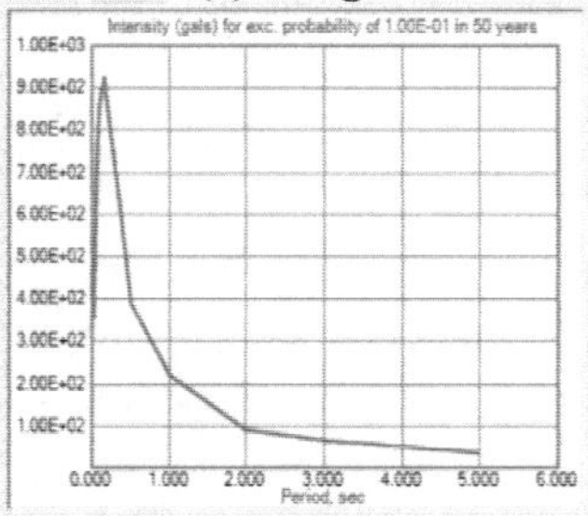

(f) Cugman

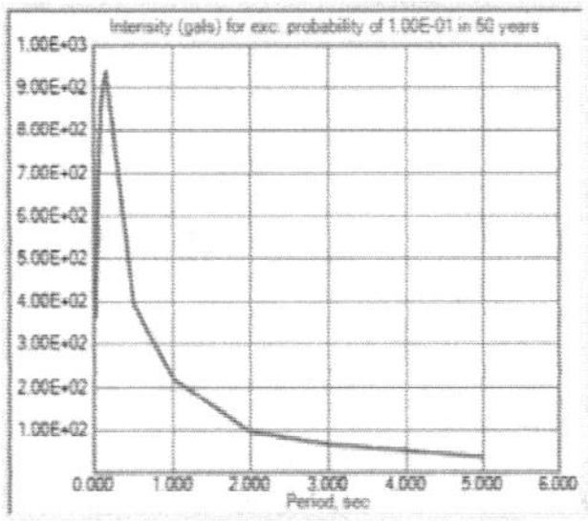

(g) FS Catanico

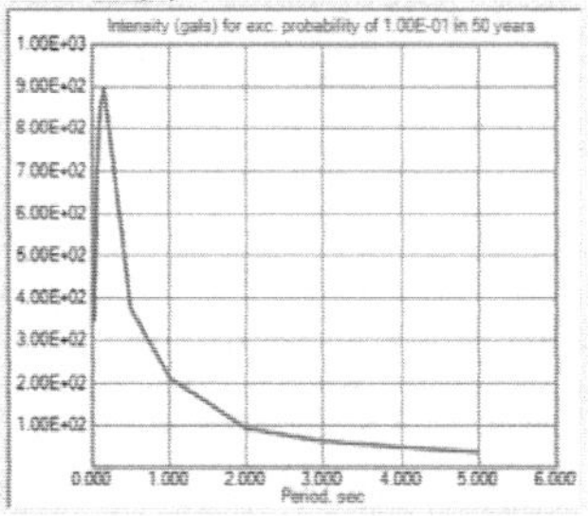

(h) Gusa

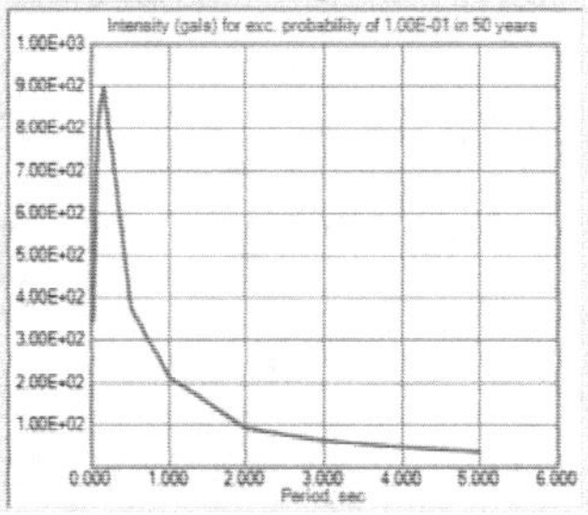

(i) Indahag

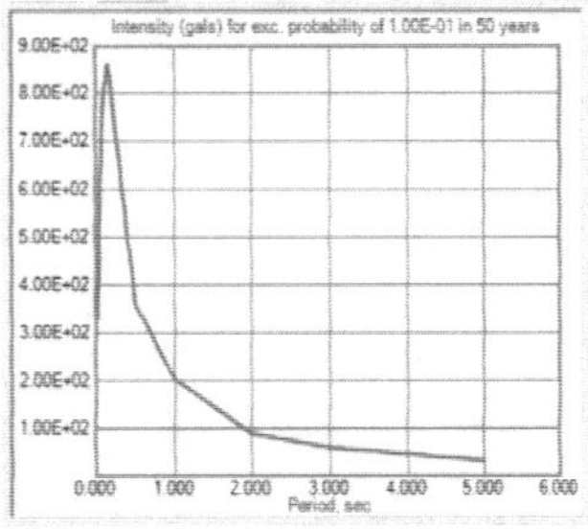

(j) Lapasan

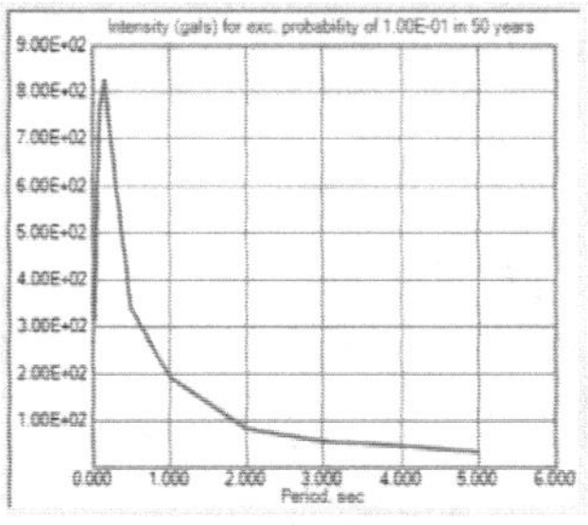

(k) Macabalan

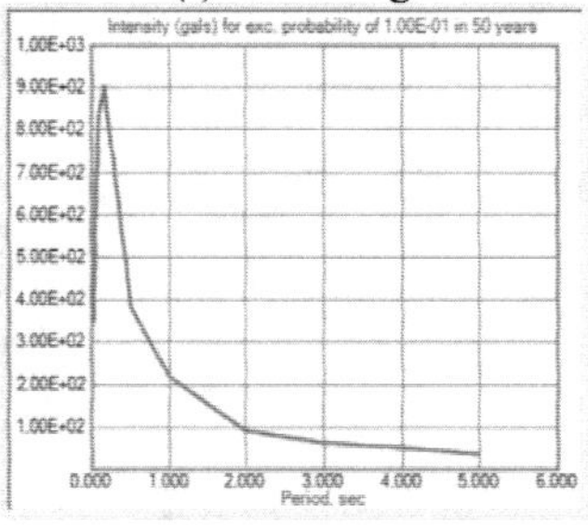

(l) Macasandig

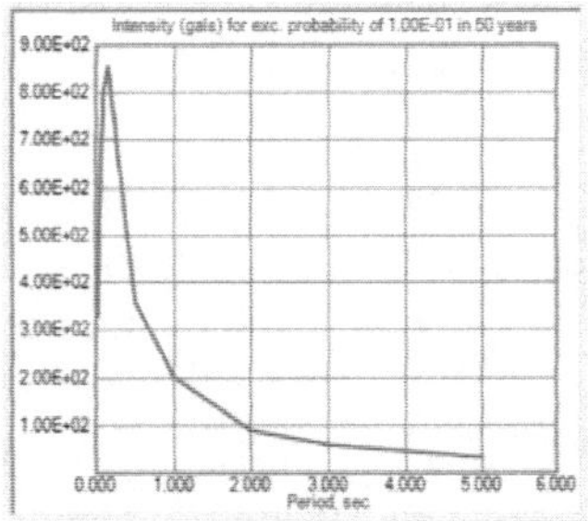

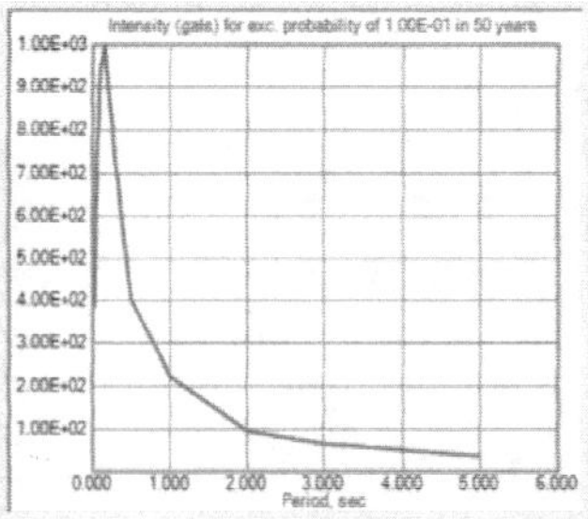

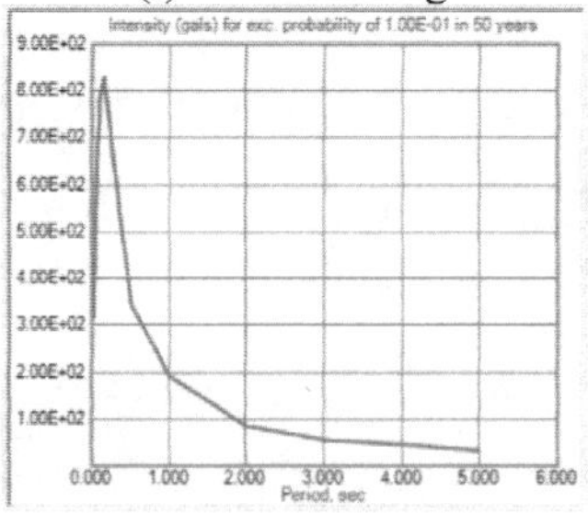

(m) Nazareth (n) Puerto (o) Puntod

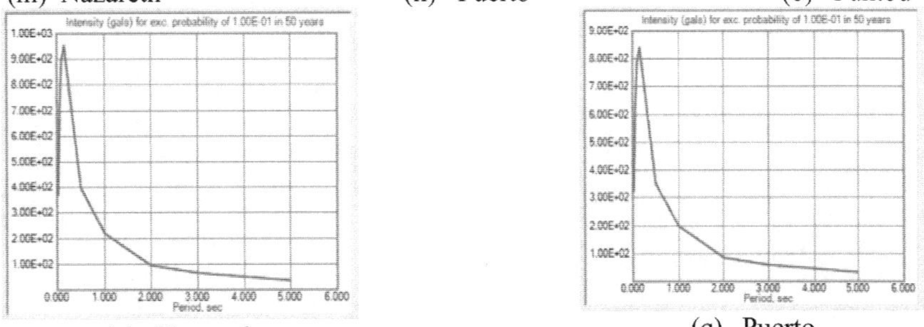

(p) Nazareth (q) Puerto

Fig. 10. Spectral Matching and Time History Generation using the SeismoMatch for each site

Figure 11 displays the matched time-histories of Agusan, Balubal, Bugo, Puerto, and Tablon using the PEER Ground Motion Database and SeismoMatch software.

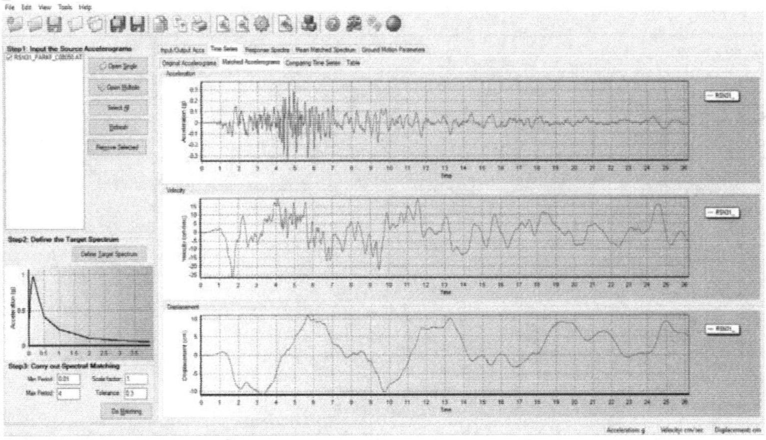

Fig.11. Matched Time-Histories of Acceleration, Velocity and Displacement for Agusan, Balubal, Bugo, Puerto and Tablon

Figure 12 displayes the matched time-histories of FS Canonico, Cugman, Indahag, Camaman-an, Gusa, Macasandig, Lapasan, Nazareth, Consolacion, Macabalan, Puntod, and Barangay 1-40 using the PEER Ground Motion Database and SeismoMatch software.

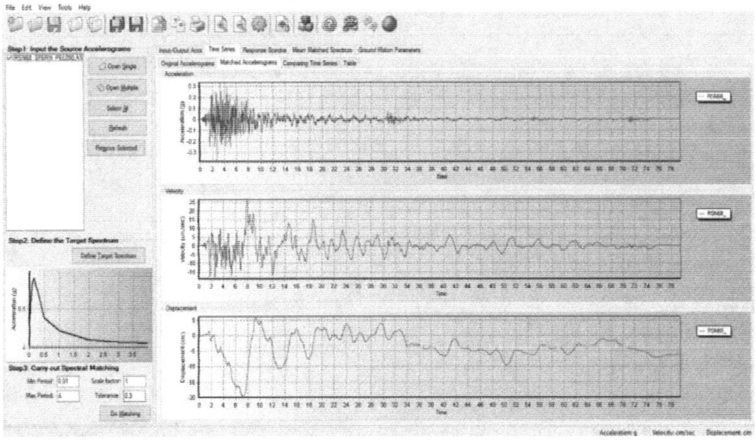

Fig. 12: Matched Time-Histories of Acceleration, Velocity and Displacement for FS Catanico, Cugman, Indahag, Camaman-an, Gusa, Macasandig, Lapasan, Nazareth, Consolacion, Macabalan, Puntod, and Barangay 1-40

Table 3 below shows the convergence on how well the generated time histories matched the target spectrum from the SeismoMatch software.

Table 3. Convergence of Spectral Matching using SeismoMatch

Sites	Spectral Matching	Iteration Needed	Average Misfit
Agusan Balubal, Bugo, Puerto and Tablon	Converged	7	3.3%
FS Catanico, Cugman, Indahag, Camaman-an, Gusa, Macasandig, Lapasan, Nazareth, Consolacion, Macabalan, Puntod, and Barangay 1-40	Converged	7	4.4%

4 Discussions

The generated site-specific response spectra of the study area in figure 10 were further summarized as shown in figure 13 below, indicating that the UBC-1997 response spectrum has conservative values in all time periods. This entails higher cost of materials when analyzing and designing a building structure.

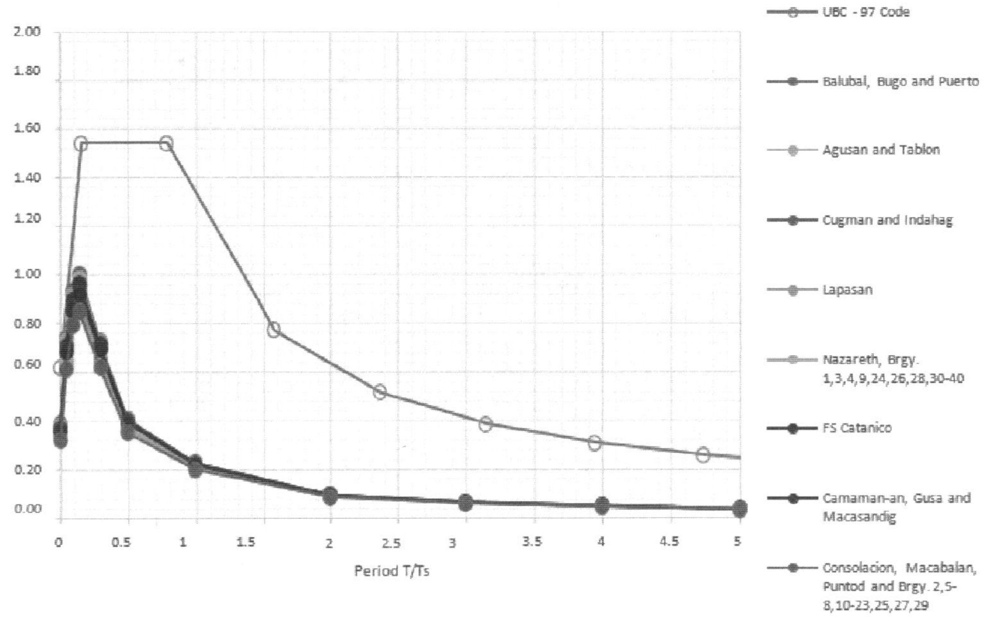

Fig. 13. Summarized Site-Specific Response Spectra of the Study Area with UBC-1997 Design Spectrum

For a five-story building, a savings of one million seven hundred thirty-one thousand six hundred twenty-five pesos & fifty-four cents (Php 1,731,625. 54) was attained when using the site-specific response spectrum in the analysis and design of a five-story commercial building (see Table 4) considering the beams and slabs at each floor levels, columns, walls, and footings. There is a significant difference of 5.83%, i.e., from thirty-one million four hundred forty-one thousand seven hundred three pesos & thirty-seven cents (Php

31,441,703.37) to twenty-nine million seven hundred ten thousand seventy-seven pesos & eighty-three cents (Php 29,710,077.83).

Table 4. Cost Comparison of Site-Specific Response Spectrum vs UBC-1997

Structural Members	**Material Cost (Php)**		**Difference (Php)**	**Difference (%)**
	Site-Specific Response Spectrum	UBC-1997		
Beams @ 3.0M	₱345,718.08	₱347,564.48	₱1,846.40	0.53
Beams @ 6.0M	₱491,122.13	₱492,427.18	₱1,305.05	0.27
Beams @ 9.0M	₱297,681.72	₱304,259.18	₱6,577.46	2.21
Beams @ 12.0M	₱300,038.79	₱265,671.05	-₱34,367.74	-11.45
Beams @ 15.0M	₱193,928.03	₱201,374.94	₱7,446.91	3.84
Slab @ 3.0M	₱2,636,860.64	₱2,537,005.64	-₱99,855.00	-3.79
Slab @ 6.0M	₱2,298,094.18	₱2,537,005.64	₱238,911.46	10.40
Slab @ 9.0M	₱2,574,740.26	₱2,617,315.88	₱42,575.62	1.65
Slab @ 12.0M	₱2,574,595.77	₱2,777,488.60	₱202,892.83	7.88
Slab @ 15.0M	₱2,770,656.71	₱2,828,732.53	₱58,075.82	2.10
Columns/Walls	₱2,380,852.13	₱2,565,720.57	₱184,868.44	7.76
Footings	₱12,845,789.39	₱13,967,137.68	₱1,121,348.29	8.73
MATERIAL COST	**₱29,710,077.83**	**₱31,441,703.37**	**₱1,731,625.54**	**5.83**

The matched time-histories, as shown in figure 11 and 12, in terms of acceleration, velocity, and displacements play an important role in the analysis and design of building structures in the Philippines. The Philippine government still lacks significant recording stations at strategic points in the country today due to budget constraints, thus, this initiative of generating time-histories is very relevant and timely especially that earthquakes nowadays are inevitable.

For the spectral matching using the SeismoMatch software, the software modifies a recorded ground motion so that its response spectrum matches a target spectrum and the algorithm iteratively adjust the accelerogram until the spectral ordinates are within an acceptable tolerance of the target. The spectral matching for all 56 sites converges, meaning the iterative matching process had successfully adjusted the accelerogram such that the response spectrum of the modified ground motion is sufficiently close to the target spectrum. In other words, the algorithm stoped because the matching criteria were satisfied, i.e., the process reached stability. As to the 3.3% and 4.4% average misfits, these indicate very good spectral matching where anything below 5% is considered excellent for engineering applications. These results mean that the adjusted accelerograms replicate the target response spectrum with high fidelity and can confidently use these results for dynamic structural analysis.

5 Conclusion

This research successfully created site-specific ground motion time histories—acceleration, velocity, and displacement—for the 2nd District of Cagayan de Oro City, Philippines utilizing a structured and data-driven methodology. Through the implementation of Probabilistic Seismic Hazard Analysis (PSHA) utilizing R-CRISIS 2015, a site-specific response spectrum was constructed, encapsulating the unique seismic demand of the location based on local seismicity, soil conditions, and attenuation qualities. Subsequent ground motion recordings supplied from the PEER NGA-West2 Database were spectrally matched using SeismoMatch, yielding a suite of time-history records customized to the defined response spectrum. The generated time histories are key inputs for nonlinear time-history calculations and provide a robust basis for evaluating structural performance under seismic stress scenarios, thereby enabling engineers to design structures that are both code-compliant and resilient against future earthquakes. The generated time histories for several barangays within the district, indicate the success of the spectral matching method in obtaining compatibility with the intended response spectrum. These data enhance the precision of dynamic structural analyses and help to more resilient, performance-based earthquake engineering designs.

6 Recommendations

It is recommended that site-specific time histories be integrated into structural analyses and local building codes to improve the safety and reliability of seismic design. Similar approaches should be extended to other districts in Northern Mindanao, supported by regular updates of seismic data and continued capacity building among engineers and policymakers. Future research should also investigate the influence of varying soil conditions and site effects to further refine seismic risk assessments and strengthen disaster resilience strategies.

Acknowledgements. The author would like to thanks the Commission on Higher Education K-12 (CHED K-12) Scholarship, Department of Public Works and Highways (DPWH) Region 10, and the Department of Science and Technology (DOST)-PHIVOLCS for the assistance in providing the data and documents needed for the completion of the study.

References

1. Ancheta, T.D., Darragh, R.B., Stewart, J.P., Seyhan, E., Silva, W.J., Chiou, B.S.J., ... Donahue, J.L.: NGA-West2 Database, Earthquake Spectra, **30**(3), 989–1005 (2014). https://doi.org/10.1193/070913EQS197M
2. Atkinson, G.M., Boore, D.M.: Earthquake Ground-Motion Prediction Equations for Eastern North America, Bull. Seismol. Soc. Am., **96**(6), 2181–2205 (2006). https://doi.org/10.1785/0120050245
3. Baguhin, I.A., Orejudos, J.N.: Site-Specific Response Spectra: 2nd District of Cagayan de Oro City, Philippines, Mindanao J. Sci. Technol., **18**(1), 93–117 (2020).
4. Bazzurro, P., Cornell, C.A.: Vector-valued probabilistic seismic hazard analysis (VPSHA), in Proc. 7th Nat. Conf. Earthquake Engineering, Boston, MA (2002).
5. Baker, J.W.: Conditional Mean Spectrum: Tool for Ground-Motion Selection, J. Struct. Eng., **137**(3), 322–331 (2011). https://doi.org/10.1061/(ASCE)ST.1943-541X.0000215
6. Boore, D.M.: Effect of baseline corrections on displacements and response spectra for several recordings of the 1999 Chi-Chi, Taiwan, earthquake, Bull. Seismol. Soc. Am., **91**(5), 1199–1211 (2001).
7. Boore, D.M., Stewart, J.P., Seyhan, E., Atkinson, G.M.: NGA-West2 equations for predicting PGA, PGV, and 5% damped PSA for shallow crustal earthquakes, Earthquake Spectra, **30**(3), 1057–1085 (2014). https://doi.org/10.1193/070113EQS184M
8. Kramer, S.L.: Geotechnical Earthquake Engineering (Prentice Hall, 1996).

9. McGuire, R.K.: Seismic hazard and risk analysis (Earthquake Engineering Research Institute, Oakland, California, 2004).
10. NEHRP: NEHRP Recommended Seismic Provisions for New Buildings and Other Structures, FEMA P-1050 (2015).
11. NSCP: National Structural Code of the Philippines, Volume 1: Buildings, Towers, and Other Vertical Structures (Association of Structural Engineers of the Philippines, 2015).
12. PEER: PEER Ground Motion Database (2010). Available at: http://ngawest2.berkeley.edu
13. PEER NGA-West2 Database (2013). Available at: http://ngawest2.berkeley.edu
14. PHIVOLCS: Philippine Earthquake Model and Seismic Hazard Maps (2020). Available at: http://www.phivolcs.dost.gov.ph
15. Seismosoft: SeismoMatch – A program for spectrum matching of earthquake records (2023). Available at: https://seismosoft.com/products/seismomatch

Seismological and Gravity Data Integration to Derive Regional Fault and Subduction Systems in Myanmar

Wahyu Sandriadi[1,3], *Umar* Muksin[1,2,3,*], *Yanis* Muhammad[1], *Nadiatul* Asra[2,3], *Siti* Maryam Purba[1,3], *Nugraha* Fathir[1,3], and *Muhammad* Haikal Gunarya[4]

[1]Department of Geophysical Engineering, Universitas Syiah Kuala, Aceh, Indonesia
[2]Department of Physics, Universitas Syiah Kuala, Aceh, Indonesia
[3]Tsunami and Disaster Mitigation Research Center, Universitas Syiah Kuala, Banda Aceh, Indonesia.
[4]Department of Earth and Planetary Science, Institute of Science Tokyo.

Abstract. Tectonic activity in Myanmar is influenced by the convergence of the Indian Plate and the Burma Plate, which form an active subduction zone in western Myanmar (Rakhine) and the Sagaing Fault system in eastern Myanmar. These two tectonic zones generate significant earthquakes with shallow to intermediate depths. This study aims to interpret the tectonic characteristics of the region through an integrated approach that combines seismicity analysis, gravity anomaly modeling based on tilt derivative and waveform inversion using a Bayesian Bootstrap approach. Earthquake data were obtained from the USGS catalog for 2010–2025, while waveform data were downloaded from the IRIS database, focusing on events with Mw > 5.5. Gravity data used in this study were sourced from the TOPEX satellite. Preliminary analysis of seismicity distribution and hypocenter depths indicates a clear plate convergence pattern and active fault systems orientation across the study area. These results were then compared with satellite gravity anomaly data, which reflected variations in crustal density. The correlation between seismicity and gravity anomalies was used to interpret the geometry of the regional geological structures. Earthquake source parameters were analyzed through waveform inversion to obtain more detailed information, such as fault plane orientation, source mechanism, and refined hypocenter location. Several earthquake events were selected for inversion based on data completeness and signal quality. The waveform inversion results show misfit values below 0.6, indicating high-quality solutions. The inversion results reveal that right-lateral strike-slip mechanisms dominate earthquakes along the Sagaing Fault, while those in the Rakhine Subduction Zone are predominantly thrust faulting events.

1. Introduction

Myanmar is one of the regions with the most complex tectonic dynamics in Southeast Asia, driven by the convergence between the Indian Plate and the Burma microplate [1]. This interaction gives rise to two contrasting active deformation systems: the Rakhine subduction zone in the west and the right-lateral Sagaing fault system in the east. Both systems play a

*Corresponding author: muksin.umar@usk.ac.id

crucial role in accommodating regional tectonic forces and serve as the primary sources of various earthquakes with diverse depths and faulting mechanisms.

One of the significant seismic events that highlights the active tectonic dynamics of the Myanmar region is the Mw 7.7 earthquake that occurred on March 27, 2025 (UTC), along the active segment of the Sagaing Fault. This earthquake caused severe damage in Mandalay and surrounding areas, resulting in thousands of fatalities and injuries [2]. This event further highlights the high seismic hazard level in Myanmar, underscoring the need for a more comprehensive understanding of its geology and tectonics to support future disaster mitigation efforts. The distribution of seismicity and Mw > 6 earthquakes in Myanmar is shown in Figure 1.

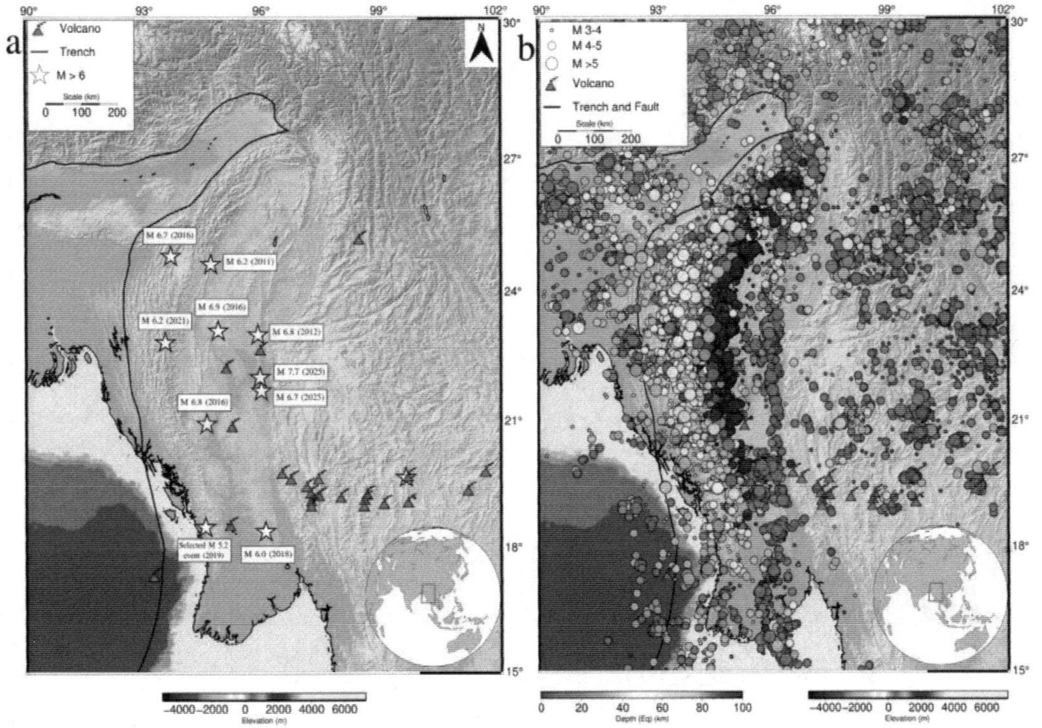

Fig. 1. (a) Distribution of earthquakes with magnitudes greater than Mw 6 in the Myanmar region. (b) Seismicity distribution in the Myanmar region showing earthquake concentration along the Sagaing Fault and the Rakhine subduction zone.

Tectonic studies in Myanmar have employed a variety of approaches. Researchers have applied earthquake source analysis with supershear models and paleoseismology [3], while subsurface imaging has relied on receiver function inversion and Rayleigh wave dispersion [4]. Structural geology investigations have included syn-kinematic sedimentation along the Sagaing Fault [5]. More recently, the Mw 7.8 supershear earthquake of 27 March 2025 (UTC) was examined in detail [6], providing new insights into large-scale fault dynamics in the region. However, most studies remain local in scope and do not fully address the broader tectonic framework. In particular, there is still a lack of research that integrates seismicity with other geophysical datasets, such as gravity anomalies, to better resolve Myanmar's tectonic geometry. Addressing this gap through integrated analysis offers the potential to strengthen current knowledge and guide future tectonic research.

Seismicity reflects crustal deformation caused by active tectonic processes; therefore, analyzing earthquake hypocenter locations and distributions is crucial for identifying active faults and subduction zones in Myanmar. Gravity data, especially Bouguer anomalies, provide complementary constraints by highlighting density variations associated with fault zones and plate boundaries. To enhance subsurface interpretations, this study also applies full-waveform inversion to refine earthquake source parameters. Unlike arrival-time methods, which rely only on first arrivals, full-waveform inversion analyzes the entire seismic waveform, improving the accuracy of source solutions and yielding deeper insights into fault characteristics. Reliable results can be obtained even with a limited number of stations, as long as the azimuthal gap remains below 180°, thereby enabling better constraints on the moment tensor, which serves as a key input for understanding fault mechanisms [7].

Based on this integrated approach, the objective of this study is to investigate Myanmar's regional tectonic framework by combining earthquake distribution, gravity anomalies, and refined source parameters. The analysis focuses on two main tectonic zones: the Rakhine subduction margin and the Sagaing Fault system. The outcomes are expected to advance understanding of Myanmar's lithospheric tectonics and provide valuable references for seismic hazard assessment in Southeast Asia.

2. Data and method

2.1 Seismicity data and analysis

The seismological analysis in this study is based on two primary datasets: seismicity data and full-waveform records. The seismicity dataset includes earthquake epicenters, focal depths, event times, and moment magnitudes (Mw), compiled from multiple catalogs, including the International Seismological Centre (ISC), the United States Geological Survey (USGS), and the Myanmar National Seismic Network, covering the period 2000–2025. This multi-catalog approach enables cross-validation of event parameters and improves the reliability of the dataset. These data were used to examine the spatial distribution of earthquakes in Myanmar, with particular attention to major tectonic features such as the Sagaing Fault and the Rakhine Subduction Zone. The results are presented as epicenter distribution maps and depth cross-sections, providing a comprehensive view of the region's tectonic framework. A similar approach has been applied in previous seismicity studies along the Sumatra Fault [8].

The waveform dataset was obtained from the open-access IRIS and GEOFON global seismic networks and consists of three-component seismograms recorded at both regional and teleseismic stations. All records were rotated into ZRT coordinates, and only traces with a high signal-to-noise ratio (SNR) were retained. Nine earthquakes with Mw > 6 were selected based on data completeness, signal quality, and representative station coverage, along with one additional Mw 5.2 event chosen for its strategic location in the central collision zone between the Indian and Burma Plates. The station distribution provided nearly complete azimuthal coverage, ensuring stable inversion results (Figure 2) and meeting the stability criteria proposed by Xu et al. [9], which require at least $N \geq 15$ channels and an azimuthal gap $\gamma \leq 180°$ for reliable source parameter estimation.

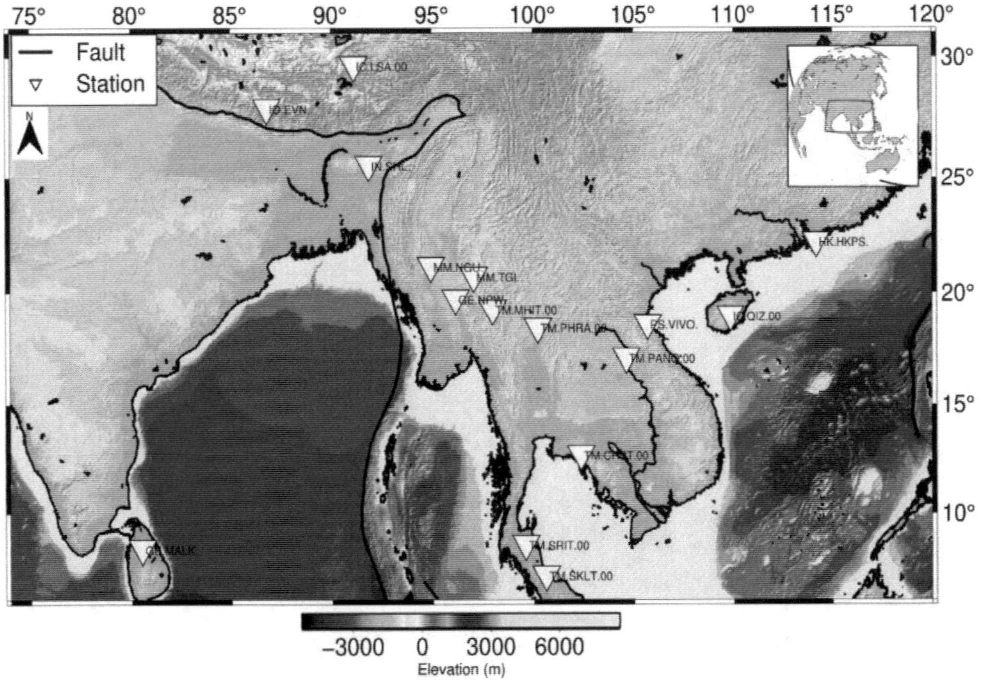

Fig. 2. Distribution of seismic stations used for waveform inversion of earthquakes in the Myanmar region. Stations from both regional and teleseismic networks (IRIS and GEOFON) are shown, covering a wide azimuthal range around the selected events.

The selected waveform data were used to refine earthquake source parameters—specifically the epicenter, focal depth, and focal mechanism—through Full Waveform Inversion (FWI). Unlike conventional catalog-based methods, FWI incorporates the complete waveform shape and explicitly accounts for subsurface velocity structures, making it a more robust approach [7], [10]. The inversions were performed using the Grond probabilistic framework, with Green's functions computed from the AK135 global velocity model via the QSEIS algorithm [11]. Frequency bands were adjusted according to event magnitude: 0.007–0.030 Hz for Mw 5.5–6.0, 0.005–0.025 Hz for Mw 6.0–7.0, and 0.003–0.025 Hz for Mw > 7. Lower frequencies were applied to larger events to stabilize the inversion and reduce the influence of unmodeled structural heterogeneities. The main objective of the inversion was to ensure that the synthetic waveforms generated from the model closely matched the observed waveforms.

The quality of the inversion results was assessed by comparing the observed and synthetic seismograms. The inversion was iterated up to 50,000 times, and if the misfit did not fall below 0.6, the frequency bands and related parameters were adjusted and the process repeated. Only solutions with an average misfit value below 0.6 were considered reliable.

2.2 Gravity data and analysis

The gravity method is a geophysical technique used to measure variations in the Earth's gravitational field caused by differences in subsurface material density. Lateral density variations between rock layers cause changes in gravitational acceleration that can be measured at the surface and used to identify geological structures such as faults, lithological boundaries, or subduction zones [12].

In this study, gravity data were sourced from the TOPEX (Topography Experiment) satellite, which provides global-scale gravity anomaly measurements. The raw TOPEX dataset consists of free-air anomalies, primarily reflecting the effects of elevation and surface relief. To obtain Bouguer anomalies that more accurately represent subsurface density variations, a series of corrections was applied. First, a free-air correction was performed to account for the observation height relative to sea level, followed by a Bouguer slab correction using a reduction density of 2.67 g/cm³, commonly adopted to represent average crustal rocks. In addition, terrain correction was implemented to minimize the influence of topographic relief, and ocean mass correction was applied to remove the gravitational effect of seawater loads. These corrections follow the standard Bouguer anomaly formulation [12]:

$$\Delta g = g_{obs} - \gamma' + \beta h - 2\pi G\rho h - \rho T \tag{1}$$

where Δg is the Bouguer anomaly in mGal, g_{obs} is the observed gravity, γ' is the normal gravity, βh is the free-air correction, G is the gravitational constant, ρ is the reduction density (2.67 g/cm³), $2\pi G\rho h$ is the Bouguer slab correction, and ρT is the terrain correction.

To highlight density contrast boundaries related to faults or lithological contacts, the tilt derivative technique was applied. This method involves computing the ratio between the vertical and horizontal derivatives of the Bouguer anomaly. The tilt derivative value can be expressed mathematically in equation 2 as follows [13]:

$$TDR = tan^{-1}\left(\frac{\frac{\partial f}{\partial z}}{\sqrt{\left(\frac{\partial f}{\partial x}\right)^2 + \left(\frac{\partial f}{\partial y}\right)^2}}\right) \tag{2}$$

where f represents the Bouguer anomaly value, $\partial f/\partial z$ is the vertical derivative, and $\partial f/\partial x$ and $\partial f/\partial y$ are the horizontal derivatives of the gravity field. The tilt derivative technique is effective in identifying subsurface structural boundaries, such as faults or lithological changes that may not be visible in the complete Bouguer anomaly analysis.

To ensure transparency and reproducibility of the analysis, the key parameters used in the tilt derivative calculation are summarized in Table 1. The table also highlights the purpose of each parameter and its sensitivity in relation to gravity anomaly interpretation.

Table 1. Tilt derivative parameters, their purposes, and sensitivities.

Parameter	Definition/formula	Function	Sensitivity
Vertical Derivative (VDR)	$\partial g/\partial z$	Emphasizes anomalies from shallow sources; highlights vertical contacts	Highly sensitive to noise; sensitivity increases with higher derivative orders
Horizontal Derivative (HDR)	$\sqrt{\left(\frac{\partial g}{\partial x}\right)^2 + \left(\frac{\partial g}{\partial y}\right)^2}$	Indicates the direction and intensity of horizontal gradients; useful for delineate lateral boundaries	More stable than VDR but less sensitive to very shallow sources
Tilt Derivative (TDR)	$\arctan \frac{VDR}{HDR}$	Combines VDR and HDR; enhances structural boundaries, faults, and lithological contacts	Relatively stable against amplitude variations; balances sensitivity to shallow and deep sources

3. Results and discussion

3.1 Identification of seismic gap and gravity anomaly variations

The Sagaing Fault is one of the major strike-slip fault system on mainland Southeast Asia. Based on historical earthquake records and instrumental data, a significant seismic gap has been identified along the fault segment between 19.2°N and 22.2°N in central Myanmar [14].

This seismic gap is clearly illustrated in Figure 3, which shows the spatial and temporal distribution of seismic activity along the Sagaing Fault. The figure indicates that, for more than a decade before the Mw 7.7 earthquake in 2025, the region exhibited relatively little significant seismic activity. The earthquakes that occurred before the mainshock were mostly small events scattered outside the Mw 7.7 (2025) segment. This pattern suggests that the segment underwent a locking period (a locked fault segment) and experienced an ongoing accumulation of tectonic strain.

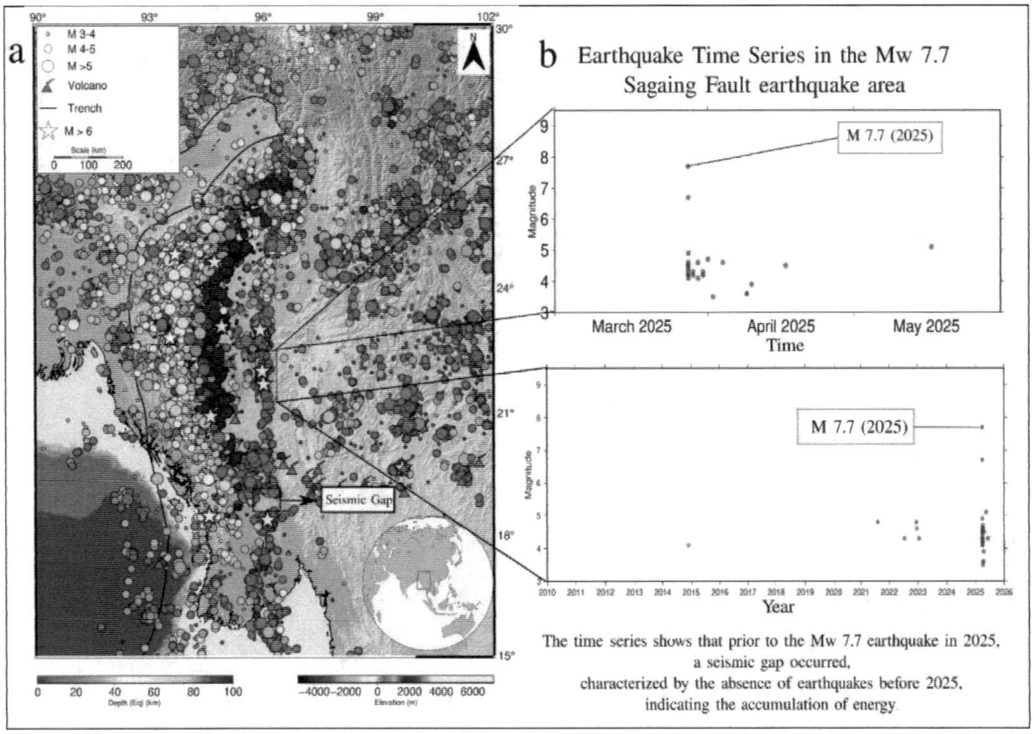

Fig. 3. (a) Spatial distribution of seismicity along the Sagaing Fault and the epicenter of the 2025 Mw 7.7 earthquake. (b) Temporal variation (time series) of seismic activity within the Mw 7.7 (2025) earthquake region.

Figure 3 shows the spatial and temporal distribution of seismic activity along the Sagaing Fault, particularly in the area of the Mw 7.7 earthquake that occurred in 2025. This earthquake was located precisely within the seismic gap zone, which had previously been identified as a fault segment with no significant seismic activity for over a century [14].

The time-series visualization in Figure 3b indicates that, during the period from 2010 to just before 2025, no earthquakes of Mw > 3 were recorded in the region. This observation further supports the indication that this segment had undergone continuous accumulation of tectonic strain. The 2025 Mw 7.7 event likely represents the release of seismic energy from the elastic strain that had been accumulating along the segment.

This event also aligns with the historical pattern of other significant earthquakes along different segments of the Sagaing Fault, such as the Mw 7.4 and Mw 7.5 earthquakes in 1930 [14]. Therefore, the 2025 Mw 7.7 earthquake not only confirms the existence of a seismic gap in this segment but also highlights that the entire Sagaing Fault system is a high-seismic potential area that warrants serious consideration in Myanmar's earthquake hazard assessment. In addition to the 2025 rupture segment, our broader seismicity analysis reveals another potential seismic gap near ~18.6°N (Figure 3). This unruptured portion of the fault may also store accumulated stress and thus should be considered in future hazard evaluations.

To support the interpretation of seismic activity and to better understand the subsurface structural controls, gravity data analysis was conducted to identify density contrast zones associated with fault presence and crustal segmentation. The gravity anomaly distribution shown in Figure 4 is used to identify subsurface density variations that are closely related to regional geological structures.

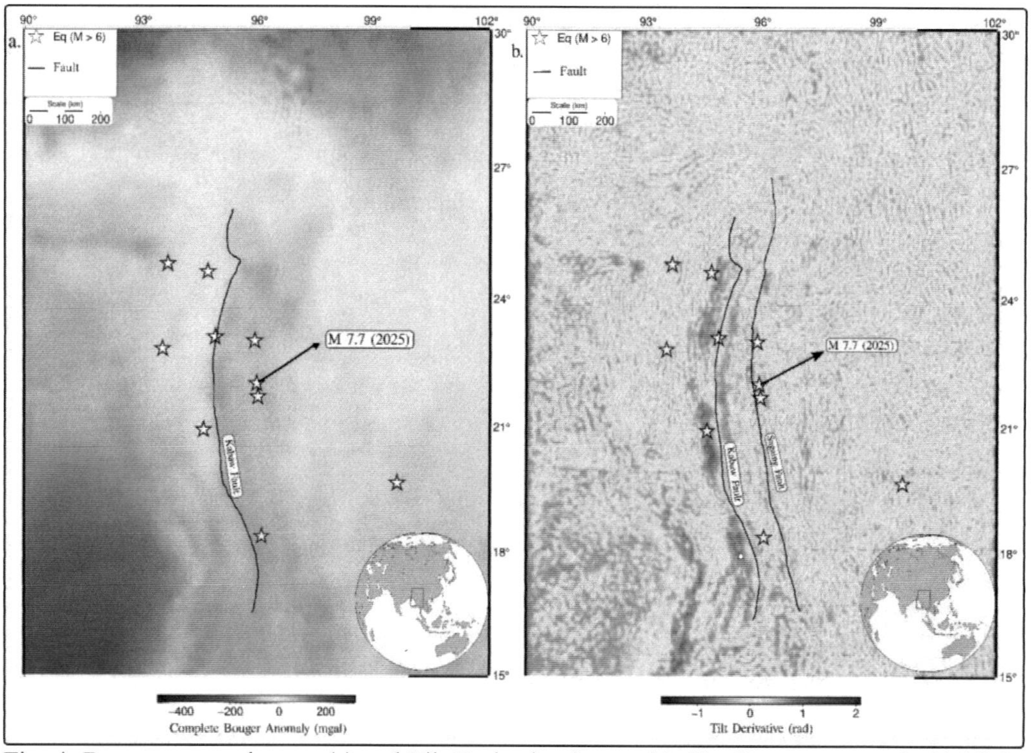

Fig. 4. Bouguer anomaly map (a) and Tilt Derivative image (b). The Bouguer anomaly highlights density contrasts indicating the presence of the Kabaw Fault, while the Tilt Derivative enhances the trace of the Sagaing and Kabaw Fault structures.

The Complete Bouguer Anomaly values (Figure 4a) in the study area exhibit prominent lateral variations. Low anomaly values are generally distributed around the Sagaing Fault and Kabaw Fault systems, indicating the presence of lower-density rocks or zones of crustal weakening. In contrast, high anomaly values are detected in the western part of the study area, particularly in the Rakhine Subduction Zone, suggesting the existence of high-density material beneath the surface. This distribution pattern reflects a strong density contrast, likely influenced by lithological differences or crustal block boundaries. The elongated contours of low anomaly values, which run nearly parallel to the fault trace, indicate the presence of the Kabaw Fault structure and are likely associated with lower-density rocks.

The Tilt Derivative image (Figure 4b) reinforces the presence of sharp structural boundaries and reveals linear patterns consistent with the orientation of the Sagaing and Kabaw Faults. These anomaly boundaries are marked by abrupt changes in gradient values, ranging approximately from -1 to 1 radian. These values reflect significant contrasts in the gravity field and typically represent the presence of active faults or major geological boundaries that separate two structurally distinct domains. The location of the 2025 Mw 7.7 earthquake lies within the transitional zone between two major fault segments and is recorded within a zone of high gradient anomaly. This finding further supports the interpretation that the major earthquake resulted from the long-term accumulation of tectonic strain in a locked fault segment, as also evidenced by the lack of significant seismic activity in the time series before 2025.

3.2 Improvement of earthquake source parameters using waveform inversion

Waveform inversion of earthquake source mechanisms was conducted to obtain more accurate source parameters, including centroid location, depth, rupture duration, and focal mechanism in the active regional tectonic zones across Myanmar. All analyzed earthquake events exhibit low misfit values (<0.6), indicating good inversion quality and reliability, as per the criteria of [7] and [10].

The moment tensor solution for the Mw 7.7 earthquake that occurred on 27 March 2025, UTC (Figure 5) shows that the source mechanism is dominated by a double-couple (DC) component of more than 90%, reflecting a purely tectonic rupture mechanism without significant volumetric deformation components (CLVD or isotropic). Based on the ensemble focal mechanism plot (Figure 5b), this earthquake is categorized as a dextral (right-lateral) strike-slip event with a north-south fault orientation.

The probability distribution of the inversion parameters (Figure 5c) indicates a dominant magnitude of approximately Mw 7.72, with a centroid depth of approximately 26 km and a source duration of around 15 seconds. The waveform fitting process (Figure 5a) demonstrates a very good match between observed and synthetic waveforms across the three rotational components (T, R, and Z), with a final misfit value of 0.59 after 50,000 iterations. The convergence plot (Figure 5d) indicates a significant decrease in misfit, stabilizing around iteration 25,000, suggesting that the derived source parameters are stable and reliable.

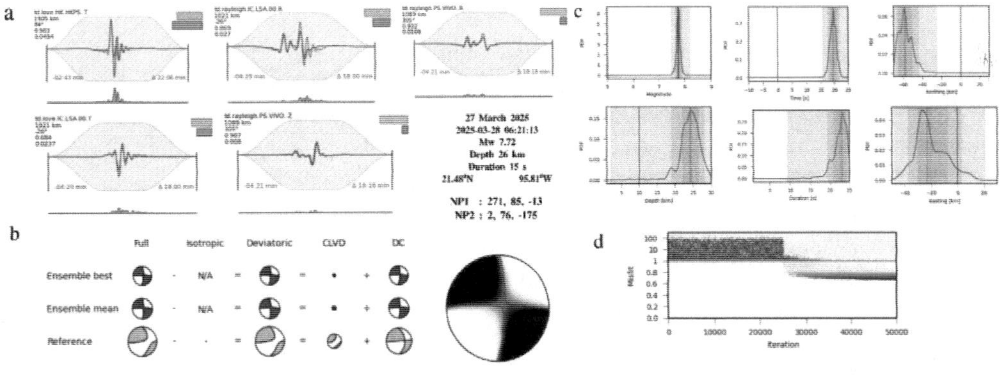

Fig. 5. Waveform inversion results for the Mw 7.7 earthquake (27 March 2025, UTC) that occurred along the Sagaing Fault. The figure includes: (a) waveform fitting, (b) focal mechanism solution with fuzzy moment tensor components, (c) probability density function distribution, and (d) misfit evolution over the iteration process.

The waveform fitting visualization on the three components (Figure 6a) shows excellent agreement between the observed and synthetic signals, with a final misfit value of 0.4 after approximately 25,000 iterations (Figure 6d). The misfit decreased significantly during the early iterations and stabilized below the 0.45 threshold, indicating strong convergence and high reliability of the inversion-derived parameters.

The fuzzy moment tensor and moment tensor decomposition (Figure 6b) indicate a subduction-related mechanism at the convergence of the Indian Plate and the Burma Microplate, where thrust faulting is the dominant mode at shallow depths along the plate interface. Meanwhile, the probability density functions (Figure 6c) display narrow and well-defined distributions, reflecting stable solutions and providing tighter constraints on the remaining source parameters.

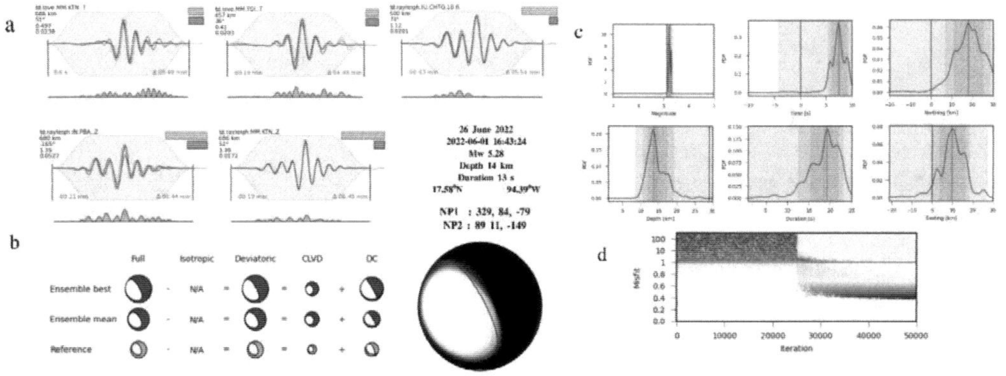

Fig. 6. Waveform inversion results for the Mw 5.2 earthquake (26 June 2022), including (a) waveform fitting, (b) focal mechanism solution with fuzzy moment tensor components, (c) probability density function distribution, and (d) misfit evolution over the iteration process.

The inversion solution for the Mw 6.82 earthquake that occurred on August 24, 2016, indicates an intraslab event, with a source depth reaching approximately 83 km. Based on the probability distribution curve (Figure 7c), the event had a source duration of 11 seconds, while the centroid location shifted by 14 km to the south and 22 km to the west from its initial catalog position.

The focal mechanism (Figure 7b) shows a dominant double-couple (DC) component of over 80%, with a thrust fault geometry. This type of mechanism is commonly observed in intraslab earthquakes triggered by internal compressional forces within the subducted oceanic plate. Our result is consistent with the focal mechanism reported by Shiddiqi et al. [15] for the August 24, 2016, Mw 6.8 Chauk earthquake in central Myanmar, which also indicated a thrust faulting solution for an intraslab event. The waveform fitting between the observed and synthetic waveforms on the three components (T, R, and Z) (Figure 7a) demonstrates a high-quality fit, with a final misfit value of 0.3 after 50,000 iterations (Figure 7d). Stability in the inversion result began to emerge after around 27,000 iterations, indicating solution convergence and high reliability of the retrieved source parameters.

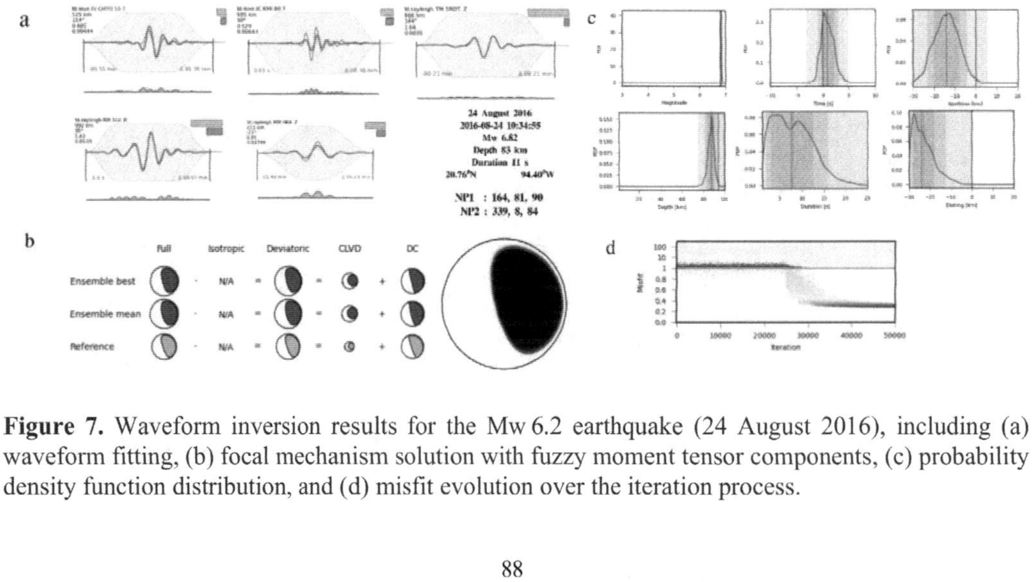

Figure 7. Waveform inversion results for the Mw 6.2 earthquake (24 August 2016), including (a) waveform fitting, (b) focal mechanism solution with fuzzy moment tensor components, (c) probability density function distribution, and (d) misfit evolution over the iteration process.

3.3 Tectonic interpretation derived from seismological and gravity data

Myanmar is an active tectonic region located at the convergence between the Indian Plate and the Burma Microplate. The interaction between oblique convergence and lateral accommodation yields a complex deformation pattern, as evidenced by seismicity, gravity anomalies, and earthquake source mechanisms recorded throughout the region.

A total of ten major earthquake events analyzed through waveform inversion reveals two dominant source mechanisms: five events with thrust faulting, typically occurring in the subduction zone, and the remaining five with dextral strike-slip faulting, characteristic of the Sagaing Fault system. To better understand subsurface deformation characteristics, cross-sectional profiles were constructed across the subduction zone and major active fault systems in Myanmar, including the Sagaing and Kabaw Faults. The distribution of hypocenters plotted along these profiles reveals earthquake clusters at various depths, indicating active tectonic segmentation. The cross-sectional view is presented in Figure 8.

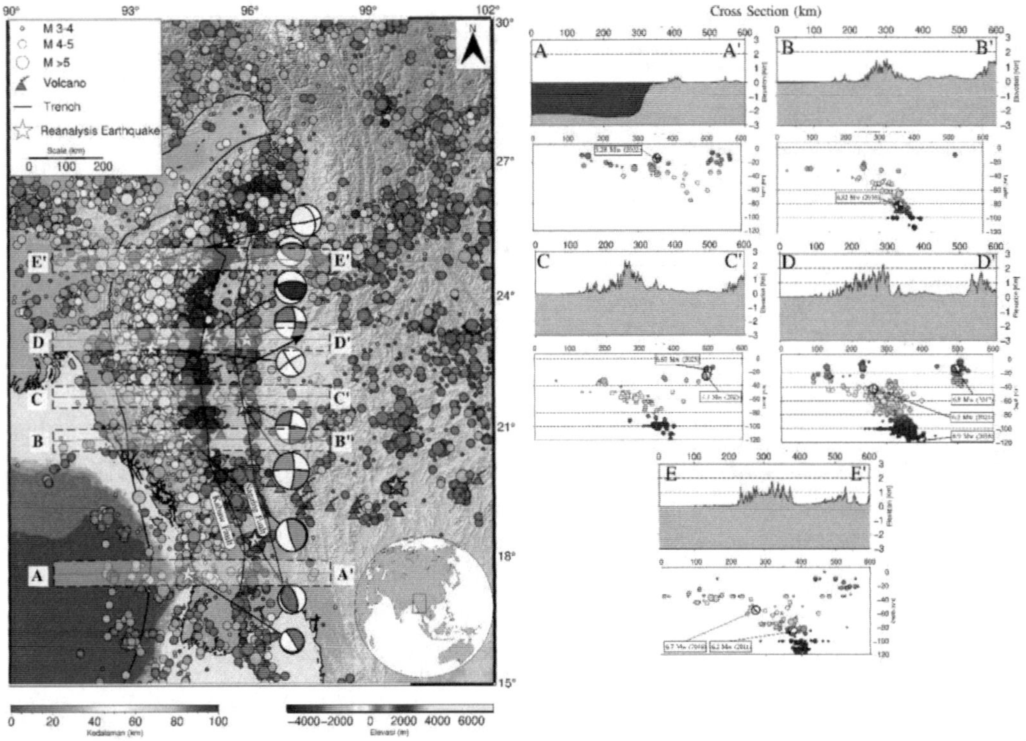

Fig. 8. Seismicity map and cross-section after earthquake parameter refinement, showing the updated distribution of event locations and focal mechanisms across Myanmar.

The cross-sectional profile in Figure 8 reveals the presence of a subduction zone in the western region. The Indian Plate is shown subducting northeastward, triggering intraslab earthquakes at depths exceeding 70 km [15]. This interplate convergence pattern is reflected in the distribution of seismicity, which dips downward from the northeast following the subduction path. On the eastern side, seismicity is concentrated at depths of less than 25 km along the Sagaing and Kabaw Fault systems. The activity of these faults, combined with interplate convergence, indicates a complex upper-crustal deformation pattern in the Myanmar region. To further investigate the subsurface structure controlling this deformation, we analyzed Complete Bouguer anomaly profiles (Figure 9), which provide additional constraints on density variations across the Myanmar region.

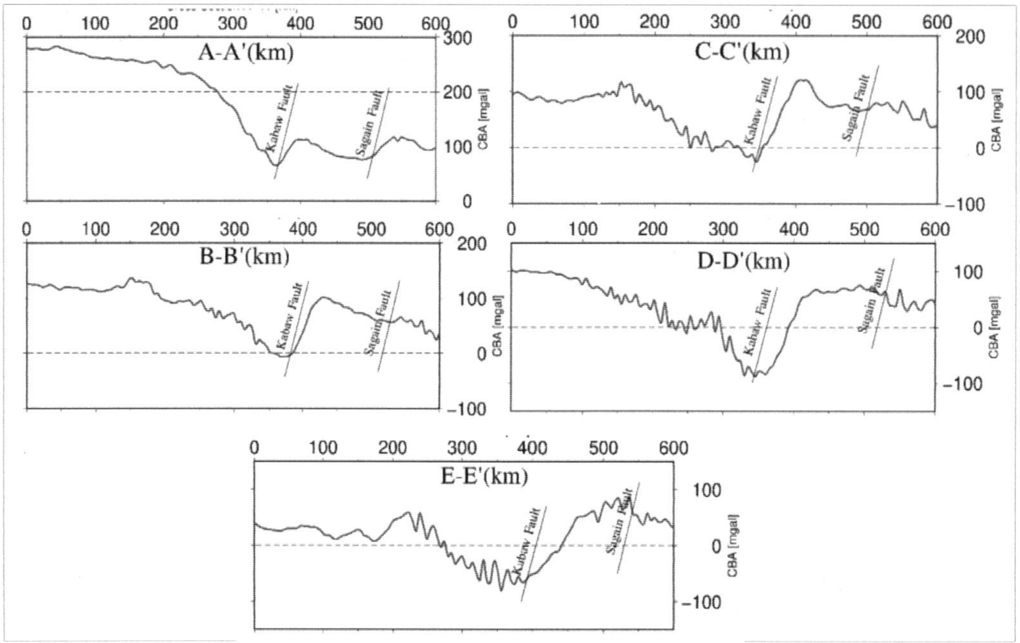

Fig. 9. Complete Bouguer anomaly profiles, illustrating subsurface density variations across the Myanmar region.

After the tectonic framework was outlined using the distribution of hypocenters and earthquake focal mechanisms, the interpretation of fault traces was further supported through cross-sectional profiles based on Complete Bouguer Anomaly (CBA) data. Figure 9 shows gravity anomaly profiles along each cross section, where high CBA values appear at the beginning of the transect and gradually decrease southward. This pattern is interpreted as a convergence zone, reflecting the interaction between tectonic plates. The interpretation is reinforced by the alignment of this anomaly pattern with seismicity distribution along the same transect. Two interpretive lines are also added to the gravity anomaly profiles, representing the presence of the Kabaw Fault and Sagaing Fault, two major tectonic structures in Myanmar that are believed to play a significant role in the region's seismic activity. The relocated earthquake events also exhibit spatial consistency with the position of these two faults. Further detailed studies are needed to improve the understanding of subsurface structures and regional seismic potential, as well as to support the development of more effective disaster mitigation strategies in the future.

4. Conclusion

This study reveals that Central Myanmar is governed by complex tectonic processes involving active fault systems and subduction dynamics. A previously identified seismic gap in the region was confirmed to be a locked fault segment that eventually ruptured during the major earthquake in March 2025. The spatial distribution of seismicity supports this interpretation, indicating concentrated stress accumulation before rupture. Gravity anomaly analysis further enhances the understanding of subsurface structures, clearly delineating variations along both the subduction interface and active faults such as the Sagaing and Kabaw Faults. Moreover, gravity anomaly profiles along cross-sections reveal consistent subsurface density contrasts that spatially align with relocated seismic events and interpreted fault traces, reinforcing the structural significance of the Kabaw and Sagaing Faults in controlling regional seismicity.

Source mechanism inversion of ten significant earthquakes yields low misfit values (<0.6), confirming the reliability of the inversion results. Among these events, five are characterized by dextral strike-slip mechanisms, while the other five exhibit thrust faulting, reflecting the diverse tectonic regime in the region. The integration of seismicity patterns, gravity anomalies, and focal mechanism solutions provides a comprehensive depiction of the tectonic framework in Central Myanmar. This multi-parameter approach not only contributes to a deeper scientific understanding of earthquake generation in the region but also offers valuable insights for future seismic hazard assessments and disaster risk reduction strategies.

References

1. Y. Wang, K. Sieh, S.T. Tun, K.Y. Lai, T. Myint, Active tectonics and earthquake potential of the Myanmar region. J. Geophys. Res. Solid Earth **119**, 3767–3822 (2014). https://doi.org/10.1002/2013JB010762
2. X. Xu, W. Kang, T. Wang, X. Zhang, Y. Liu, Y. Zhang, J. Zhao, K. Li, Q. Wang, J. Cheng, J. Ren, The Mw7.7 Myanmar earthquake: a continental longest surface-rupturing supershear cascading event. npj Nat. Hazards **2**, 73 (2025). https://doi.org/10.1038/s44304-025-00125-z
3. D. Melgar, R. Weldon, Y. Wang, M.G. Bato, L.T. Aung, X. Shi, T. Dawson, Supershear source model of the 2025 M7.8 Myanmar earthquake and paleoseismology of the Sagaing Fault: regions of significant overlap with past earthquakes. EarthArXiv (2025). https://doi.org/10.31223/X5QT6G
4. J. Bai, Y. He, X. Yuan, F. Tilmann, Y. Ai, M. Jiang, G. Hou, C.T. Mon, M. Thant, K. Sein, Seismic structure across central Myanmar from joint inversion of receiver functions and Rayleigh wave dispersion. Tectonophysics **818**, 229068 (2021). https://doi.org/10.1016/j.tecto.2021.229068
5. C.K. Morley, Syn-kinematic sedimentation at a releasing splay in the northern Minwun Ranges, Sagaing Fault zone, Myanmar: significance for fault timing and displacement. Basin Res. **29**, 684–700 (2017). https://doi.org/10.1111/bre.12197
6. L. Ye, T. Lay, H. Kanamori, The 28 March 2025 Mw 7.8 Myanmar earthquake: preliminary analysis of an ~480 km long intermittent supershear rupture. Seism. Rec. **5**, 260–269 (2025). https://doi.org/10.1785/0320250021
7. S. Heimann, M. Isken, D. Kühn, H. Sudhaus, A. Steinberg, S. Daout, S. Cesca, H. Bathke, T. Dahm, Grond: a probabilistic earthquake source inversion framework. GFZ Data Services (2018). https://doi.org/10.5880/GFZ.2.1.2018.003
8. U. Muksin, K. Bauer, M. Muzli, T. Ryberg, I. Nurdin, M. Masturiyono, M. Weber, AcehSeis project provides insights into the detailed seismicity distribution and relation to fault structures in Central Aceh, Northern Sumatra. J. Asian Earth Sci. **171**, 20–27 (2019). https://doi.org/10.1016/j.jseaes.2018.11.005
9. X. Zhao, Z. Duputel, Z. Yao, Regional W-phase source inversion for moderate to large earthquakes in China and neighboring areas. J. Geophys. Res. Solid Earth **122**, 10052–10068 (2017). https://doi.org/10.1002/2017JB014950
10. M.H. Gunarya, A.V. Simanjuntak, U. Muksin, Characteristics of earthquake mechanisms along the Andaman-Nicobar region derived from Bayesian moment tensor inversion. IOP Conf. Ser. Earth Environ. Sci. **1373**, 012064 (2024). https://doi.org/10.1088/1755-1315/1373/1/012064
11. S. Heimann, H. Vasyura-Bathke, H. Sudhaus, M.P. Isken, M. Kriegerowski, A. Steinberg, T. Dahm, A Python framework for efficient use of pre-computed Green's functions in seismological and other physical forward and inverse source problems. Solid Earth **10**, 1921–1935 (2019). https://doi.org/10.5194/se-10-1921-2019

12. M. Yanis, I. Marwan, R. Idroes, N. Zaini, A.Y. Paembonan, R. Ananda, A.A. Ghani, A pilot survey for mapping the fault structure around the Geuredong volcano by using high-resolution global gravity. Acta Geophys. **70**, 2057–2075 (2022). https://doi.org/10.1007/s11600-022-00860-1
13. M. Yanis, F. Abdullah, N. Zaini, N. Ismail, The northernmost part of the Great Sumatran Fault map and images derived from gravity anomaly. Acta Geophys. **69**, 795–807 (2021). https://doi.org/10.1007/s11600-021-00574-5
14. N. Hurukawa, P. Maung Maung, Two seismic gaps on the Sagaing Fault, Myanmar, derived from relocation of historical earthquakes since **1918**. Geophys. Res. Lett. 38, L01303 (2011). https://doi.org/10.1029/2010GL045932
15. H.A. Shiddiqi, P.P. Tun, T.L. Kyaw, L. Ottemöller, Source study of the 24 August 2016 Mw 6.8 Chauk, Myanmar, earthquake. Seismol. Res. Lett. **89**, 1773–1785 (2018). https://doi.org/10.1785/0220170278

"URGENSI" Platform Development to Enhance Disaster Preparedness Among the Elderly

Rika Yuliwulandari[1,*], *Zulfan* Febriawan[2], *Rifda* El Mahroos[2], *Debrina* Kusuma Devi[3], *Eko* Teguh Paripurno[4], *Awang* Hendrianto Pratomo[4], *Tedy* Agung Cahyadi[4], *Yohana* Noradika Maharani[4], *Mohammad* Idhom[5], *Gideon* Setyo Budi Witjaksono[6], *Kysara* Aureldy Davonso Matualage[7], *Hafiz* T.A. Khan[8,9]

[1]Department of Pharmacology, Faculty of Medicine, Universitas Pembangunan Nasional Veteran Jawa Timur, Surabaya, Indonesia
[2]Department of Physiology, Faculty of Medicine, Universitas Pembangunan Nasional Veteran Jawa Timur, Surabaya, Indonesia
[3]Department of Internal Medicine, Faculty of Medicine, Universitas Pembangunan Nasional Veteran Jawa Timur, Surabaya, Indonesia
[4]Master Program of Disaster Management, Faculty of Mineral and Energy Technology, Universitas Pembangunan Nasional Veteran Yogyakarta, Yogyakarta, Indonesia
[5]Department of Data Science, Faculty of Computer Science, Universitas Pembangunan Nasional Veteran Jawa Timur, Surabaya, Indonesia
[6]Department of Accounting, Faculty of Economics and Business, Universitas Pembangunan Nasional Veteran Jawa Timur, Surabaya, Indonesia
[7]Student Faculty of Medicine, Universitas Pembangunan Nasional Veteran Jawa Timur, Surabaya, Indonesia
[8]Public Health Group, University of West London, London, United Kingdom
[9]Oxford Institute of Population Ageing, University of Oxford, Oxford, United Kingdom

Abstract. Surabaya, Indonesia's second-largest city, lies between the Waru and Surabaya faults, making it prone to earthquakes and frequent flooding due to its dense population and inadequate drainage system. Among its residents, approximately 350,000 elderly individuals are particularly vulnerable during disasters because of limited mobility, physical frailty, and difficulty accessing information. However, research on disaster preparedness among the elderly and the use of technology-based support remains limited. This study aimed to develop and test "URGENSI", a web-based platform designed for the elderly around UPN "Veteran" Jawa Timur, following the ADDIE Model (Analysis, Design, Development, Implementation, Evaluation). Stakeholder discussions informed the design process. The application integrates two main components: 1) KESTANA Assessment, a self-assessment tool to evaluate disaster and health preparedness, and 2) educational modules (KESTANA education and KITANA education) offering self-learning materials on health, disaster readiness and emergency kit preparation. Functionality testing was

* Corresponding author: rika.fk@upnjatim.ac.id

conducted with 25 elderly participants living near UPN "Veteran" Jawa Timur. Results showed most respondents were neither healthy nor resilient when facing disasters. Nevertheless, participants rated the application as user-friendly, visually engaging, and easy to understand. Overall, URGENSI demonstrates strong potential to enhance disaster preparedness, support elderly health, and contribute to broader disaster risk reduction and community resilience efforts in Indonesia.

1 Background

Natural disasters are triggered by natural phenomena such as floods, landslides, coastal abrasion, tornadoes, droughts, forest and land fires, earthquakes, tsunamis, and volcanic eruptions [1]. According to data from the 2023 World Risk Index (WRI), ten countries rank highest in terms of disaster risk globally, the Philippines ranks first with the highest disaster risk index at 46.86%, followed by Indonesia at 43.5%, and India in third place with an index of 41.57% [2]. Indonesia experiences a particularly high frequency of natural disasters, especially volcanic eruptions and earthquakes. Located along the Pacific Ring of Fire, the country is among the most disaster-prone regions in the world [3]. Data from Indonesia's National Disaster Management Agency (Badan Nasional Penanggulangan Bencana or BNPB) indicate that a total of 4,852 natural disaster events occurred across the country in 2023. Floods were the most frequent, accounting for 1,801 incidents, followed by extreme weather (1,135 incidents), forest and land fires (1,117), landslides (568), droughts (168), tidal waves or coastal abrasion (31), earthquakes (29), and volcanic eruptions (3) [4].

These disasters resulted in approximately 370 fatalities, 39 missing persons, and 536 individuals injured [5]. Older adults represent one of the most vulnerable population groups before, during, and after disasters. This is primarily due to the high prevalence of chronic illnesses, physical decline, cognitive impairments, dementia, and general frailty among this age group [6]. Physical limitations, coupled with diminished social support, often leave older adults living alone. This situation further increases their vulnerability to disaster impacts, as they may lack the physical capacity and family assistance necessary for timely response and recovery [7]. Moreover, the immune system in older individuals tends to function less effectively, reducing their ability to fight off infections caused by bacteria or viruses. These factors collectively contribute to making older adults one of the most at-risk groups during disasters, as their physical, cognitive, and social limitations significantly increase their likelihood of becoming victims in emergency situations [8].

Preparedness is an essential component of disaster management. Disaster preparedness enables older adults to formulate and plan necessary actions to be taken when a disaster occurs [9]. Older adults at the Puntodewo Tanjungsari Posyandu in Surabaya showed adequate disaster-related knowledge; however, their actual disaster experience and response were found to be inadequate [10]. In the current digital era, information and communication technology (ICT) holds significant potential in enhancing public literacy and preparedness for disasters. The use of digital platforms such as mobile applications, websites, or community-based information systems can serve as effective tools for education, early warning, and coordination in disaster response [11]. However, most existing platforms have not been specifically designed to reach older adults, particularly in terms of language, user interface, and modes of information delivery. This poses a substantial barrier to improving preparedness within this demographic group. Therefore, this study aims to develop a digital platform and field trials to determine people's experience in using this digital platform.

2 Methods

2.1 Development Model

The development of the "URGENSI" platform was guided by the ADDIE model, which includes five systematic stages: Analysis, Design, Development, Implementation, and Evaluation. This model was selected to ensure a structured and user-centered approach in creating a digital platform tailored to enhance disaster preparedness among the elderly.

2.1.1 Analysis

In the analysis phase, a needs assessment was conducted to identify the specific challenges and preparedness gaps faced by elderly individuals in disaster situations. Data were gathered through literature review, expert consultation and informal interviews with elderly residents in the area surrounding UPN "Veteran" Jawa Timur. Key findings highlighted limited access to disaster-related information, low digital literacy, and a lack of elderly-specific preparedness tools.

2.1.2 Design

The design phase involved translating these needs into specific platform features and learning objectives. The platform was planned to include educational modules, preparedness checklist, and a consultation feature. User interface design prioritized accessibility, with large icons, simple language, and audio-visual support to accommodate sensory or cognitive limitations common among the elderly. A content structure was developed based on KESTANA (Elderly Health Resilience in Disasters) principles and government disaster preparedness guidelines.

2.1.3 Development

During the development phase, the platform prototype was built collaboratively by software developers and experts in gerontology and disaster management. Educational content was produced in multiple formats including text, infographics, and short videos. The KESTANA self-assessment tool was embedded to allow elderly to evaluate their personal disaster and health preparedness. The platform also included a disaster and health consultation feature, enabling users to submit questions directly to professionals.

2.1.4 Implementation

In the implementation phase, the prototype was tested with 25 elderly participants living near UPN "Veteran" Jawa Timur. A short orientation was conducted to introduce participants to the platform and assist them in navigating its features, particularly the KESTANA assessment. Participants were encouraged to use the platform independently or with optional researcher assistance for those unfamiliar with mobile technology.

2.1.5 Evaluation

In the evaluation phase, data were collected to assess both platform usability and its effectiveness in increasing knowledge and preparedness through user satisfaction questionnaire. The satisfaction instrument evaluated content clarity, ease of use, visual design, and perceived usefulness. Feedback was analyzed to inform revisions and future improvements to the platform.

2.2 Platform Trial

The "URGENSI" platform was trialed with 25 elderly participants residing in the area surrounding UPN "Veteran" Jawa Timur. Before the trial, the researcher conducted an orientation session to guide participants on how to use the platform, with particular emphasis on how to complete the KESTANA assessment, an instrument designed to evaluate elderly health preparedness in the face of disasters. During the session, the researchers also explained how to navigate and utilize the platform's features, including disaster and health education modules, as well as a consultation feature that allows users to ask questions related to disaster preparedness and health management. Throughout the trial, participants interacted with the platform either independently or with assistance, depending on their technological proficiency. After the usage period, participants were asked to provide feedback on their user experience by completing an online questionnaire. This feedback focused on their satisfaction with the platform's usability, content, and perceived usefulness in improving their preparedness and health awareness.

2.3 Data Collection

Data collection in this study was conducted among 25 elderly individuals residing in the area surrounding UPN "Veteran" Jawa Timur. To assess their level of disaster preparedness and health awareness, participants were asked to engage with the "URGENSI" platform using their own mobile devices. For participants who faced difficulties in operating the device, assistance was provided by the research team to ensure equitable access and usability. The platform included a preparedness checklist, interactive modules containing educational content, which participants were to complete over a set period. Following this, to evaluate user satisfaction, participants were asked to complete an online questionnaire that captured their experience using this platform. The questionnaire consisted of Likert-scale items assessing four key dimensions: content quality, ease of use, visual and interface design, and perceived usefulness.

Table 1. User Satisfaction Scoring.

No.	Type of User Satisfaction	Score	Information
1.	Content quality : How easy was it for you to understand the information presented in the "URGENSI" platform?	1	Not understandable
		2	Difficult to understand
		3	Easy to understand
		4	Very easy to understand
2.	Ease of use : How easy was it for you to operate the platform on your own?	1	Could not use
		2	Difficult to use
		3	Easy to use
		4	Very easy to use
3.	Interface and visual design : How would you rate the readability and appearance of the text, icons, and overall layout?	1	Uncomfortable to view
		2	Slightly hard to read
		3	Clear and comfortable
		4	Very clear and comfortable
4.	Perceived Usefulness : To what extent did the platform help you feel more prepared for disasters?	1	Not helpful
		2	Slightly helpful
		3	Quite helpful
		4	Very helpful

2.4 Data Analysis

Data were analyzed using both quantitative and qualitative methods. The analysis of data focused on two main components: (1) the level of health and disaster preparedness among the elderly, and (2) user feedback related to satisfaction with the platform. Quantitative data obtained from data on health and disaster preparedness among the elderly were analyzed using descriptive statistics to determine differences in preparedness in the elderly. Whereas, qualitative data was obtained from user satisfaction data. User satisfaction data were analyzed using percentage distributions to assess the level of satisfaction in areas such as content clarity, ease of use, visual design and perceived usefulness. The responses were categorized into score intervals to determine satisfaction levels (very dissatisfied to very satisfied).

3 Results

3.1 Description of the "URGENSI" Platform

The findings of this study led to the development of a digital platform named **URGENSI** (https://www.lansiatangguh.com), which is designed to support disaster preparedness specifically among older adults and families. URGENSI provides accessible, practical information aimed at promoting healthy aging and strengthening family resilience in the face of disasters (Figure 1). The platform integrates several key features tailored to the needs of its target users. One of the core components is the Family and Elderly Health Resilience Assessment Tool (KESTANA), which enables users to evaluate their health status and preparedness level. In addition, the platform includes the Disaster Resilience Kit (KITANA), which offers essential guidelines and tools for disaster readiness. Educational modules are also embedded, including KESTANA Education, which provides structured learning content on maintaining physical, mental, and social well-being of older adults to improve their capacity for disaster response and recovery, and KITANA Education, which delivers guidance on assembling and using essential emergency supplies tailored to the needs of older adults and their families.

Furthermore, URGENSI features a Disaster-Related Health Consultation Service, allowing users to seek professional advice on health concerns in the context of emergency preparedness. Collectively, these components make URGENSI a comprehensive, elderly-friendly platform that aims to bridge the digital gap and improve disaster resilience through inclusive, health-oriented technology.

Fig 1. Homepage of "URGENSI" Platform.

3.2 Description of the "URGENSI" Platform Features

To address the specific needs of older adults and families in disaster preparedness, the URGENSI platform incorporates several integrated features designed to enhance both Awareness and practical readiness. These features are systematically organized to support health resilience and disaster response capabilities. An overview of the platform's main components is illustrated in Figure 2.

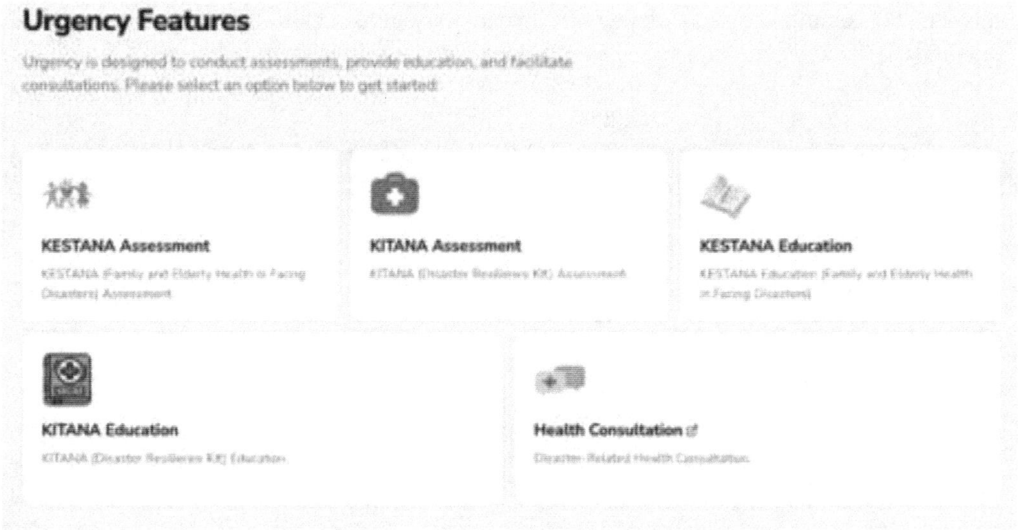

Fig 2. "URGENSI" Platform Features.

3.2.1 *KESTANA (Family and Elderly Health Resilience) Assessment*

The KESTANA (Kesehatan Keluarga dan Lansia Tangguh Bencana) assessment is designed to evaluate various aspects of elderly individuals' preparedness and resilience in the context of health and disaster situations. This assessment focuses on four key areas. First, it examines the overall health status of older adults, including the presence of chronic conditions and access to routine medical care. Second, it explores practices to maintain health quality, such as nutrition, physical activity, and access to healthcare services. Third, the assessment measures the elderly's ability to use technology, particularly in accessing health information, communication tools, and emergency alerts. Lastly, KESTANA captures personal experiences of older adults during past disasters, including their coping strategies, challenges faced, and the support systems that were available to them. Through these dimensions, KESTANA provides a comprehensive understanding of how well-prepared elderly individuals and their families are in maintaining health and ensuring resilience during emergencies (Figure 3).

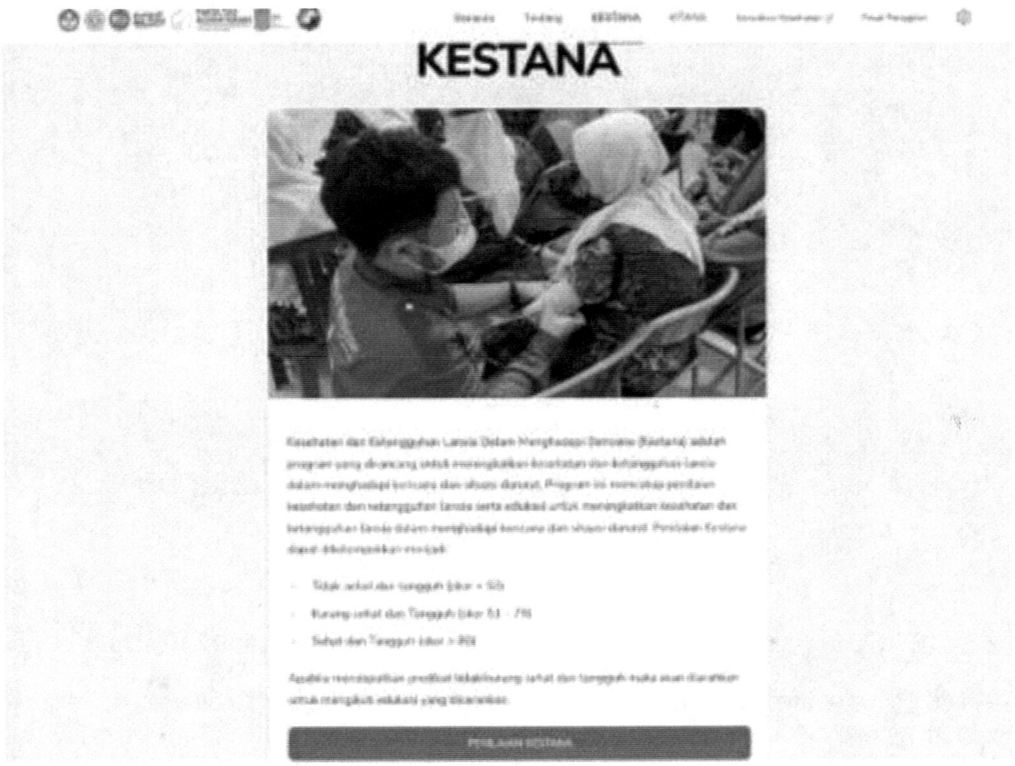

Fig 3. "URGENSI" Platform Features.

3.2.2 *KITANA (Disaster Resilience Kit) Assessment*

The KITANA (Disaster Resilience Kit) Assessment is a tool developed to evaluate the availability, completeness, and readiness of essential supplies and equipment needed by families, especially those with elderly members, to respond effectively in disaster situations. This assessment focuses on the presence of critical items such as food, clean water, first aid kits, medications, important documents, flashlights, batteries, communication tools, and personal protective equipment. It also considers whether these items are stored in an accessible, organized, and easily transportable manner. Furthermore, the KITANA assessment evaluates the family's awareness and routine in checking and maintaining the contents of their disaster kit to ensure its functionality when needed. By identifying gaps and strengths in household preparedness, the KITANA assessment serves as a vital tool in promoting disaster resilience, ensuring that families, particularly those with vulnerable members like the elderly, are better equipped to face emergencies with confidence and reduced risk (Figure 4).

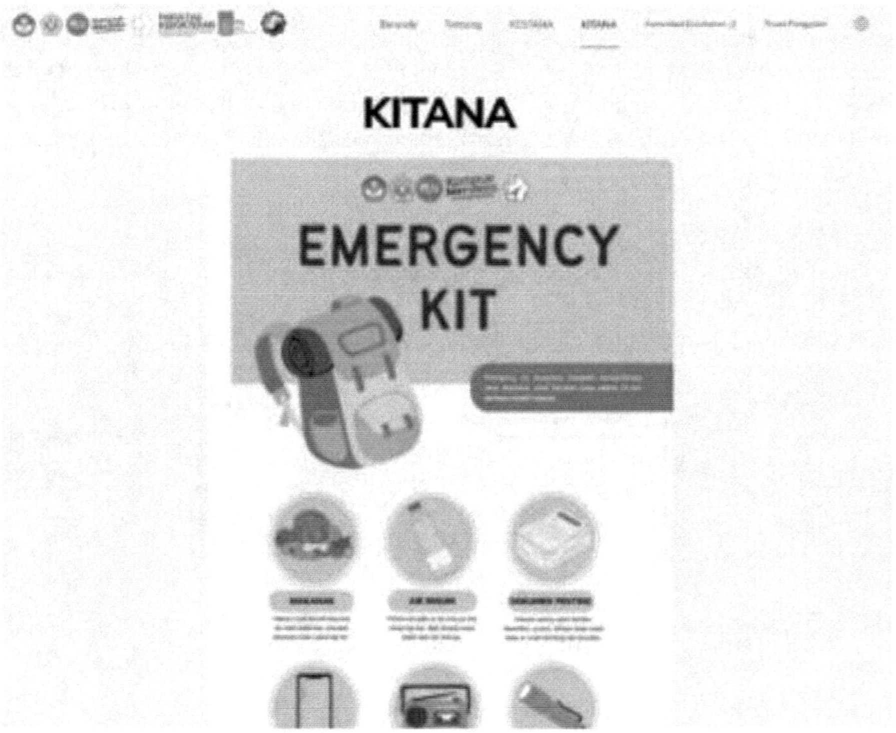

Fig 4. "URGENSI" Platform Features.

3.2.3 *KESTANA (Family and Elderly Health Resilience) Education*

The KESTANA (Family and Elderly Health Resilience) Education is an educational initiative aimed at increasing awareness, knowledge, and skills related to health maintenance and disaster preparedness among elderly individuals and their families. This program focuses on key topics such as promoting healthy lifestyles, managing chronic illnesses, understanding age-related vulnerabilities, and practicing preventive healthcare. In addition, the education includes practical guidance on disaster preparedness, such as how to develop emergency plans, identify risks, and maintain communication during crises. It also introduces basic digital literacy to help older adults access reliable health and disaster-related information through technology. By integrating health education with disaster resilience strategies, KESTANA Education empowers families to actively support the well-being and safety of elderly members, strengthening their capacity to respond effectively to emergencies and reduce health-related risks during disasters (Figure 5).

Fig 5. "URGENSI" Platform Features.

3.2.4 *KITANA (Disaster Resilience Kit) Education*

The KITANA (Disaster Resilience Kit) Education is a learning program designed to increase awareness and knowledge about the importance of preparing essential items needed during emergencies or disasters. This educational component teaches individuals and families, especially those with elderly members, how to assemble, manage, and maintain a comprehensive disaster resilience kit. It covers the identification of critical supplies such as food, water, medications, personal hygiene items, important documents, first aid materials, and communication tools. Participants are also guided on how to properly store and regularly update the kit to ensure accessibility and usability when disasters strike. Additionally, KITANA Education emphasizes the role of preparedness in reducing panic and increasing survival chances, especially for vulnerable populations like older adults. Through this program, families become better equipped with the practical skills and understanding needed to build resilience and respond effectively to emergency situations (Figure 6).

Fig 6. "URGENSI" Platform Features.

3.2.5 *Disaster-related Health Consultation*

Disaster-related Health Consultation is a service aimed at providing targeted medical advice, guidance, and support to individuals particularly the elderly and vulnerable groups before, during, and after a disaster. This consultation focuses on addressing health risks associated with disasters, such as the management of chronic diseases, mental health support, access to medications, and the prevention of disease outbreaks in emergency settings. It also includes personalized health assessments, recommendations for maintaining well-being during crises, and education on recognizing early warning signs of health deterioration. Furthermore, disaster-related health consultations help individuals understand how to navigate healthcare services during emergencies, including where to seek help and how to communicate specific medical needs. By offering timely and relevant health information, this service plays a crucial role in enhancing community resilience, minimizing health complications, and supporting recovery efforts in the aftermath of a disaster (Fig 7.).

Fig 7. "URGENSI" Platform Features.

3.3 Disaster Preparedness for the Elderly

Data on disaster preparedness and health were collected using the "URGENSI" platform among 25 elderly individuals living around UPN "Veteran" Jawa Timur Surabaya. The results are presented in Figure 8. Figure 8 illustrates the distribution of elderly individuals across three categories of disaster preparedness and health: *Not healthy and resilient*, *Less healthy and resilient*, and *Healthy and resilient*. The data reveals a striking imbalance, with the vast majority (92%) of older adults falling into the *Not healthy and resilient* category. This indicates that most elderly individuals in the sample lack both adequate health status and the capacity to withstand or respond effectively to disaster situations. Interestingly, there were no respondents (0%) categorized as *Less healthy and resilient*, suggesting that the participants were either in a severely unprepared condition or already met both health and resilience standards.

Only 8% of the elderly were classified as *Healthy and resilient*, representing a small fraction of the population who possess both good health and the necessary preparedness to face disasters. These findings highlight a critical need for targeted interventions to improve both health literacy and disaster readiness among older adults, particularly through accessible and tailored support systems such as the URGENSI platform.

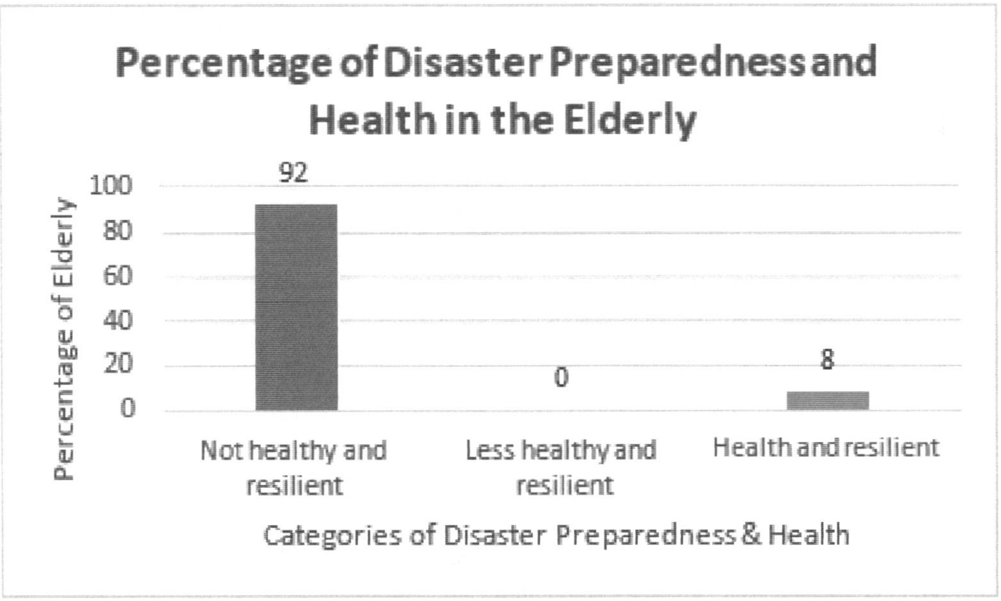

Fig 8. Percentage of Disaster Preparedness and Health in the Elderly.

3.4 Assessment of Elderly Satisfaction with Platform Usage

An evaluation of user satisfaction regarding four key aspects quality of content, ease of use, interface and visual design, and perceived usefulness was conducted among 25 older adults around UPN "Veteran" Jawa Timur. Data were collected through an online questionnaire specifically designed for elderly users. The results of this assessment are presented in Figures 9 to 12. Figure 9 illustrates user satisfaction with the content quality of the *URGENSI* platform as reported by 25 elderly respondents. The graph shows that 100% of participants rated the content quality with a score of 4 on the satisfaction scale, indicating a very high level of satisfaction.

No respondents gave lower scores (1, 2, or 3), suggesting that the platform's content was perceived as relevant, accurate, and appropriate for the needs and comprehension levels of older adults. This finding reflects the effectiveness of the educational material and information provided through *URGENSI*, particularly in its clarity, usefulness, and alignment with disaster preparedness goals for the elderly population.

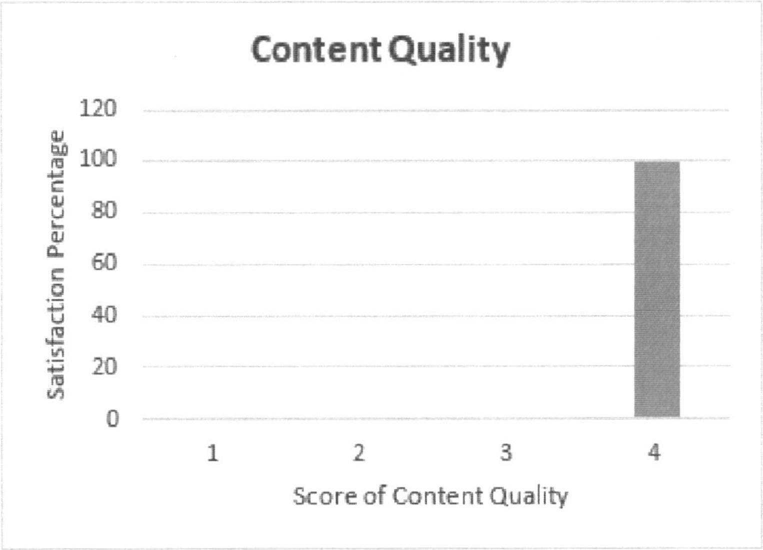

Fig 9. Percentage of Satisfaction in Content Quality.

Figure 10 presents the level of user satisfaction regarding the ease of use of the *URGENSI* platform among 25 elderly participants. The results indicate that 100% of respondents rated the platform with a score of 4 on the ease-of-use scale, reflecting a very high degree of usability. No participants assigned lower scores (1, 2, or 3), suggesting that the platform interface and navigation were considered intuitive, accessible, and manageable even by older adults who may have limited experience with digital technology. This outcome highlights the platform's success in incorporating user-centered design principles that accommodate the specific physical and cognitive needs of elderly users, thereby minimizing barriers to access and encouraging continued engagement.

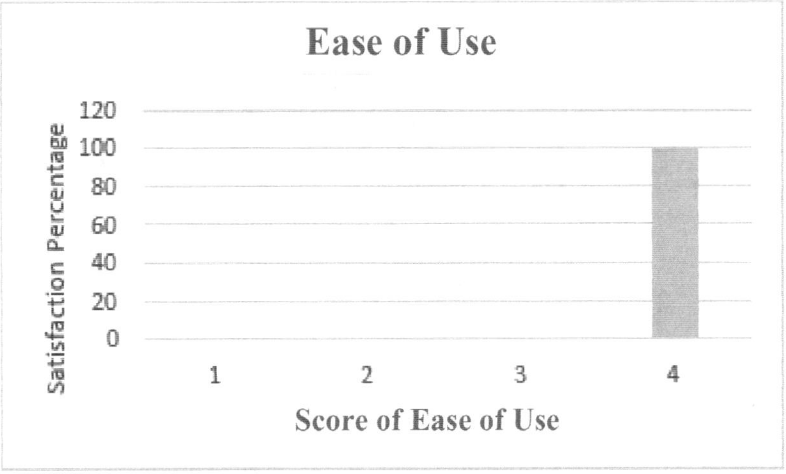

Fig 10. Percentage of Satisfaction in Ease of Use

Figure 11 illustrates user satisfaction with the interface and visual design of the *URGENSI* platform, based on feedback from 25 elderly participants. The results show that 100% of respondents assigned a score of 4, representing the highest level of satisfaction on the evaluation scale. No lower scores (1–3) were recorded, indicating a universally positive response to the platform's aesthetic quality and visual usability. This result suggests that the design elements, including layout, color scheme, font size, and iconography, were perceived as clear, visually appealing, and accessible by older adults. The consistent satisfaction also reflects the importance of age-sensitive design considerations in digital platforms, especially when targeting users with potential visual or cognitive limitations. The outcome reinforces that a well-designed interface plays a crucial role in supporting user engagement and learning, particularly in disaster preparedness contexts.

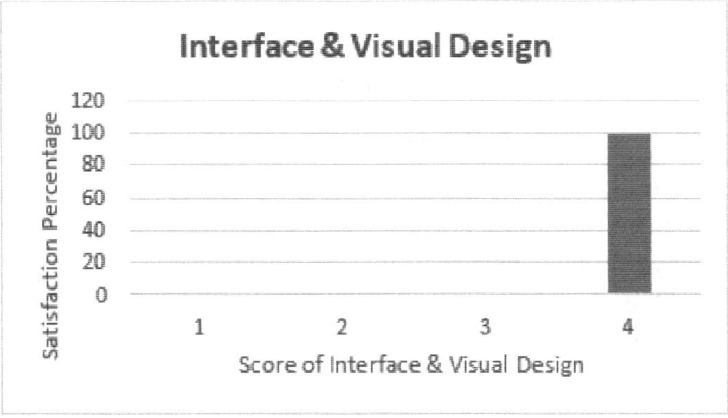

Fig 11. Percentage of Satisfaction in Interface & Visual Design.

Figure 12 illustrates user satisfaction with the perceived usefulness of the platform as reported by the respondents. The graph shows that 100% of participants rated the perceived usefulness with a score of 4 on the satisfaction scale, indicating a very high level of satisfaction. No respondents assigned lower scores (1, 2, or 3), suggesting that the platform was consistently viewed as highly beneficial and effective in meeting user expectations. This unanimous rating reflects the platform's success in delivering features or content that are practical, valuable, and aligned with user needs, thereby reinforcing its overall usefulness and relevance in the intended context.

Fig 12. Percentage of Satisfaction in Perceived Usefulness.

4 Discussion

This study highlights a critical gap in disaster preparedness and health resilience among elderly individuals, as evidenced by the fact that 92% of respondents fell into the "not healthy and resilient" category. This finding underscores the compounded vulnerability of older adults in disaster contexts, which is frequently attributed to age-related declines in physical health, presence of chronic diseases, limited mobility, and insufficient knowledge of preparedness strategies [20]. Such vulnerability is further exacerbated by poor access to health services, technology, and information, especially in lower-income or rural areas. The absence of respondents in the "less healthy and resilient" category suggests a polarization, either individuals are completely unprepared or fully prepared, emphasizing the need for intermediate, progressive educational interventions that can gradually build resilience and health capacity. The introduction of the URGENSI platform represents a strategic response to this issue. Its integrated features, including KESTANA (Family and Elderly Health Resilience) Assessment, KITANA (Disaster Resilience Kit), educational modules, and disaster-related health consultations, are tailored specifically for older adults and their families. This aligns with research that advocates for holistic, person-centered approaches in health education and emergency preparedness [21]. The high satisfaction ratings in all aspects evaluated content quality, ease of use, interface design, and perceived usefulness suggest that the platform succeeded in meeting the needs of its target audience. Notably, 100% of respondents gave the maximum score across all categories, indicating exceptional levels of user engagement and perceived relevance.

From a usability perspective, the platform's success can be linked to the application of age-friendly digital design principles, such as large fonts, simple navigation, minimal cognitive load, and clear visual hierarchy. These elements are essential for improving accessibility for users with age-related visual, motor, or cognitive limitations. Additionally, the incorporation of disaster consultation services is vital, as elderly individuals often have limited access to timely medical guidance during emergencies. This feature ensures continuity of care and enhances confidence among elderly users when navigating disaster scenarios. Another important aspect is the emphasis on digital literacy and family involvement, which is embedded in both KESTANA and KITANA educational modules.

Studies have shown that disaster resilience in older populations is not only influenced by individual preparedness but also by the quality of family and community support networks. By encouraging family-based assessments and preparedness actions, the URGENSI platform fosters a collaborative approach that distributes the burden of readiness and ensures that elderly individuals are not left to manage disaster risks alone. In terms of broader implications, the URGENSI platform can serve as a scalable model for elderly disaster preparedness education in other regions, especially within the ASEAN context, where aging populations are rapidly increasing.

As digital transformation becomes more inclusive, platforms like URGENSI could be integrated into public health strategies and local government disaster response programs, especially in areas prone to natural hazards such as floods, earthquakes, or pandemics. This approach is consistent with the global call for integrating disaster risk reduction into health and aging agendas, as outlined by the WHO and UNDRR. Future research should explore the longitudinal impact of platform usage on disaster outcomes, health behavior change, and resilience-building. Additionally, examining the role of socioeconomic status, education level, and previous disaster experience could enrich the understanding of the determinants of preparedness among older adults.

5 Conclusion

The results of this study underscore the critical need for disaster preparedness initiatives tailored specifically to the elderly population. Data collected from 25 older adults living around UPN "Veteran" Jawa Timur revealed that a significant majority (92%) were categorized as not healthy and not resilient in the face of disasters, highlighting considerable gaps in both health status and preparedness. Only a small proportion (8%) met the criteria for being healthy and resilient. These findings emphasize the importance of targeted interventions that promote health literacy, technological readiness, and disaster response capacity among older adults. To address these gaps, the URGENSI platform was developed as a comprehensive, elderly-friendly digital tool that integrates assessment, education, and consultation services. Key components such as the KESTANA and KITANA assessments and educational modules help users evaluate their resilience, improve their health practices, and prepare essential disaster kits.

Moreover, the platform's disaster-related health consultation service ensures accessible support for medical and emergency-related concerns.

User satisfaction evaluations showed overwhelmingly positive results. Across four measured aspects like content quality, ease of use, interface and visual design, and perceived usefulness, 100% of elderly respondents gave the highest satisfaction rating (score of 4). This indicates that the platform was perceived as highly accessible, relevant, and beneficial to its intended users. In conclusion, URGENSI not only addresses an urgent need among elderly communities but also demonstrates strong potential as an effective, user-approved tool for enhancing disaster preparedness and health resilience.

References

1. T.N. Hidayati, A. Yolla, R. Febrialdi. Correspondence Analysis to Determine the Relationship Between Disaster Types and House Damage in Indonesia in 2019. *Bulletin of Scientific Contribution Geology*. **18**, 2 (2020).
2. M. Ira, N.D. Erika. Overview of the Level of Community Knowledge Regarding Preparedness in Facing Volcanic Eruption Disasters in Jrakahselo Village, Boyolali. *Indonesian Journal of Public Health*. **2**, 4 (2024).
3. O. Yenny, W.K. Nabila. Literature Studies on Natural Disasters and Communication in Indonesia. *E3S Web of Conference*. 604, 02008 (2025). https://doi.org/10.1051/e3sconf/202560402008
4. BNPB. *Rekapan Data Bencana di Indonesia 2023*. Badan Nasional Penanggulangan Bencana. (2023).
5. M. Zana, H.H. Ali, A.A. Haltam. Prevalence of Cognitive Impairment Amongst Older Adults and the Associated Risk Factors. Kurdistan. *Journal of Applied Research*. **9**, 1 (2023).
6. N. Phraknoi, J. Susanto, Y. Hu, Y.S. Goh, C.E.C. Lee. Older People's Needs in Urban Disaster Response: A Systematic Literature Review. *International Journal of Disaster Risk Reduction*. **96** (2023).
7. N.M. Risdiana. Kesiapsiagaan Lansia Dalam Menghadapi Bencana Angin Puting Beliung di Kecamatan Sitinjo. *Jurnal Keperawatan dan Fisioterapi*. **5**, 2 (2022).
8. C.G. Marcella, M.D. Walter, N. Cahyo. Field Insights and Analysis of the 2018 Mw 7.5 Palu, Indonesia Earthquake, Tsunami and Landslide. *Pure Appl. Geophys*. **178** (2021).

9. I. Nurhidayati, E. Ratnawati. Kesiapsiagaan Keluarga dengan Lanjut Usia pada Kejadian Letusan Merapi di Desa Belerante Kecamatan Kemalang. *Jurnal Keperawatan dan Kesehatan Masyarakat.* **7,** 1 (2018).
10. B. Biswajit. Role of Information and Communication Technology (ICT) in Disaster Management: A Study in India. *International Journal of Political Science and Governance.* **6**, 2 (2024).
11. F. Johan, E. Anna, B. Frida, E. Tilda. Digital Health Platforms for the Elderly? Key Adoption and Usage Barriers and Ways to Address Them. *Technological Forecasting and Social Change.* **189** (2023).
12. S-G. Zahra. Challenges in Health and Technological Literacy of Older Adults: A Qualitative Study in Isfahan. *BMC Geriatr.* **24**, 247 (2025).
13. Guddo, B. Ramesh. The Vulnerability of Older Adults and Natural Disaster. *Disaster Advances.* **13**, 11 (2020).
14. S.A. Regina, A.D. Joie, C.K. Emily, C.G. Eric, X. Lea, A.M. David, C. Anita. Improving Disaster Resilience Among Older Adults. *Rand Health Q.* **8**, 1 (2018).
15. L. Jiacheng, Z. Ru, X. Junping, P. Younghwan. Positive Strategies for Enhancing Elderly Interaction Experience in Smart Healthcare Through Optimized Design Methods: An INPD-Based Research Approach. *Sustainability.* **16**, 20 (2024).

Towards Sustainable Schools: Applying New European Bauhaus Principles in Banda Aceh's Educational Infrastructure

Rifqi Irvansyah[1,2,*], *Tiziana* Rossetto[2], *Yunita* Idris[1], and *Nazriatun* Nisa[1,3]

[1]Tsunami and Disaster Mitigation Research Center (TDMRC), Universitas Syiah Kuala, Indonesia
[2] EPICentre, University College London, United Kingdom
[3] Department of Geography Education, Universitas Samudra, Indonesia

Abstract. This study assesses the extent to which existing school buildings in Banda Aceh align with the New European Bauhaus (NEB) principles of Sustainability, Inclusiveness, and Beauty. While large-scale reconstruction after the 2004 Indian Ocean tsunami restored access to education, many schools built in the post-recovery phase prioritized speed and basic functionality, with limited consideration for long-term sustainability, cultural relevance, and community integration. Using a tailored assessment framework, the study evaluates energy efficiency, sustainable materials, community engagement, accessibility, and aesthetic quality. The findings reveal significant gaps, particularly in renewable energy adoption, inclusive community spaces, and the integration of local cultural identity. The overall compliance with NEB principles is modest, with Sustainability at 46%, Inclusiveness at 48%, and Beauty at 38%. The results underscore the need to retrofit existing schools and update design practices to fully integrate NEB values, ensuring safer, more resilient, and inspiring educational environments in disaster-prone contexts.

1 Introduction

Safe, functional, and sustainable school buildings are essential for creating conducive learning environments that support student well-being and educational outcomes [1]. Research consistently shows that the quality of school facilities plays a critical role in shaping students' academic performance, emotional development, attendance rates, and overall engagement in learning activities [2]. Schools serve as ideal settings to promote health and well-being, especially because children spend a substantial portion of their developmental years within school environments [3]. Therefore, investing in school infrastructure is not only about providing shelter but also about creating spaces that support holistic development and equitable education.

In disaster-prone regions such as Banda Aceh, Indonesia, the importance of resilient and sustainable school infrastructure becomes even more critical. The 2004 Indian Ocean Tsunami caused unprecedented devastation across Aceh Province, including the destruction

* Corresponding author: rifqirvansyah@gmail.com

of thousands of schools. During the post-disaster reconstruction phase between 2005 to 2009, the Indonesian Government established the *Badan Rehabilitasi dan Rekonstruksi Aceh dan Nias* (BRR – Aceh and Nias Rehabilitation and Reconstruction Agency), which led large-scale efforts to rebuild the education sector. According to BRR records [4], 1,759 schools were reconstructed during this period out of a total of 3,415 schools that had been damaged by the disaster. These efforts marked a crucial step in restoring access to education. Reconstruction continued beyond 2009 as part of longer-term development and infrastructure recovery.

As of 2025, data from the Indonesian Ministry of Education's online portal (*Kementerian Pendidikan Dasar dan Menengah*) show that Banda Aceh currently has 218 existing school buildings across elementary, junior high, and senior high school levels [5]. Despite the substantial progress in physical reconstruction, questions remain about whether these schools were rebuilt with adequate attention to long-term sustainability. Many of the buildings constructed during the early reconstruction phase prioritized speed and basic functionality, and may not reflect current principles of sustainable, adaptive, and human-centered design.

In 2020, sixteen years after the tsunami, the Ministry of Public Works and Public Housing (PUPR) issued updated national guidelines for school design, known as *Petunjuk Teknis Bangunan Gedung Sekolah* (Juknis) [6]. These guidelines introduced new minimum standards for structural safety, accessibility, and spatial functionality. While these updates mark a significant policy step, they largely emphasize technical compliance and operational safety. They do not explicitly integrate broader sustainability concepts such as environmental performance, material reuse, climate responsiveness, and community well-being. As such, there remains a gap between national building codes and emerging global frameworks for sustainable education infrastructure.

The New European Bauhaus (NEB) initiative, launched by the European Commission in 2020, presents a holistic framework that integrates sustainability, inclusiveness, and beauty in the design and use of built environments. The NEB connects the goals of the European Green Deal with everyday spaces, schools for instance, by promoting thoughtful, human-centric design that enhances both environmental quality and quality of life [7]. NEB promotes an innovative and holistic approach to design (see Fig 1) with the aim of creating environments that are not only functional but also enhance well-being and social cohesion [8] [9]. This framework aligns strongly with the goals of resilient and adaptive school infrastructure.

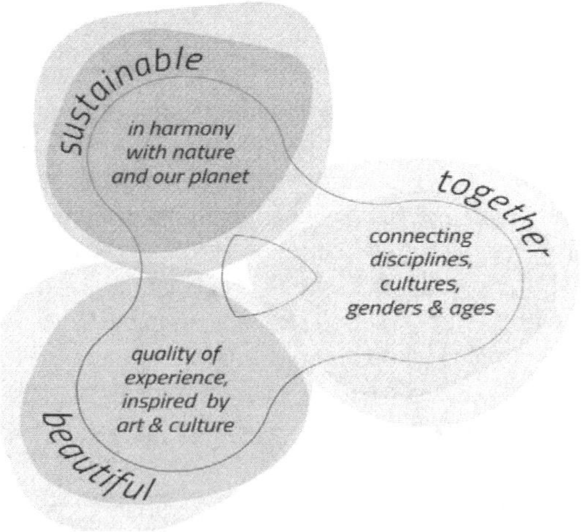

Fig. 1. New European Bauhaus from concept to action [10]

In the NEB framework, sustainability involves designing buildings that minimize ecological impact and are resilient to future challenges. For school buildings, this includes incorporating energy-efficient technologies, using sustainable materials, and implementing water-saving systems. Features like solar panels, natural ventilation, and rainwater harvesting can reduce energy use and create school environments that are not only safer but more sustainable in the long term.

The NEB "Together" pillar focuses on creating inclusive spaces that strengthen a sense of community. Together or inclusiveness ensures physical access for all, including persons with disabilities, togetherness goes a step further by encouraging interaction, collaboration, and shared experiences. In school settings, this could involve promoting layouts that invite participation and engagement. Schools built with this spirit become not just centres of learning, but places where people connect and belong.

Beauty in the NEB framework is about creating environments that inspire emotional well-being, creativity, and connection. It includes both aesthetic design and what is referred to as "quality of experience." This broader interpretation of beauty captures critical aspects such as health, safety, accessibility, and resilience to hazards. A beautiful school environment not only integrates elements such as natural light, ventilation, and cultural identity but also promotes functionality and emotional comfort. Beauty is no longer limited to surface-level aesthetics but is deeply connected to how spaces make people feel safe, welcome, and engaged. Schools that are thoughtfully designed with both aesthetics and functionality in mind can foster creativity, emotional well-being, and a sense of pride among students and staff.

This study aims to evaluate the compliance of existing school buildings in Banda Aceh with the three core NEB principles: Sustainability, Togetherness, and Beauty. Through this assessment, the research also seeks to highlight opportunities for improving school infrastructure to become more environmentally responsible, inclusive, culturally rooted, and resilient to future challenges. The findings will provide a clear roadmap for improving the implementation of building codes in Banda Aceh by identifying specific areas of deficiency, such as the lack of renewable energy adoption, cultural relevance, and accessible design for people with disabilities.

2 School Building Assessment Approach

2.1 School buildings data

In this study, the focus is on evaluating how school buildings align with the NEB principles of sustainability, inclusivity, and beauty. Towards this aim, an initial inventory of schools in Banda Aceh was developed using data the Indonesian Ministry of Education's online portal. This data contains information on three types of schools:

- Elementary Schools (SD/MI): For students aged 6–12 years, covering grades 1 through 6, consisting of 114 schools.
- Junior High Schools (SMP/MTs): For students aged 12–15 years, covering grades 7 through 9, consisting of 53 schools.
- Senior High Schools (SMA/MA/SMK): For students aged 15–18 years, covering grades 10 through 12, consisting of 51 schools.

The following images (Fig 2 to 4) illustrate the three types of school buildings based on PUPR Guideline 2020.

Fig. 2. Elementary school 3D view

Fig. 3. Junior high school 3D view

Fig. 4. Senior high school 3D view

The schools are assessed through comprehensive and detailed visual inspections carried out on-site. Data collected from previous surveys conducted during the INSPIRE Project in 2018 [11] and the AAUI-Maipark research grant in 2021 [12] are re-assessed and added to through further visual inspections. The assessment was conducted on a sample of 69 schools, comprising 119 individual buildings, which were part of previous surveys and were subject to further on-site inspections. The schools assessed include both those that were newly constructed as part of the post-tsunami reconstruction effort and those that were pre-existing facilities. This diverse sample allows for a comprehensive evaluation against modern standards, regardless of their initial design period.

The unit of analysis for this study is the individual school building rather than the entire school complex. Hence, in one complex school area, there are several buildings that can be assessed. This approach allows for a precise evaluation of specific design features and their compliance with NEB principles. In each case, structural design and layout of the buildings is recorded and photographs are taken to document the building's condition and record architectural elements. Elements such as plumbing and electrical systems are not considered in this study. The distribution of surveyed schools is shown in Fig. 5.

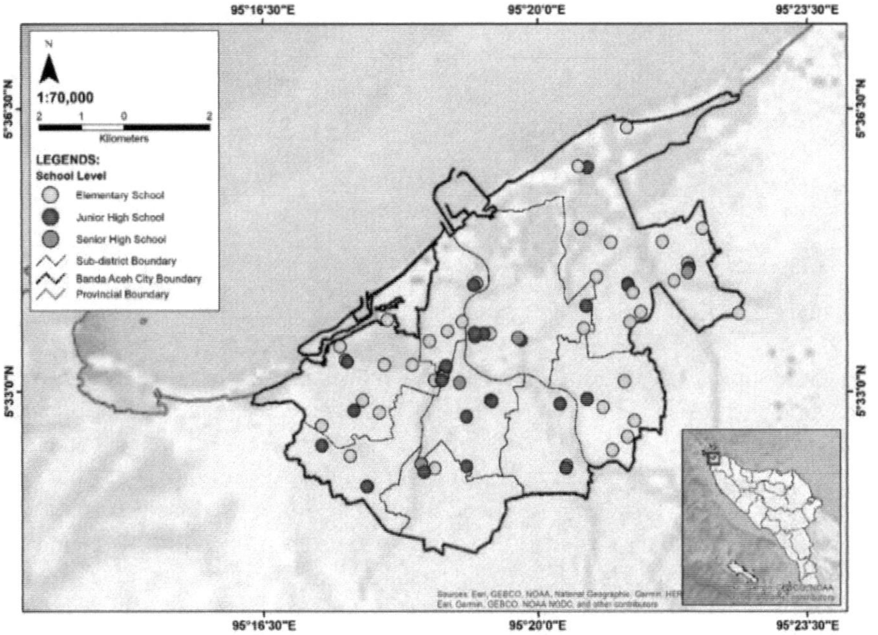

Fig. 5. Distribution of surveyed schools

The process of assessing existing school buildings in Banda Aceh follows a systematic evaluation based on the three core principles of the New European Bauhaus (NEB) framework: Sustainability, Togetherness, and Beauty. The criteria used for this study are adapted from *A Practical Guide to the New European Bauhaus Self-Assessment Method and Tool* [13]. While the original guide provides an extensive list of indicators for each principle, these are highly detailed and primarily intended for the design phase of individual projects. Applying them directly to a large population of existing buildings, such as schools, would be impractical.

To address this, the assessment was optimized by selecting the most relevant and measurable parameters that align with the environmental conditions and local context of Banda Aceh. These parameters serve as practical indicators to represent each NEB principle while ensuring that the evaluation remains applicable and focused on the realities of existing school infrastructure.

As illustrated in Fig. 6, each NEB principle is represented by three carefully chosen assessment criteria, providing a clear and manageable framework for evaluating how the principles are reflected in the design, function, and experience of school buildings. These criteria were adapted from the NEB practical guide but simplified to suit the evaluation of an existing building stock, rather than being applied during the design phase. For Sustainability, the selected criteria focus on *energy efficiency*, *sustainable materials*, and *water conservation*. For Togetherness, the assessment considers *community engagement*. The Beauty principle is divided into five aspects such as *design appeal, cultural relevance, environmental harmony, accessibility,* and importantly, *health and safety.*

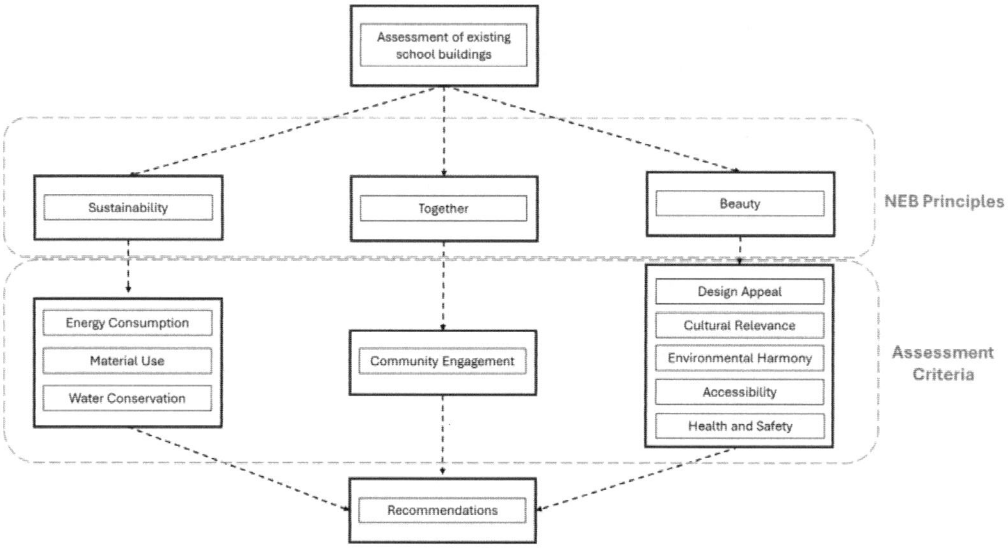

Fig. 6. The flowchart of school buildings assessment according to NEB principles

The flowchart illustrates how each NEB principle is linked to its respective measurable criteria, ensuring a consistent and transparent method for assessing the extent to which Banda Aceh's schools align with the NEB vision. This approach also allows for a more balanced comparison across principles and helps identify targeted areas for improvement.

For the evaluation process, each surveyed school building is assessed against the selected NEB parameters. Compliance is recorded when a building meets the specified guideline (assigned a value of 1). Non-compliance is recorded when a building does not meet the requirement (assigned a value of 0), with specific justifications noted for each case. This method ensures that the results highlight both strengths and deficiencies, providing a clear basis for recommendations, whether through retrofitting, redesign, or future construction improvements.

2.2 Application of NEB principles in building assessment

The NEB framework offers a holistic approach that connects the goals of the European Green Deal with everyday spaces, promoting thoughtful, human-centric design. This differs from the local Juknis guidelines and building codes, which focus more on technical and operational safety. This study justifies the use of the NEB framework by arguing that it aligns strongly with the goals of resilient and adaptive school infrastructure, especially in a disaster-prone region like Banda Aceh.

This study adapted the New European Bauhaus (NEB) Self-Assessment Guide to evaluate existing school buildings in Banda Aceh. Because the original framework was developed for a European context, several items were excluded as they were either irrelevant in a tropical climate or not feasible for existing infrastructure.

For Sustainability, insulation measures (foam, double glazing) and heating systems were excluded as they address temperate-climate needs. Renewable energy was noted but not scored, since no schools had installed such systems. For Inclusiveness, housing-related indicators (affordability, rent regulation, de-commodification) were excluded as they apply to residential projects, not schools. The focus instead was on community engagement, such as shared spaces and programmes. For Beauty, the original NEB guide contains indicators for "aesthetic pioneering" and advanced design styles like deconstructivism and regionalism.

For a post-disaster context where functional and culturally relevant design is a more urgent priority, these advanced aesthetic concepts were considered less applicable.

2.2.1 Sustainability

Sustainability within the NEB framework focuses on minimizing environmental impact through energy efficiency, the use of sustainable materials, and water conservation. The sustainability assessment helps ensure that school buildings contribute to a lower environmental footprint. To assess sustainability, the evaluation looks at several factors in the following Table 1.

Table 1. NEB Parameter for Sustainability

Aspect	Criteria	Practical Implementation
Energy Consumption	Uses LED or energy-efficient lighting	LED bulbs
	Classrooms have sufficient windows for natural daylight	Large windows, skylights
	Has energy-efficient heating or cooling systems	Natural ventilation
	Building insulation in walls, roof, and windows present	Foam insulation, double-glazed windows
	Uses renewable energy sources (solar panels, etc.)	Solar PV panels on roof
	Energy consumption is monitored or managed	Energy meters, smart energy systems
Material Use	Uses locally sourced construction materials	Bamboo, local stone, clay bricks
	Incorporates recycled or reused materials	Recycled steel, reclaimed wood
	Uses low-toxicity paints and finishes	Water-based paints
	Materials are durable and require low maintenance	Concrete, treated timber
	Building materials can be recycled or reused after use	Steel frames, modular panels
Water Conservation	Has low-flow faucets and toilets	Dual flush toilets, sensor taps
	Rainwater harvesting system installed	Rooftop rainwater tanks
	Greywater reuse or recycling system in place	Greywater from sinks reused for irrigation
	Efficient irrigation for plants	Drip irrigation
	School promotes water conservation awareness	Signage, student programs

2.2.2 Together / Inclusiveness

The Together principle in the NEB framework focuses on strengthening social cohesion and fostering meaningful connections between the school and the wider community. In this study, the evaluation of inclusiveness is limited to community engagement. This principle also examines how well the schools foster a sense of belonging and engagement by involving local communities in the design process and ensuring the spaces are used collectively. The aim is to create schools that serve not only as educational facilities but also as vibrant community hubs. Key aspects of this evaluation can be shown in the following Table 2.

Table 2. NEB Parameter for Inclusiveness

Aspect	Criteria	Practical Implementation
Community Engagement	Spaces available for community events or meetings	Multipurpose halls, outdoor amphitheatre
	School runs programs that involve local community	After-hours classes, local workshops
	Facilities accessible outside school hours	Community centres, library access
	Evidence of community involvement in building upkeep	Volunteer groups, parent associations
	Multipurpose halls or shared spaces	Gymnasiums, auditoriums

2.2.3 Beauty

Beauty in the NEB framework is not just about aesthetic appeal but about creating spaces that foster emotional well-being, creativity, and connection with nature. The beauty of a building can significantly impacts the learning experience and social interactions. In the context of school buildings, this principle includes both aesthetics and quality of experience. The aesthetic aspect considers harmonious design, cultural relevance, and integration with the natural environment. While quality of experience, which covers physical accessibility, health, and safety. Accessible designs ensure that all students, including those with disabilities, can navigate the school environment with ease. Health and safety further enhance the overall quality of the learning experience. By focusing on beauty, schools can become not only functional spaces but also emotionally supportive environments that promote creativity and well-being. Key factors in this core include in the following Table 3.

Table 3. NEB Parameter for Beauty

Aspect	Criteria	Practical Implementation
Design Appeal	Building has pleasant and inspiring interior design	Colourful walls, murals, natural materials
	Use of colour, art, or murals to enhance learning spaces	Student artwork, cultural motifs
	Layout promotes openness and calmness	Open-plan spaces, natural light
	Exterior design fits well within its surroundings	Harmonizes with neighbourhood style
Cultural Relevance	Design incorporates local architectural styles	Traditional roof shapes, carved woodwork
	Use of traditional materials or motifs	Bamboo, terracotta tiles
	Dedicated spaces for cultural or traditional activities	Community halls, ritual spaces

Aspect	Criteria	Practical Implementation
	Design considers local climate and environment	Shading devices, cross ventilation
Environmental Harmony	Presence of landscaped gardens or green spaces	School gardens, playground trees
	Use of native plants in landscaping	Local flowers, shrubs, and trees
	Design allows connection with nature (views, outdoor spaces)	Outdoor classrooms, nature trails
	Sustainable landscaping practices employed	Rain gardens
Accessibility	Ramps or elevators for wheelchair access	Concrete ramps, electric lifts
	Doors and corridors wide enough for mobility aids	Minimum 90 cm door width
	Accessible toilets and washrooms	Grab bars, wide stalls
	Clear signage for visually or hearing-impaired students	Braille signs, auditory signals
	Handrails and non-slip surfaces on stairs and ramps	Textured tiles, sturdy handrails
	Inclusive design of classrooms and common areas	Adjustable desks
Health and Safety	Good ventilation and fresh air circulation	Operable windows, HVAC systems
	Natural daylight in classrooms	Large windows, light wells
	Noise reduction measures implemented	Acoustic panels, landscaping buffers
	Use of non-toxic materials and finishes	Low-VOC paints
	Safety protocols and regular emergency drills	Fire drills, evacuation plans

3 Results and Discussion

3.1 NEB Compliance

The NEB framework encompasses three core principles: Sustainability, Togetherness, and Beauty, each representing essential aspects of high-quality, future-ready school infrastructure. The evaluation results for the surveyed schools in Banda Aceh reveal notable gaps in meeting these standards.

As shown in Fig. 7, overall compliance is relatively modest, with Sustainability scoring 46%, Togetherness 48%, and Beauty 38%. The result indicates that less than half of the surveyed schools meet the selected criteria for each principle. The low compliance rates suggest that current school designs, while functional, often do not fully consider environmental sustainability, community involvement, and attractive, healthy learning environments.

The near-similar values for sustainability and togetherness suggest that while some initiatives, such as basic energy efficiency measures and occasional community involvement, are present, they are not yet consistently incorporated across the school stock. The lowest score in the beauty category highlights a persistent gap in design elements that enhance user experience, cultural relevance, accessibility, and health and safety.

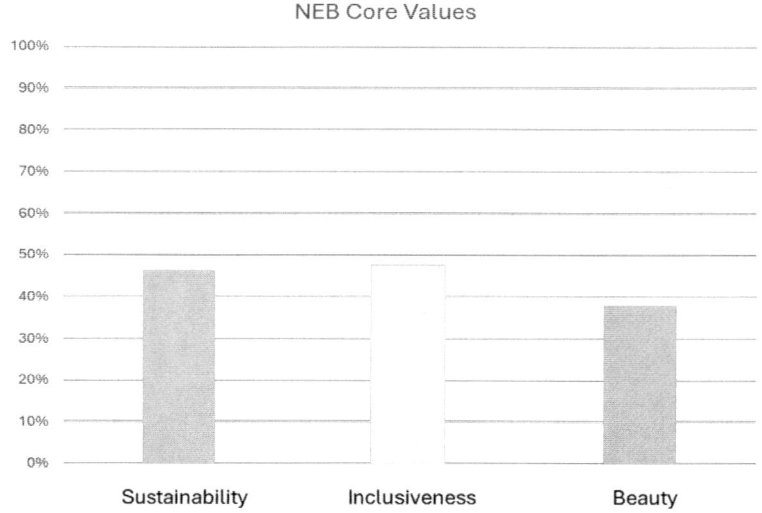

Fig. 7. The compliance of NEB concepts

3.1.1 Sustainability

The Sustainability principle of the New European Bauhaus (NEB) emphasizes reducing the environmental footprint of buildings by improving energy efficiency, using sustainable materials, conserving water, and enhancing climate resilience. In this study, the assessment of existing school buildings in Banda Aceh considered multiple sustainability aspects, including energy consumption patterns, the type and sourcing of construction materials, and the implementation of water-saving measures. The overall compliance for sustainability is illustrated in Fig. 8, while Fig. 9 provides examples of observed field conditions from the surveyed schools.

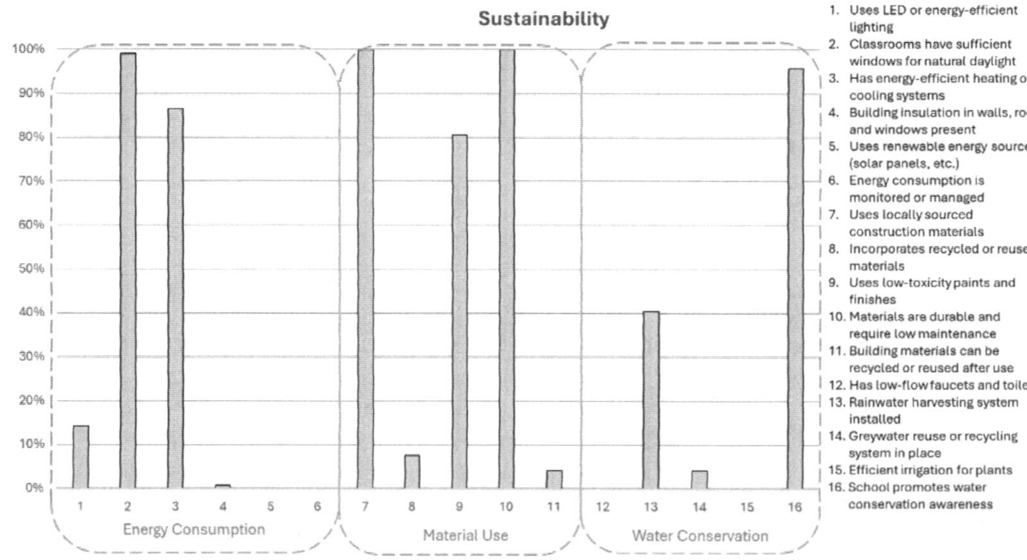

Fig. 8. Compliance of surveyed schools with NEB Sustainability criteria

a. Energy consumption

The majority of surveyed schools (99%) had classrooms with sufficient windows for natural daylight, which supports passive energy use by reducing reliance on artificial lighting. However, only 14% of the buildings implemented LED or energy-efficient lighting systems, and a small fraction (1%) incorporated building insulation in walls, roofs, and windows. Additionally, none of the buildings utilized renewable energy sources, such as solar panels. Energy consumption management systems were also not commonly in place.

b. Material use

Regarding material use, 100% of the schools utilized locally sourced construction materials, with clay bricks. Recycled or reused materials were incorporated in 8% of the schools, and 81% used low-toxicity paints and finishes. However, only 4% of the schools used materials that could be recycled or reused after the building's lifecycle. Furthermore, while 100% of the schools used durable materials like concrete, only a small percentage implemented low-maintenance building materials such as steel frames and modular panels.

c. Water conservation

Water conservation was a significant concern in the surveyed schools. Only 40% of the schools had rainwater harvesting systems installed, and 4% had greywater reuse or recycling systems. Moreover, no schools had efficient irrigation systems for plants, though 96% of schools actively promoted water conservation awareness through signage and student programs. Additionally, none of the schools featured dual-flush toilets and sensor taps to reduce water consumption.

a. Concrete structure b. Clay bricks material c. Rainwater tank

d. Water conservation program e. Large window f. Water-based painting

Fig. 9. Field examples of Sustainability criteria implementation in surveyed schools

3.1.2 Together / Inclusiveness

The "Together" principle in the New European Bauhaus (NEB) framework focuses on fostering community engagement and social cohesion, ensuring that schools serve as shared spaces for interaction, collaboration, and collective activities. This principle emphasizes the active participation of local communities in school life. As shown in Fig. 10, compliance with community engagement indicators varies widely across the surveyed schools. The detailed breakdown is presented in Fig. 11.

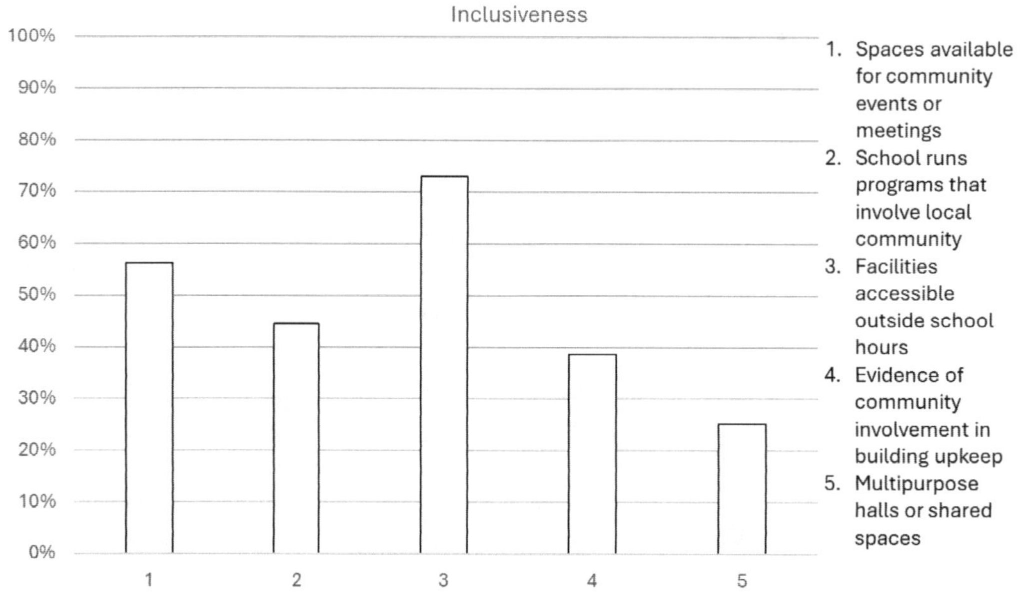

Fig. 10. Compliance of surveyed schools with NEB Inclusiveness criteria

a. Library access b. Parent association c. Multipurpose Hall

Fig. 11. Field examples of Inclusiveness criteria implementation in surveyed schools

Facilities accessible outside school hours scored the highest (73%), indicating that a significant proportion of schools provide spaces such as community centres or libraries for public use beyond regular school activities. Spaces available for community events or meetings also showed relatively strong performance (56%), often in the form of multipurpose halls.

However, only 45% of schools run programs that actively involve the local community, such as after-hours classes or local workshops, highlighting a missed opportunity for deeper engagement. Community involvement in building upkeep, such as through volunteer groups or parent associations, was recorded in 39% of schools, suggesting that maintenance and improvement efforts could benefit from stronger community participation. The availability of multipurpose halls or shared spaces, including gymnasiums and auditoriums, was the lowest (25%), showing that facilities specifically designed for collective activities remain limited.

3.1.3 Beauty

The Beauty core value in the New European Bauhaus (NEB) framework goes beyond aesthetics. It encompasses functionality, accessibility, and health and safety to create environments that inspire, support well-being, and foster meaningful connections between people and their surroundings. In the context of school buildings, this means integrating visually appealing design with practical features that ensure comfort, inclusivity, and safety for all users. The compliance results for each aspect are shown in Fig. 12, with examples from the surveyed schools in Fig. 13.

- *Design appeal*

Design appeal considers how the overall appearance and spatial arrangement of the school contribute to an inspiring and pleasant learning environment. Elements such as colourful walls, murals, natural materials, and student artwork can enhance engagement and creativity. The survey results show that 49% of schools have colourful walls, murals, and natural materials, while 35% incorporate student artwork or cultural motifs. Notably, 92% of schools feature open-plan layouts that promote openness and natural light, which supports a calm and engaging learning atmosphere. However, exterior design integration with the surrounding neighbourhood remains absent (0%).

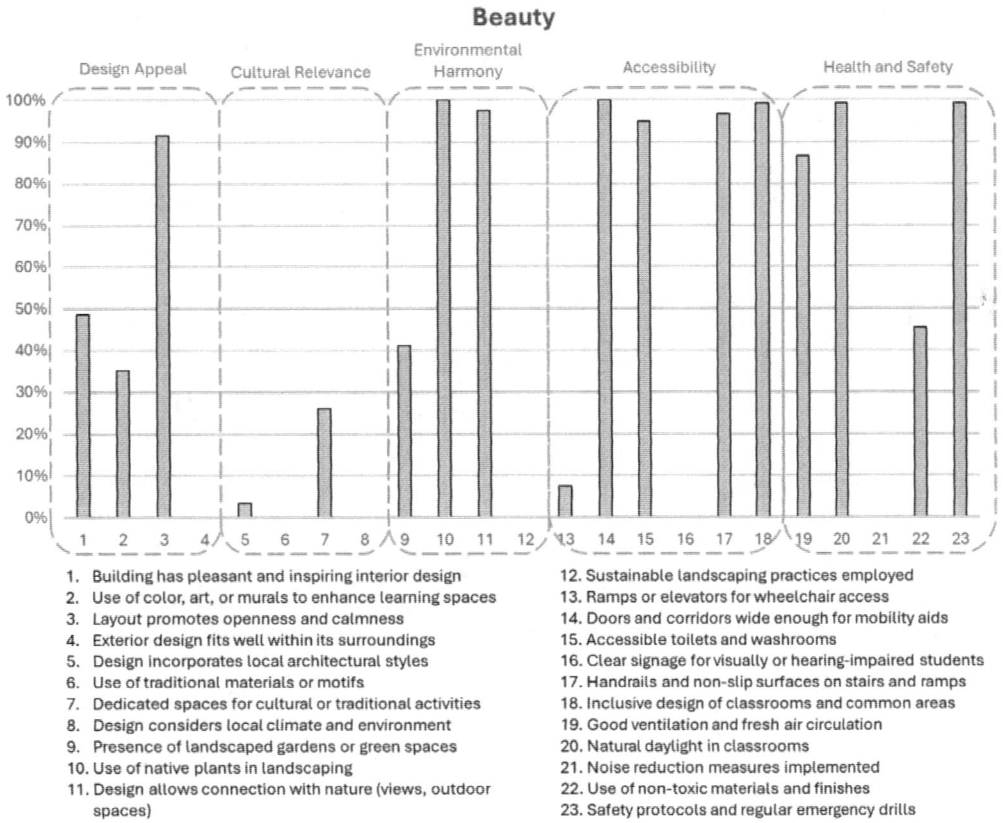

Fig. 12. Compliance of Surveyed Schools with NEB Beauty Criteria

- *Cultural Relevance*

Cultural relevance ensures that school buildings reflect and celebrate the local identity, traditions, and community values. This could involve using traditional materials, motifs, architectural styles, or providing dedicated spaces for cultural activities. The survey reveals very low integration of cultural relevance, with no schools using traditional materials (0%), only 3% incorporating local architectural styles, and 26% providing dedicated spaces for cultural or traditional activities. Additionally, no schools were found to consider local climate in the cultural design approach (0%), highlighting a significant gap in integrating local heritage into school design.

- *Environmental Harmony*

Environmental harmony focuses on connecting the built environment with nature. It includes landscaped gardens, the use of native plants, and sustainable landscaping practices. The findings show strong compliance in some areas: all surveyed schools (100%) use local flowers, shrubs, and trees, and 97% allow for direct connection with nature through outdoor classrooms. However, only 41% have landscaped gardens or green spaces, and none implement sustainable landscaping practices such as rain gardens (0%).

a. Colorful murals b. Local flowers c. Praying room
d. Outdoor classroom e. Student artwork f. School garden
g. Ramp h. Sturdy handrails i. Evacuation plan

Fig. 13. Field examples of Beauty criteria implementation in surveyed schools

- *Accessibility*

Accessibility in the NEB framework ensures that school facilities are usable by all, including students with disabilities. The assessment reveals high compliance in basic accessibility features: 100% of schools have doors and corridors wide enough for mobility aids, 95% have accessible toilets, and 97% have handrails and non-slip surfaces. However, ramps or elevators for wheelchair access are present in only 8% of schools, and no schools provide clear signage for visually or hearing-impaired students (0%). Additionally, while 99% offer inclusive classroom layouts, adjustable desks and quiet zones for students with specific needs are available in only 56% of the buildings.

- *Health and Safety*

Health and safety are crucial for creating secure and supportive learning environments. The survey shows that 99% of schools follow safety protocols and conduct regular emergency drills, an essential practice in a disaster-prone area like Banda Aceh. Natural daylight is available in almost all classrooms (99%), and 87% of schools have good ventilation systems. However, noise reduction measures are entirely absent (0%), which could affect concentration and comfort. Furthermore, only 45% of schools use non-toxic materials and finishes, leaving room for improvement in ensuring healthier indoor environments.

3.2 The Link to Disaster-Resilience

The connection between the NEB principles and disaster resilience is a central theme of this research. In a region like Banda Aceh, which is highly prone to natural disasters, the NEB principles offer a framework for building schools that are not only structurally sound but also contribute to a community's overall resilience. For example, the Sustainability principle, through its focus on energy independence and water conservation, can ensure schools remain functional even when public utilities are disrupted. The Togetherness principle promotes schools as community hubs with spaces for meetings and emergency shelters, while the Beauty principle, which includes aspects of health and safety, ensures that evacuation routes and safety protocols are part of a thoughtful and human-centric design. In the future, Banda Aceh can build school infrastructure that serves as a foundation for a safer and more resilient future for the entire community by adapting these principles.

3.3 Recommendations

Based on the survey data and findings, several key recommendations can be made to enhance the resilience, safety, and sustainability of school buildings in Banda Aceh, particularly alignment with the New European Bauhaus (NEB) principles. These recommendations are intended for stakeholders, policymakers, and decision-makers involved in future school construction and renovation projects in Banda Aceh.

Additionally, the recommendations can guide updates to the current school building guidelines, ensuring they better account for multi-hazard risks and sustainability while incorporating NEB principles. As emphasized by the World Bank's 2025 Global Assessment of Building Codes [14], it is important to focus on integrating green, resilient, and inclusive building strategies, which should be taken into consideration when developing or updating building codes for school infrastructure. Several recommendations can be implemented:

3.3.1 Sustainability:

- Energy efficiency: introduce solar panels and energy-efficient lighting systems (e.g., LED lights) in all schools to minimize energy consumption. Furthermore,

incorporate natural ventilation in classrooms to reduce reliance on air conditioning and decrease energy use.
- Water management: implement rainwater harvesting systems and low-flow toilets in all school buildings. Schools should also consider greywater recycling systems to reduce water consumption and increase self-sufficiency in water management.

3.3.2 Together:

- Community engagement: design spaces that foster community engagement, such as multipurpose halls, outdoor amphitheatres, and community gardens. Schools should also encourage local participation in school upkeep and community activities.

3.3.3 Beauty:

- Aesthetic and emotional well-being: schools should incorporate colourful murals and use natural materials in interiors to enhance the emotional well-being of students.
- Cultural integration: reflect local cultural identity by incorporating traditional materials and architectural styles that resonate with the community. This could also include the creation of dedicated spaces for cultural and traditional activities that align with the local context.
- Environmental harmony: encourage the integration of green spaces such as school gardens, playgrounds, and outdoor classrooms. Native plants should be prioritized to foster a harmonious relationship between the built environment and nature.
- Universal design features: ensure accessibility for people with disabilities, such as ramps, wheelchair-friendly toilets, and adjustable desks. This would ensure inclusivity in all aspects of school design, promoting equal access to education for all students.

While there has been substantial progress in physical reconstruction (not only schools but also several structures) since the 2004 tsunami, the findings of this study suggest that the quality of buildings, has not improved significantly in terms of long-term sustainability, cultural relevance, and community integration. A key reason for this is that much of the post-disaster reconstruction prioritized speed and basic functionality over comprehensive, long-term sustainable design. The national guidelines and building codes emphasize technical compliance and operational safety, but they do not explicitly integrate broader sustainability or community-focused concepts. This creates a gap where buildings can be structurally safe but still fail to meet the standards of modern, resilient, and human-centered design.

The findings of this research provide a clear roadmap for improving the implementation of building codes in Aceh by highlighting specific areas of deficiency. The low compliance scores in key areas, such as renewable energy adoption, cultural relevance, and accessible design for people with disabilities, indicate where current guidelines fall short and where future policy interventions should be focused.

For instance, this study reveals that no schools utilized renewable energy sources or incorporated noise reduction measures, and only a small percentage had features like wheelchair ramps. By specifically identifying these gaps, policymakers can be guided to revise the *Petunjuk Teknis Bangunan Gedung Sekolah (Juknis)* to include these criteria as mandatory standards for all new and retrofitted school buildings. This will ensure a better integration of green, resilient, and inclusive building strategies, consistent with global recommendations. The NEB framework provides a useful model for how to move beyond basic safety and functionality towards a more holistic approach that considers environmental, social, and aesthetic factors.

3.4 Future Research

Future research should address key gaps identified in this study to further enhance the resilience, functionality, and sustainability of school infrastructure in Banda Aceh. The following areas are proposed for deeper investigation:

3.4.1 Multi-hazard vulnerability assessment

Given that Banda Aceh is highly prone to natural disasters such as earthquakes, tsunamis, and liquefaction, future studies should conduct detailed vulnerability assessments of existing school buildings against multiple hazards. This includes structural integrity analysis, evacuation route safety, and disaster preparedness measures. Such assessments can guide the development of targeted retrofitting strategies and disaster-resilient designs tailored to the local hazard profile.

3.4.2 Functionality compliance with updated building guidelines

An in-depth review is needed to evaluate whether schools constructed after the 2004 tsunami meet the latest Ministry of Public Works and Public Housing (PUPR) Building Guidelines in 2020. This includes assessing classroom size, circulation spaces, accessibility, structural elements, lighting, and other functional requirements. The study should also examine whether the 2020 guidelines themselves require revisions to better address the evolving needs of educational infrastructure in disaster-prone areas.

In addition, future research should also be considered more comprehensive justification of a comparative matrix highlighting the key differences between the NEB and the local guidelines, such as ministerial regulations (*Peraturan Menteri*), and/or suitable building codes that consist of the infrastructure regulations.

3.4.3 Integrated design recommendations

Future work should develop prototype school designs that integrate three critical aspects:

- Safety: Compliance with multi-hazard resilience standards for earthquakes, tsunamis, and liquefaction.
- Functionality: Alignment with the most recent building codes and functional requirements for educational facilities.
- Sustainability: Application of the NEB principles to promote energy efficiency, environmental harmony, and community inclusiveness.

These integrated designs could serve as a reference model for policymakers, architects, engineers, and other stakeholders, providing a practical blueprint for constructing safe, functional, and sustainable schools in disaster-prone regions.

4 Conclusions

The assessment of school buildings in Banda Aceh reveals key gaps in the integration of the New European Bauhaus (NEB) principles into existing infrastructure. While many schools have made progress in accessibility and community engagement, the application of sustainability and aesthetic principles remains limited. Natural lighting and some use of locally sourced materials are common, yet there is significant scope for improvement in energy efficiency, renewable energy adoption, and water conservation.

Although inclusiveness is partially achieved through community engagement, the integration of beauty requires greater attention. The beauty aspects ensure schools inspire creativity, emotional well-being, and a sense of belonging. Cultural relevance and overall aesthetic quality should be strengthened to reflect local identity and enhance the learning environment.

To fully align with the NEB framework, future efforts should prioritize rehabilitation school buildings with energy-efficient technologies, sustainable materials, and improved water management systems. These enhancements will not only improve the resilience of schools in Banda Aceh but will also foster long-term sustainability and social inclusivity, creating environments that are functional, inspiring, and well-adapted to the community's needs.

Consistent with the World Bank's 2025 Global Assessment of Building Codes, school infrastructure policy in Banda Aceh must evolve to integrate green, inclusive, and resilient building strategies. Future projects should emphasize multi-hazard–resistant structures, renewable energy integration, sustainable materials, and effective water management, ensuring that schools are prepared to meet both current challenges and the long-term impacts of climate change and natural disasters.

References

1. E. Alasino, A. Martinez, P. Barrett, F. Ramirez, T. Shmis and J. Teixeira, RIGHT+ framework for physical learning environments (PLEs): Guidance for resilient, inclusive, green, healthy, and teaching- and learning-conducive (RIGHT) PLEs effectively implemented (+). World Bank, Washington, DC (2025).
2. M.C. Coronado et al., The impact of school facilities on student learning and engagement. NetZED Laboratory, University of Oregon (2021).
3. M. Pulimeno, P. Piscitelli, S. Colazzo, A. Colao, A. Miani, School as ideal setting to promote health and wellbeing among young people. *Health Promot. Perspect.* **10**(4), 316–324 (2020). https://doi.org/10.34172/hpp.2020.50
4. BRR (Badan Rehabilitasi dan Rekonstruksi Aceh-Nias), Book 0: Tsunami. Limited publication to mark the closing of the BRR, Jakarta (2009).
5. Kemendikdasmen, Data Induk Satuan Pendidikan Kota Banda Aceh. Portal Data Kemendikdasmen. https://data.kemendikdasmen.go.id/datainduk/satpen/060000/066100
6. Ministry of Public Works and Housing (PUPR), Petunjuk teknis dan standardisasi desain dan penilaian kerusakan sekolah dan madrasah. PUPR, Jakarta (2020).
7. European Commission, Communication… The European Green Deal (COM (2019) 640 final). European Commission, Brussels (2019)
8. European Commission, NEB Compass — a guiding framework (2022).
9. C. Coujard, L. De Rosso, A. Yurchyshyna, A. Gandini, V. Garcia Cortes, S. Urra Uriarte and C.H. Keller, New European Bauhaus: A handbook. NEBULA Project (2023).
10. Built4People, Everything you ever wanted to know on the New European Bauhaus is in NEBULA's handbook! Built4People (2023).

11. R. Gentile, C. Galasso, Y. Idris, I. Rusydy, E. Meilianda, From rapid visual survey to multi-hazard risk prioritisation and numerical fragility of school buildings. *Nat. Hazards Earth Syst. Sci.* **19**, 1365–1386 (2019). https://doi.org/10.5194/nhess-19-1365-2019
12. R. Irvansyah, I. Rusydy, A.V. Simanjuntak, A comparative analysis of seismic and tsunami fragility curves for school buildings in Banda Aceh coastal area. *E3S Web Conf.* **340**, 02003 (2022). https://doi.org/10.1051/e3sconf/202234002003
13. P.B. Lourenço, T. Maloutas, E. Romano, M. Santamouris, P. Negro, B. Widera, F. Ansaloni, C. Balaras, I. Katurić, D. Kolokotsa, T. Rossetto, G. Senatore, A. Tomaszewicz, E. Medeiros, K. Gkatzogias, D. Pohoryles, M. Acri, B. Becerik-Gerber, A. Bisello and M. Vecco, A practical guide to the New European Bauhaus self-assessment method and tool. Publications Office of the European Union (2024) https://data.europa.eu/doi/10.2760/9581060
14. World Bank, Global Facility for Disaster Reduction and Recovery (GFDRR) and Inter-American Development Bank (IDB), A global assessment of building codes: Current status and evolving needs to promote resilient, green and inclusive buildings. World Bank, Washington, DC (2025). http://hdl.handle.net/10986/43310

Revealing Submarine Landslide Signatures in South Western Palu Bay: Insights from New Shallow Bathymetric Data

Haekal A. Haridhi[1,2,3,*], *Syamsidik*[1,4], *Nadri* Pratama[1], *Haddad* Rahmat[1], *Zawil* Kiram[1], *Fahrur* Rozy[2], *Dimas S. J.* Sianipar[5], *Marzuki* Sinambela[5], *Muzli*[5], *Munawar*[5], *Indah* Fajerianti[5], *Muhammad* Nanda[1,2,3], *Yudi* Haditiar[1,2,3]

[1]Tsunami and Disaster Mitigation Research Center (TDMRC), Universitas Syiah Kuala, Banda Aceh, 23111, Indonesia

[2]Department of Marine Sciences, Faculty of Marine and Fisheries, Universitas Syiah Kuala, Banda Aceh, 23111, Indonesia

[3]Research Center for Marine Sciences and Fisheries, Universitas Syiah Kuala, Banda Aceh, 23111, Indonesia

[4]Civil Engineering Department, Faculty of Engineering, Universitas Syiah Kuala, Banda Aceh, 23111, Indonesia

[5]State College of Meteorology, Climatology, and Geophysics of the Republic of Indonesia (STMKG), Tangerang, 15221, Indonesia

Abstract. The 2018 Palu earthquake and tsunami was a rare event where tsunami generation was partly driven by coastal and submarine landslides triggered by a supershear Mw 7.5 strike-slip earthquake. While much attention has focused on the seismic source, the role of submarine slope failure remains less understood. To investigate this, a shallow bathymetric survey was conducted in 2024 by TDMRC and STMKG, covering over 500 km^2 in the central to southern parts of Palu Bay using a portable multibeam echo sounder (MBES). The bathymetric data were analyzed using slope angle and Factor of Safety (FS) modeling via Scoops3D. Several potential submarine slide features were identified, including steep scarps, chaotic seafloor patterns, and low FS zones, particularly along the offshore extension of the Palu–Koro Fault. Key slide locations, such as Benteng (A) and further south (D), are consistent with earlier post-tsunami findings. These results support the interpretation that multiple submarine landslides likely contributed to tsunami generation. The study highlights the need to incorporate landslide hazards into tsunami models and demonstrates the effectiveness of low-cost, portable MBES systems for submarine hazard assessment in Indonesia's tectonically active coastal areas.

1 Introduction

The Palu earthquake and tsunami on 28 September 2018 stand out as one of the most complex and unexpected disasters in modern Indonesian history [1–5]. Unlike most tsunamis in

*Corresponding author: haekal.azief.haridhi@usk.ac.id

subduction zones, which typically triggered by seafloor deformation from megathrust earthquakes, the Palu tsunami was caused by a combination of sources, including coastal and submarine landslides induced by intense ground shaking from a supershear strike-slip earthquake (Mw 7.5) [3]. Such non-seismic tsunami sources remain poorly understood, mainly due to limited data and the inherent challenges of detecting and analyzing submarine dynamics in detail [5].

While many post-disaster studies have focused on the seismic aspects and impacts of the earthquake [1,6], the contribution of submarine landslides, i.e. either as primary or secondary tsunami triggers, remains an open question [1,7]. Gaining a better understanding of seafloor morphology and slope stability is crucial, especially in the central to southern parts of Palu Bay, which are particularly vulnerable to submarine mass movements.

In 2024, the Tsunami and Disaster Mitigation Research Center (TDMRC) of Universitas Syiah Kuala, in collaboration with the Meteorology, Climatology, and Geophysics College (STMKG), conducted a shallow bathymetric survey using a multibeam echo sounder (MBES). This new bathymetric dataset opens up new opportunities for detailed post-tsunami forensic analysis of the seafloor morphology. Using numerical and geospatial approaches, the bathymetric data can be processed to identify key features of submarine landslides, such as scarps, sediment lobes, and slope instabilities [8].

This study aims to investigate the characteristics and distribution of submarine landslides in the south western parts of Palu Bay based on the latest shallow bathymetric data. By analyzing submarine slopes and calculating the Factor of Safety (FS), the research seeks to provide a more comprehensive understanding of potential non-seismic tsunami sources.

2 Methods

2.1 Acquisition and mapping shallow bathymetric data

Shallow bathymetric data (0–250 m depth) were collected over a 12-day period, from October 5 to 17, 2024, focusing on the central to southern parts of Palu Bay. The survey was conducted using a traditional fishing boat equipped with a differential GPS, motion sensors, and a multibeam echo sounder (MBES; Geoswath 4R, 125 kHz). The MBES is capable of recording depths up to 200 m with a swath width of up to 780 m. The GPS and motion sensors were co-located, and real-time corrections for vessel movement were automatically applied during data acquisition, with the corrected motion data fed directly into the MBES system. Figure 1 shows the installation setup of the survey equipment.

Fig. 1. Installation of equipment for the shallow bathymetric survey. The transducer was lifted when approaching the shoreline or during high-speed transit to the survey area. At the survey location, the transducer was submerged and securely locked in place to ensure high-accuracy data recording.

The survey followed a track line layout designed parallel to the coastline, with an estimated 40–50% overlap between swaths to ensure complete coverage. Data acquisition began from deeper waters and proceeded toward shallower areas. Figure 2 illustrates the survey lines covering the central to southern regions of Palu Bay.

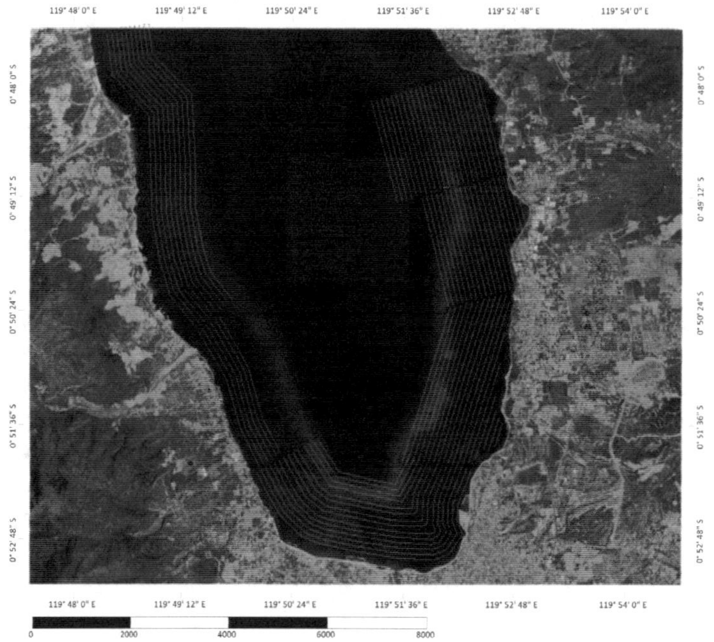

Fig. 2. Survey lines for shallow bathymetric data acquisition in the central to southern part of Palu Bay. The survey tracks are represented by thin solid grey lines.

During data collection, the boat maintained a speed of 2.5–3 knots during daylight hours. Night time acquisition was not feasible due to interference with local fishing activities. Fishermen in the area use light to attract fish, which could disrupt or be disrupted by the survey operations. Figure 3 shows the shallow bathymetric data acquisition in action.

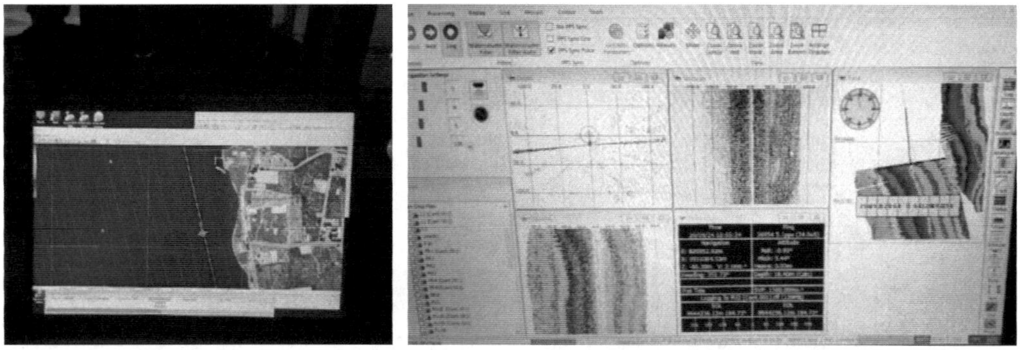

Fig. 3. Documentation of shallow bathymetric data acquisition. Left: Current track location indicated by the thick red line, with the boat position marked by a crosshair. Right: Quality control during data recording, showing the waterfall display, swath width, navigation, and ping information along the survey track.

The collected data were processed using Hypack® software, which included filtering to remove erroneous recordings such as depth spikes, as well as de-tiding and datum corrections. The processed data were then finalized using Agisoft for 3D modeling, and QGIS was used to generate map visualizations.

2.2 Analysis on the submarine landslide feature

The shallow bathymetric dataset was analyzed to identify features indicative of submarine landslides. Characteristic indicators include scarp formations, chaotic or disturbed seafloor textures, and debris lobes. Significant changes in slope contours may also suggest previous landslide events. By combining these morphological features with visual analysis, potential submarine landslide locations can be effectively identified.

2.3 Factor of Safety (FS)

The Factor of Safety (FS) was calculated to assess slope stability using a simplified approach, adapted to the limitations of available subsurface physical data. The general equation used in the morphological approach is:

$$FS = \frac{c+(\rho \cdot z \cdot cos^2\theta - u) \cdot \emptyset}{\rho \cdot z \cdot sin\theta \cdot cos\theta} \quad (1)$$

where:
c = sediment cohesion (obtained from literature or previous studies),
ϕ = internal friction angle,
ρ = wet bulk density of the sediment,
z = depth or slope height, and
θ = slope angle (derived from bathymetric data).

Scoops3D (https://www.usgs.gov/software/scoops3d, last accessed August 7, 2025) was used to analyse slope angles and calculate the FS, using the bathymetric data as input. The results from Scoops3D were consistent with observed submarine landslide locations and have previously been successfully applied to identify submarine slide sources in tsunami scenarios offshore northern Aceh [8].

3 Result and Discussion

Out of the total 500 km² of collected shallow bathymetric data, one specific area, at the south western area of Palu Bay as the offshore extension of the Palu-Koro Fault, is the focus of this study. Figure 4 Left shows the overall data coverage, with the slope angle analysis is shown on the right. In this area, the offshore extension of the Palu-Koro Fault appears as a steep, narrow feature marked by an abrupt change in slope angle.

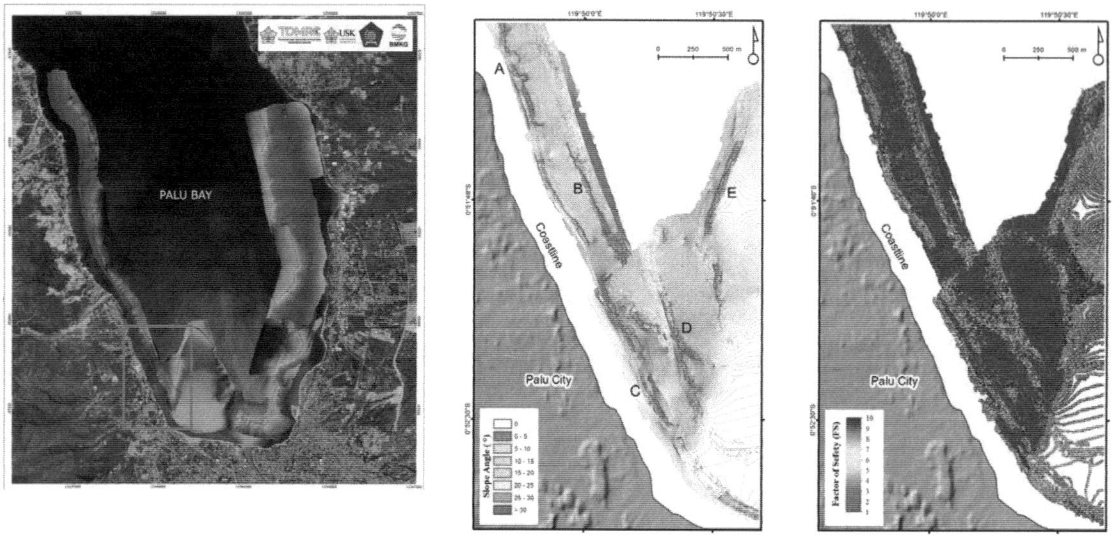

Fig. 4. Left: Overall coverage of shallow bathymetric data collected during the survey, with the red rectangle indicating the area analyzed and discussed in this study. Center: Slope angle map of the selected area, highlighting significant slope changes that coincide with the inferred location of the offshore extension of the Palu–Koro Fault. Right: Factor of Safety (FS) map of the same area, showing zones of low FS values corresponding to potential submarine landslide locations.

An abrupt change in slope angle is observed along the northeastern end of the study area (Figure 4, center, annotated as A), corresponding to a coastal landslide zone characterized by at least three channel-like features arranged sequentially. This area has previously been identified as Slide F (Benteng) by [5], interpreted it as a key contributor to localized tsunami waves during the 2018 event [7]. Our findings confirm similar morphological patterns, indicating either a reactivation or ongoing instability of this zone. The observed features align well with previously mapped coastal failure zones.

A narrow, coast-parallel slide is also evident at location B (Figure 4, center), located near the offshore trace of the Palu–Koro Fault. This feature closely borders the fault zone and likely represents fault-controlled slope instability. Additionally, large-scale submarine landslide features are identified in the southern part of the study area (locations C and D), characterized by scarps and downslope debris lobes. These features are aligned with the offshore continuation of the Palu–Koro Fault and reflect a structural control on slope failure distribution.

Our results are consistent with those reported by [9] in their first results of a bathymetric survey of Palu Bay, which documented multiple seafloor disturbances, particularly in the central and southern sectors of the bay is interpreted as evidence of submarine mass-wasting events. Their bathymetric difference analysis also highlighted bathymetric depressions and disrupted seafloor surfaces that coincide spatially with our slope angle anomalies and low FS zones.

Notably, the area corresponding to slide G [5] (Figure 4, center, annotated as D) also shows steep slope gradients and low FS values in our Scoops3D simulations. This further supports the interpretation of ongoing or recent submarine slope instability at that site.

All identified landslide features in our study exhibit low FS values, determined through box search simulations using Scoops3D. The spatial consistency between our results and those of previous studies underscores the reliability of integrating shallow bathymetric data with geomechanical modeling. Moreover, it emphasizes the need to consider fault-controlled and non-seismic landslide sources in tsunami hazard assessments for tectonically complex coastal zones such as Palu Bay.

4 Conclusions

This study presents a shallow bathymetric analysis of the central to southern parts of Palu Bay, focusing on identifying submarine landslide features and evaluating slope stability using the FS derived from numerical modeling with Scoops3D. The newly acquired high-resolution bathymetric data reveal multiple distinct submarine slope failure features, including scarps, debris lobes, and chaotic seafloor textures.

Several key landslide locations identified in this study, particularly in the northeastern and southwestern parts of Palu Bay, closely align with previously reported slides. Specifically, slide F (Benteng) corresponds to location A in this study, and slide G aligns with location D, as documented in [5] and [9]. These areas are characterized by steep slope gradients and low FS values, indicating current or potential slope instability. The spatial correlation between the mapped morphological features and FS anomalies strengthens the interpretation that these zones are prone to submarine mass movements and may pose ongoing tsunami hazards.

The findings highlight the critical role of submarine and coastal landslides in tsunami generation, especially in tectonically complex settings like Palu Bay. They also demonstrate the effectiveness of integrating bathymetric surveys with slope stability modeling to identify potential non-seismic tsunami sources. Continued monitoring, combined with sedimentological and geotechnical investigations, is essential for improving tsunami hazard assessments and early warning capabilities in the region.

Acknowledgements

Authors are grateful to Sekolah Tinggi Meteorologi, Klimatologi, dan Geofisika (STMKG) of Indonesia for a collaborative research project under Indonesia Disaster Resilience Initiative Project (IDRIP) 2024 funded by World Bank Contract No. 595/PL.101/PPK-IDRIP/XII/STMKG/2024 (Physical Modelling of Non-Seismic Tsunamis, Development of Formulation of Modelling Parameters for Multi-Source Tsunami for Palu Region). The grant has enabled this research to be published. This research is also funded by PTNBH USK under the scheme Penelitian Lektor Kepala Contract No. 1251/UN11/KPT/202 titled "Karakteristik Longsoran Bawah Laut di Teluk Palu Tengah–Selatan: Studi Forensik dengan Data Batimetri Dangkal Terbaru".

References

1. M. Heidarzadeh, A. Muhari, and A. B. Wijanarto, Pure Appl. Geophys. **176**, 25 (2019)
2. D. H. Natawidjaja, M. R. Daryono, G. Prasetya, Udrekh, P. L. F. Liu, N. D. Hananto, W. Kongko, W. Triyoso, A. R. Puji, I. Meilano, E. Gunawan, P. Supendi, A. Pamumpuni, M. Irsyam, L. Faizal, S. Hidayati, B. Sapiie, M. A. Kusuma, and S. Tawil, Geophys. J. Int. **224**, 985 (2021)
3. A. Socquet, J. Hollingsworth, E. Pathier, and M. Bouchon, Nat. Geosci. **12**, 192 (2019)
4. A. R. Gusman, P. Supendi, A. D. Nugraha, W. Power, H. Latief, H. Sunendar, S. Widiyantoro, Daryono, S. H. Wiyono, A. Hakim, A. Muhari, X. Wang, D. Burbidge, K. Palgunadi, I. Hamling, and M. R. Daryono, Geophys. Res. Lett. **46**, 8721 (2019)
5. P. L. F. Liu, P. Higuera, S. Husrin, G. S. Prasetya, J. Prihantono, H. Diastomo, D. G. Pryambodo, and H. Susmoro, Landslides **17**, 2085 (2020)
6. Syamsidik, Benazir, M. Umar, G. Margaglio, and A. Fitrayansyah, Int. J. Disaster Risk

Reduct. **38**, 101229 (2019)

7. K. Pakoksung, A. Suppasri, F. Imamura, C. Athanasius, A. Omang, and A. Muhari, Pure Appl. Geophys. **176**, 3323 (2019)
8. H. A. Haridhi, B. S. Huang, K. L. Wen, A. Mirza, S. Rizal, S. Purnawan, I. Fajri, F. Klingelhoefer, C. S. Liu, C. S. Lee, C. R. Wilson, T. Wu, I. Setiawan, and V. B. Phung, Nat. Hazards Earth Syst. Sci. **23**, 507 (2023)
9. M. C. G. Frederik, Udrekh, R. Adhitama, N. D. Hananto, Asrafil, S. Sahabuddin, M. Irfan, O. Moefti, D. B. Putra, and B. F. Riyalda, Pure Appl. Geophys. **176**, 3277 (2019)

Nature and Partnerships: Building Global Coastal Resilience

Nizam Nizam[1,*], *Benazir Benazir*[1], *Sri Puji Saraswati*[1]

[1]Department of Civil & Environmental Engineering, Faculty of Engineering, Universitas Gadjah Mada, Indonesia

Abstract. The 2004 Indian Ocean Tsunami catalyzed global disaster risk reduction, yet accelerating sea-level rise, coastal urbanization, and escalating losses demand new approaches. This keynote argues that future resilience requires coupling engineered systems with nature-based solutions (NbS) and global partnerships integrating science, policy, finance, and community stewardship. Synthesizing evidence from the Indian Ocean basin, Indonesia's mangrove rehabilitation, and hybrid projects worldwide, we demonstrate that ecosystems attenuate waves, shrink flood footprints, trap sediments, and generate co-benefits for carbon and livelihoods. Critical governance enablers include people-centred early warnings, long-term finance, and nature-positive land-use planning, aligned with global frameworks like Sendai and the Paris Agreement. We conclude with a call for a Blue-Green Resilience Compact: open data, investable NbS pipelines, community stewardship funds, and scale-able risk transfer instruments. Converting disaster memory into partnership investments can lower long-term risk and build coastal societies that live in harmony with water.

1 Introduction

The morning of December 26, 2004, changed forever how humanity understands coastal vulnerability and disaster risk. The Indian Ocean Tsunami, with its unprecedented human and economic toll—over 230,000 lives lost across 14 countries—became a defining moment that reshaped global approaches to disaster risk reduction. The disaster have brought global transformation in disaster risk reduction. It catalyzed the establishment of the Indian Ocean Tsunami Warning System, strengthened the Hyogo Framework for Action [1], and fundamentally altered how nations approach coastal risk management. Most importantly, it demonstrated that disasters, while devastating, can become catalysts for building back better based on social capital imbedded in many traditional communities—a principle deeply rooted in Indonesian philosophy of gotong royong (mutual cooperation) and the understanding that disaster can become a momentum for a nation's resilience revitalization [2]. This profound concept recognizes that disasters, while bringing immense suffering, also create extraordinary opportunities for transformation.
Indonesia's experience with the 2004 tsunami, the 2006 Yogyakarta earthquake, and countless other disasters has demonstrated that communities can emerge from catastrophe

* Corresponding author: nizam@ugm.ac.id

not merely recovered, but fundamentally stronger, more unified, and more resilient than before. The principle is embedded in Indonesia's rich cultural heritage—from the Javanese concept of *tanggap, tangguh, tanggon, dan trengginas* (responsive, resilient, dependable, and agile) to the broader understanding that disasters serve as tests that reveal and strengthen the character of individuals, communities, and nations (ibid.). This wisdom, developed through millennia of living with geological and climatic hazards, offers lessons for global coastal resilience building.

In 2025, as we commemorate the 21st anniversary of the tsunami, coastal communities worldwide face an even more complex risk landscape. Over 2.75 billion people now inhabit coastal areas—more than 40% of the global population—while climate change accelerates sea-level rise, intensifies storms, and increases the frequency of extreme events. The Intergovernmental Panel on Climate Change (IPCC) warns of accelerating sea-level rise and storm intensification, while global disaster losses now exceed US$200 billion annually. This evolving context demands a fundamental shift from reactive disaster response to proactive resilience building. The future of coastal protection lies not merely in stronger seawalls or more sophisticated early warning systems, but in harnessing the power of nature itself—through partnerships between human communities and coastal ecosystems that have protected shorelines for millennia.

2 The Evolving Coastal Risk Landscape

2.1 Climate Change Amplified Coastal Vulnerabilities

The risk landscape facing coastal communities has fundamentally transformed since 2004. Global mean sea level has risen approximately 0.2 meters since 1900 and is projected to rise another 0.43-0.84 meters by 2100 under high-emissions scenarios. This is not merely a future threat—current rates of sea-level rise have accelerated to approximately 4.7 mm per year, three times faster than the 20th-century average. As a consequence, coastal erosion has become a global phenomenon affecting all continents. Coastal erosion rates have accelerated dramatically, a global study of coastal erosion over 32 years (1984 – 2015) shows the overall surface of eroded land is about 28,000 km^2, twice the surface of gained land with some regions experiencing shoreline retreat in the order of kilometers. Extreme weather events, including hurricanes, typhoons, and storm surges, are becoming more frequent and intense, with devastating consequences for coastal infrastructure and communities.

2.2 Ecosystem Degradation Increases Vulnerability

Mangrove forests provide valuable habitat for living organisms, provide ecosystem services through provisions of nutrients, coastal protection, carbon sequestration, habitat provision, and support for fisheries and livelihoods. Parallel to climate change, widespread degradation of natural coastal buffers has dramatically increased vulnerability. The Ramsar Convention estimates that 35% of wetlands were lost between 1970 and 2015—a rate three times faster than forest loss. Mangrove forests, which once protected vast stretches of tropical coastlines, have declined by over 50% globally, with annual loss rates of 1-2% [3]. Coral reefs, which provide natural wave protection to over 100 million people worldwide, face unprecedented threats from ocean acidification, warming, and pollution. Studies indicate that coral reefs can reduce wave energy by up to 97%, yet continued degradation removes this critical natural defense system [4].

2.3 Urbanization and Development Pressures

Rapid coastal urbanization compounds these challenges. Low-elevation coastal zones (LECZs), which comprise only 2% of the global land area, contain 10% of the world's population and 13% of urban populations. Urban expansion into wetlands, coastal plains, and natural buffer zones not only increases exposure but also destroys the very ecosystems that provide natural protection. In many developing countries, informal settlements in coastal areas lack adequate infrastructure and are particularly vulnerable to coastal hazards. Agriculture and aquaculture, as the economic backbone for many coastal communities, add pressure to the coastal environment. The intersection of poverty, environmental degradation, and climate change creates compounding vulnerabilities that traditional engineering approaches alone cannot address.

3 Nature-Based Solutions: Scientific Evidence and Global Potential

Recent scientific literature underscores the increasing attention being paid to nature-based solutions for coastal protection. A growing body of evidence, synthesized through systematic reviews and mapping exercises, demonstrates the performance of these solutions within shallow coastal ecosystems. These ecosystems include salt marshes, mangroves, coral reefs, seagrass beds, seaweeds, and shellfish reefs. Meta-analyses further contribute to this understanding by quantitatively assessing the effectiveness of specific NBS, such as salt marsh planting, and by evaluating the ecological consequences of shoreline hardening. This surge in comprehensive literature reviews signifies a field that is increasingly focused on consolidating existing knowledge to provide a robust foundation for future research and application. The emphasis on shallow biogenic ecosystems likely reflects the critical role these environments play in natural wave attenuation and shoreline stabilization across many coastal regions.

Research in this domain has predominantly focused on evaluating the ecological outcomes, such as changes in species abundance and biodiversity, and the physical outcomes, like reductions in wave energy and mitigation of flood risk (ibid.). However, the economic and social performance of Nature based Solutions (NBS) have received comparatively less attention. This disparity suggests a crucial need for future research to more comprehensively assess the broader impacts of NBS, particularly their socio-economic implications, as these factors are vital for the long-term success and societal acceptance of these solutions. Systematic maps of the existing literature reveal specific areas where evidence is concentrated, as well as significant gaps in our understanding. Evidence clusters exist around habitat restoration, structural interventions that incorporate natural, artificial, or hybrid elements, and the performance of NBS in salt marsh and mangrove ecosystems. These well-studied areas offer opportunities for more in-depth analysis through targeted systematic reviews. Conversely, substantial knowledge gaps remain regarding the effectiveness of NBS in ecosystems like seagrass and kelp forests. Furthermore, there is a limited understanding of the economic and social performance of many NBS, their long-term effectiveness, and the feasibility of scaling up certain solutions to meet broader coastal protection needs.

3.1 The Science of Natural Coastal Protection

Nature-based solutions represent a paradigm shift from fighting against natural forces to working with them. Extensive scientific research now demonstrates the quantifiable protective benefits of coastal ecosystems:

Mangrove Forests: Research across multiple studies shows that mangrove forests can reduce wave energy by 50-72% and wave heights by 32-77%. During extreme events, they can reduce wave heights by 85-90% at low tide and 96-97% at high tide. A global economic analysis estimates that mangroves prevent over US$65 billion annually in flood damages and reduce flood risks for over 15 million people [5].

Coastal Forests: Apart from mangrove, coastal forests, such as casuarina can act as buffer against tsunami, storm surges and other coastal disasters. Coastal forest can reduce tsunami runup velocity by 70% and force by 80% [6]. However, deforestation of coastal forests for agriculture, aquaculture, urban development and tourism have significantly reduce its effectiveness as coastal protection.

Coral Reefs: Coral reefs act as natural breakwaters, dissipating up to 97% of incoming wave energy. Recent studies show that coral reefs provide flood protection worth over US$1.8 billion annually to US coastlines alone, with benefits extending to tourism, fisheries, and coastal stability. Accretion of coral reef will increase coastal protection. Indonesia as part of coral triangle has among the highest net accretion rate of 11.7 kg/m^2 annually, however global sea temperature rise and coastal pollution lead into coral bleaching that reduce the effectiveness of coral reef protection [7].

Salt Marshes and Coastal Wetlands: Salt marshes can reduce wave energy by up to 72% over relatively short distances. They also provide critical flood storage capacity—a single hectare of wetland can store 1-1.5 million gallons of floodwater [8].

Shore Protection Vegetation: 31% of the world's shoreline are sandy beach. The presence of shore vegetation protects the sand and dunes from erosion. Shore vegetation protects the shoreline by decreasing wave runup, holding sand loss from backwash, and stabilizing the sandy sediments from water and aeolian erosion. Beach grasses, meadows, and pandanus are among shore vegetation that have been proven to protect shorelines from wave erosion.

Seagrass Meadows and Kelps: Often overlooked, seagrass beds contribute significantly to coastal protection by reducing wave energy by up to 50% and stabilizing sediment through their root systems. They can reduce wave heights by up to 90% in optimal conditions.

3.2 Hybrid and Living Shoreline Approaches

The most innovative advances in coastal protection now integrate natural and engineering approaches, representing a paradigm shift from traditional "grey" infrastructure toward more resilient, adaptive systems ("grey-green" infrastructure). Recent meta-analysis of 304 academic studies demonstrates that hybrid measures perform best for climate adaptation and mitigation, combining the immediate protective benefits of engineered structures with the long-term resilience and co-benefits of natural systems [9]. Living shorelines are nature-based shoreline erosion control techniques that combine natural coastal habitats with engineered means of breaking wave energy. Unlike traditional hardened shorelines like bulkheads or seawalls that create barriers between land and water, living shorelines incorporate plants, coral, rocks, and oysters to naturally stabilize shorelines while maintaining habitat connectivity.

Hybrid coastal defenses represent a broader category that strategically integrates both nature-based components and hard engineering structures, such as placing concrete breakwaters in front of mangroves, low-crested breakwaters in front of seagrass, or combining seawalls with vegetated foreshores. Hybrid measures show exceptional effectiveness in low wave energy conditions (SMD = 9.01), significantly outperforming other approaches through multiple defense lines :

1. Offshore structures provide initial wave breaking
2. Vegetated zones dissipate remaining energy through drag forces
3. Engineered or nature-based backing provides final protection during extreme events.

Fig. 1. Conceptual illustration of hybrid grey-green coastal protection.

Hybrid systems harness the advantages of both hard and soft measures, providing immediacy of engineered barriers while maintaining ecological functionality. All coastal defence measures show positive economic returns over 20 years, but hybrid measures are more cost-effective than traditional hard measures. One important advantage of hybrid systems is ecosystem services restoration, otherwise difficult to achieve using "grey" measures. Another advantage of a hybrid system is its self-repair capabilities. Natural component recovers through biological processes, sediment accumulation naturally raises bed elevation over time, and vegetation or coral regrowth restores protective capacity after storms. Hybrid systems also have cost advantages arising from lower maintenance requirements, multiple co-benefits, and avoided ecosystem service replacement costs.

3.3 Co-benefits Beyond Coastal Protection

Nature-based and hybrid coastal protection provide multiple co-benefits that traditional "grey" infrastructure cannot match:

- Climate Mitigation: Coastal ecosystems are among the most carbon-rich environments on Earth. Mangroves alone account for 14% of global oceanic carbon sequestration despite covering less than 1% of coastal areas. Salt marshes and seagrass beds provide additional significant carbon storage capacity.
- Biodiversity Conservation: Coastal ecosystems support disproportionately high biodiversity. Mangroves provide nursery habitat for over 75% of commercially important fish species in tropical regions, while salt marshes support critical bird migration routes and spawning areas.
- Economic Benefits: Beyond flood protection, coastal ecosystems support tourism, fisheries, and other economic activities. Global estimates suggest that ecosystem services from coastal habitats provide over US$25 trillion annually in economic benefits.

Water Quality Improvement: Coastal vegetation filters nutrients and pollutants from runoff, improving water quality in nearshore environments. Oyster reefs can filter millions of gallons of water daily, removing excess nutrients and suspended sediments.

Table 1. Effectiveness of Nature-based Coastal Protection

Nature-Based Solution	Wave Attenuation	Erosion Control	Flood Protection	Habitat Provision	Environmental Benefits	Climate Adaptation
Mangroves	**High**	**High**	**High**	**Very High**	**Very High**	**Very High**
	Dense root systems and canopy reduce wave energy by 66% (McIvor et al., 2012)	Extensive root networks stabilize sediments and shoreline (Gijsman et al., 2021)	Act as natural barriers against storm surge and coastal flooding (Luo et al., 2024)	Critical nursery habitats for fish, birds, and marine life (Perricone et al., 2023)	Excellent carbon sequestration, water filtration (Morris et al., 2023)	Highly adaptable to sea level rise, self-sustaining (Chen et al., 2025)
Coastal Forests	**Medium-High**	**High**	**Medium**	**Medium-High**	**High**	**Medium-High**
	Tree canopy and trunks reduce wind and wave energy (Perricone et al., 2023)	Deep root systems prevent soil erosion and landslides (Huynh et al., 2024)	Moderate storm surge protection (Morris et al., 2024)	Habitat for terrestrial wildlife and some marine species (Paxton et al., 2024)	Carbon storage, air quality improvement (Paxton et al., 2023)	Can adapt to changing conditions with proper management (Morris et al., 2024)
Coral Reefs	**Very High**	**Medium**	**High**	**Very High**	**Very High**	**Medium**
	Effective at breaking wave energy offshore - reduce wave energy by 97% (Ferrario et al., 2014)	Some sediment trapping but limited direct erosion control (Carlot et al., 2023)	Excellent protection against ocean swells and storm waves (Norris et al., 2024)	Among the most biodiverse ecosystems on Earth (Geldard et al., 2023)	Tourism value, fisheries support, water quality (Storlazzi et al., 2024)	Vulnerable to ocean acidification and warming - 94% (Carlot et al., 2023)
Salt Marsh, Wetlands	**Medium-High**	**High**	**Very High**	**Very High**	**Very High**	**High**
	Vegetation dampens wave energy by up to 60% and reduces current velocity (Möller et al., 2014)	Plant roots bind sediments and trap suspended particles (Marin-Diaz et al., 2023)	Excellent flood water absorption - avoided $625M in damages during Hurricane Sandy (Narayan et al., 2017)	Critical habitat for migratory birds, fish spawning (Cassalho et al., 2024)	Superior nutrient cycling, pollutant filtration (Zhu et al., 2020)	Can migrate inland with sea level rise if space available (Familkhalili et al., 2023)
Shore Vegetation	**Medium**	**High**	**Low-Medium**	**Medium**	**Medium-High**	**Medium**
	Moderate wave energy reduction near shore (Morris et al., 2024)	Excellent root binding of coastal soils (UK Government, 2025)	Limited flood protection capacity (Huynh et al., 2024)	Habitat for coastal birds and small mammals (Morris et al., 2024)	Soil stabilization, some carbon storage (NCCOS, 2025)	Requires management for climate adaptation (Morris et al., 2024)
Seagrass Meadows	**Medium**	**Medium-High**	**Low**	**Very High**	**Very High**	**Medium**
	Underwater vegetation reduces current velocity and wave energy - drag coefficients 0.02-5.12 (Paul & Gillis, 2020)	Trap sediments and reduce bottom scouring up to threefold (van der Heide et al., 2013)	Minimal direct flood protection (Magdalena & Magdalena, 2025)	Critical nursery areas and food source for marine life (James et al., 2020)	Excellent carbon sequestration ("blue carbon") (Mazarrasa et al., 2023)	Sensitive to temperature and water quality changes but can provide resilience during hurricanes (James et al., 2020)

4 Community-Centered Resilience: The Heart of Sustainable Solutions

4.1 The Philosophy of Disaster as A Momentum

A disaster could provide a transformative framework for understanding how disasters can catalyse community resilience. This philosophy, deeply rooted in Indonesian cultural values and tested through centuries of living with natural hazards, recognizes that while disasters bring devastation, they also create unprecedented opportunities for renewal, unity, and strength-building. The philosophical foundation draws from multiple cultural sources. In Javanese tradition, the story of Jabang Tetuko being tested in Kawah Candradimuka before becoming the mighty Gatotkaca illustrates how trials forge strength. Hindu concepts of Dewa Siwa as both destroyer and creator remind us that destruction can clear the way for new and better life. Islamic teachings emphasize that Allah does not burden souls beyond their capacity, suggesting that surviving disaster reveals inner strength previously unknown [2]. This framework manifests in four key dimensions of resilience that mirror the Javanese concepts of resilience: t*anggap, tangguh, tanggon, dan trengginas* (ibid.)*:*

- *Tanggap* (Responsive): The capacity to quickly understand and respond to changing circumstances. In disaster contexts, this means communities that are aware of risks, prepared for emergencies, and able to adapt quickly to new challenges. In broader terms, it represents the ability to be responsive to global changes, competitive pressures, and emerging opportunities.
- *Tangguh* (Resilient): The fundamental strength to withstand shocks and maintain core functions under stress. This encompasses both physical infrastructure that can survive disasters and social systems that maintain cohesion during crises.
- *Tanggon* (Dependable): The reliability that others can count on during difficulties. Communities and nations that are tanggon become sources of strength for others, providing assistance and leadership when disasters strike.
- *Trengginas* (Agile): The capacity for quick, smart, and precise action. This represents the ability to make rapid decisions and implement them effectively, whether in disaster response or in seizing opportunities for advancement.

4.2 The Principle of Gotong-Royong in a Global Context

The Indonesian concept of *gotong royong* – mutual cooperation and community self-reliance — offers profound insights for global coastal resilience building. This principle recognizes that sustainable resilience cannot be imposed from above but must emerge from empowered communities working collectively to protect and restore their environment.

The 2006 Yogyakarta earthquake provided a powerful demonstration of this principle in action. While international best practices suggested relocating survivors to temporary housing complexes away from damaged areas, the affected communities chose to build simple shelters from rubble near their destroyed homes [2]. This decision, initially questioned by disaster management experts, proved remarkably effective. By maintaining social bonds and community networks, survivors were able to support each other psychologically, organize collective rebuilding efforts, and accelerate recovery. Within one year, over 138,000 homes had been rebuilt through *gotong royong* and community self-help—far faster than the three years required for the more technologically advanced but socially disruptive approach used in Kobe, Japan, after the 1995 earthquake.

This experience revealed the critical importance of what Indonesian scholars term "*modal sosial*" (social capital)—the community's capacity to develop cooperation and partnerships to achieve common goals, grounded in mutual support, religious values, and local wisdom. Social capital represents the ability of communities to work together effectively, maintaining strong social bonds while demonstrating tolerance and adaptability. Community-based disaster risk reduction (CBDRR) approaches have proven most effective when local actors are empowered with:

- Access to early warning information and decision-making authority,
- Technical training and resources for ecosystem restoration,
- Long-term stewardship responsibilities and incentives, and
- Integration of traditional ecological knowledge with modern science.

4.3 Women's Leadership in Coastal Resilience

Evidence consistently shows that women's leadership enhances the effectiveness of coastal resilience initiatives. Women often possess specialized knowledge of coastal resources and play critical roles in household and community adaptation strategies. Programs that prioritize women's participation and leadership show [11]:

- Higher rates of long-term project sustainability
- Better integration of social and environmental objectives
- More equitable distribution of benefits
- Enhanced community cohesion and resilience

Research across various coastal seascapes reveals clear gendered divisions of labor, with women maintaining extensive knowledge of coastal forests, shallow areas, and nearshore resources through activities like collecting wood, invertebrates, and farming seaweed. Women often demonstrate higher environmental stewardship, showing greater concern for preserving coastal and marine ecosystems for future generations, particularly related to their family nurturing roles. Studies across Fiji, Papua New Guinea, Solomon Islands, Indonesia, Kenya, and Madagascar involving 3,063 people found significant gender differences in how coastal management impacts are perceived, with women articulating impacts differently, particularly regarding economic, governance, and health aspects [12].

The evidence overwhelmingly demonstrates that women's leadership in coastal resilience is not merely beneficial but essential for effective climate adaptation and sustainable coastal management. As the Gender Snapshot 2024 report indicates that climate change may push up to 158 million more women and girls into poverty by 2050, prioritizing women's leadership becomes both a rights imperative and a practical necessity for coastal resilience [11].

Success requires systematic approaches that address structural barriers, build institutional capacity, and create enabling environments for women's meaningful participation in coastal governance and management. The growing body of evidence and successful models provides clear pathways for scaling up women's leadership to enhance coastal resilience globally.

4.4 Youth Engagement and Intergenerational Knowledge Transfer

Successful coastal resilience programs increasingly recognize the importance of engaging youth as agents of change while preserving traditional ecological knowledge. Educational programs that combine traditional knowledge with modern science create:

- Stronger community ownership of restoration projects
- Continuity of management across generations
- Innovation in adaptation strategies
- Enhanced scientific literacy and environmental awareness

Recent research demonstrates that youth-led coastal restoration initiatives can significantly empower environmental stewardship at community scales. Thailand's largest youth-led marine conservation initiative, "Care for Coral," exemplifies this approach through immersive restoration and educational experiences that inspire youth to become active stewards across various habitats and species. The program has promoted lifelong ocean literacy, together with skills and knowledge transfer, providing meaningful opportunities for local communities to engage in direct restoration experiences.

Scaling and Multiplier Effects: In Peru, the Planeta Océano organization's Marine Educators Network, involving over 50 schools, has reached over 500,000 people and engaged over 400 young people in youth-led initiatives ranging from mangrove reforestation to advocating for sustainable fisheries. This approach has sparked multiplier effects and ownership in local communities, since messaging comes from youth community members themselves. Many universities in Indonesia have student community empowerment program (*Kuliah Kerja Nyata* – KKN), as part of their curriculum, by sending students for one to two months doing community empowerment program especially related to SDGs achievement. Every year, this program has engaged tens of thousands of university students to reach millions of rural communities across Indonesia.

Enhanced Project Sustainability: Youth engagement creates lasting institutional memory and adaptive capacity. Research shows that when young people lead and participate in research with tangible outcomes—such as developing new conservation strategies or improving local environmental health—it affects current policy and lays the groundwork for ongoing environmental advocacy. These experiences often inspire a lifelong commitment to environmental stewardship as young researchers see the direct impact of their work.

4.5 Indigenous and Traditional Knowledge Systems

Indigenous and traditional communities possess millennia of experience in sustainable coastal management, rooted in a deep, place-based understanding of ecological processes and intergenerational stewardship. These communities have developed sophisticated systems of environmental governance, often encoded in cultural norms, customary laws (adat), taboos, and spiritual beliefs. Their Traditional Ecological Knowledge (TEK) offers not only insights into species behavior and habitat dynamics but also frameworks for managing risk, ensuring equity, and fostering resilience. Integrating traditional ecological knowledge with modern science has proven essential for:

- Selecting appropriate species and restoration techniques, informed by historical baselines, seasonal knowledge, and traditional resource use patterns;
- Understanding long-term ecosystem dynamics and variability, including how local ecosystems respond to climate shifts, sedimentation, and extreme events;
- Developing culturally appropriate management systems, such as sasi laut, awig-awig, or pa*nglima laot*, which balance ecological sustainability with social legitimacy and local governance;
- Building community support and participation, by reinforcing local ownership, pride, and continuity in conservation efforts, thereby enhancing compliance and monitoring.

The Indonesian experience illustrates how disasters can also act as powerful catalysts for rediscovery and revitalization of these knowledge systems. Following events like the 2004 Indian Ocean tsunami and more recent coastal flooding and erosion, communities in Aceh, Mentawai, Nusa Tenggara, and elsewhere have turned to ancestral practices for guidance. In many cases, they have re-established traditional zoning, replanted protective mangroves and pandanus belts, and revived rituals that reinforce a collective sense of responsibility toward coastal ecosystems.

These blended approaches—uniting indigenous wisdom and scientific innovation—are now recognized as vital to advancing nature-based solutions (NbS) for coastal resilience. They not only enhance ecological outcomes but also ensure that solutions are socially embedded, culturally resonant, and politically inclusive.

5 Governance and Policy Integration: Building Enabling Frameworks

5.1 International Policy Alignment

The integration of nature-based coastal protection into international frameworks has accelerated significantly. Sendai Framework for Disaster Risk Reduction (2015-2030) explicitly promotes ecosystem-based approaches and emphasizes the importance of inclusive governance and community participation in building resilience [13]. Paris Agreement on Climate Change recognizes nature-based solutions as essential for both mitigation and adaptation, with many countries including ecosystem restoration in their Nationally Determined Contributions [13]. Kunming-Montreal Global Biodiversity Framework establishes targets for ecosystem restoration (30% by 2030) and emphasizes the integration of biodiversity conservation with climate action and disaster risk reduction [14]. UN Early Warnings for All Initiative seeks universal coverage of multi-hazard early warning systems by 2027, with emphasis on community-centered, locally relevant warning systems that integrate ecosystem-based risk reduction.

5.2 National Policy Innovation

Leading countries are pioneering policy frameworks that integrate nature-based solutions into national coastal management. Indonesia's National Action Plan on Climate Change

Figure 2. a. Coral reef replantation (Bali), b. living shoreline (North Carolina), c. mangrove rehabilitation (Makassar), d. mixed coastal plantation - casuarina and pandanus (Pacitan). (source: a. Reef builder, b: NOAA, c: Makita, d: Benazir)

Adaptation prioritizes ecosystem-based adaptation and establishes institutional frameworks for coordinated implementation across sectors and scales. In restoring coral reef, more than 1 million fragmented hard coral have been planted, and more than 120,000 units of artificial reef have been deployed. Under the plan, Indonesia has targeted to restore 600,000 hectare of mangrove forest by 2024 [15]. Living shoreline, practiced had been successfully implemented to mitigate coastal erosion in many states in the USA. Singapore's Coastal Adaptation Studies demonstrate how small island states can integrate nature-based and engineered solutions for comprehensive coastal protection. The Netherlands' Room for the River Programme illustrates large-scale implementation of nature-based flood management, providing models for adaptive delta management worldwide.

5.3 Local Governance and Implementation

Effective implementation of nature-based coastal protection requires strong, inclusive, and adaptive local governance institutions. These institutions serve as the backbone for planning, executing, and sustaining coastal resilience strategies rooted in natural processes and local ecosystems. Their effectiveness depends on several critical capacities:

- *Coordinating across sectors and scales:*
  Nature-based solutions (NbS) often span multiple sectors—such as environment, fisheries, public works, tourism, and disaster management—as well as governance levels, from village authorities to provincial and national agencies. Strong local institutions are needed to facilitate inter-agency collaboration, harmonize overlapping mandates, and align local initiatives with national policies and international commitments (e.g., SDGs, Sendai Framework, NDCs).
- *Integrating scientific information with local knowledge:*
  Sound decision-making requires blending empirical research (e.g., on erosion rates, habitat functions, climate projections) with place-based knowledge systems that have evolved over generations. Local governance bodies must have mechanisms to translate technical data into actionable community plans, while also validating and incorporating traditional ecological knowledge (TEK) through participatory processes, co-production of knowledge, and inclusive dialogues.
- *Ensuring equitable participation and benefit-sharing:*
  For nature-based coastal protection to be socially sustainable, governance systems must ensure that all stakeholders—especially marginalized groups such as women, indigenous peoples, and small-scale fishers—are meaningfully involved in planning, implementation, and monitoring. Equity also requires fair distribution of benefits, such as access to restored mangrove resources, ecotourism opportunities, or protection from storm surges, and transparent mechanisms for resolving conflicts.
- *Adapting to changing conditions and new information:*
  Coastal environments are dynamic, and climate risks are uncertain. Effective local governance must embrace adaptive management—regularly updating strategies based on new data, community feedback, and observed ecosystem responses. This includes maintaining institutional flexibility, cultivating a culture of learning, and investing in local capacity-building and monitoring systems to detect change and inform timely responses.
- *Securing long-term financing and stewardship:*
  Nature-based solutions require sustained investment—not just for initial implementation, but for long-term maintenance, community engagement, and ecosystem monitoring. Local institutions must be capable of mobilizing diverse funding sources, including public budgets, private partnerships, environmental trust funds, and payment

for ecosystem services (PES). Equally important is cultivating a sense of local stewardship, ownership, and accountability to ensure the durability of interventions beyond short-term project cycles.

6 Building the Blue-Green Resilience Compact

6.1 Vision for Global Cooperation

The concept of a Blue-Green Resilience Compact represents a new framework for international cooperation that recognizes the inseparable connections between ocean health, coastal ecosystem integrity, and human resilience. This compact would establish:

- Shared Principles for sustainable coastal management that prioritize ecosystem-based approaches while recognizing the need for hybrid solutions in high-risk areas.
- Common Standards for monitoring, assessment, and reporting on coastal ecosystem health and protection services.
- Technology and Knowledge Sharing platforms that facilitate South-South cooperation and transfer of innovative approaches.
- Coordinated Financing mechanisms that leverage public, private, and philanthropic resources for large-scale implementation.

6.2 Key Components of the Compact

6.2.1 Open Data and Information Sharing:

- Standardized coastal risk and ecosystem datasets accessible to all stakeholders
- Real-time monitoring networks that support adaptive management
- Traditional ecological knowledge databases that complement scientific information
- Early warning systems that integrate ecosystem and social indicators

6.2.2 Investment and Finance Innovation:

- Standardized project development tools for nature-based coastal protection
- Blended finance facilities that reduce investment risks
- Insurance products that recognize and protect natural infrastructure
- Payment for ecosystem services mechanisms that reward stewardship

6.2.3 Community Empowerment and Capacity Building:

- Training programs that build local technical capacity
- Leadership development initiatives that prioritize women and youth
- Governance frameworks that ensure meaningful participation
- Livelihood support that aligns community welfare with ecosystem health

6.2.4 Policy Alignment and Integration:

- Legal frameworks that recognize and protect ecosystem services
- Planning processes that integrate ecosystem considerations
- Regulatory reforms that enable innovative financing and implementation

- International cooperation mechanisms that support transboundary ecosystem management

6.3 C. Implementation Pathways

- Regional Cooperation Networks would facilitate knowledge sharing, joint planning, and coordinated implementation of large-scale coastal protection initiatives.
- City-to-City Learning programs would enable urban coastal areas to share experiences and innovations in nature-based coastal protection.
- Scientific Collaboration would advance understanding of coastal ecosystem functions and optimize restoration and management techniques.
- Private Sector Engagement would leverage business innovation and investment to scale successful approaches.
- Revitalization of local wisdom and indigenous knowledge, ensuring that traditional ecological practices, customary governance, and intergenerational learning are integrated into policy, planning, and on-the-ground action.

7 Conclusion: Toward a Resilient Coastal Future

Twenty one years after the 2004 Indian Ocean Tsunami, we stand at another inflection point in coastal risk management. The convergence of accelerating climate change, continued ecosystem degradation, and growing coastal populations demands transformative approaches that go beyond traditional disaster risk reduction.

The path forward lies not in choosing between natural and engineered solutions, but in creating synergistic partnerships that harness the best of both. Nature-based coastal protection, supported by innovative governance, finance, and technology, offers unprecedented opportunities to build resilience while providing multiple co-benefits for biodiversity, climate, and human well-being. The lesson of the tsunami—that disasters can catalyze transformation—remains as relevant today as it was two decades ago. But now we have the knowledge, tools, and global cooperation mechanisms to build forward better before the next disaster strikes.

The Blue-Green Resilience Compact represents more than a new approach to coastal protection—it embodies a fundamental recognition that human security and ecosystem health are inseparable. In the words often attributed to Nietzsche, "What does not destroy us makes us stronger." The tsunami did not destroy the spirit of coastal communities, but revealed their extraordinary capacity for renewal and transformation.

As we look toward the future, let us remember that resilience should not be built in the wake of disaster, but in the years preceding it—through partnerships between communities and nature, between nations and ecosystems, between traditional knowledge and modern science. The coastal ecosystems that have protected humanity for millennia are ready to continue that service, but only if we commit to protecting and restoring them in return.

The choice before us is clear: we can continue to fight against the ocean's power with ever-higher walls and stronger barriers, or we can work with natural systems to create coastlines that are not just protected, but alive, productive, and resilient. The science, the technologies, and the partnerships needed for this transformation exist today. What remains is the collective will to implement them at the scale and speed that our coastal future demands.

In this new era of coastal resilience, let us ensure that the legacy of the 2004 tsunami is not measured only in the lives lost, but in the lives protected, the communities empowered, and the ecosystems restored for generations to come. The ocean that brought such devastation twenty one years ago can become our greatest ally in building a resilient, sustainable coastal future—if we have the wisdom to listen to its lessons and the courage to transform our relationship with the sea.

Acknowledgements. The authors would like to express their sincere gratitude to Universitas Gadjah Mada for entrusting us with the leadership of the infrastructure disaster response team in both the Aceh tsunami, Yogyakarta earthquake, Mt Merapi volcanic eruption, Lombok earthquake, and Palu multi-disasters recovery efforts. This opportunity has been invaluable in contributing to these significant disaster recovery initiatives and in gathering the insights presented in this paper. We also wish to acknowledge the countless individuals, community members, and local leaders in those disaster areas who shared their wisdoms, show their resilience, knowledge, and the spirit to recover and raise from disasters. Finally, we thank our colleagues at Universitas Gadjah Mada and other partnering institutions for their expertise, dedication, and support throughout these challenging but rewarding projects.

References

1. UNSDR, Hyogo Framework for Action 2005-2015: Building Resilience of Nations and Communities to Disasters, UNSDR (2005)
2. N. Nizam, Bencana Alam Sebagai Momentum Revitalisasi Ketangguhan Bangsa, (Natural Disasters as A Momentum For Revitalizing National Resilience), UGM's 61st Inaugural Speech. (in Bahasa, translated into English) (2010)
3. H. Robertson, et.al., Convention on Wetlands (2025) Global Wetland Outlook 2025: Valuing, conserving, restoring and financing wetlands. Gland, Switzerland: Secretariat of the Convention on Wetlands. DOI: 10.69556/GWO-2025-eng. (2025)
4. F. Ferrario, M.W. Beck, C.D. Storlazzi, F. Micheli,C.C. Shepard, L. Airoldi, L., The effectiveness of coral reefs for coastal hazard risk reduction and adaptation. Nature Commun. 5 (3794), http://dx.doi.org/10.1038/ ncomms4794. (2014)
5. G. Mancheño, et.al., Integrating mangrove growth and failure in coastal flood protection designs. Sci Rep. 2024 Apr 4;14(1):7951. doi: 10.1038/s41598-024-58705-4. (2024)
6. K. Harada, F. Imamura, Study on the evaluation of tsunami reducing by coastal control forest for actual conditions. In: Proc. APAC 2003 Makuhari; Japan. (2003)
7. B. Benazir, R. Triatmadja, S. Syamsidik, N. Nizam, W. Warniyati, Vegetation-based approached for tsunami risk reduction: Insights and challenges, Progress in Disaster Science, 23 (2024), https://doi.org/10.1016/j.pdisas.2024.100352. (2024)
8. O. Hoegh-Guldberg, L. Pendleton, A. Kaup, People and the changing nature of coral reefs, Regional Studies in Marine Science, Volume 30, 2019, 100699, ISSN 2352-4855, https://doi.org/10.1016/j.rsma.2019.100699. (2019)
9. A. De Souza de Lima, F. Cassalho, T.W. Miesse, M. Henke, M.R. Canick, C.M. Ferreira, Assessing the potential long-term effects of sea-level rise on salt marsh's coastal protective capacity under different climate pathway scenarios. Environ Monit Assess. Aug 16;196(9):817. doi: 10.1007/s10661-024-12961-z. (2024)
10. L. Huynh, J. Su, Q. Wang, L. Stringer, A. Switzer. Meta-analysis indicates better climate adaptation and mitigation performance of hybrid engineering-natural coastal defence measures. Nature Communications. 15. 10.1038/s41467-024-46970-w. (2024)
11. UN Women, Gender Snapshot 2024: How gender inequality and climate change are interconnected. (2024)
12. S.J. Harper, et al. Gender differences in the perceived impacts of coastal management and conservation. npj Ocean Sustainability, June 19, 2024. https://doi.org/10.1038/s44183-024-00070-w. (2024)

13. United Nation, Sendai Framework for Disaster Risk Reduction 2015-2030. (2015)
14. UNEP, Kunming-Montreal Global Biodiversity Framework, Decision Adopted By The Conference Of The Parties To The Convention On Biological Diversity, UNEP. (2022)
15. D. Murdiyarso & R. Ambo-Rappe. Coastal area rehabilitation for climate change mitigation: The role of mangroves and subnational emissions reductions. Working Paper 15. Bogor, Indonesia: CIFOR. (2022)

https://doi.org/10.1051/e3sconf/202565102001

Bridging the Living Income Gap for Cocoa and Palm Oil Farmers in Rural Post-Disaster Aceh

Kurnia Asni[1] and *Syafruddin* Chan[2,*]

[1]Master of Management Study Program, Faculty of Economics, Abulyatama University, Indonesia.
[2]Management Program, Faculty of Economics and Business, Syiah Kuala University, Indonesia.

Abstract. This study examines how cocoa and oil palm farmers in Aceh Timur and Aceh Tamiang, two disaster-prone districts in Indonesia's westernmost province respond to economic and environmental vulnerabilities through adaptive livelihood strategies. Using a mixed-methods approach combining household surveys and Focus Group Discussions (FGDs), the research analyses income gaps, expenditure sufficiency, and strategic responses among smallholder farmers. Findings reveal that a significant portion of households operate under financial stress even in stable conditions, with 41% unable to meet basic nutritional and non-food needs. The economic fragility is compounded by exposure to recurring floods and commodity price volatility. Despite these challenges, farmers are not passive recipients of risk. Adaptive strategies such as intercropping, scion sales, and certification-based partnerships (RSPO for palm oil, Rainforest Alliance for cocoa) emerged as viable solutions to boost household income and enhance resilience. RSPO-certified palm oil, when integrated into underutilized mills via cooperatives, could raise income by 15–20% and reduce the share of high-deficit households. Similarly, cocoa farmers practicing intercropping and seedling sales reported projected annual income gains of IDR 30 million, with certification adding a 10–15% price premium. The study applies an integrated analytical framework combining the Sustainable Livelihoods Framework and Economic Vulnerability and Resilience Theory, emphasizing household agency in managing risk. By linking economic strategies to disaster preparedness, this research contributes to a deeper understanding of how rural households can serve as key agents of resilience in ecologically fragile regions. Policy implications include the need for localized certification-led support systems and cooperative-based development models.

1 Introduction

Aceh, the westernmost province of Indonesia, remains heavily reliant on agriculture, with farming serving as the primary source of income for most rural households. Districts such as East Aceh and Aceh Tamiang exemplify the vulnerability of agrarian communities, particularly smallholder farmers of cocoa and oil palm, who face entrenched structural challenges. These include erratic commodity prices, limited infrastructure, poor access to credit and markets, and exposure to climate-related hazards [2, 9]. As a result, many farming households are unable to attain a decent standard of living. The situation is especially dire in Aceh Tamiang, where more than 22% of the rural population falls below the poverty line, compared to 18% in East Aceh (Badan Pusat Statistik Provinsi Aceh, 2024). Economic vulnerability is further exacerbated by low agricultural productivity and minimal support for risk mitigation, leaving households with few buffers against both economic and environmental shocks.

This economic fragility extends beyond daily hardship. It contributes to broader vulnerability landscapes, especially in regions like Aceh that are repeatedly affected by floods, droughts, and climate variability. Aceh is one of Indonesia's most disaster-prone provinces, not only because of its seismic history but also due to increasing hydro-meteorological risks exacerbated by climate change. In this context, the linkage between household economic security and disaster risk reduction (DRR) becomes especially salient. Research has emphasized the

* Corresponding author: syafruddin.chan@usk.ac.id

importance of integrating livelihoods into DRR efforts, arguing that economic resilience is a crucial pillar of sustainable disaster mitigation [12]. Without strengthening economic security at the household level, particularly among vulnerable rural groups, macro-level DRR interventions may fail to produce lasting impacts. Farming households in Aceh, however, are not passive victims of these conditions. They employ various adaptive strategies such as livelihood diversification, engaging in non-farm employment, seasonal migration, and petty trade alongside land management innovations and social capital mobilization [10]. These strategies reflect an effort to manage risk, smooth income, and build a form of endogenous resilience. Nonetheless, a clear understanding of how these strategies are constructed, what constraints shape them, and how they relate to broader development and disaster frameworks is still lacking. Particularly absent is a systematic view of how adaptive capacities at the household level can inform community-wide resilience building.

This study applies the Sustainable Livelihoods Framework (SLF), which offers a multidimensional perspective on how households navigate vulnerability through the deployment of various capitals: human, natural, financial, physical, and social [14]. SLF is especially relevant for examining household decision-making in fragile contexts. However, to capture the dynamic interplay between shocks and recovery, this framework is complemented with the Economic Vulnerability and Resilience theory, which emphasizes the importance of absorptive and adaptive capacities in responding to external pressures [4]. The integration of these two frameworks enables a robust examination of how smallholder farmers in disaster-prone Aceh adapt economically while simultaneously engaging in informal risk reduction.

The urgency of this inquiry is underscored by recent economic trends. Between 2022 and 2024, the price of fresh fruit bunches (FFB) from oil palm fell by over 25%, while cocoa prices stagnated despite rising production costs (BPS Aceh 2024). These economic shocks occurred alongside recurrent flooding events in East Aceh that damaged farmlands and disrupted market access. The compounding effect of market volatility and environmental instability creates a dual burden on households, threatening not only their income but also their long-term well-being and sustainability. Despite these realities, most current studies either treat disaster vulnerability and rural poverty as separate phenomena or frame households as passive recipients of policy rather than active agents of adaptation. Van Schie et al. (2020) argue that such depoliticized framings can obscure the creative and strategic actions that households undertake to protect their futures. Moreover, Birkmann et al. (2022) note that traditional vulnerability assessments often neglect the role of informal institutions, social networks, and indigenous knowledge in shaping adaptive responses. These gaps point to the need for more grounded, household-level research that can inform cross-sectoral policies linking economic recovery with disaster risk governance.

The novelty of this study lies in its integrated theoretical model and its empirical grounding in one of Indonesia's most ecologically and socioeconomically vulnerable regions. Few studies have attempted to bridge SLF and economic resilience theory in the context of household adaptation to disaster risk in Southeast Asia. In doing so, this research aims to generate practical insights on how livelihood strategies contribute to disaster resilience, with implications for policy, planning, and grassroots development. The use of qualitative data gathered through interviews and field observations, augmented by limited quantitative measures of vulnerability, ensures both contextual richness and analytical rigor.

This research is expected to benefit a range of stakeholders. For local governments and national disaster agencies, it offers a detailed understanding of community-level resilience pathways, enabling better targeting of social protection and rural development initiatives. For NGOs and development partners working on post-disaster recovery, the findings provide operational insights on how to design adaptive livelihood support programs. Academic communities will also benefit from its methodological contributions, particularly in the areas of resilience studies, rural livelihoods, and disaster governance. Most importantly, farming households themselves may indirectly benefit through better-informed interventions that reflect their lived realities and resource constraints.

2 Literature Review and Theoretical Framework

2.1 Household Livelihood Strategies of Smallholder Farmers

Household livelihood strategies refer to the diverse combinations of on-farm and off-farm activities that rural families pursue to ensure economic survival and adapt to external shocks. Among smallholder farmers, this includes efforts such as crop diversification, agroforestry, informal labor, and seasonal migration [7]. In Aceh Timur and Aceh Tamiang, reliance on a single commodity like oil palm or cocoa, coupled with poor infrastructure and limited institutional support, makes diversification critical. Recent studies highlight how adaptive behaviors ranging from petty trade to engagement with cooperatives are vital not only for income smoothing but also as informal mechanisms of disaster risk reduction [10, 8]. These strategies are often embedded in social networks and community practices, reinforcing their importance in fragile, hazard-prone regions.

2.2 Household Economic Resilience

Economic resilience at the household level is defined as the ability to absorb and recover from economic or environmental shocks while maintaining core functions [5]. This concept extends beyond income alone to include access to savings, social capital, and the capacity for strategic adaptation. In regions like Aceh, frequent flooding and volatile commodity prices increase the urgency of strengthening household resilience. Sarr (2021) argues that flexible livelihood planning and access to community support are key characteristics of resilient households. He further emphasizes that rural households with diversified income sources and adequate financial buffers are less likely to fall into disaster-induced poverty. Viewing resilience as a dynamic and evolving process, rather than a fixed condition, is therefore essential for the development of effective, household-centered disaster response policies.

2.3 Sustainable Livelihoods Framework (SLF)

The Sustainable Livelihoods Framework (SLF), developed by the Department for International Development (1999), offers a structured approach to analysing how households use different forms of capital—human, natural, physical, financial, and social to pursue sustainable livelihoods. Abazue et al. (2015) emphasize the interaction between these capitals within specific institutional and environmental contexts. In disaster-prone areas like Aceh, this framework is useful in identifying how shocks to one capital (e.g., loss of land or reduced market access) can undermine overall household security. At the same time, strengthening another capital, such as social ties through cooperatives or informal networks, can improve resilience. SLF thus serves as both a diagnostic and strategic tool to understand the adaptive capacities of rural households facing chronic vulnerability.

2.4 Economic Vulnerability and Resilience Theory

Briguglio et al. (2009) distinguish between economic vulnerability, referring to a household or region's exposure to external risks, and economic resilience, which denotes the ability to recover and adapt. In Aceh, economic vulnerability is heightened by commodity dependence, poor access to credit, and infrastructure deficits. However, resilience manifests through behaviours such as informal cooperation, household re-budgeting, and flexible income strategies [13]. This theory underscores the role of household agency and the importance of behavioural adaptation in the face of shocks. Applying this lens to cocoa and oil palm farmers allows the study to capture not only structural weaknesses but also the creative ways in which families build coping mechanisms to reduce long-term risk.

2.5 Analytical Framework of the Study

This research adopts an integrated analytical framework by combining the SLF with economic vulnerability and resilience theory to explore how households in disaster-prone areas manage income deficits while enhancing resilience. The framework positions smallholder households as active agents who deploy livelihood assets strategically to cope with overlapping challenges ranging from income shocks to climate-related hazards [12]. This approach also supports a localized understanding of disaster risk reduction by highlighting the connection between economic insecurity and environmental exposure. By analysing how decisions around land use, labour, and cooperative engagement contribute to both economic stability and hazard mitigation, the study provides grounded insights for designing inclusive, context-sensitive resilience policies.

2.6 Review of Previous Studies

Several foundational and recent studies inform the conceptual basis of this research. Ellis (2000) provided early insights into livelihood diversification as a household response to risk and resource scarcity. Cutter et al. (2008) emphasized the multidimensional nature of resilience, including social and economic assets that influence recovery capacity. Briguglio et al. (2009) contributed theoretical clarity by separating vulnerability from resilience, introducing operational indicators that are still widely used today. More recently, Wise et al. (2021)

and Yang et al. (2021) have explored how smallholder households in Indonesia adapt to intersecting environmental and economic stressors. However, few studies have attempted to explicitly bridge SLF and resilience theory to analyse household responses to disaster risks. This study addresses that gap by offering an empirically grounded, theoretically integrated analysis of farmer adaptation in Westernmost Indonesia.

3 Methodology

3.1 Sample Size

This study focuses on heads of farming households in Aceh Timur and Aceh Tamiang, with 75 respondents selected purposively from each district (totalling 150) to capture the actual structure of income, consumption, and household expenditure. Given the study's exploratory mixed-methods approach, the emphasis is on the depth and contextual relevance of information, not statistical generalization. The sample size was deemed sufficient as the data collected reached a point of saturation, where additional responses no longer provided new insights. This aligns with common practices in qualitative and exploratory research, where information richness and analytical relevance outweigh sample size maximization. The sample size also reflects practical considerations related to data validation and the development of policy recommendations.

3.2 Data Collection Method

This study employed a mixed-methods data collection strategy aligned with the Anker Methodology, which is widely recognized for estimating living wages and incomes based on local socio-economic contexts. Primary data were gathered through quantitative surveys with farming household heads to assess income, expenses, and other economic indicators, and complemented by qualitative interviews to explore consumption patterns, spending priorities, and livelihood challenges. Additionally, secondary data from institutions such as BPS and WHO were analysed to establish benchmarks for decent living standards. Focus Group Discussions (FGDs) with local stakeholders were also conducted to validate findings and contextualize economic realities. In accordance with Anker's framework, this approach integrates local price data, household expenditure surveys, and community insights to determine living costs covering nutritious diets, housing, non-food essentials, and emergency margins while adjusting for local conditions through both regression analysis and contextual interpretation.

To minimize recall bias in reporting income and expenditure, the survey employed a short recall period (monthly reporting) rather than annual estimates. In addition, we conducted focus group discussions to validate income and yield figures, where discrepancies could be cross-checked among peer farmers. Self-reported commodity prices and yield estimates were triangulated with secondary data obtained from local agricultural offices, cooperatives, and certification bodies. These steps reduced the likelihood of significant inaccuracies and increased confidence in the reliability of the data.

3.3 Focus Group Discussion (FGD) Planning for the Study in Aceh Timur and Aceh Tamiang

To gain a deeper understanding of farmers' welfare challenges and income strategies in Aceh Timur and Aceh Tamiang, Focus Group Discussions (FGDs) are a key component of this study. FGDs are essential not only for capturing the lived experiences of farmers but also for identifying practical strategies to close income gaps. They provide a platform for farmers to share contextual insights, challenges, and coping mechanisms, offering a richness that complements quantitative findings. The discussions aim to uncover adaptive strategies such as income diversification, harvest optimization, collective marketing, and access to government support. Each FGD will involve a mix of participants, including cocoa and palm oil farmers, government representatives (e.g., from Bappeda, the Plantation Office, and the Trade and Cooperative Office), buyers, and relevant NGOs. This multi-stakeholder approach ensures that a wide range of perspectives is considered in designing sustainable solutions for improving farmer livelihoods.

3.4 Data Analysis Technique

The data analysis involves calculating the cost of living based on WHO standards and the Anker Methodology, which includes components such as nutritional needs, adequate housing, basic education, and healthcare expenses. This cost-of-living estimate is then compared to the take-home income of each respondent to assess economic sufficiency. The analysis also identifies the income gap, the difference between actual earnings and the required living costs, and explores the strategies employed by households to bridge this gap.

4 Results

This section presents the key findings derived from both the household survey and Focus Group Discussions (FGDs), highlighting the income disparities of both palm oil and cacao farmers. The analysis outlines how certification schemes, cooperative partnerships, and crop diversification strategies contribute to narrowing these income gaps in disaster-prone regions.

4.1 Respondent Profile and Plantation Area

Table 1 explains the profile of respondents in this study and also the plantation area.

Table 1. Respondent Profile.

Profile Category	Subcategory	Frequency	Percentage
Gender	Male	117	47%
	Female	26	53%
	Total	143	100%
	Bachelor	5	3%
Education Level	Senior High School	61	43%
	Below SMA	77	54%
	Total	143	100%
Household Size	3 Members	27	19%
	4 Members	78	55%
	5 Members	27	19%
	6 Members	11	8%
	Total	143	100%
Types of plantations	Only Cocoa	33	23%
	Only Palm Oil	87	61%
	Palm Oil and Cocoa	23	16%
Total		**143**	**100%**

Most of the farmers are female (53%), and the majority have an education below senior high school (54%). The typical household size is 4 members, representing 55% of the sample. Regarding plantation types, palm oil dominates, with 61% of farmers cultivating only palm oil, while 23% focus on cocoa and 16% grow both. Overall, the sample consists of 143 respondents with diverse backgrounds in education, household size, and plantation activities.

Table 2. Plantation Area.

Land Area Category	Palm Oil (Total)	Palm Oil (%)	Cocoa (Total)	Cocoa (%)
≤ 0.50 Ha	20	18%	24	43%
> 0.50 – 1.00 Ha	41	37%	16	29%
> 1.00 – 2.00 Ha	31	28%	13	23%
> 2.00 – 3.00 Ha	11	10%	-	-
> 3.00 Ha	7	6%	3	5%
Total	**110**	**100%**	**56**	**100%**

Table 2 shows the distribution of plantation area for palm oil and cocoa across different land size categories. Most palm oil farmers (37%) cultivate on plots between 0.50–1.00 hectares, followed by 28% on 1.00–2.00 hectares, while only 6% operate on land larger than 3.00 hectares. In contrast, cocoa farmers tend to work on smaller plots, with 43% farming on ≤ 0.50 hectares and another 29% on 0.50–1.00 hectares. Very few cocoa farmers operate on land above 1.00 hectares, and none are recorded above 2.00 hectares. This suggests palm oil farming is generally done on larger plots, while cocoa cultivation is concentrated on smaller holdings.

4.2 Income

This section explains the farmers' income, both for Palm Oil and Cocoa production.

Table 3. Income Drive from Palm Oil and Cocoa Plantation

Nett Income from Palm Oil and Cocoa (IDR)	Aceh Tamiang		Aceh Timur		Total	%
	Palm Oil	Cocoa	Palm Oil	Cocoa		
1. ≤ 1 millions	14	16	14	12	56	34%
2. > 1 – 2 millions	19	3	20	12	54	33%
3. > 2 – 4 millions	13	0	10	6	29	18%
4. > 4 – 6 millions	4	0	7	3	14	9%
5. > 6 millions	4	2	3	2	11	7%
Total	**54**	**21**	**54**	**35**	**164**	**100%**

Table 3 presents the net income distribution of palm oil and cocoa farmers in Aceh Tamiang and Aceh Timur. Most farmers earn modest incomes, with 34% earning ≤ 1 million IDR and 33% earning between 1–2 million IDR. Palm oil is the more prevalent income source in both regions, especially in higher income brackets. In Aceh Tamiang, 54 farmers earn from palm oil, but only 21 from cocoa, while in Aceh Timur, 54 farmers also earn from palm oil and 35 from cocoa. Cocoa income is generally lower, with no farmers in Aceh Tamiang earning above 2 million from cocoa. Overall, palm oil generates higher and more widespread earnings, while cocoa income is concentrated among fewer farmers and mostly within lower income brackets.

Table 4. Other Incomes.

Regencies	Average Other Income/Month	Respondent
Aceh Tamiang	IDR 2,103,867	60
Aceh Timur	IDR 1,499,616	59

In addition to income from palm oil and cocoa, farmers in both Aceh Tamiang and Aceh Timur also rely on other sources of income. On average, respondents in Aceh Tamiang reported higher additional monthly income, amounting to IDR 2,103,867, compared to IDR 1,499,616 in Aceh Timur. This suggests that farmers in Aceh Tamiang have more diversified or lucrative supplementary income streams. A total of 119 respondents, 60 from Aceh Tamiang and 59 from Aceh Timur, reported other income sources, highlighting the importance of non-plantation income in supporting rural livelihoods in both regencies.

4.3. Household Expenditures of Palm Oil and Cocoa Farmers

This section presents the analysis of household expenditures among palm oil and cocoa farmers. The discussion is divided into two parts. The first part explores the gap between actual income and actual expenditure. Actual income refers to the average monthly earnings reported by farming households, as obtained through the survey. Meanwhile, actual expenditure reflects the average monthly household spending, also based on survey data. This comparison aims to highlight whether farming households are operating with a financial surplus or deficit under current economic conditions.

The second part of the analysis evaluates the sufficiency of household income in relation to the cost required to meet ideal nutritional standards. Specifically, it compares actual income to estimated household expenditures based on a benchmark dietary requirement of 7,300 kilocalories per household per day. This standard is used to assess whether the income earned by palm oil and cocoa farmers is adequate to ensure basic nutritional needs are met for their families. The findings from this comparison provide deeper insights into the farmers' economic vulnerability and their capacity to maintain minimum living standards.

Table 5. Actual Expenditure of Palm Oil and Cocoa Household.

Expenditure for Food & Non-Food Cost (IDR)	Regencies		Total	%
	Aceh Tamiang	Aceh Timur		
≤ IDR 2,500,000	46	14	60	42%
IDR 2,500,001 – IDR 3,500,000	23	21	44	31%
IDR 3,500,001 – IDR 4,500,000	4	18	22	15%
IDR 4,500,001 – IDR 5,500,000	0	14	14	10%
> IDR 5,500,000	0	3	3	2%
Total	**73**	**70**	**143**	**100%**

Table 5 outlines household expenditure on both food and non-food items across Aceh Tamiang and Aceh Timur. Overall, 42% of households spend less than IDR 2.5 million per month, with a significant majority of these (46 out of 60) coming from Aceh Tamiang. In contrast, higher expenditure brackets are more common in Aceh Timur particularly in the IDR 3.5 to 5.5 million range where 32 households fall into these categories compared to only 4 in Aceh Tamiang. The highest spending group, those spending over IDR 5.5 million monthly, is found exclusively in Aceh Timur. This suggests a noticeable difference in consumption patterns between the two regencies, with households in Aceh Timur tending toward higher overall expenditure.

Table 6. Average Actual Expenditure of the Household per Month

Regencies	Average Food Cost/Month	Average Non-Food Cost/Month	Average Total Expenditure	Ratio
Aceh Tamiang	IDR 1,030,711.96	IDR 1,340,422.83	IDR 2,371,134.79	40%
Aceh Timur	IDR 1,370,481.43	IDR 2,222,572.33	IDR 3,593,053.76	60%

The average monthly expenditure data further highlights the disparity between Aceh Tamiang and Aceh Timur. Households in Aceh Timur spend significantly more overall, averaging IDR 3,593,053.76 per month compared to IDR 2,371,134.79 in Aceh Tamiang. Notably, non-food expenses make up the bulk of spending in both regions. Still, the gap is especially pronounced in Aceh Timur, where non-food costs average over IDR 2.2 million more than 60% of total expenditure. In contrast, Aceh Tamiang households allocate only 40% of their budget to non-food needs. This suggests that residents of Aceh Timur may have higher living costs or place greater emphasis on non-food necessities such as education, transportation, and healthcare.

Table 7. Income Gap: The Income vs Actual Expenditure.

Range	Categories	Regencies		Total	%
		Aceh Tamiang	Aceh Timur		
<= -750,000	High Deficit	19	14	33	23%
-749,000< Value <= -1,000	Low Deficit	2	20	22	15%
1,000 < Value <= 500,000	Low Surplus	9	10	19	13%
501,000 < Value <= 1,500,000	Medium Surplus	10	12	22	15%
> 1,500,000	High Surplus	33	14	47	33%
Total		**73**	**70**	**143**	**100%**

The analysis of household income and expenditure in Aceh Tamiang and Aceh Timur shows a varied financial condition. About 38% of households experience a deficit, with 23% in a high deficit and 15% in a low deficit category, meaning their spending exceeds income. On the other hand, 61% of households enjoy a surplus, consisting of 13% with a low surplus, 15% with a medium surplus, and 33% with a high surplus. Overall, the largest group is households with a high surplus, while nearly two-fifths of households still struggle with income deficits.

Household expenditure among surveyed farmers can be broadly categorized into non-food and food-related spending. Non-food expenses include essential needs such as housing, transportation, children's education, health care, household utilities, clothing, and allocations for savings or emergency funds. Meanwhile, food expenditures cover both staple and supplementary items, including staple foods, animal-based proteins, plant-based foods, vegetables, fruits, eggs, milk and processed dairy products, instant noodles, cooking oil, sugar, snacks, and various beverages. This breakdown reflects both subsistence priorities and lifestyle needs within farming households.

Table 8. The Requirements of Nutrition for a 4-member Household.

Family Members	**Age Assumption**	**Gender**	**Activity**	**Energy Requirements (kcal/day)**
Father	35 years	Male	Moderate	2,750
Mother	33 years	Female	Moderate	2,250
Child 1	10 years	Male	Light-Moderate	2,000
Child 2	13 years	Female	Light-Moderate	1,950
Total				**7,300**

The ideal daily nutritional requirement for a four-member household, based on energy needs, totals 7,300 kilocalories per day. This estimation is derived from age, gender, and activity levels as follows: a 35-year-old father (male, moderate activity) requires 2,750 kcal/day; a 33-year-old mother (female, moderate activity) requires 2,250 kcal/day; a 10-year-old son (light to moderate activity) needs 2,000 kcal/day; and a 13-year-old daughter (light to moderate activity) requires 1,950 kcal/day. These nutritional needs are calculated according to the Indonesian Ministry of Health's Recommended Dietary Allowances (RDA), as outlined in Peraturan Menteri Kesehatan RI No. 28 Tahun 2019 tentang Angka Kecukupan Gizi [14].

Table 9. Food and Non-Food Cost Based on 7300 kcal/day.

Category	**Value (IDR)**
Non-Food	IDR 2,348,917
Food	IDR 1,620,000
Food & Non-Food	IDR 3,968,917

Based on the ideal nutritional requirement of 7,300 kcal/day for a four-member household, the monthly expenditure needed to meet both food and non-food needs totals IDR 3,968,917. Of this, IDR 1,620,000 is allocated to food, which reflects the minimum cost required to provide nutritionally adequate meals for the family. The remaining IDR 2,348,917 covers essential non-food expenses such as housing, transportation, education, healthcare, clothing, and emergency savings. These figures provide a benchmark for evaluating household economic sufficiency in meeting both dietary and living.

Table 10. Income Gap: The Income vs Ideal Food Nutrition Expenditure

Range	WHO Income & Expenditure Bracket	Regencies		Total	%
		Aceh Tamiang	Aceh Timur		
<= -750,000	High Deficit	32	26	58	41%
-749,000< Value <= -1,000	Low Deficit	8	17	25	17%
1,000 < Value <= 500,000	Low Surplus	9	6	15	10%
501,000 < Value <= 1,500,000	Medium Surplus	8	8	16	11%
> 1,500,000	High Surplus	16	13	29	20%
Total		**73**	**70**	**143**	**100%**

The second deficit table (Table 10) evaluates household income against a more stringent benchmark the cost of meeting ideal food nutrition (7,300 kcal/day) and non-food expenses. Under this standard, financial strain becomes more apparent. A significant 41% of households fall into the high deficit category, with 32 cases in Aceh Tamiang and 26 in Aceh Timur, indicating that nearly half of the respondents cannot meet basic ideal needs. An additional 17% fall into the low deficit range. In contrast, only 20% of households achieve a high surplus, and another 21% fall within low to medium surplus brackets. Compared to the first deficit table, this analysis reveals a much higher level of economic vulnerability when nutritional adequacy is considered, highlighting the gap between actual income and the cost of achieving a healthy and sustainable standard of living.

4.4 Strategy to Overcome Income Deficit

The strategy to address income deficits was formulated through two Focus Group Discussions (FGDs), held separately in Aceh Timur and Aceh Tamiang. Each FGD brought together a comprehensive range of stakeholders, including representatives from three key regional government agencies Bappeda, the Department of Plantations and Agriculture, and the Department of Industry, Trade, and Cooperatives. In addition, the discussions included participants from palm oil and cocoa buyer groups, farmer cooperatives, individual cocoa and palm oil farmers, as well as representatives from non-governmental organizations and independent observers specializing in plantation issues. This diverse representation ensured that the proposed strategies were grounded in practical realities and informed by multiple perspectives.

4.5 Strategy for Palm Oil

The Aceh Tamiang and Aceh Timur FGD identified RSPO certification particularly through the Identity Preserved (IP) model as the most viable lever for closing household income gaps among smallholder palm-oil farmers. Participants emphasized that RSPO certified fresh fruit bunches (TBS) currently command an average premium of ID 95 per kg over conventional sales, while many farmers still defect to traders offering only IDR 50 per kg higher spot prices because of pre-existing debt ties. Cooperative leaders estimated that if 500 hectares of smallholder plots (≈ 10% of the district's independent estates) attained IP-level certification, participating households could boost net monthly farm income by 15–20 percent, shifting roughly one-third of high-deficit households (≥ IDR 750,000 shortfall) into at least a low-surplus category.

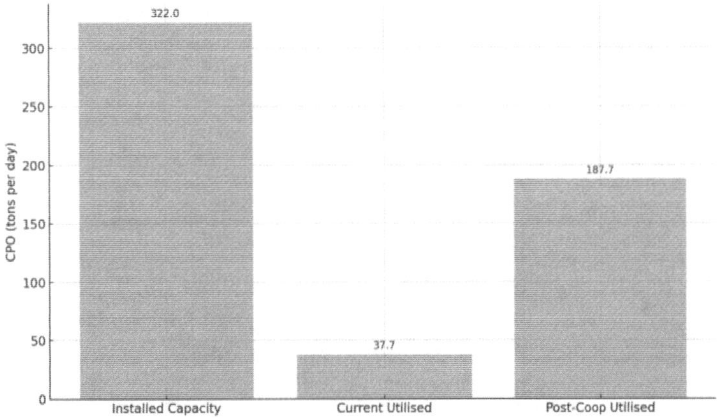

Figure 1. Typical Mill Utilisation Scenario.

PT Socfindo's mill data reinforced the business case for such partnerships. With an installed capacity of 322 tons CPO per day, the mill currently processes only ≈ 37.7 tons from its 3,000-ha nucleus estate just 11.7 percent utilization leaving 284 tons of idle capacity daily. FGD simulations showed that integrating certified smallholder cooperatives supplying an additional 150 tons TBS per day (rendement 23 percent) would lift utilization to nearly 60 percent and generate an extra IDR 450 million in annual cooperative premiums, assuming current IP differentials. This idle-capacity argument persuaded mill representatives to commit to year-end performance bonuses and per-delivery quality premiums, contingent on verifiable RSPO compliance.

Table 11. Key Indicators for RSPO-IP Certification Scenario.

Indicator	Figure
Current RSPO IP premium offered by cooperative (Rp/kg)	95
Typical spot-price premium from traders (Rp/kg)	50
Target smallholder plots for 1st-wave certification (ha)	500
Additional certified TBS supply targeted (tons/day)	150
Mill installed CPO capacity (tons/day)	322
Current CPO output from nucleus estate (tons/day)	37.7
Projected CPO output with cooperative supply (tons/day)	187.7
Projected mill utilisation after integration (%)	58
Average projected household income boost (%)	15–20
High-deficit households potentially moved to surplus (share)	≈⅓

Finally, the discussion underscored the enabling roles of Forum Konservasi Leuser (FKL) and the PUPL. Their joint roadmap targets 100 farmer households for first-wave certification, introducing GPS-based geo-traceability and cooperative pre-financing to break debt cycles with informal agents. By bundling legal land verification, capacity-building modules, and digital record-keeping, FKL–PUPL forecasts that smallholders could meet RSPO audit requirements within 18 months, at a cost of roughly IDR 12 million per cooperative, a figure judged "economically justified" given projected income gains. Collectively, these FGD findings position RSPO-IP integration as both a mill-level efficiency solution and a household-level poverty-reduction strategy in Aceh's palm-oil sector.

4.6 FGD Findings: Adaptive Income Strategies Among Cocoa Farmers in Disaster-Prone Areas

In line with the paper's core theme—Disaster-Prone but Not Defenceless—Focus Group Discussions (FGDs) with certified cocoa farmers in Aceh Tamiang reveal that smallholders are not passive victims of environmental shocks, but active agents of economic adaptation. Farmers outlined several strategies they have adopted to increase household income and build resilience in the face of recurring climatic and market-related disruptions.

The most prominent adaptation strategy discussed was intercropping, which farmers identified as a critical buffer against both seasonal uncertainty and price volatility. By planting short-cycle food or horticultural crops such as bananas, chili, and vegetables alongside cacao, farmers reported earning up to IDR 24 million per year, comprising 75% of the projected additional income in post-certification scenarios. This practice not only diversifies income streams but also enhances food security and land-use efficiency, key priorities in disaster-prone rural economies.

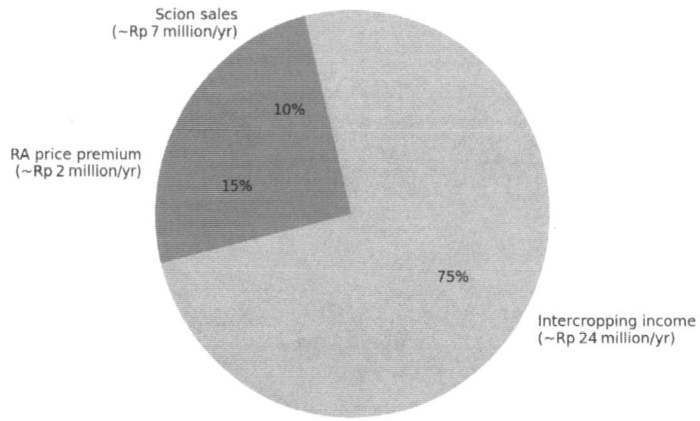

Figure 2. Additional Income of Cocoa Household.

Certification premiums from the Rainforest Alliance (RA) estimated at around Rp 2 million per year (15%) were also acknowledged as a valuable, though secondary, income source. Farmers appreciated the price uplift, but noted that on its own, the premium was insufficient to close income gaps without complementary strategies. A third adaptive mechanism highlighted was scion sales, in which farmers propagate and sell high-quality cacao seedlings to peers. This approach, contributing approximately IDR 7 million per year (10%), was viewed as a sustainable entrepreneurial pathway enabled by capacity-building efforts from local NGOs, the Doktor Kakao network, and demonstration plot initiatives.

These findings underscore that cocoa farmer in Indonesia's westernmost, disaster-exposed regions are not defenseless; rather, they are actively leveraging certification schemes and agroecological knowledge to stabilize and grow their incomes. The shift toward integrated, market-linked practices represents a significant evolution in local adaptation strategies one that bridges sustainability, economic security, and disaster resilience at the household level.

The livelihood strategies identified intercropping, scion sales, and certification-based partnerships—not only enhance income diversification but also strengthen resilience to disaster shocks. For instance, focus group participants reported that intercropping with food crops (e.g., cassava, maize) provided a safety net during flood-related income disruptions, reducing dependence on external food aid. Likewise, certification schemes and cooperative membership improved access to financial resources and recovery networks after extreme weather events. By stabilizing income flows and embedding farmers within collective support structures, these strategies contribute directly to disaster risk reduction and enhance household adaptive capacity in hazard-prone environments.

5 Conclusions

The household survey confirmed that even before natural-hazard shocks strike, many farm families in Aceh Tamiang operate on an economic knife-edge. When actual income is set against actual spending, 23% of households run a high deficit, and when the benchmark rises to a nutritionally adequate food basket plus essential non-food costs, the share in high deficit climbs to 41% (58 of 143 households). More than two-thirds of cocoa-growing families earn ≤ IDR 2 million per month, and 60% of palm-oil households fall short of the IDR 3.5 million threshold needed to meet the 7,300 kcal/day dietary standard for a four-member family. These gaps underscore the article's thesis—"Disaster-Prone but Not Defenceless"—by showing that financial stress is already acute in calm years; cyclones, floods, or market shocks would simply widen an existing chasm.

Focus-group discussions then shifted from diagnosis to adaptation. Stakeholders agreed that certification-led supply-chain upgrades offer the fastest, climate-smart route to close the gap. For palm oil, an RSPO-Identity Preserved partnership with PT Socfindo would lift farm-gate prices by ≈ IDR 95 kg^{-1} and utilise an extra 150 t TBS day^{-1} of idle milling capacity enough to pull one-third of high-deficit households into surplus. In cocoa, Rainforest Alliance certification coupled with intercropping and scion sales can raise household cash flow by IDR 30 million yr^{-1} and deliver a 10–15% price premium on beans. Both roadmaps rest on cooperative Internal Management Systems, GPS-based geo-traceability, and blended finance to cover transition costs, turning smallholders from passive hazard victims into proactive stewards of resilient, biodiversity-friendly landscapes. In doing so, the findings illustrate how Western Indonesia's farmers, though situated in a disaster zone, can defend their livelihoods through integrated economic and environmental strategies.

6 Research Limitations

Several caveats apply to this research. The income–expenditure survey, based on 143 households, relied on self-reported data collected at a single point in time. Without corroboration from independent financial records, recall errors or misreporting may have affected the precision of deficit estimates. In addition, seasonal fluctuations in crop prices and household spending were not captured, making the analysis a snapshot rather than a year-round profile. The purposive sampling approach ensured coverage of cocoa and palm oil smallholders in Aceh Timur and Aceh Tamiang, but limits the generalizability of results beyond this target population.

The focus group discussions primarily involved participants in Aceh Tamiang who were already connected to NGO or corporate programmes. As a result, the strategies identified may reflect the perspectives of better-networked farmers and service providers, while overlooking challenges faced by more isolated producers. Furthermore, the projected income gains from RSPO and RA certification assume stable premium pricing, reliable technical support, and cooperative cohesion factors that remain vulnerable to shifts in market conditions, policies, or stakeholder dynamics. Taken together, these limitations suggest that the findings should be interpreted as indicative rather than definitive. Further longitudinal and multi-site research with broader and more representative sampling would help validate and refine the insights presented here.

Acknowledgements. This study was made possible through the financial support of the Leuser Conservation Forum (Forum Konservasi Leuser/FKH), the Sustainable Trade Initiative (IDH), and MARS. The authors also wish to express their gratitude to the Tsunami Disaster Mitigation and Research Center (TDMRC), Universitas Syiah Kuala (USK), for its valuable support and expertise in the field of disaster risk and mitigation.

References

1. Sarr, M., Ayele, M. B., Kimani, M. E., & Ruhinduka, R. (2021). Who benefits from climate-friendly agriculture? The marginal returns to a rainfed system of rice intensification in Tanzania. *World Development*, ***138***, 105160. https://doi.org/10.1016/j.worlddev.2020.105160
2. Abazue, C. M., Er, A. C., Ferdous Alam, A. S. A., & Begum, H. (2015). Oil palm smallholders and its sustainability practices in Malaysia. *Mediterranean Journal of Social Sciences*, ***6***(6 S4), 482-488.
3. Birkmann, J., Cardona, O. D., Carreño, M. L., et al. (2022). Framing vulnerability, risk and societal responses: The MOVE framework revisited. *Natural Hazards*, ***112***, 151–177. https://doi.org/10.1007/s11069-022-05344-5
4. Briguglio, L., Cordina, G., Farrugia, N., & Vella, S. (2009). Economic vulnerability and resilience: Concepts and measurements. *Oxford Development Studies*, ***37***(3), 229–247. https://doi.org/10.1080/13600810903089893
5. Cutter, S. L., Burton, C. G., & Emrich, C. T. (2010). Disaster resilience indicators for benchmarking baseline conditions. *Journal of Homeland Security and Emergency Management*, ***5***(1). https://doi.org/10.2202/1547-7355.1732
6. Department for International Development. (1999). Sustainable livelihoods guidance sheets. Department for International Development. https://www.livelihoodscentre.org.uk
7. Ellis, F. (2000). *Rural livelihoods and diversity in developing countries*. Oxford University Press.
8. Yang, X., Guo, S., Deng, X., Wang, W., & Xu, D. (2021). Study on livelihood vulnerability and adaptation strategies of farmers in areas threatened by different disaster types under climate change. Agriculture, **11**(11), 1088.

9. Ratner, B. D. (2015). Building resilience in rural livelihood systems as an investment in conflict prevention. In *Livelihoods, natural resources, and post-conflict peacebuilding* (pp. 327-340). Routledge.
10. Russell M. Wise, James R. A. Butler, Wayan Suadnya, Ketut Puspadi, Imam Suharto, and T. D. Skewes, *Clim. Risk Manag.* **12**, 100–114 (2016).
11. Danielle Van Schie, A. B. Mirza, Ranon Jahan Khan Ranon, Masfiqua Malek, N. Naushin, Farid Hossain, … and Simon Anderson, *Clim. Dev.*, 1–13 (2025).
12. Rahman, M. A., Akter, S., & Das, N. C. (2023). Linking livelihood diversification with disaster risk reduction: Evidence from flood-prone areas in South Asia. *Environmental Hazards, 22*(1), 42–60. https://doi.org/10.1080/17477891.2023.2172333
13. Zhou, Y., Li, Y., Zhang, L., Su, Q., & Huang, X. (2023). Spatial–temporal characteristics of geological disaster resilience in poverty and disaster-prone areas in China: a case study of Ganzi Prefecture. *Environmental Science and Pollution Research,* **30**(35), 83401-83420.
14. Ministry of Health of the Republic of Indonesia (2019). *Regulation of the Minister of Health of the Republic of Indonesia Number 28 of 2019 concerning Recommended Dietary Allowances* Jakarta: Ministry of Health of the Republic of Indonesia.
15. Badan Pusat Statistik Provinsi Aceh (2024). *Statistik perkebunan kelapa sawit dan kakao 2022–2024* [Palm oil and cocoa plantation statistics 2022–2024]. Badan Pusat Statistik Provinsi Aceh. https://aceh.bps.go.id

Entrepreneurial activities post-disaster: How to respond to the changing environment

Michi Fukushima[1,*]

[1]Tohoku University, Graduate School of Economics and Management, 27-1 Kawauchi, Sendai, Japan.

Abstract. It is well-documented that entrepreneurial activities increase after disasters. However, as time passes and the effects of the disaster fade, it becomes increasingly difficult to ascertain how entrepreneurs who emerge in the aftermath respond to the changing environment and achieve business grow. To address this gap, this study examines the factors that support expansion and the strategies they employed to sustain such growth. In our research a combination of methods was adopted, namely a questionnaire survey and a case study. A questionnaire was administered to individuals who started businesses after the Great East Japan Earthquake in Japan. The survey suggests that there may have been a shift in motivation among individuals embarking on new business creation in the aftermath of a disaster. Their focus appears to have transitioned from being primarily social to being more self-oriented. Additionally, it seems that entrepreneurs who expand their businesses may also exhibit a tendency towards altruism. A case study highlights how one entrepreneur expanded his business by remaining altruistic and leveraging its unique conceptual ability to adapt to and transform its environment. The findings suggest that entrepreneurs who expand their companies tend to show social orientation and are supported by a return to the fundamentals of management, while offering new interpretations of their products and services with cooperative relationships with others.

1 Background of study and awareness of the problem

Research supports the idea that entrepreneurship increases immediately after a disaster. Several empirical studies have been conducted on such phenomena. Just after the Great East Japan Earthquake in devastated areas many entrepreneurs were shown to have emerged [10]. The Black Saturday fires were the worst in Australia's history. In the immediate aftermath of the tragedy, many ventures were created [1]. Moreover, it has been argued that the rate of new businesses launched adjacent to New York City exceeded the rate prior to the September 11 attacks [2]. The reasons cited for the phenomena include economic factors, increased prosocial activities, and social networks of aspiring entrepreneurs [3]. Others argue that disasters destroy existing customs and networks and replace them with free resources to facilitate innovation under certain conditions [4][5].

* Corresponding author: michi@tohoku.ac.jp

A common theme across these studies is that entrepreneurs who emerge immediately after a disaster establish businesses and conduct activities under the unique circumstances of a post-disaster environment. Grounded in an investigation of five major disasters, the phrase "paradise built in hell" describe how cooperation and mutual aid flourish precisely in devastated areas [9]. Other scholars argue that new ventures have functions to alleviate victims' suffering after the disaster [1].

However, research investigates that, in countries with strong governance natural disasters promote entrepreneurial activity but only temporarily and this effect is often short lived [6]. As time passes and the post-disaster disruption and instability subside and the condition gradually stabilize, entrepreneurs must adapt their operations to match a normalized environment. Yet, most research has been limited to short-term outcomes, with few studies tracking post-disaster entrepreneurs over the long term (ten years or more).

This study aims to investigate how entrepreneurs and firms that emerge immediately after a disaster react to a gradually stabilizing environment. It also clarifies the characteristics of managers who continue to expand their businesses in the face of normalization and how they manage their operations.

To address these issues, we investigated the entrepreneurs who started their businesses after the 2011 Great East Japan Earthquake in Tohoku regions in, northern Japan. Fourteen years later their trajectories vary: some expanded their businesses, some remained small, others disappeared. Our primary research question is to examine the causes of their differences.

The first half of this study presents findings from multiple questionnaire surveys, clarifying (1) how entrepreneurs have shifted their motivation of starting businesses under evolving conditions and (2) what types of entrepreneurs are expanding their businesses and what their behavioral characteristics are. The second half provides a case study of an entrepreneur who is expanding his businesses.

2 Survey of new entrepreneurship

2.1 Overview

The first part of this study draws on a survey of new entrepreneurs in Sendai City conducted by the Sendai City Industrial Promotion Organization and the author. After the Great East Japan Earthquake, the Sendai City Industrial Promotion Organization established the ASHI ☆SUTA program to provide counseling to both new and aspiring entrepreneurs. The survey targeted individuals who started their businesses and received support from ASHI☆SUTA in each year. The survey was administered three times online (see Table 1): in 2017 (103 respondents), 2019 (133 respondents), and 2021 (81 respondents). Questions in the survey were as follows; sex and age of respondents, their previous careers, their motivation of starting businesses, the years and the places of their starting their businesses and reasons, the number of employees at starting business and present, their business performance comparing with those at starting businesses and present, fund-raising, challenges they face and the supports they wanted from the city. Additionally, follow-up surveys were conducted in 2019 (70 respondents) and 2021 (121 respondents) for new businesses that had responded earlier.

Table 1 . Numbers of respondents by category

	Survey in 2017	Survey in 2019	Survey in 2021	Total
Number of entrepreneurs participating in this survey for the first time	103	133	81	317

Number of entrepreneurs who participated in the previous survey		70 (who participate in 2017&2019)	121 (who participate in 2017&2021 or 2019&2021 or 2017&2019&2021)	191
Total	103	201	202	508

2.2 Changes in motivation

First, we examined the motivations of new entrepreneurs to start a business. The survey asked respondents to indicate all applicable motivations for starting a business (multiple responses were permitted). The horizontal axis presents the items, while the vertical axis shows the percentage of "yes" responses to the total number of respondents (Fig. 1).

Based on a survey of entrepreneurs who started businesses after the Great East Japan Earthquake [9], a clear tendency toward more altruistic motivations emerged. Consistent with the aforementioned, altruism was found to heighten after a disaster [8].

However, in this study, there were consecutive declines in the number of new entrepreneurs who cited the following reasons for starting a business: "because I want to contribute to earthquake recovery," "because I want to create a place of employment." Overall, the number of new entrepreneurs citing social motives such as earthquake recovery and job creation decreased. By contrast, an increasing number of respondents answered in favor of self-actualization motives such as "because I was interested in the job of business management" and "because I want to make use of my abilities and experience." These data show that the motivation of entrepreneurs with respect to establishing new companies has gradually changed.

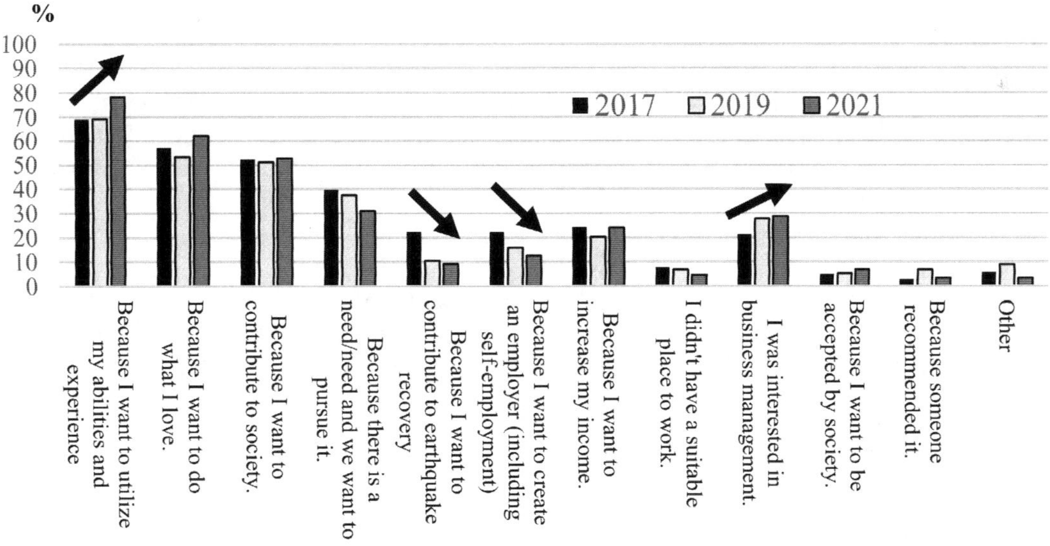

Fig. 1. Changes in the motivations of new entrepreneurs to start a business by year.

Next, for the 121 respondents who undertook the follow-up survey in 2019 or 2021 or both, we examined the relationship between entrepreneurial motivation at the time of opening and subsequent firm growth. As shown in Fig. 2, there was a difference in the content of the entrepreneurial motive between firms that "expanded" and those that responded, "about the same."

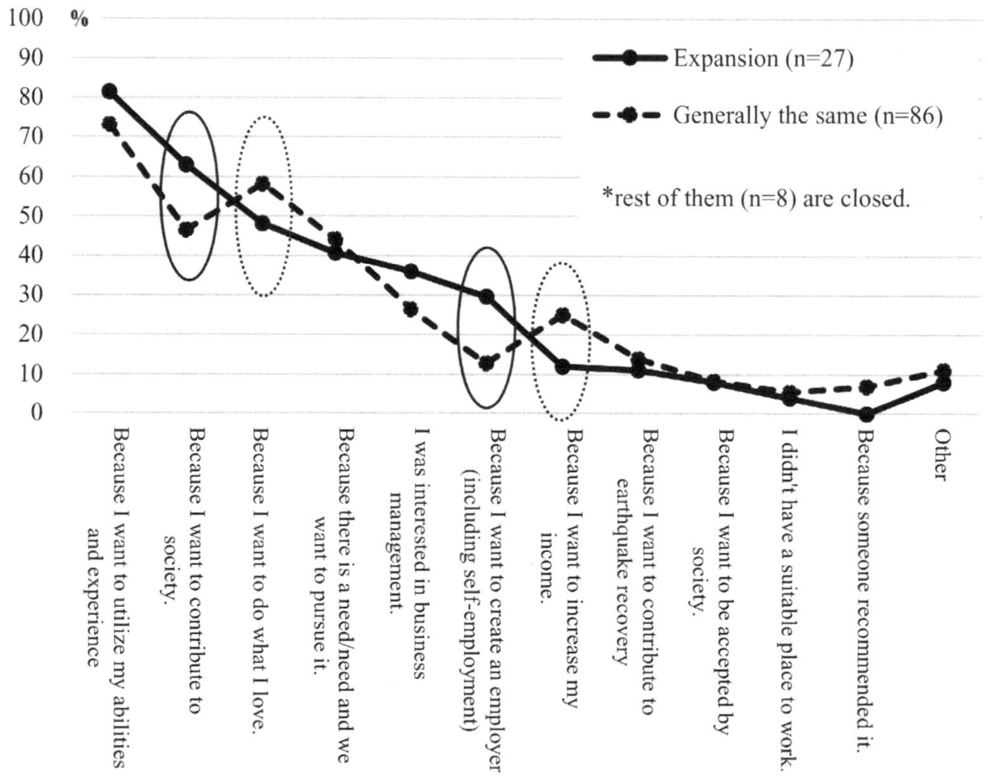

Fig. 2. Motivation of new entrepreneurs to start a business and growth of their firms.

The items that exhibited a discrepancy of more than 10% between the companies that experienced growth and those that remained the same were "Because I want to contribute to society." "Because I want to create a place of employment.", "Because I want to do what I love.", and "Because I want to increase my income." The "expanded companies were more likely to cite social contribution as their motivation, while the "generally similar" companies were more likely to cite self-interested motivations.

The above-stated results indicate not only that "contribution to earthquake recovery" as a motive for starting a business has been decreasing year by year, but also that entrepreneurial motives related to social contribution may be positively related to firm expansion.

3 Case Study: Akiu Winery

3.1 Methodology

In the next section, a case study was conducted on an entrepreneur that has started his businesses after the Great East Japan Earthquake and successfully expanded it. We selected Akiu Brewery Co. (officially Akiu Brewery, but commonly known as Akiu Winery) because of its relevance. This winery was established out of the founder's deep concern for the disaster-affected areas, and it has grown through the concept of "terroirge", collaborating with local farmers and wineries.

The primary data source was in-depth interviews with the founder of Akiu Winery, Chikafusa Mouri. He wasprimarily asked about his career, business activities, motivations of founding the winery, the strategy for the winery's growth, and his philosophy and prospects. Two of the interviews were conducted face-to-face and one was conducted online. The transcripts were analyzed using an inductive approach and Gioia methodology [7]. In addition to the interviews publicly available secondary data about the Akiu Winery was collected and used in the analysis.

3.2 Founding Akiu Winery

Akiu Brewery Co. was established in March 2014 by Mouri in Akiu, Sendai, and its core business is wine production and sales.

Mouri had no intention of becoming an entrepreneur. Prior to the earthquake, he worked for an architectural firm and oversaw the design of the hot spring facility Yupopo in Onagawa town, Miyagi Prefecture. After the earthquake, through his work of inspecting and designing the facility, he participated in local reconstruction meetings and made many acquaintances. He proposed the idea of reconstruction through wine tourism and marriage to the local government, suggesting to the municipality that the winery serve as the nucleus for establishing restaurants and bakeries, and for the gradual formation of a town. However, this idea was not adopted by any of the municipalities; therefore, Mouri decided to set up his own winery to realize the objective.

Subsequently, Mouri established a new network of contacts during the reconstruction process, including with general contractors. Mr. Mouri cleared the rubble in Yamamoto town and began experimenting with grape cultivation in preparation for establishment of a winery therein. However, he faced obstacles regarding commercialization. Grape cultivation was rather unsuccessful, and working for a design firm hampered his availability to spend time on the cultivation process and other activities. Furthermore, he did not obtain cooperation from the government or local community in Yamamoto town. Mouri explains the main reasons for this as follows:

> "They probably thought I was an outsider and that there was no way someone from a design firm could do what I was doing. Also, they probably couldn't support an outsider when they had to rebuild their existing businesses. I felt that the ladder had been removed." (Interview with Mouri, January 31, 2019[8])

The project ended in 2013 without the cooperation of the local government. However, things then began to take a completely different turn. Through a friend in Akiu town, Taihaku ward, Sendai, Mouri was able to rent an abandoned plot of land. In 2014, with the support of the local people of Akiu, he planted 1,200 saplings and began growing grapes in Akiu. Furthermore, the Mitsubishi Corporation Reconstruction Assistance Foundation supported the project by providing 50 million yen toward reconstruction assistance, while financial institutions provided financing to the amount of 90 million yen. The city of Sendai and Miyagi Prefecture also provided significant cooperation. Grapes had never been cultivated in Akiu before, and Mouri, an amateur, worked on the project while reading books by foremost wineries. He also received help from a university regarding land conditions and grape cultivation. Mouri was unable to take on the important task of vinification because he rarely drank alcohol. A former winemaking student from Italy responded to Mr. Mori's request on social networks and agreed to help him address this problem.

With the help of this new connection, the construction of a winery in Akiu was well underway, and the harvesting of in-house gown grapes began in 2016; in 2022, the company's wine won a gold medal in the "Sakura Awards 2021."

3.3 Incubation role

A major characteristic of Akiu Winery is that it does not only produces its own products but also serves as an incubator for various wineries. Since the earthquake, the number of people entering the wine business has increased in the Tohoku prefectures, especially in the disaster hit areas of Miyagi and Iwate. Various wineries have sprung up, driven by the growing popularity of "Japanese wine" worldwide and as a part of broader reconstruction efforts. Some of these wineries may be able to grow grapes with labor but unable to conduct in-house winemaking which requires capital investment. Akiu Winery serves as a hub among such wineries, providing not only employee training in winemaking and other related aspects, but also product development support, such as test brewing. The winery holds occasional workshops on the production process, and participants come from within the prefecture as well as from Yamagata, Iwate, and Fukushima.

In addition, Akiu Winery helps start-up wineries get their businesses off the ground. Not only by conducting test vintages but also by commissioning vintages from wineries that have successfully cultivated grapes ahead of others [8]. This kind of project typifies more as a form of support for new wine businesses in Tohoku rather than a commercial venture.

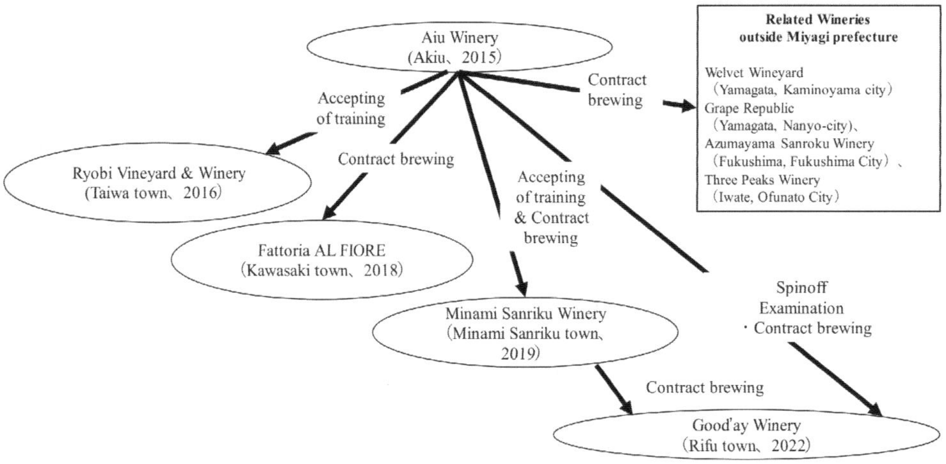

Fig. 3. Newly created wineries supported by Akiu Winery [8]

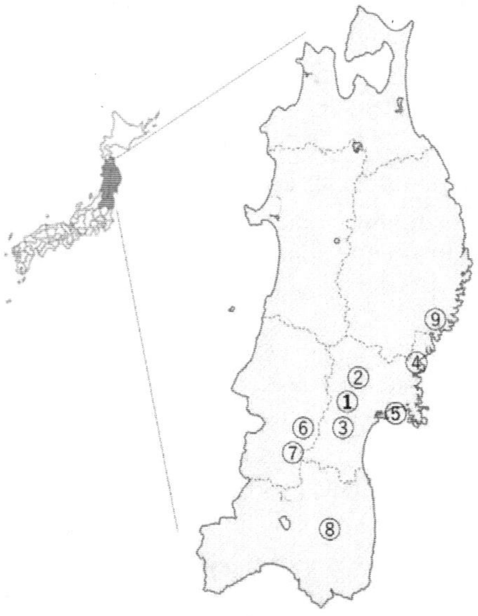

Fig. 4. Locations of the wineries supported by Akiu Winery [8]

3.4 Community Cooperation

Mouri, who had no ties to Akiu, was helped by various people. Specifically, when Mouri decided to start his business, the first person to support him was a well-liked local neighborhood association president. This person was extremely helpful in advancing the business and educating Mouri about the current situation in Akiu and helping him with human relations. Mouri describes the "receptive capacity" of the community as follows:

> "After the earthquake, Akiu became a frontline base for people from all over the country, including various police departments and water authorities. I heard that the people of Akiu really wanted to support the coastal areas. So, Akiu has that kind of constitution to accept people from outside like that." (Interview with Chikafusa Mouri, January 31, 2019,[8]).

This support from the local community has strongly linked Mouri's business to the "promotion of the region." From there, the idea was born to not only create a winery that would connect with others, but also use wine to attract people to Akiu and serve the food made in Miyagi Prefecture. The initial idea of "supporting producers through the marriage of wine and food" developed into the coined term "terroirge,". The term is a portmanteau of the words "terroir" and "marriage" that refers to the idea of enjoying wine while enjoying Tohoku food, tourist attractions, and sports.

This emphasis on community relations evolved into a new company that promoted Akiu as a whole and developed the town, while focusing on "wine tourism." In 2017, Mouri founded Akiu Tourism Factory, Inc. with two entrepreneurs to promote the region. One of them is Daiki Chiba, a serial entrepreneur living in Sendai with multiple entrepreneurial experiences in the food industry, and the other is Kanzaburo Sato, the owner of Sakan, one of the largest local traditional hotels in Akiu. Akiu Tourism Factory runs Akiu Sha, an old private house restaurant that serves local food and organizes various food events. Currently, it is beginning to become a mainstay of facilities that attract tourists who enjoy "walking around town" outside of the hot springs. In this way, Mouri's business aligns with the broader regional strategy of the Akiu region. When "the linkage of food and tourism with the region creates a story," being able to visit a region like Akiu, enjoy its atmosphere, and drink local wine becomes a key factor in delivering a satisfying experience.

Furthermore, as Akiu is increasingly featured in the press, including as a winery, the number of young people moving to Akiu is increasing. Chiba, the head of Akiu Tourism Factory, says, "Designers, photographers, creators, and others have started working in Akiu, and an increasing number of people who admire them are moving to Akiu."

The abovementioned efforts, in conjunction with the establishment of a microbrewery for local beer in the summer of 2023, will further contribute to the branding of the region.

4 Conclusion and limitations

In the first half of this paper, a hypothesis was presented regarding the relationship between sociability and the expansion of a company's size. This hypothesis was confirmed by the case study presented in the second half.

Firstly, through survey research, the findings suggest that entrepreneurs motivated by social contribution are more likely to expand their businesses. Subsequently, focus was directed toward Akiu Winery, a representative company that was established with altruistic motives immediately after the Great East Japan Earthquake and has undergone significant growth since its foundation. The company has maintained its original altruistic nature to this day. With its unique concept of "terroirge" at its core, the company attracted stakeholders, garnered community support and empathy, and established a system of coexistence and co-prosperity for the community. In other words, the company grew by remaining altruistic and leveraging its unique conceptual ability to adapt to and transform its environment.

Mouri's goal was to create a wine cluster, while trying to create jobs and restore and revitalize the local industry. This strong social orientation is at the core of his businesses, which are still going strong 10 years later.

Products and services were introduced in the aftermath of the earthquake. However, Mouri did not stick to the context of the earthquake disaster but added new interpretations to the products and services, and brought them to the market by expressing their value in a new dimension.

Furthermore, Mouri has the same "attitude and level-headedness of honestly pursuing the fundamentals of management." As a manager, Mouri had been overseeing numerous activities daily. His intention was to "just do what I have to do as a manager, no matter what the earthquake or what the situation is."

Entrepreneurs who emerge after a disaster will be affected by the incident immediately after starting their businesses; however, as time passes, they will be required to respond flexibly and adapt to the changing environment. At this time, the social aspects that were the initial motivations for starting a business and the attitude of not losing sight of the fundamentals of management may play a vital role in their adaptation.

This study has certain limitations that need to be taken into consideration when analyzing the findings. Regarding the questionnaire survey, not all entrepreneurs aim to expand in the first place, and they may have different interpretations of the social nature of their businesses. For the case study sample bias is also notable. While keeping these issues in mind, we would like to take this survey as a first step toward deepening our knowledge about the phenomena in the future.

Acknowledgements. This research was supported by the Japan Academic Society of Ventures and Enterprises. We would also like to thank Chikafusa Mouri (CEO, Akiu Winery) and ASHI☆SUTA, Sendai City Industrial Promotion Organization, for their cooperation in conducting the survey. We would like to express our gratitude to them.

References

1. D. A. Shepherd & T. A. Williams, Local venturing as compassion organizing in the aftermath of a natural disaster: The role of localness and community in reducing suffering. J. of Manag. Stud. **51**(6),952-994 (2014). https://doi.org/10.1111/joms.12084
2. S. Paruchuri, P. Ingram, Appetite for destruction: The impact of the September 11 attacks on business founding. Ind. Corp. Change. **21**(1), 127–149 (2012). https://doi.org/10.1093/icc/dtr075
3. S. Shinada, Entrepreneurial activity after disasters: Why do catastrophic events cause an increase in entrepreneurship? Venture Review. **22,** 43–57 (2013). https://doi.org/10.24717/jasve.22.0_43
4. M. Skidmore & Toya, H, Do natural disasters promote long-run growth?. Economic Inquiry. **40**(4), 664-687(2002).
5. Y. Sawada, R.Bhattacharyay & T. Kotera. Aggregate impacts of natural and manmade disasters: A quantitative comparison. The Research Institute for Economy, Trade, and Industry, Discussion Paper Series 11-E, 23. (2011).
6. C. J. Boudreaux, A. Jha. & M. Escaleras. Natural disasters, entrepreneurship activity, and the moderating role of country governance. Small Bus. Econ. **60**, 1483–1508 (2023). https://doi.org/10.1007/s11187-022-00657-y
7. D. A. Gioia, K.G. Corley & A. L. Hamilton. Seeking qualitative rigor in inductive research: Notes on the Gioia methodology. Organ.Res.Methods, **16**(1), 15-31. (2013). https://doi.org/10.1177/1094428112452151
8. M. Fukushima & S. Shinada. The Growth and Strategy of Companies Born After the Great East Japan Earthquake: Akiu Brewery's Altruism and Visionary Concept of "Terroir. Shoko Kinyu,**79**(4),5-19. (2024).
9. R. Solnit, *A paradise built in hell: The extraordinary communities that arise in disaster* (Viking Press, New York, 2009).
10. M. Fukushima, Entrepreneurships after a disaster, in ISPIM Conference Proceedings, International Society for Professional Innovation Management (ISPIM), Melbourne, Australia, December 10-13 (2017),pp.1-9.

Resilience in Action: A Study of Disaster Volunteers Using Play Therapy in the 2018 Lombok and Palu Earthquakes

Alice Arianto[1,*], *Lily* Salim[1], and *Amabelle* Vanalyssa[1]

[1]Department of Applied Counselling, Politeknik Bentara Citra Bangsa, 10130 Jakarta, Indonesia

Abstract. This paper explores the transformative experience of disaster volunteers with the involvement of professionals who applied play and creative art therapy during the post-earthquake recovery efforts in Lombok and Palu, Indonesia in 2018. Drawing on participatory action research, the study involved 24 volunteers over six months, uncovering a mixed method of quantitative and qualitative insights into disaster volunteers' motivation, resilience, and the therapeutic impact of creative interventions in disaster settings. This participatory action research involved twenty four disaster volunteers within the span of 24 weeks. The study utilises narrative inquiry, questionnaire, observations, and the resilience scale. Fostering a culture of volunteerism, developing strategies for resilience to further navigate through disasters, this study suggests a practical implementation to help children, families and communities experiencing disasters in Indonesia or globally.

1. Introduction

Lombok and Palu are situated within the Pacific Ring of Fire, widely known for its geologically active areas. Lombok Island is located near the border between the Sunda Arc and the Banda Arc tectonic systems that was likely affected by fluid suspected from Mount Rinjani magmatic activity or Flores Oceanic Crust [1]. In 2018, earthquakes in Lombok Island struck up to M_W 6.9, that triggered landslides and several structural damages that significantly impacted the lives of children and families. While thousands were injured, more than 509,000 people have been affected by the earthquakes. Central Sulawesi was struck by an earthquake of M_W 7.4, followed by 4-7 meters tsunami, coastal collapse and liquefaction in Palu city [12, 13]. The disasters resulted in 400,000 people affected in Lombok, and among 140,000 were children who faced psychological distress and disrupted education. 1.5 to 1.9 million people in Palu were affected, with over 500,000 children lost access to safe shelter and education. Reports from UNICEF, IFRC of earthquakes and tsunamis have shown that these intertwined geophysical and humanitarian impacts underscore the acute vulnerability of children in disaster zones and highlight an urgent need for child-centred psychosocial support. Some disaster volunteers learn to help children cope with their mental health and recovery after disasters. Play therapy is an intervention that can help restore the psychological

* Corresponding author: alice.arianto@bentaracampus.ac.id

well-being of children affected by disasters. The child-centred play therapy approach may be modified to provide a safe space according to the needs [5]. Disaster volunteers play a significant role, engaging closely with social communities, health agencies, government, and individuals according to their respective expertise. This research highlights the role of professional play and creative art therapists and their involvement within the disaster volunteer groups. The volunteers contributed based on their willingness and community spirit, exemplified by the culture of mutual collaboration within society, also known as '*gotong royong*' in the Indonesian language. Although theories affirm the need for child-centred, play-based approaches for disaster recovery, the gap persists in the application of such interventions and in providing research of post-intervention strategies. This study emphasises the importance of resilience for disaster volunteers and extends findings to address the need to equip individuals in mental health and well-being, and identify strategies explicitly tailored to their context. This study examines the experience of disaster volunteers in conducting play therapy and creative art as part of a recovery program in Lombok and Palu in 2018. This participatory action research (PAR) involves 24 disaster volunteers across five groups in Lombok and Palu, spanning approximately six months of field volunteering, or 24 weeks. These are collaborative efforts between HOPE Worldwide Indonesia, a non-governmental organisation (NGO), and Play Therapy Indonesia (Association of Indonesian Play Creative Arts Therapists).

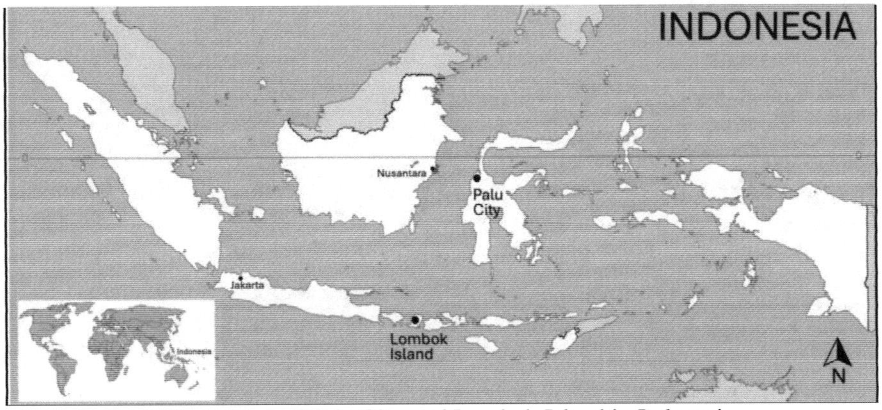

Fig 1. Map of Palu City and Lombok Island in Indonesia

1.1. Lombok Earthquake and Palu Earthquake, Tsunami, Liquefaction 2018

The 2018 Lombok earthquakes were a series of powerful seismic events that struck Lombok Island, Indonesia, between July and August 2018, causing widespread destruction, fatalities, and displacement, severely impacting children and leading to trauma and long-term psychosocial challenges. The disasters may cause a mild to severe reaction in the form of anxiety, stress and other mental health problems [10]. The 2018 Palu earthquake, tsunami, and liquefaction disaster in Indonesia left profound psychological scars on children and families, creating a complex mental health crisis that required extensive intervention such as psychosocial support [14]. Similarly, Lombok and Palu disasters underscored how catastrophic events can simultaneously shatter lives but strengthen human resilience. Resilience within disaster risk reduction is focused in the psychological and social aspects and described as the ability to adapt [3]. While the psychological wounds ran deep, the responses amidst unimaginable loss, recovery and growth remain a possibility when appropriate support and interventions are available. While Indonesia continues to grapple with high disaster risks, the mental health conditions that occurred from Lombok and Palu served both as warnings and hope for protecting vulnerable developing minds in future crises.

1.2. Disaster Volunteers

Volunteerism is often defined in the context of emergencies and disasters with an emphasis on volunteer activities and outcomes rather than on individual motivations, underlying factors and characteristics of the volunteers [8]. Disaster volunteers are individuals or groups who provide immediate and long-term assistance to communities affected by disasters (e.g., floods, earthquakes, tsunamis, liquefaction). They work with non-profit, non-governmental organisations (NGO), government agencies, or grassroots organisations to support survivors, particularly vulnerable groups like children and families, by offering medical aid, psychological support, shelter, food, and education. Disaster volunteers play a crucial role in helping children and families recover from disasters. Their work is physically and emotionally taxing, but their resilience, teamwork, and compassion save lives. The volunteers create safe spaces, consistency and safety for children to play and recover from their trauma. By improving support systems for disaster volunteers and strengthening disaster preparedness, we can ensure better protection in future crises. Twenty-four disaster volunteers were involved in assisting children and families in Lombok and Palu.

1.3. Play and Creative Art Approach

Disasters can leave deep emotional scars, especially on children who experience traumatic events during their young lives, lose their home, possessions and struggle to convey their thoughts [2]. Play and creative art therapy are powerful, trauma-informed approaches that help both child survivors and volunteers to process grief, fear and stress which may lead to healing in a non-verbal process. Through many theories of John Bowlby, Virginia Axline, Jean Piaget and Vygotsky, along with other theorists, we conclude that children often have difficulty expressing traumatic events using words, but play and creative art approach allow safe emotional release, expression and restore a sense of normalcy, control, and joy amidst chaos. Moreover, the neurobiological effects of trauma within the brain and body will affect their emotional regulation, stress responses, coping mechanisms, well-being and overall development. The use of play, art, music and movement to help children process, express their trauma and experiences within a safe and supportive environment are emphasised in this research [2]. This play and creative art approach as well as other skills, have provided disaster volunteers with competence to facilitate interventions, develop coping strategies, regulate emotions, adapt in difficult or unfamiliar settings and promote resilience. Furthermore, playful and creative environment allows children and families to communicate, express using non-verbal, also a universal language to help reduce traumatic stress. This helps them show through games or drawings, rather than tell their feelings and memories using verbal languages [10]. For volunteers, using play and creative art helps them decompress after witnessing disasters and strengthen their team bonding through shared activities. Play and creative arts therapy can turn pain into expression, fear into courage, confusion and loss into connection. For volunteers, these methods are not just tools, they serve as lifelines, helping them heal, while in turn, offer help to others. The play and creative art therapists as disaster volunteers prepared seven tools for the Lombok and Palu resiliency programme: art, clay, therapeutic story, movement, music, creative visualisation, puppet, and sand. They empower children to rebuild their resilience by expressing, regulating, processing their trauma in a developmentally appropriate manner, using self-expression and creative approaches.

1.4. Resilience

Resilience refers to the ability to adapt, withstand, and recover from adversity, trauma, or prolonged stress. In disaster volunteering, resilience includes: a) psychological endurance that involves managing stress, grief, secondary trauma; b) physical stamina for coping with harsh environments, tasks and fatigue; c) social adaptability including working effectively in chaotic, fast-paced environments, diverse, multicultural teams; d) post-traumatic growth such as emerging stronger after crises and self-regulation during field work. Resilience is dynamic; volunteers must adapt to changing disaster conditions. The psychological state of volunteers has a significant impact on children in disaster recovery settings, particularly when utilising play and creative art therapy and facilitating the resiliency program, commonly recognised as psychosocial support. The emotional contagion through the mirroring effect and atonement affects the children subconsciously. This reflects the ability for play therapists to recognise the mind-body connection and underlying conditions of trauma survivors and their role as part of the disaster volunteers [2]. A grounded resilience promotes a safe and healthy social state. Having self-awareness and regulation, clear boundaries and clarity in role delegations, proper training and supervision, consistent, predictable routines are helpful in the process to lead to positive change. Studies show that disaster volunteers often report greater life meaning after a mission. Results show a significant influence on their well-being, consideration of social relationships, daily life and their values [15]. Among the disaster volunteers, professional play and creative art therapists, working collaboratively in teams with other volunteers from NGO and local volunteers over the span of 24 weeks.

2. Methods and Tools

This action research explores the motivations that propel disaster volunteers into action, particularly involving a strong sense of duty, shared cultural values, and impact of mental health factors which drive their willingness. Furthermore, it underscores the importance of investigating the resilience, mental health and well-being of the volunteers. The merger between play and creative art therapists as informal volunteers, who were assisted by trained disaster volunteers allowed sufficient coordination on the field [11]. The coordination and structure of the team consisted of a combination of professional play and creative art therapists with trained disaster volunteers. The therapists served as facilitators of psychosocial interventions, while trained volunteers managed logistics, safety and field coordination, and local volunteers partake as community liaisons to ensure cultural relevance, etiquette and trust. Each field unit typically included one lead therapist, two to three trained disaster volunteers, and one local volunteer. Assigned roles were given in daily briefings and coordination of tasks that included adaptation of field conditions. Disaster volunteers who are professional therapists focused on creating and delivering interventions, while other volunteers support operations, documentation and strengthen the community engagement through shared values such as *gotong royong*. This layered structure enhanced clarity of roles, efficiency and fostered collaboration, practiced skills and sustainability, enabling this model for future PAR and disaster interventions. The participatory action research (PAR) utilises narrative inquiry, questionnaire, observations, and the resilience scale, which has been adapted for this study to provide translations in English and Indonesian language [7]. This study allows disaster volunteers to identify their strengths, skills, and resources. The processes of this study enabled disaster volunteers to equip children and families with coping strategies that would provide a longer-term ability for them to prepare for life challenges.

The research data is collected from 24 participants who volunteered in the Lombok and Palu. The themes reveal a range of motivations which are thematically grouped. The themes of volunteers' motivations draw on mostly internal reasonings: compassion (n = 8), personal growth and healing (n = 4), moral duty and faith (n = 3), organisational duty (n = 2), skill-based contributions (n = 4), willingness to learn and gain field understanding (n = 3). Table 1 reveals the frequency of disaster volunteer responses that range from 1 week to 2 years, showing short-term and long-term commitments. The smaller number of group volunteers (n = 6) contributed between 1 to 2 weeks due to professional and work-related commitments. The ranging numbers of 1 to 6 months of commitment involve play and creative art therapists as well as NGO volunteers. The long-term commitment of 24 months (n = 2) involves local volunteers and NGO volunteers. Other factors that affect the duration of volunteers in the field may vary as disaster volunteers within the smaller number of months are mostly self-funding, willing to contribute or they may receive limited support of fundings. These factors of duration are not the main focus of the study. Although it contributes to the interest in researching motivations and decision-making when they are called to help people in need almost immediately within days to weeks after the disasters in Lombok and Palu occurred in 2018.

Table 1. Standardised Duration and Frequency of Responses

Standardised Duration (Months)	**Frequency of Disaster Volunteer Responses**
0.25	2
0.33	2
0.5	2
1	4
2	4
3	2
4	1
6	4
10	1
24	2

This study combines mixed-methods participatory action research (PAR), where volunteers acted as co-researchers, contributing to insights and meaning-making through narrative inquiry and through personal stories and reflection. Narrative inquiry is a qualitative method that involves the study of experiences and allows participants to be storytellers. Narrative inquiry explores a particular phenomenon from the point of view of participants, in which the researcher form stories in a systematic structure [6]. This method integrates qualitative and quantitative data for insights into experiences, emotional states, thoughts and resilience that focused on four key aspects: a. Volunteer Experience and Motivation: What motivated you to volunteer? What were the most rewarding and most challenging aspects of your volunteering experience?; b. Emotional and Psychological Impact: How has this volunteering experience affected you mentally and emotionally? Did you experience stress, anxiety, or emotional exhaustion? If so, how did you cope? Did you feel a sense of fulfilment or personal growth from helping others?; c. Resilience and Coping Mechanisms: What personal strengths helped you during this experience? Did volunteering change your perspective on resilience in difficult situations? What support systems (family, community, organisation) helped you during or after volunteering?; and d. Future Suggestions: Would you volunteer again for disasters or similar programmes? What kinds of training or support would help volunteers manage stress better? How can communities better prepare volunteers for future disasters? Validated instruments such as the Connor-Davidson Resilience Scale (CD-RISC-10), Brief Resilience Scale (BRS), and WHO-5 Well-being Index, alongside the open-ended narrative inquiry and reflections were integrated to assess resilience among a sample of 24 disaster volunteers. CD-RISC-10 provides a focused assessment that only takes 5 minutes or less, in consideration of disaster volunteers and professional therapists who experience time-consuming, high-stress, overlapping responsibilities in various environments. The CD-RISC-25 has been an improvement and known to be well-established in structure, however this study uses CD-RISC-10 for a reliable measurement that is globally fit, validated within numerous studies, an instrument with clear language that helps volunteers understand the questions, as some disaster volunteers are non-English speakers. The study integrates various methodological approaches to collect empirical evidence through narrative inquiry methods, combined with tools to obtain quantitative and qualitative insights for this research. To highlight the PAR method, the questionnaire and narrative inquiries are gathered post-earthquake in recovery efforts after disasters in Lombok and Palu, mostly collected from our actively engaged disaster volunteers. To strengthen the mixed-methods design, quantitative results were integrated with qualitative narratives, to show consistent patterns across tools. CD-RISC-10 measures adaptability, perseverance and recovery as shown in Figure 2. The Brief Resilience Scale (BRS) assesses the ability to bounce back, or recover from stress, while WHO-5 Well-Being Index evaluates the emotional well-being of disaster volunteers post-volunteering. The CD-RISC-10 revealed scores ranging from 23 to 49 ($M = 41.13$, $SD = 6.45$) showing most volunteers clustered around high resilience levels. The BRS showed a supportive result, with mean of 3.12 ($SD = 0.49$) indicating normal to high resilience. The WHO-5 Well-Being Index showed an average of 16.87 ($SD = 2.47$) that reflects generally positive emotional health despite trauma exposure. The scores indicate that higher resilience is directly associated with one's ability to regulate for emotional well-being. Significant correlation ($r = 0.515$, $p = 0.009$) demonstrated higher resilience that linked to stronger belief of the effectiveness of play and creative art interventions. The results align with volunteers' narratives of growth, recovery, perseverance, values of unity such as *gotong royong* and conclusively highlights resilience as a measurable trait and ongoing process fostered by mutual support along with development of professional skills. Disaster volunteers described strategies of facing challenges, including the adaptation for long hours of preparation and injury during field work, collaboration, cooperation and professional support with an aligned purpose. Many disaster volunteers shared through qualitative data how they developed into resilient individuals, growing in engagement within community services or child and family related work as they felt the long-term effect that carried forward.

3. Findings

The data revealed transformative journeys through their experiences and service. One disaster volunteer wrote after reflecting on the experience, "I came to help others, but I ended up healing myself." This response reflects how engagement in volunteering service activated a reflective process that catalysed identity growth and emotional insight. Stories are containers of both trauma and recovery. Volunteers shared profound self-awareness and emotional integration during post-disaster interventions, indicating that narrative itself becomes a mechanism of healing. They described emotions ranging from grief to hope, exhaustion to resilience. One disaster volunteer shared, "Some nights, I cried alone—but reminded myself that if children can survive, so can I. They brought courage and comfort for me to move on with my life in times of trouble." Emotional intensity was not viewed as weakness but as a path to find meaning. The volunteers became emotionally present, aware of pain and grief in others, transforming their suffering into resilience. Through this they shifted their thoughts, emotional growth and their purpose fostered by their disaster volunteer experience [15]. Through thematic analysis, searching for themes across codes and finding context from data collection [9]. The researchers summarised themes of resilience and compassion within the mental health and well-being categories, such as "purpose fuels perseverance" and "the community heals the helper". Indonesian cultural values commonly known as '*gotong royong*' or cooperation. Results show the term was cited several times by volunteers. It was observed that mothers in the neighbourhoods supported and encouraged one another in bearing the burden of loss. Connections built on inclusiveness, support, being listened to and cared for provided comfort. The volunteers were immersed in a unique routine such as morning songs, group debriefs and adopting a shared habit. One volunteer described the team as "a family bound by compassion". The saying goes, "Ringan sama dijinjing, berat sama dipikul", which translates to "light are carried together, heavy are shouldered together" or in simple terms, a problem shared is a problem halved. The saying emphasis on mutual support and shared responsibility within relationships, as seen in the values of togetherness in Indonesia. Another volunteer shared that an earthquake happened in the midst of an intervention. They noticed the children were calm and enjoyed playing during that moment, they seemed less anxious or scared and the children along with their families learned to accept the disasters that occurred.

In other words, in the wake of a disaster, the loss was not carried alone. Whether the devastation was great or small, it was collectively borne. The shared weight of sorrow became bearable when held in the hands of many. This mutual understanding shows the path through suffering in a positive light. The implementation of play and art therapy approach offered a glimpse into the therapeutic process. It became an innovative solution during a crisis in both locations in Lombok and Palu. Human beings, as inherently social creatures, are in constant need of one another. When we create connections, the oxytocin hormone is released and mend our mind and body as pain-relief among other benefits. Neuroscience shows that humans are wired from birth to connect, to care for others and provide a healthy environment that shows healing power [4]. The process happens with a presence, a listening ear, nonjudgmental heart, shared laughter, freedom of play, freedom to express thoughts and emotions through creative interventions, whether individually or collaboratively. These seemingly simple gestures affirm a profound point that healing often arises from connection between human beings in mutual support. Despite the different settings in Lombok and Palu, both cases displayed a shared outcome. The data points towards resilience as a collaborative, shared experience to build empathy. The study has given volunteers the opportunity to reflect and walk into their personal journey, enabling transformation for the disaster volunteers themselves.

3.1. The Connor-Davidson Resilience Scale (CD-RISC-10) and Scores

CD-RISC-10 is a widely accepted instrument for assessing an individual's capacity to cope with adversity and build adaptability. In this study, the CD-RISC-10 was administered to disaster volunteers who engaged in play and creative art programs for intervention during the 2018 Lombok and Palu earthquakes in Indonesia. The quantitative data provides a compelling narrative of inner strength and adaptive functioning. As illustrated in Figure 2, the distribution of total CD-RISC-10 scores among 24 volunteers ranges from 23 to 49. The mean resilience score shows 41.13 (n = 24, SD = 6.45), and the median score is 42.00, indicating moderate variability within the sample. The distribution of scores is positively skewed toward the upper end of the scale (maximum score = 50), suggesting most participants demonstrated a high level of resilience.

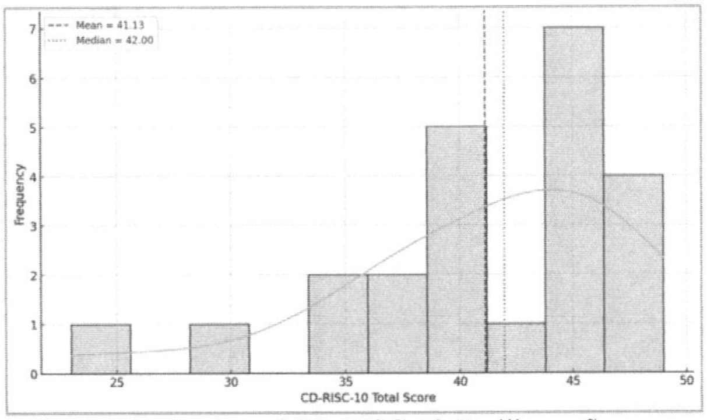

Fig 2. Distribution of CD-RISC-10 Resilience Scores

Table 2. CD-RISC-10 Z-Scores

CD-RISC-10 Total Score	z-Score
48	1,08818110961016
36	-0,812692221101258
45	0,612962776932305
45	0,612962776932305
40	-0,179067777530785
46	0,771368878249230
41	-0,0206616666381675
45	0,612962776932305
49	1,24658722050277
35	-0,971098331993877
41	-0,0206616666381675
20	-1,92153499734958
23	-2,87197166270529
30	-0,337473888423404
44	0,454556666039687
47	0,929774998717541
35	-0,971098331993877
37	-0,65428611020864
46	0,771368887824923
40	-0,179067777530785
42	0,13774444425445
49	1,24658722050277
44	0,454556666039687

This visual distribution corroborates with the volunteers' self-reported experiences during fieldwork, as evidenced by narratives that reflect grit, emotional awareness, and the ability to adapt as well as rebound from emotional stress. The qualitative reflections are compatible and synchronized with the elevated CD-RISC-10 scores, reinforcing the strength of the volunteers not merely as helpers but as co-survivors in the shared landscape of post-disaster healing. Furthermore, Z-scores were calculated to determine each individual's deviation from the group mean. Most volunteers are within ±1 SD of the mean, with a few high outliers indicating exceptional personal resilience. This statistical sequence not only validates the consistency of resilience across the group but also suggests that the experience of disaster volunteering itself may serve as a resilience-enhancing practice. In conclusion, this data underscores that resilience, as measured through the CD-RISC-10, is a dynamic construct shaped not only by internal traits but also by external contexts and community bonding. When volunteers are both emotionally and mentally prepared and supported, they become more than volunteers in the field, but they have proven to be vessels of hope and resilience. Results of Z-scores calculated for CD-RISC-10 are shown in Table 2.

3.2. Brief Resilience Scale (BRS) and Score Analysis

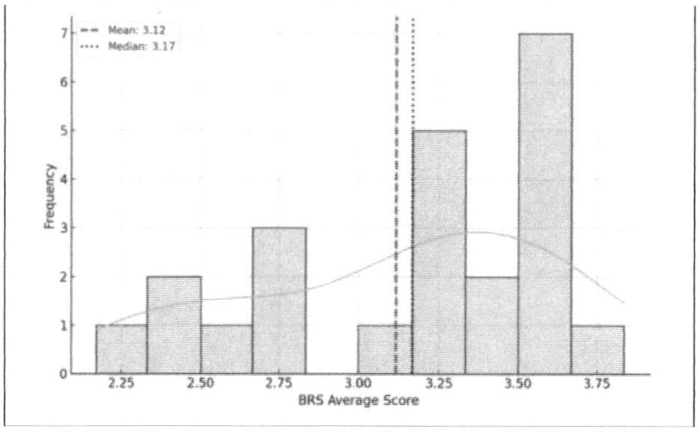

Fig 3. Distribution of Brief Resilience Scale (BRS) Scores

The results for Brief Resilience Scale (BRS) are presented in Figure 3. It reveals a modest right-skew in the distribution of scores, indicating generally high resilience within the sample. The mean score for the sample is 3.12 (SD = 0.49), while the median score shows 3.17. The small standard deviation indicates low variability, closely clustered in the mean, showing a concentrated distribution of responses. This proximity between the mean and median signifies a relatively symmetrical and well-balanced pattern, implying that resilience is a shared and evenly distributed trait among the group. To better understand the individual differences within this data set, a Z-score analysis was conducted and presented in Table 3. Most volunteers scored within ±1 standard deviation of the mean, emphasising a cohesive group profile with minimal deviation. This pattern highlights the volunteers' strengths in emotional resilience and well-being, likely cultivated through their shared experience, professional training, awareness and coping mechanisms.

Table 3. Brief Resilience Scale (BRS) Z-Scores

BRS Average Score	Z-Score
12.00	0,804138695970585
3,33	0,455172846775803
3,166	0,10620699758102
3,66	1,15310454516536
3,5	0,804138695970585
3,5	0,804138695970585
3,166	0,10620699758102
3,5	0,804138695970585
2,5	-1,2896563991981
2,33	-1,63862224839289
3,33	0,455172846775803
3	-0,242758851613761
2,66	-0,940690550003327
3,66	1,15310454516536
2,66	-0,940690550003327
3,83	1,50207039436015
2,66	-0,940690550003327
3,16	0,10620699758102
2,3	-1,63862224839289
3,16	0,10620699758102
3,16	0,10620699758102
2,16	-1,98758809758767
3,66	1,15310454516536

The finding that only a small number of participants presented as statistical outliers, either significantly higher or lower than the mean, underscores an important insight: the volunteers' resilience is not merely a function of personality traits but a dynamic quality shaped by community, preparation, and purpose. These results reinforce the idea that when individuals engage in meaningful, group-based, and goal-driven creative interventions such as play therapy and creative art approach, they are more likely to develop or sustain psychological resilience even in challenging environments. In humanitarian work, where emotional endurance is as essential as technical skill, this evidence confirms that resilience can be fostered, supported, and shared. The volunteers' relatively high and stable resilience scores provide scientific backing to their capacity for sustained emotional engagement, reinforcing the importance of integrating reflective and creative methods within a field of study into the volunteer training and support programmes for psychosocial support.

3.3. WHO-5 (Well-being Index) and Scores

In most post-disaster interventions where the emotional state is frequently scarred with grief, uncertainty, shock and exhaustion, the psychological well-being of volunteers is in critical conditions, yet they are often overlooked. The WHO-5 Well-Being Index was developed by the World Health Organization (WHO), provides a snapshot of an individual's emotional state. As shown in Figure 4, the distribution of WHO-5 scores collected from 24 disaster relief volunteers actively involved in psychosocial interventions using play therapy and the arts. The scores range between 12 and 20 with a maximum score of 25 ($M = 16.87$, $SD = 2.47$) and a median of 17.00. The calculated standard deviation indicates moderate variability within the group. With a clinical cut-off of 13, the volunteers' average WHO-5 scores remained well above it, implying good emotional well-being within the group of disaster volunteers. This is a threshold commonly used to identify risk of depression or emotional distress. The benchmark is internationally recognised, however direct application to Indonesian adult volunteers may vary and differ with cultural adaptation, community context, background, values along with other factors that may influence the cut-off for volunteers in Indonesia. However, the consistently elevated scores did not show signs of distress. The disaster volunteers were functioning adaptively, reinforcing usefulness and efficacy as the indicator of emotional health in this context. Despite their proximity to trauma-affected populations, the volunteers' scores remain well above the clinical cut-off of 13, which shows the threshold typically used to screen for the risk of depression or emotional distress. This suggests that participants maintained a generally positive affective baseline, even amidst potentially critical and unexpected environments. Moreover, it has been reported by volunteers who co-research through the PAR method, that some earthquakes were experienced while they actively volunteered on the field.

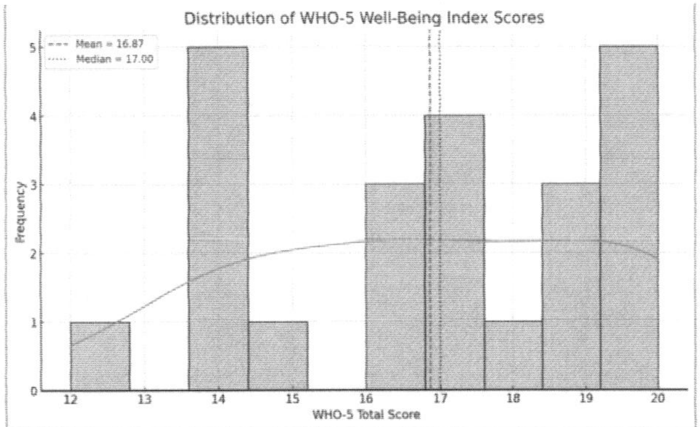

Fig 4. Distribution of WHO-5 Well-being Index Scores

The near alignment of mean and median values points to a symmetrical and emotionally consistent profile across the group. This pattern reflects not only the absence of distress but also the presence of effective coping mechanisms. Disaster volunteers appeared to function with emotional clarity and stability. These findings align with the previous data gathered, reinforcing the volunteers' well-being as they contributed to the recovery of others. The average WHO-5 score of 16.87 (SD = 2.47) indicates that volunteers maintained good well-being above the clinical cut-off. One volunteer reflected, "Seeing the children smile again made my own heart lighter." Such statements capture the nature of compassion, inspiration, shared by the children led to joy and emotional health. This aligns with the quantitative results, showing that well-being was both measurable and reflected in moments of recovery and built connections. The WHO-5 results provide quantitative validation for these qualitative sentiments, anchoring the narratives in measurable psychological outcomes. Notably, no extreme outliers were recorded in the Z-score analysis, which confirms the relative cohesion and psychological parity within the group. This suggests a well-supported team structure that likely fostered emotional regulation, containment, connection and interpersonal encouragement. The practical implications of this analysis are clear: the disaster volunteers require not only logistical preparation and aid, but also emotional support. Integrating expressive arts, narrative therapy elements, and structured group reflection may serve as protective factors against emotional fatigue and secondary trauma. Future programme planning and design should therefore include well-being checklist or tracking system alongside performance outcomes. In conclusion, the WHO-5 findings affirm that well-being is not incidental, it is essential for preventing harm to both disaster volunteers and children in the future. Emotional resilience and strength observed in the volunteers display their capacity to act as emotional anchors in times of crisis, offering not only their skills, but also their humanity. Their well-being is both a metric of individual resilience and a marker of collective hope. In other cases where disaster volunteers have a range of professional backgrounds, and limited training or experience with children, the results may vary and suggest another finding to this study.

Table 4. WHO-5 Z-Scores

WHO-5 Total Score	Z-Score
19	0,880634314494746
14	-1,18616050523782
17	0,053916386601719
20	1,29399327844126
20	1,29399327844126
20	1,29399327844126
14	-1,18616050523782
16	-0,359442577344794
20	1,29399327844126
14	-1,18616050523782
18	0,467275350548233
12	-2,01287843313085
14	-1,18616050523782
19	0,880634314494746
14	-1,18616050523782
19	0,880634314494746
17	0,053916386601719
16	-0,359442577344794
16	-0,359442577344794
17	0,053916386601719
15	-0,772801541291308
20	1,29399327844126
17	0,053916386601719

To assess the resilience of disaster volunteers helping children through play and creative art approach, we collect the calculations using the following statements:

1. Volunteer Resilience Score: "Saya memiliki motivasi pribadi yang kuat untuk menjadi relawan" ["I have a strong personal motivation to be a volunteer"], "Saya merasa memiliki kesiapan emosional saat memulai penugasan" ["I felt emotionally prepared when starting the assignment"], "Saya merasa kegiatan ini meningkatkan ketahanan (resilience) saya sendiri" ["I feel this activity has enhanced my own resilience"], and "Saya mampu menjaga kesehatan mental saya selama penugasan" ["I was able to maintain my mental health during the assignment"].
2. Perceived Effectiveness of Play and Creative Art Interventions: "Saya memahami konsep dasar terapi bermain dan seni…" ["I understand the basic concepts of play and art therapy…"], "Saya dapat mengidentifikasi kebutuhan emosional anak-anak…" ["I can identify the emotional needs of children…"], and "Bagaimana Anda menilai kontribusi play therapy terhadap pemulihan anak-anak?" ["How do you assess the contribution of play therapy to children's recovery?"].

The significant correlation ($r = 0.515$, $p = 0.009$) between volunteers' resilience and perceived effectiveness of play and creative art interventions should be interpreted with caution, as correlation does not establish causality. It is unclear whether scores for higher resilience lead disaster volunteers to view the positive outcome of play and creative art interventions, or whether meaningful interventions reinforce resilience. Factors such as professional skills, background, prior volunteering experience, mental and emotional well-being, personalities and other possibilities have an influence on the results. In this study these confounders were not controlled. Therefore, acknowledging it underscores the need for further research using larger samples and the need to clarify the correlation that points towards direct effectiveness of resilience.

Table 5. WHO-5 Z-Scores

Variable	Mean	Median	Std Dev	Min	Max
Volunteer Resilience Score	4.63	4.63	0.36	4.00	5.00
Perceived Intervention Score	4.25	4.00	0.64	3.00	5.00

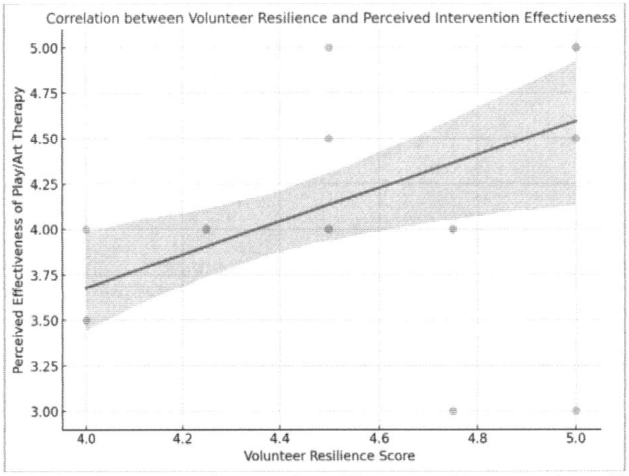

Fig 5. Correlation between volunteers' resilience and intervention effectiveness

Most volunteers who reported higher levels of internal resilience also tended to express greater belief in the positive impact of play and art therapy on disaster survivors, in both children and adults. This confirms that resilience is not just a personal trait but is also interlinked with empathy, an understanding of healing, and the perceived effectiveness of creative interventions using the play and creative art approach.

The open-ended survey responses and reflective prompts draw on four main themes that emerged. Firstly, personal growth and transformation such as describing, "I discovered parts of myself I never knew existed". Second, the path to healing as stated by one of the volunteers, "The child who had not spoken in days began to draw a sunrise. That was our breakthrough", and next, the shared struggles described, "We cried alone at night but smiled in front of the children". Lastly, the personal and cultural resilience in sayings such as, "Gotong royong gave me the strength to keep going". These narratives confirm that play and creative art not only helped children express their trauma but also served as a reflective mirror, a useful process for the volunteers themselves, facilitating emotional regulation and resilience.

4. Limitations

A number of improvements can be implemented to this study, notably an important factor is the gap of 6 years after the interventions in Lombok and Palu, between the year 2019 to early 2025. This gap may suggest disaster volunteers may have forgotten some information, events may be blurred and may be less accurate. The use of mixed methods provided generous data to this participatory action research (PAR). However, researchers should integrate quantitative and qualitative data, to use tests such as t-test, ANOVA and redesign the development of further action research in the future. A comparison of differences or correlation between genders, as well as comparison between three groups of disaster volunteers: 1) professional play and creative art therapists; 2) trained nonprofit disaster volunteers; 3) local or community volunteers should be studied. Results of data should be calculated in comparison, to find association between different categories and variables, such as professional skills vs. gender, age, or experience and factors that may correlate. The duration of volunteer work has not shown significant differences, although variables show a trend that is close to significance using Mann-Whitney U test. Results of 'I can adapt well' and 'handle stress well' ($p = 0.063$, $p = 0.068$) may suggest longer duration of field volunteering work helped them to be more resilient and adaptive. These may indicate that disaster volunteers who spent more than one month for psychosocial interventions have higher resilience and are more able to cope with stress, although this slight trend has not been statistically proven. Researchers found the p-value test showed no significant differences between the levels of resilience for both genders, male and female. The data showed only a small number of differences, as it was found that female groups can cope better after difficulties, adapt to stress, yet the male group had higher scores for preparation of strength, excitement and bringing energy. Also, another limitation lies in how t-values are taken from a small sample consisting of only 24 disaster volunteers. Among these limitations, data gathered reflected back to another research. Most volunteers from professional play and creative art therapy background have stated that they were able to experience a different work environment compared to their client therapy settings. This experience allowed professional therapists to work with other volunteers in an open, unfamiliar and challenging environment. Most volunteers expressed that they would like to be involved and engage in more interventions such as disaster psychosocial support. This research points towards another data gathered in Lombok Island from one of our researchers, entitled, *Measuring Post-Disaster Resilience Perception in Small Island: Lessons from 2018 Lombok Island Earthquakes in Indonesia,* which presented the highest concerning results for perceived family or household resilience (75.6%). The results showed low perceived local government and low perceived family resilience. Limitations have been considered during the execution of this research, with a smaller number of samples, yet it leads to the need to study how disaster volunteers' resilience may impact and possibly associate with resilience among families in Lombok, Palu city and several disasters impacted locations in the future.

5. Conclusion

This study has shown the powerful and impactful role of volunteerism in disaster contexts, highlighting its contribution in the strengths, resilience, transformation and empathy that emerges when individuals are open to learn, grow, collaborate, cooperate and adapt in every situation, focusing on human connection as the core and support for resilience. When survivors and volunteers become emotionally and relationally connected, mutual acceptance creates a foundation for inner strength. Transformation is often described as arising from within, requiring self-awareness and readiness [9]. Disaster volunteers reported that emotional and relational bonding with local survivors in Lombok and Palu have fostered mutual acceptance as they co-create a safe, comforting environment. It is suggested for emotional processing sessions or programmes to be led by trained mental health professionals using tools such as group painting, group sandtray therapy, guided storytelling, and expressive movement. Effectiveness of psychosocial support using creative interventions have been studied to provide mutual regulation in post-disaster environments and enable emotional expressions for children and families [2, 5]. The implementation of play and creative art approach allow emotional containment, creativity, narrative integration, and mutual support.

Disaster response organisations should adopt formal policies for mental health and well-being monitoring using standardised tools such as the WHO-5 Well-Being Index and Brief Resilience Scale (BRS), which demonstrated reliability in tracking emotional health and adaptive functioning in high-stress contexts in volunteerism. By embedding these tools and good practices into future disaster preparedness programmes, we not only address the psychosocial needs of affected communities but also ensure that the very volunteers who serve the people, are equipped to thrive and develop. Resilience is not merely an individual trait—it is a collective and shared capacity cultivated through presence, purpose, and community engagement [4]. These interventions are not abstract ideals—they were actively practiced on the ground. Planning, execution, and evaluation took place daily in the field to ensure that volunteers remained emotionally grounded and spiritually strong enough to hold space for survivors whether they are children, young people or adults—to listen, affirm, be present and rebuild lives together. The mixed methods data highlights the need for a long-term intervention for the children and families, increased professional development through training, workshops in play and creative art or skills to enhance competence and disaster volunteers' readiness and resilience. The study has shown that psychosocial support with play and creative art interventions have given a significant impact to disaster volunteers and through their help to children and families in Indonesia. They described personal growth, positive team collaborations with other professionals, to support one another with flexibility, creativity, adaptability and affirm boundaries, to ensure safety for everyone.
Ultimately, this research invites us to rethink disaster response and interventions that utilise the strengths and resilience in volunteerism. The relational and expressive, collaborative methods using PAR with play and creative art approach demonstrated the importance of bridging humanitarian responses, such as trauma, fostering connection [7]. This includes not only focusing on logistical efforts but shifting to a deeply, profoundly human process of co-regulation, empathy, empowerment, and mutual transformation. This integrative analysis confirms that volunteers engaged in creative psychosocial interventions demonstrate both resilience and emotional well-being. The application of five qualitative frameworks (PAR, ABR, IPA, Narrative Inquiry, and Mixed Methods) helped validate and deepen the insights. Although limitations persist from the small number of samples collected and limitations in the research findings have been addressed, it brings to light the importance of resilience, maintaining a healthy well-being and continuous development with professional support for disaster volunteers as they facilitate and build a safe environment for psychosocial support and interventions. The findings support a development of future frameworks in applying creative, culturally grounded approaches in humanitarian contexts, focused on compassion, empathy, mutual understanding that is carried within the culture of the country, Indonesia. The findings may suggest a practical implementation into further projects and programmes to help children, families and communities facing disasters or groups of children at-risk of trauma and conditions that may affect their well-being. Researchers in this study hope to contribute to the body of knowledge and offer a helping hand to many.

References

1. A. T. Sasmi et al., *Shear wave splitting of the 2018 Lombok earthquake aftershock area, Indonesia*, Geosci. Lett., **10**, Article 7 (2023). https://doi.org/10.1186/s40562-022-00258-3
2. C.A. Malchiodi, *Creative Interventions with Traumatized Children* (Guilford Press, 2015)

3. D. E. Alexander, *Resilience and disaster risk reduction: an etymological journey,* Nat. Hazards Earth Syst. Sci., **13**, 2707–2716 (2013). https://doi.org/10.5194/nhess-13-2707-2013
4. D. Rakel, *The Compassionate Connection: The Healing Power of Empathy and Mindful Listening* (W.W. Norton & Company, 2018)
5. E. Cohen, R. Gadassi, *The Function of Play for Coping and Therapy with Children Exposed to Disasters and Political Violence,* Curr. Psychiatry Rep., **20**, Article 31 (2018). https://doi.org/10.1007/s11920-018-0895-x
6. G. B. Chand, *Narrative inquiry: A critical examination of its theoretical foundations and methodological applications,* Far West. Rev. **2**, 135–152 (2024). https://doi.org/10.3126/fwr.v2i1.70514
7. J.M. Chevalier, D. Buckles, *Participatory action research: theory and methods for engaged inquiry* (Routledge, 2019). https://doi.org/10.4324/9781351033268
8. J. Whittaker, B. McLennan, J. Handmer, *A review of informal volunteerism in emergencies and disasters: Definition, opportunities and challenges*, Int. J. Disaster Risk Reduct., **13**, 358–368 (2015). https://doi.org/10.1016/j.ijdrr.2015.07.010
9. K. Charmaz, *Constructing Grounded Theory: A Practical Guide through Qualitative Analysis* (SAGE Publications, London, 2006)
10. L. Triasari et al., *Play therapy to reduce traumatic stress in earthquake-affected children in North Lombok, West Nusa Tenggara*, Int. J. Nurs. Health Serv. **3**, 133–140 (2020)
11. M. Aminizade et al., *The role of volunteer citizens in response to accidents and disasters*, Health Emerg. Disasters Q., **2**, 107–124 (2017). https://doi.org/10.18869/nrip.hdq.2.3.107
12. M. Ferrario, *Landslides triggered by multiple earthquakes: insights from the 2018 Lombok (Indonesia) events*, Nat. Hazards, **98**, 575–592 (2019). https://doi.org/10.1007/s11069-019-03718-w
13. R. Omira, et al., *The September 28th, 2018, Tsunami in Palu–Sulawesi, Indonesia: A Post-Event Field Survey*, Pure Appl. Geophys., **176**, 1379–1395 (2019). https://doi.org/10.1007/s00024-019-02145-z
14. Y.A. Sokang et al., *Lesson Learned from Post-Disaster Psychosocial Support for Children in Palu, Central Sulawesi—Indonesia*, J. Disaster Res., **19**, 678–690 (2024). https://doi.org/10.20965/jdr.2024.p0678
15. Y. Saito, Y. Ishida, *The effect of disaster volunteer experience on the well-being of young people in the Great East Japan Earthquake*, J. Disaster Res., **16**, 957–961 (2021). https://doi.org/10.20965/jdr.2021.p0957

Identification of factors in the implementation of the primary health care service integration system to support disaster risk management

Sri Kusumadewi*, *Rahadian* Kurniawan[1], *Elyza Gustri* Wahyuni[1], *Aridhanyati* Arifin[1], and *Linda* Rosita[2]

[1]Department of Informatics, Faculty of Industrial Technology, Universitas Islam Indonesia, Yogyakarta, Indonesia
[2]Department of Clinical Pathology, Faculty of Medicine, Universitas Islam Indonesia, Yogyakarta, Indonesia

Abstract. Integration of primary health services (IPHS) is an Indonesian government program to provide comprehensive, easily accessible, and sustainable health services for the community through the coordination of various health programs at the primary health care facilities. As a country with high disaster potential, the Indonesian government needs disaster risk management supported by IPHS. This study aims to identify factors that should be considered so that IPHS can play a maximum role in supporting disaster risk management. The initial identification included eight factors. The study employs the concept of a group support system. There are six stages: data collection, ranking aggregation, factor feasibility test, determination of factors and weights, revision of weights, and analysis and recommendations. A total of 26 respondents were involved in this study. The study's results demonstrated that the respondents reached a consensus. The availability of data on vulnerable groups is the main factor that needs to be considered. By using the voting by plurality at each rank aggregation method, a score of 21 was obtained, and the final weight of this factor was 0.3209. In the future, a multi-criteria group decision support system (GDSS) will be needed to select the most appropriate strategy based on these eight factors.

1 Introduction

Disasters still pose a serious hazard to people and the environment, and they cause huge financial losses despite improvements in disaster risk reduction strategies and technologies [1]. The possibility of death, injury, or other health conditions, as well as long-term economic, social, environmental, and cultural harm, is known as disaster risk. Vulnerable people are primarily affected in terms of health [2,3]. People with disabilities, older age, gender, pre-existing chronic health issues, and socially marginalized groups are all factors that affect population vulnerability [4]. Disruptions in the delivery of health services may

* Corresponding author: sri.kusumadewi@uii.ac.id

exacerbate pre-existing conditions [5]. Access to healthcare may be further restricted for these persons in the post-disaster period due to structural obstacles, health system failure, and heightened demand for medical services. Effective disaster preparedness should be prioritized more and developed in partnership with vulnerable people in order to address these problems [4]. Therefore, as part of disaster mitigation measures, it is crucial to gather and manage data on vulnerable groups, such as pregnant women, children under five, the elderly, and people with disabilities [6].

The health care system of a nation, which includes clinics, hospitals, and primary health care (PHC) facilities, is essential to managing and reducing the risk of disasters [7]. For PHC centers to continue operating both during and after a disaster, disaster risk assessment and emergency preparedness are essential. In particular, PHC facilities should be prepared for disasters, according to the World Health Organization (WHO) [8].

Evidence and practice regarding PHC and disaster management organization collaboration are limited [4]. PHC's relationship with clients and responsibility for their care in both normal and emergency situations make it essential to bolstering disaster preparedness for vulnerable populations. PHC plays a significant role in the community's disaster risk management system. In order to improve community resilience and speed up disaster response, health and disaster information systems must be integrated [9]. To assist decisions at the PHC level in community health management, the information system must provide an information dashboard that integrates health and disaster risk data [9]. An early warning system that is connected with PHC can enhance disaster response, and a rapid reporting system that uses digital technology to speed up reporting in the field is crucial [10].

Currently, health data management in Indonesia is conducted via the SatuSehat platform. SatuSehat is a digital health ecosystem platform created by the Ministry of Health of the Republic of Indonesia. SatuSehat operates as a cohesive system that links diverse health services, including PHC, hospitals, clinics, pharmacies, and laboratories, to enhance access to public health data. Through SatuSehat, users can digitally record their health history, enabling access to this data by multiple registered health organizations. The integration of primary health services (IPHS) constitutes a component of the SatuSehat ecosystem, emphasizing PHC. IPHS emphasizes three primary aspects: the life cycle as the central theme for integrating health services via networks at the village and hamlet levels, and enhancing regional oversight through the health situation dashboard. In other words, IPHS must be substantiated by data originating from the grassroots level, specifically the village, via integrated service activities (IHS). The deployment of applications among IHS volunteers to enhance the collection of public health data is crucial [11] The information technology infrastructure requires enhancement, and training for volunteers and health staff is essential [11].

Because Indonesia is a disaster-prone nation, the IPHS information system is crucial to disaster risk reduction. Several factors must be considered to ensure that IPHS can optimally contribute to disaster risk management. This study aims to determine the important factors for consideration. This study is significant as it offers three primary contributions: identifying key factors that affect the efficacy of IPHS in disaster risk management, compiling recommendations to enhance information systems through early detection features, data integration, and rapid response, and establishing a scientific foundation for formulating health policies aimed at reinforcing primary service systems in disaster contexts.

2 Method

In general, this study will identify factors that need to be considered so that IPHS can play a maximum role in supporting disaster risk management, which will be obtained through the group support system.

2.1 Data collection

The data collection technique was carried out using a questionnaire. The demographic data of respondents included gender, age, education, and role (volunteer/cadre, village staff, or health professional). We present eight facets that must be considered in the implementation of IPHS in IHS in conjunction with disaster risk management. The eight criteria were derived from previous studies [6,9,10,11] and the results of interviews with village officials. Participants were asked to establish the priority ranking of each factor. A value of 1 indicates the highest priority factor, whereas a value of 8 indicates the lowest priority factor. Table 1 presents the eight factors.

Table 1. Factors to consider when implementing IPHS in IHS

Code	Factor
C1	Availability of vulnerable group data (pregnant women, toddlers, elderly, disabled)
C2	Integration of IPHS system with disaster information system
C3	Ability of IHS volunteers/cadres to input data into IPHS system or use mobile application
C4	Availability of health and disaster risk information dashboard in health centers
C5	Early warning system connected to primary services
C6	Internet access and devices (smartphones, tablets) in IHS and health centers
C7	Training of cadres and health workers in using IPHS information system
C8	A rapid reporting protocol when a disaster occurs

The participants were volunteers, village personnel, and health professionals from Tirtorahayu village, DIY Province, Indonesia. Twenty-six respondents completed the questionnaire. The location was selected due to its proven success in deploying the IHS information system [11]. All participants were verified to have utilized the IHS information system. The process continued with data validity assessment, during which outlier data was eliminated and excluded from subsequent procedures.

2.2 Rank aggregation

We process the ranking results of each respondent and aggregate them into group decision support. We used four ranking aggregation methods, namely Borda Count, Median Ranking, and voting by plurality at each rank.

The expert preference rankings must be obtained before factors can be ranked using the Borda rule. Suppose there are m factors to be selected. If factor C_i is ranked k-th, the Borda count of C_i is $B(C_i) = m - k + 1$, which means the number of factors defeated by C_i. The greatest Borda count factor is ranked first, and the lowest Borda count factor is ranked last.

In the mean ranking method, we aggregate all ratings provided by respondents for each factor and subsequently compute the average of those scores as the overall rating. Upon calculating the average of each component, the factors are arranged in ascending order according to that average value, from the lowest (best rating) to the highest.

voting by plurality at each rank is a ranking aggregation method in which for each ranking position (rank 1, rank 2, etc.), the alternative receiving the most number of votes from respondents at that place is declared the winner of that ranking. Upon the voting of an alternative, it is excluded from the computation, and the procedure is continued for the subsequent position until all alternatives are ranked. This method is simple, fast, and focuses on the majority of choices in each position.

2.3 Determining weight

We calculate the weight of each factor using the pairwise comparison technique. We will compare the rating outcomes from the previous stage with each other. For example, r_i is the ranking of the i-th factor, then the weight of the *i*-th factor is:

$$a_i = \frac{1}{r_i} \quad (1)$$

Next, we carry out normalization to ensure that the total weight equals one. The formula used is:

$$w_i = \frac{a_i}{\sum_j^n a_j} \quad (2)$$

Where w_i is the normalized weight of the i-th factor, and n is the number of factors.

2.4 Factors feasibility test

After obtaining the weight of each factor, the next step is to test whether it is worthy of consideration or not. We use two methods: general criteria and the ratio test [12,13]. The general criteria used is that a factor can be selected for consideration if it has a weight of at least 5%. The ratio test is performed using the following formula:

$$\propto_i = \frac{w_i}{w_{max}} \quad (3)$$

The *i*-th factor can potentially be chosen for consideration if $\propto_i > 0.1$.

2.5 Determining factors and revising weights

We will identify the relevant factors to examine. Upon the exclusion of factors that fail to meet the eligibility criteria, we will then adjust the weight assigned to each element. However, if all factors are considered significant, no adjustment to the weight will be made.

The factors mentioned previously and their associated weights may subsequently serve as factors for identifying optimal recommendations related to the deployment of IPHS that improve a disaster risk management. The Multi-Attribute Decision Making (MADM) framework is applicable for this purpose. Nonetheless, this work does not address the application of the MADM model.

We cross-validated our findings using qualitative data. This process included interviews with village officials to ensure the relevance and accuracy of the results. The interviews provided strong support and confirmation for the quantitative analysis.

2.6 Analysis and recommendation

The last step is to examine each factor while taking into consideration the actual conditions in Indonesia. This study has been supported by empirical evidence derived from observations and conversations with stakeholders.

Subsequently, we will be providing recommendations to enhance the effectiveness of the IPHS deployment in support of disaster risk management. These recommendations will incorporate feedback from stakeholders. Recommendations are provided for both general policies and those relevant to information technology applications.

3 Results and Discussion

3.1 Factors identification

This study involved 26 respondents. However, only 22 respondents provided valid data. Figure 1 illustrates the respondent data categorized by gender, age, education, and status. The rankings provided by each respondent are compiled to obtain the final ranking. Table 2 presents the results of the ranking aggregation using the Borda Count method.

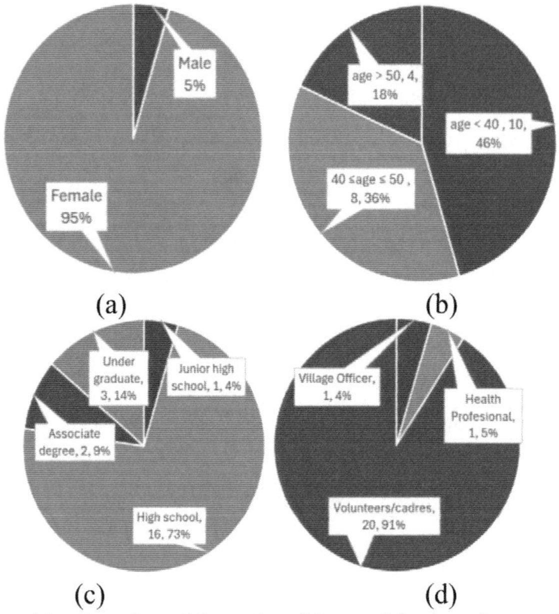

Fig. 1. Respondent demographics based on: (a) gender; (b) age; (c) education; and (d) status.

Table 2. Ranking aggregation using the Borda Count method

Rank	Factor	Score
1	C7	215
2	C1	214
3	C3	208
4	C6	205
5	C5	200
6	C8	198
7	C4	194
8	C2	180

Table 3 presents the results of the ranking aggregation using the mean ranking approach. According to Table 2 and Table 3, both methods produced similar ranking aggregate results.

Table 3. Ranking aggregation using the mean ranking method

Rank	Factor	Mean
1	C7	1.227
2	C1	1.273
3	C3	1.545
4	C6	1.682
5	C5	1.909
6	C8	2.000
7	C4	2.182
8	C2	2.818

Table 4 presents the results of the ranking aggregation using the voting by plurality at each rank method. This table indicates that C3 and C6 obtained identical scores of 10, so they were assigned the same ranking of third position. Similarly for C5, C4, and C8 obtained identical scores of 15, resulting in an equivalent ranking of fourth place.

Table 4. Ranking aggregation using the voting by plurality at each rank method

Rank	**Factor**	**Score**	**Ideal rank**
1	C1	21	1
2	C7	19	2
3	C3	15	3
4	C6	15	3
5	C4	10	4
6	C5	10	4
7	C8	10	4
8	C2	6	5

The ranking aggregate results in Table 4 support the findings of the previous two tables. This signifies that, in this instance, the Borda Count method and the Mean Ranking approach yield a consensus among decision makers.

Subsequently, we utilize the results presented in Table 4 to compute the weight value of each factor. Table 5 presents the weight of each factor. The weight is calculated using Eq. 1. For the next step, it is necessary to normalize these weights so that their cumulative sum equals one. Normalization is performed with Eq. 2.

Table 5. Weight of each factor

Factor	**Rank**	**Weight**	**Norm weight**
C1	1	1.000	0.321
C2	5	0.200	0.064
C3	3	0.333	0.107
C4	4	0.250	0.080
C5	4	0.250	0.080
C6	3	0.333	0.107
C7	2	0.500	0.160
C8	4	0.250	0.080

The factor feasibility test was conducted using a threshold, $\theta = 0.05$. It was found that each factor has a weight of more than 0.05. Thus, all factors are worthy of consideration.

In the ratio test, the threshold will be used $\theta = 0.1$. First, the ratio of each factor is calculated using Eq. 3. It turns out that the ratio of all factors exceeds 0.1. The result indicates that all factors are worthy of consideration. Table 6 shows the results of the ratio test.

Table 6. Ratio test results

Rank	**Norm Weight**	**Ratio**	**Decision**
C1	0.321	1.000	feasible
C2	0.064	0.200	feasible
C3	0.107	0.333	feasible
C4	0.080	0.250	feasible
C5	0.080	0.250	feasible
C6	0.107	0.333	feasible
C7	0.160	0.500	feasible
C8	0.080	0.250	feasible

We accepted all factors as considerations; therefore, we did not alter or revise the weights. We continued in use the prior weights.

3.2 Analysis and recommendations

Table 7 provides a summary of the factors, priority scale, and weight of each factor. According to the Table 7, the database of vulnerable groups should be strengthened, the capability of health professionals should be increased, and infrastructure and operational systems should be provided as the first priority when merging primary and disaster care. The best technology integration between systems will only occur when the fundamental data and human aspects have been addressed. Therefore, the key to this integration process is a phased strategy that emphasizes field preparedness.

Table 7. Weight of each factor

Rank	Factor		Weight
1	C1	Availability of vulnerable group data (pregnant women, toddlers, elderly, disabled)	0.3209
2	C7	Training of cadres and health workers on using IPHS information system	0.1604
3	C3	Ability of IHS volunteers/ cadres to input data into IPHS system or use mobile application	0.1070
4	C6	Internet access and devices (smartphones, tablets) in IHS and Health Center	0.1070
5	C4	Availability of health and disaster risk information dashboard in Health Center	0.0802
6	C5	Early warning system connected to primary services	0.0802
7	C8	A rapid reporting protocol when disaster occurs	0.0802
8	C2	Integration of IPHS system with disaster information system	0.0642

Integrating primary care systems with disaster management systems is likely to encounter several significant challenges. These include insufficient training for healthcare workers who lack disaster-related competencies, as well as the uneven availability and high cost of necessary digital infrastructure. Another major concern is the potential violation of data privacy when sensitive health information is used for disaster response. Trade-offs must be made between strengthening disaster preparedness and maintaining routine health services, as well as between data transparency and individual privacy. Furthermore, sectoral egos and overlapping institutional authority often obstruct inter-agency coordination, complicating the integration process.

The government must improve the availability and accuracy of data on vulnerable groups, including pregnant women, children, the elderly, and people with disabilities, within the IPHS. This data is critical for identifying at-risk populations and planning appropriate disaster preparedness and response strategies. Such data should be acquired through systematic data collection conducted by PHC, IHS, and community volunteers, and connected with other demographic and social systems. It must also be provided with a decision support system capable of identifying health issues in the aged, including disease complications [14,15]. During a disaster, the process of determining requirements and allocating relief can be completed swiftly and precisely with the use of accurate and centralized data.

The training of personnel, volunteers/cadres, and healthcare professionals in the utilization of the IPHS system and mobile applications is crucial. The government must provide technology-driven and disaster-oriented training programs that are comprehensible and customized to local competencies. This training should be accompanied by consistent mentoring and incentives for participants to ensure their motivation to improve their proficiency in technology utilization. Community support is essential, particularly through partnerships with local organizations to facilitate informal training initiatives.

Access to digital devices and internet connectivity is a crucial factor. The government should supply devices such as tablets, smartphones, and internet access at primary care centers, particularly in disaster-prone regions. Conversely, the community can actively participate in the regular maintenance of the provided facilities and pursue partnerships with the business sector or corporate social responsibility initiatives to enhance the accessibility of resources in their regions. The preparedness of this digital infrastructure will enhance the effectiveness of the information system during emergencies.

To utilize the obtained information effectively, it is essential to create a health and disaster risk information dashboard at PHC. This dashboard will serve as an important tool for vulnerability mapping, disaster response planning, and health resource distribution. The IPHS system must be integrated with early warning systems from institutions such as the Meteorology, Climatology, and Geophysics Agency, as well as the National Disaster Management Agency, to ensure that disaster information is received by primary services in real-time and addressed promptly in accordance with established protocols.

The integration of IPHS with national disaster information systems, such as InaRISK and DIBI, is crucial for establishing effective cross-sector cooperation. The government must establish an expedited reporting system via IPHS, enabling cadres or health professionals to communicate field circumstances during a crisis. The community is expected to comprehensively grasp the functionality of this system, engage in disaster simulations, and offer input to ensure the system evolves in alignment with local requirements. The collaboration between the government and the community is essential for developing a robust and adaptable primary service information system for disaster management.

Interviews with representatives of the village officials supported these findings. They emphasize the potential of the IPHS to strengthen community-level disaster preparedness. By offering relevant information on disaster risks and recommended actions, the IPHS can be used to design targeted training programs. This approach helps ensure that communities are better informed and more resilient during emergencies.

3.3 Future research

Since IPHS is closely related to public health preparedness, response, and recovery, choosing the best implementation method is essential for supporting disaster risk management. Disaster scenarios necessitate a robust and coordinated healthcare system; therefore, a decision-making framework must account for critical factors such as resource availability, service capacity, accessibility to affected regions, and community involvement. Consequently, research using the Multiple Attribute Decision Making (MADM) model is crucial for determining the optimal strategy based on these diverse factors.

MADM allows a systematic assessment of diverse alternative IPHS implementation strategies, considering multiple attributes that affect the effectiveness of disaster risk management, including response speed, intersectoral integration, service flexibility, and communication effectiveness. In complex and evolving scenarios like disasters, data-driven decision-making and multi-criteria analysis help mitigate subjective bias and enhance the precision of strategy selection. This is essential for IPHS to function optimally in detecting risks early and providing adaptive services according to needs.

The MADM model can also promote the engagement of diverse stakeholders from the health sector, disaster management, and local communities in the decision-making process. The selected method should both technically efficient and possesses significant social legitimacy and feasibility for implementation. Consequently, research utilizing the MADM approach is essential to ensure that the application of IPHS in facilitating disaster risk management is conducted in a systematic, evidence-based, and community-focused manner.

4 Conclusion

There are eight factors that need to be considered so that IPHS can play a maximum role in supporting disaster risk management. The availability of vulnerable group data (such as pregnant women, toddlers, the elderly, and people woth diasabilities) is the main factor that needs to be considered. Using the voting by plurality at each rank ranking aggregation method, a score of 21 was obtained, and the final weight of this factor was 0.3209. Synergy between the government and the community is key to creating a primary service information system that is resilient and adaptive to disasters.

We need a multi-criteria group decision support system (GDSS) in the future to select the most appropriate strategy based on these eight factors. The MADM group model can facilitate the involvement of various stakeholders in the decision-making process.

Acknowledgements. The author would like to thank the Directorate of Research and Community Service, Universitas Islam Indonesia, for funding the research through the Institutional Development Research Grant with contract number 001/Dir/DPPM/70/Pen.PengembanganInstitusi/ IV/2025.

References

1. U. Lagap, S., Ghaffarian. Digital post-disaster risk management twinning: A review and improved conceptual framework. IJDRR, **110**, 104629 (2024). https://doi.org/10.1016/j.ijdrr.2024.104629
2. M. A. Benevolenza, and L. DeRigne. The impact of climate change and natural disasters on vulnerable populations: A systematic review of literature. J Hum Behav Soc. Environ., **29**, 2, 266–281 (2019). https://doi.org/10.1080/10911359.2018.1527739
3. R. S. C. Friedman, D. M. Carpenter, J. M. Shaver, S. C. McDermott, and J. Voelkel. Telemedicine familiarity and post-disaster utilization of emergency and hospital services for ambulatory care sensitive conditions. Am. J. Prev. Med., **63**, 1, e1–e9, (2022). https://doi.org/10.1016/j.amepre.2021.12.025
4. P. Şimşek, M. Kako, N. Harada, J. Abrahams, and I. Tayfur. Scoping review of exploring the roles of primary care providers to increase disaster preparedness of vulnerable populations. Prog. Disaster Sci., **23**, 100339 (2024). https://doi.org/10.1016/j.pdisas.2024.100339
5. G. C. Alexander, and D. M. Qato. Ensuring access to medications in the US during the COVID-19 pandemic. JAMA., **324**, 1, 31 (2020). https://doi.org/10.1001/jama.2020.6016
6. K. H. E. Lin, S. Khan, L. A. Acosta, R. Alaniz, D. R. Olanya, The dynamism of post disaster risk communication: A cross-country synthesis. IJDRR, **48**, 101556 (2020). https://doi.org/10.1016/j.ijdrr.2020.101556
7. E. Melnychuk, T. D. Sallade, C. K. Kraus, Hospitals as disaster victims: Lessons not learned?. JACEP Open, **3**, 1, e12632 (2022). https://doi.org/10.1002/emp2.12632
8. F. Shojaei, S. Sohrabizadeh, P. Qaraeian, H. Yousefi Khoshsabeghe, J. Jandaghi, F. Farivar, Comparative analysis of two disaster risk assessment tools for primary health care centers: The case of Iran. IJDRR, **107**, 104466 (2024). https://doi.org/10.1016/j.ijdrr.2024.104466

9. L. Rao, M. McNaughton, A knowledge broker for collaboration and sharing for SIDS: the case of comprehensive disaster management in the Caribbean. Inf. Technol. Dev., **25**, 1, 26–48 (2019). https://doi.org/10.1080/02681102.2018.1510363
10. R. Šakić Trogrlić, M. van den Homberg, M. Budimir, C. McQuistan, A. Sneddon, B. Golding, Early warning systems and their role in disaster risk reduction," in towards the "perfect" weather warning, Cham: Springer Int. Publishing, 11–46 (2022). https://doi.org/10.1007/978-3-030-98989-7_2
11. C. Hardiyanti, S. Kusumadewi, R. Kurniawan, Evaluation of success and failure factors for maternal and child health in integrated healthcare center information systems (IHCIS) using the HOT-fit method. JISEBI, **10**, 1, 152–166 (2024). https://doi.org/10.20473/jisebi.10.1.152-166
12. M. A. Hatefi, Indifference threshold-based attribute ratio analysis: A method for assigning the weights to the attributes in multiple attribute decision making. Appl. Soft. Comput., **74**, 643–651 (2019). https://doi.org/10.1016/j.asoc.2018.10.050
13. D. Boix-Cots, F. Pardo-Bosch, P. Pujadas, A systematic review on multi-criteria group decision-making methods based on weights: Analysis and classification scheme. Inf. Fusion, **96**, 16–36 (2023). https://doi.org/10.1016/j.inffus.2023.03.004
14. S. Kusumadewi, L. Rosita, E. Gustri Wahyuni, Implementation of fuzzy associative memory toward optimizing a neural network model to predict total iron binding capacity. Biomed. Signal Process. Control, **86**, 105297, (2023). https://doi.org/10.1016/j.bspc.2023.105297
15. S. Kusumadewi, L. Rosita, E. G. Wahyuni, Stability of classification performance on an adaptive neuro fuzzy inference system for disease complication prediction. IJ-AI, **12**, 2, 532–542 (2022). https://doi.org/10.11591/IJAI.V12.I2.PP532-542

The 2024 flood in Demak Regency, Indonesia: A retrospective qualitative investigation into Indonesian Red Cross response

Yuli Arinta Dewi1[1,2,*], *Koentjoro* Soeparno[2], *Mizan* Bustanul Fuady Bisri[3, 4], and *Praditya* Putri Pertiwi[2]

[1]Faculty of Psychology, Universitas Islam Sultan Agung, Semarang, Indonesia
[2]Faculty of Psychology, Universitas Gadjah Mada, Yogyakarta, Indonesia
[3]Graduate School of International Cooperation Studies, Kobe University, Kobe, Japan
[4]Asian Disaster Reduction Center, Kobe, Japan

Abstract. In early 2024, Demak Regency, Central Java, Indonesia, experienced severe flooding that affected 13 of 14 sub-districts, 70 villages, and over 131,703 residents. This study aimed to assess the Indonesian Red Cross (PMI) Demak Branch response by identifying the key actions undertaken and highlighting the integration of psychosocial aspects. Using a retrospective qualitative approach, the study analyzed PMI operational reports and news coverage from February and March 2024. PMI Demak provided healthcare services across nine camps, attending to 358 individuals with conditions such as cough and skin irritations, alongside logistical support, including meal distribution and clean water supply, and post-disaster sanitation of wells and public facilities. A total of 123 personnel, comprising volunteers and staff, were mobilized, with additional support from external PMI branches. Operational challenges were observed, including weak coordination, spontaneous volunteer mismanagement, and fragmented data systems. The findings also revealed significant psychological distress among victims, particularly children and the elderly, as well as stress signs among volunteers. Strengthening coordination, integrating systematic psychosocial support, and improving volunteer management are recommended to enhance disaster response resilience in flood-prone areas.

1 Introduction

Floods are among the most destructive natural disasters, often caused by heavy rainfall, tidal surges, and poor land management practices. They result in numerous fatalities, mainly due to drowning, and lead to various health issues, including injuries, waterborne diseases, and mental health problems [1]. Economically, floods damage infrastructure, homes, and agriculture, causing significant losses, especially in rural areas [2]. Environmental impacts include soil erosion, water contamination, and biodiversity loss, while essential services such

* Corresponding author: yuliarintadewi@mail.ugm.ac.id

as clean water and healthcare are often disrupted. These impacts highlight the importance of coordinated efforts in disaster preparedness and response, while also acknowledging that not all consequences can be entirely prevented.

Flooding is the most frequent natural disaster in Indonesia, significantly affecting the socio-economic and developmental landscape of the country [3]. Due to rapid urbanization, land subsidence, and inadequate drainage infrastructure, flood risk has increased substantially in many lowland and coastal regions along the northern coast of Java Island, including Demak Regency [4]. In early 2024, Demak Regency experienced one of its worst flood events in recent history, inundating 13 sub-districts and 70 villages, and displacing more than 131,703 residents [5]. The map of Demak Regency is shown in Figure 1.

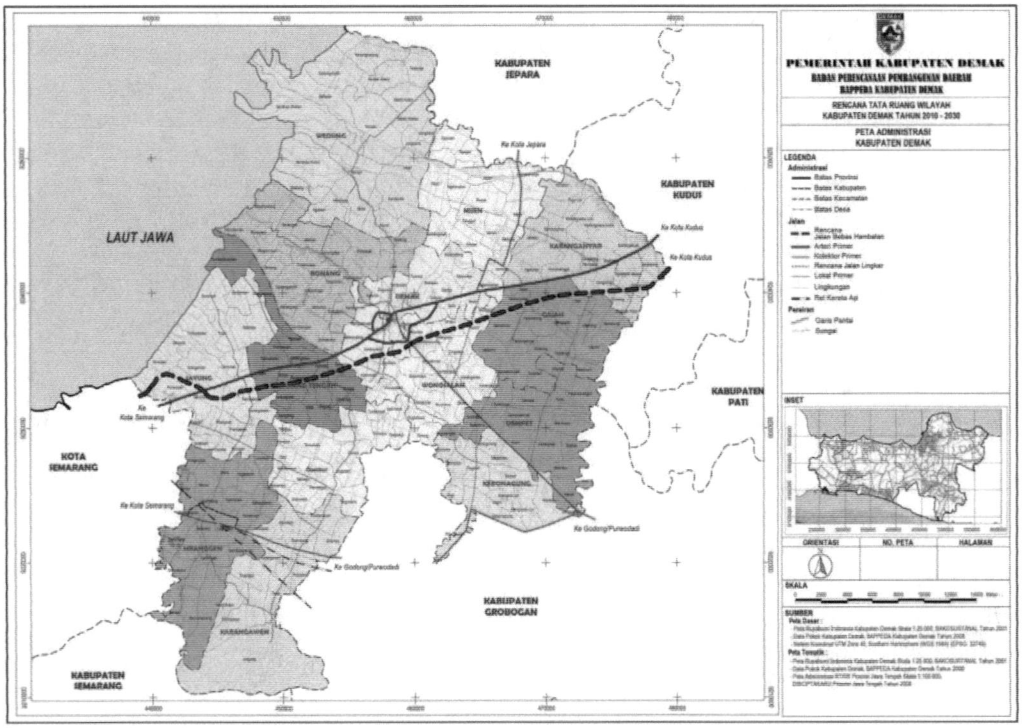

Fig 1. Map of Demak Regency, Central Java (source: Demak Government).

In response to the 2024 flood in Demak, the Indonesian Red Cross (Palang Merah Indonesia/PMI) Demak Branch deployed 123 volunteers, 4 professional humanitarian workers, and 3 board members to conduct emergency operations such as healthcare delivery, logistics distribution, and environmental cleanup. However, preliminary reviews of operational documents and field reports suggested that psychosocial support was not systematically integrated into the response, particularly for the volunteers themselves [6]. Without structured psychological support systems in place, these volunteers may face long-term consequences that could undermine both individual resilience and organizational sustainability.

PMI, particularly the Demak Branch, played a central role in the 2024 flood response. However, to date, there has been limited documentation on the specific types of interventions carried out, the scope of their implementation, and the extent to which psychosocial needs—particularly among volunteers—were identified and addressed. In addition, anecdotal reports highlighted operational constraints such as weak coordination, unmanaged spontaneous volunteers, and fragmented data systems, which require further empirical investigation.

This study employed a retrospective qualitative design to examine PMI Demak's disaster response and psychosocial dimensions of the emergency. We analyzed two primary sources—internal operational reports and publicly available media coverage—published between February and March 2024. The goal was to investigate how psychosocial needs were addressed, what operational challenges affected both community and responder resilience, and what systemic gaps persist in volunteer management. Through this lens, the study aimed to generate practical recommendations for integrating psychosocial components more effectively into future disaster response strategies in Indonesia.

This paper presents a situational analysis of the 2024 Demak floods, organized into three parts. It begins by outlining the pre-disaster conditions that shaped the region's chronic vulnerability, followed by a sequential account of the two flood events, and concludes with an assessment of PMI's operational response. The analysis emphasizes critical gaps in community-based action, inter-agency coordination, and psychosocial interventions, contributing to the broader discourse on integrated disaster resilience in Southeast Asia.

The objective of this study was to evaluate the PMI Demak Branch's response to the 2024 flood disaster in Central Java, with particular attention to the psychosocial dimensions of the emergency response. This qualitative investigation seeks to identify the main actions undertaken during the disaster, analyze the challenges encountered—especially in relation to volunteer well-being and service coordination—and highlight areas for improvement in integrating psychosocial support into operational frameworks. Drawing on the Psychological Resilience and Stress-Coping Frameworks, the study aims to inform evidence-based recommendations for enhancing both community and responder resilience in future disaster contexts.

2 Methods

2.1 Research design

This study adopted a retrospective qualitative design to examine the operational response and psychosocial dimensions of the 2024 Demak flood emergency. using document analysis as the primary qualitative research method [7].

2.2 Data source

A total of fourteen documents were reviewed, including two operational situation reports issued by the PMI Demak Branch and eleven news articles from verified national and regional media outlets (Kompas, Detik, Suara Merdeka, Antara, and others) published between February and March 2024. These news articles were identified through a Google search using the keyword combination "banjir Demak 2024", which initially returned approximately 280 results. From this pool, eleven articles were selected based on credibility, relevance to the study objectives, and appearance in the first two pages of search results. The PMI reports provided chronological accounts of volunteer deployment, relief services, and operational challenges, while media reports offered external perspectives on community impacts and coordination issues. These documents were treated as primary data sources for qualitative analysis, as outlined in the Appendix.

2.3 Data analysis

Data were analyzed using thematic analysis, following the six-phase method developed by Braun and Clarke (2006). The process began with familiarization, involving multiple close readings of all documents to gain an initial understanding of the disaster context and PMI's operational activities. This was followed by generating initial codes that captured recurring patterns across the dataset. Documents were coded to identify themes related to four major categories: (1) response strategies, including healthcare delivery, logistics distribution, evacuation services, water distribution, disaster assessment, and environmental sanitation; (2) psychological impacts, including emotional and cognitive reactions, behavioral and social impacts, and psychosocial support; (3) operational challenges, including poor inter-agency coordination, unstructured spontaneous volunteers, limited transportation access, and inadequate safety equipment; and (4) psychological resilience factors, including the absence of formal psychosocial support for volunteers, and psychological support interventions for victims.

Coding was iterative and recursive, allowing for codes and categories to evolve in response to emerging insights. These were then clustered into broader themes that reflected systemic issues surrounding preparedness and recovery, guided by the Psychological Resilience Framework. The framework enabled interpretation of both individual and institutional capacities to adapt under adversity, with particular focus on gaps in support systems that hindered emotional and organizational recovery during and after the flood events.

To assist with data organization and ensure analytical traceability, all coding was conducted using ATLAS.ti software. Triangulation was applied by comparing PMI's internal reports with media accounts, allowing the analysis to capture both internal operational perspectives and public-facing narratives. This strengthened the credibility and contextual depth of the findings.

2.4 Ethical considerations

Since the study used publicly available data and internal PMI reports with no individual identifiers, no formal ethical review was required. However, care was taken to anonymize potentially sensitive operational details and to align with ethical standards for qualitative document analysis.

3 Result and Discussion

3.1 Contextual overview: Demak region and the 2024 flood

3.1.1 Pre-disaster situation

Demak Regency is a northern coastal regency in Central Java Province, strategically located 26 km from the provincial capital of Semarang. It spans an area of 99,532 hectares and is administratively divided into 14 sub-districts, 243 villages, and 6 urban wards, with a total population of 1,223,217 in 2022. The population was concentrated in urbanizing zones like Mranggen, Demak, and Sayung, with the highest density recorded in Mranggen at 2,275 people/km^2. Geographically, Demak lies within a low-lying coastal and deltaic plain, featuring flat topography and multiple alluvial river systems, notably the Tuntang, Jragung, and Serang–Lusi Rivers. These rivers were prone to overflow during periods of heavy rainfall, especially when upstream discharge from neighboring districts exceeded local channel capacity. The area was also characterized by sedimentation, tidal intrusion, and land subsidence, making it a chronic flood hotspot.

The region was primarily composed of paddy fields (59.72%), non-rice agriculture (21.1%), and rapidly expanding residential zones (19.17%). Residential settlements were organized in both urban and rural clusters, with significant concentrations in Sayung, Karanganyar, and Karangawen, areas that historically experience frequent flooding. According to the Demak Regency in Figures 2024 report, urban residential development is notably concentrated in flood-prone sub-districts such as Mranggen (7.8%), Karangawen (8.2%), and Demak City (6.3%) of the regency's total land area. These regions have been prioritized in the regional spatial plan (RTRW) as urban settlements due to their population density and economic activity, based on Demak's Regional Regulation No. 1 of 2020, a revision of the 2011–2031 spatial plan.

Historical disaster records from 1815 to 2015 indicate that the Demak Regency experienced 36 major disaster events, with extreme weather and floods accounting for approximately 65.79% of all occurrences. These events caused substantial human and material losses, including over 7,000 damaged houses and more than 14,000 hectares of land affected. More recent data from 2015 to 2021 showed an intensification of disaster frequency, with Demak recording an average of 98 disaster events annually and a peak of 174 incidents in 2017. The most recurring hazards during this period were droughts, extreme weather events, and annual floods [8]. Based on the Disaster Risk Assessment for 2016–2020, Demak remained highly vulnerable to hydrometeorological hazards such as floods, droughts, and coastal abrasion, underscoring the urgent need for sustained disaster risk reduction efforts [8].

River embankments in Demak, especially those along Wulan, Lusi, and Jeratun Rivers, had already shown vulnerabilities in the years prior to 2024. Community-based risk assessments pointed to repeated small-scale breaches and overtopping during the rainy season [8]. However, no large-scale structural retrofitting or community drills were conducted in the years before and up to early 2024, leaving the region highly exposed to hydrometeorological extremes. This systemic exposure would ultimately amplified the impact of the two flood events in February and March 2024.

3.1.2 The disaster

In February-March 2024, Demak Regency experienced two major flood events triggered by intense upstream rainfall and failing river infrastructure. The first occurred on February 5, 2024, when continuous heavy rain led to the collapse of 10 embankments, primarily affecting the sub-districts of Karangawen, Kebonagung, Wonosalam, Gajah, Dempet, Mranggen, and Karanganyar. These breaches involved both major rivers, such as the Serang, Tuntang, and Wulan Rivers, and tertiary irrigation channels, inundating hundreds of hectares of residential, agricultural, and peri-urban areas.

Field assessments recorded inundation depths of 50–150 cm in several locations, with stagnant waters persisting for up to three weeks. Public infrastructure, including schools, mosques, health centers, and national roads, was rendered unusable. By mid-February, 70 villages across 13 sub-districts were submerged, displacing over 130,000 residents and severely disrupting transportation, water supply, and public services. Most evacuation points, such as village halls and schools, quickly exceeded capacity, while sanitation systems collapsed, triggering spikes in respiratory and waterborne illnesses.

The second flood struck on March 13, 2024, at 9:00 PM, following another episode of torrential rain and strong winds. This time, seven additional embankments failed, spreading damage to both previously affected zones and new areas, including Grobogan Regency. In total, 17 embankment breaches were recorded in less than two months, underscoring the escalating pressure on Demak's already fragile flood control infrastructure.

According to the local disaster agency and *Rencana Rekonstruksi dan Rehabilitasi Paska Bencana* (R3P) documentation, this disaster caused damage and losses across multiple sectors, particularly affecting government functions, public access, and essential services. The total economic loss and damage were estimated at Rp 2.049 trillion, with damages accounting for Rp 1.432 trillion and losses amounting to Rp 617 billion. The most affected sectors included housing (Rp 1.091 trillion), public infrastructure such as clean water and roads (Rp 616 billion), and the economic sector, particularly agriculture and trade (Rp 1.146 trillion). Disruptions also extended to health, education, and public security services. These widespread impacts highlighted the vulnerability of critical public functions and economic systems in the face of extreme hydrometeorological events [9].

The combination of river overflow, inadequate drainage capacity, and limited community preparedness transformed this event from a seasonal flood into a protracted humanitarian crisis. The compound nature of the floods—intense rainfall, embankment failure, and prolonged inundation—exposed critical governance and resilience gaps in both urban and rural areas of Demak.

3.2 Disaster impacts

The 2024 flood in Demak Regency unfolded in two major waves, with the first flood in early February affecting 29,228 people across several sub-districts. The second and more severe flood in March impacted a total of 131,703 people spanning 13 sub-districts and 126 villages, severely disrupting daily life and overwhelming public infrastructure. In total, 40,354 households were affected in March 2024, with many residents forced to evacuate to temporary shelters due to rising floodwaters. Despite the scale of the disaster, there were no reported fatalities, though the physical and psychological toll was substantial. Critical community facilities—including 430 places of worship, 143 educational buildings, 15 health centers, and 15 government offices—were damaged or rendered inaccessible. The flood also inundated 9,442 hectares of rice fields and 529 hectares of aquaculture ponds, significantly affecting livelihoods. These extensive disruptions created a highly stressful situation.

3.2.1 Psychological impacts

The psychological impact of the 2024 flood in Demak was observed among both survivors and specific community groups, particularly school-aged children. A qualitative discourse analysis conducted by Junyanti (2024) revealed multiple emotional reactions experienced by students at SMAN 1 Karanganyar following the disaster. These included anxiety, depressive symptoms, and trauma-related responses such as flashbacks and loss of motivation, as reported by Antara News (2024) in the article *"Banjir Demak Terparah dalam 30 tahun terakhir – Apakah bencana ini berkaitan dengan Selat Muria?"*

Additionally, field-based psychosocial interventions were documented during emergency response efforts. According to the Ministry of Health report *Banjir di Demak, Jawa Tengah, 05-02-2024* (2024), trauma healing sessions were carried out by Central Java Police and local counselors in shelters across Demak and Kudus. Children were engaged in play therapy and psychological first aid (PFA) to mitigate distress. The program aimed to restore emotional stability in informal and child-friendly settings:

> *"We carried out psychosocial activities. Anxiety emerged as the main impact, where students felt fear, worry, and restlessness about the prospect of flooding recurring [...] the psychological consequences of the flood could include signs of depression, manifested in feelings of melancholy, hopelessness, and lack of interest in educational activities." (Junyanti, 2024, p. 204).*

> *"This is so that the children affected by the flood can stay motivated, not experience stress, and reduce anxiety and pain." (Susanto in Jatengdaily.com, 2024).*

Despite these initiatives, there is limited publicly available evidence documenting systematic assessments of psychological distress among disaster responders or broader community members. Most of the available data came from isolated observations or single-event reports. This suggests that while there was recognition of psychological vulnerability among certain groups, comprehensive data collection and follow-up psychosocial support mechanisms remained limited during the emergency response phase.

3.2.2 Operational Challenges

The 2024 Demak flood response revealed several operational challenges. One of the most critical issues was the coordination gap between local government levels. As noted in PMI's 2024 field report, these gaps specifically hampered water distribution and sanitation efforts during the second response phase:

> *"The sub-district and the village didn't coordinate well during the second phase of clean water distribution and well-draining."*

Access to remote areas was severely limited by high floodwaters and inadequate transport equipment—some boats used lacked engines or were damaged, as noted in the PMI Demak report *Emergency Response to the February–March 2024 Floods*. Additionally, the absence of an integrated data system hindered real-time identification of at-risk groups and limited informed decision-making during field operations, as documented in the PMI Demak report *Posko TDB Banjir Kabupaten Demak* (2024).

3.3 PMI response strategies

PMI Demak Branch demonstrated a geographically extensive response to the March 2024 flood, with services concentrated in the most severely affected sub-districts such as Karanganyar, Gajah, Sayung, Bonang, and Mranggen. Based on the map (Figure 2), PMI deployed key services including logistical distribution (marked by green dots), assessment teams (red dots), WASH (blue dots), and limited psychosocial support (purple dots), with the latter primarily focused in Karanganyar.

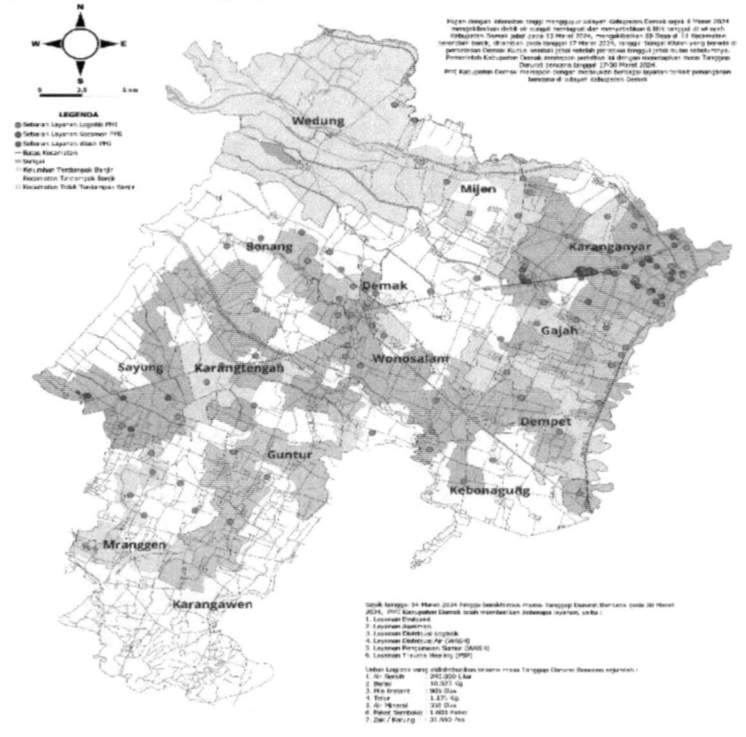

Fig 2. PMI Demak response map.

While logistical and water distribution reached a wide area across almost all flood-affected zones, psychosocial services were confined to select shelters, indicating uneven coverage. This spatial distribution highlights PMI's capacity to mobilize critical resources across regions with varied flood intensity, though it also reflects service concentration in high-density, high-impact areas rather than uniform access across the district.

PMI Demak mobilized 123 personnel, including 117 volunteers and 6 staff, to implement emergency responses. Key interventions are shown in Figure 3. The diagram outlines the structure of services provided by the Indonesian Red Cross (PMI) during the disaster response. These services are categorized into three main clusters: Disaster Services, Health Services, and Psychological Services, as described in PMI Demak's reports (*Posko TDB Banjir Kabupaten Demak, 2024,* and *Emergency Response to the February–March 2024 Floods*). Under Disaster Services, PMI delivered a range of support, including logistical services, disaster relief kitchen services, disaster response post services, evacuation services, and disaster assessment. Logistic services also encompassed water distribution, reflecting the critical role of clean water supply during emergencies. The Health Services cluster included general health service delivery as well as public facility sanitation services, which aimed to prevent disease outbreaks in affected areas. Meanwhile, psychological services focused on mental health support, particularly through targeted interventions such as psychosocial support for children affected by the floods.

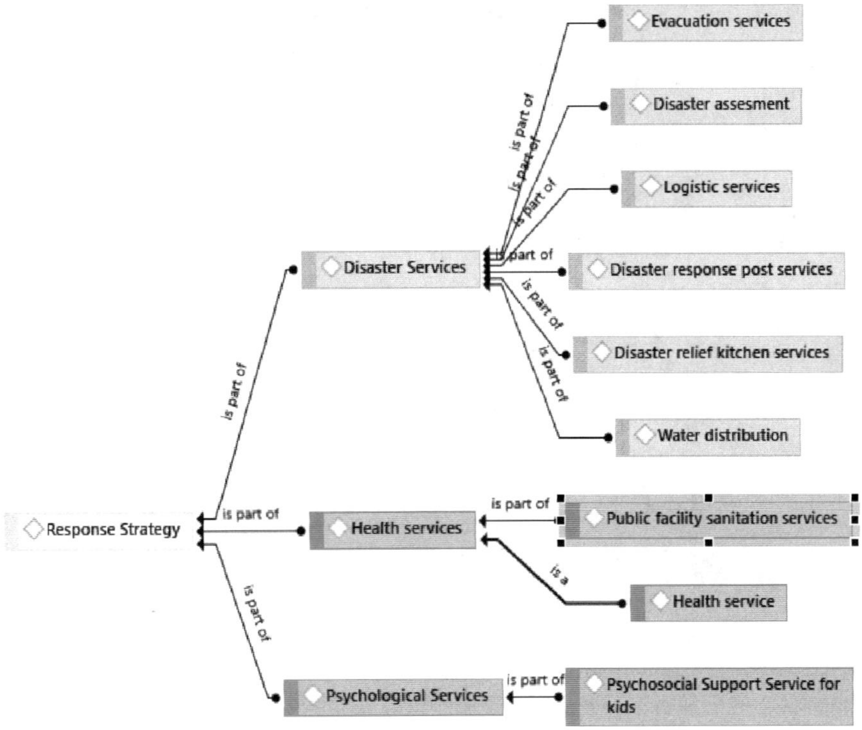

Fig 3. PMI response strategies on the 2024 Demak flood.

3.3.1 Disaster services

During the 2024 flood disaster in Demak Regency, the PMI implemented a comprehensive range of disaster services across affected areas. After the flood, PMI mobilized teams to conduct mass evacuations in at least 13 sub-districts, including Sayung, Karanganyar, Wonosalam, Dempet, and Mranggen, assisting residents from inundated villages to safe shelters. Simultaneously, PMI carried out rapid disaster assessments in locations such as

Karangtengah and Sayung to identify needs and guide further interventions. In terms of logistics distribution, PMI provided essential relief supplies—including food, hygiene kits, baby kits, cleaning equipment, and emergency sacks—across numerous sub-districts and even reached neighboring Kudus. Additionally, PMI established and operated emergency community kitchens, distributing thousands of ready-to-eat meal packages to displaced residents. Clean water provision was also a major component, with over 400,000 liters of clean water distributed in two phases, primarily benefiting residents in Karanganyar, Gajah, and Sayung (Figure 4). These coordinated efforts exemplify PMI's frontline role in safeguarding communities and addressing basic survival needs during the critical emergency phase.

As part of its emergency relief operations during the 2024 flood in Demak Regency, the PMI distributed a wide range of essential supplies to affected communities, as reported in the PMI Demak document *Emergency Response to the February–March 2024 Floods*. The aid packages included basic food items (raw food ingredients), ready-to-eat meals, hygiene kits, baby kits, medicines, sacks for flood response needs, and cleaning equipment. These items were vital in addressing the immediate needs of displaced residents, ensuring food security, sanitation, and health support in evacuation shelters and flood-affected areas.

Fig 4. Water and sanitation services (Source: PMI Demak, internal documentation, 2024).

To address the urgent need for clean water during the 2024 flood disaster in Demak, the PMI carried out large-scale water distribution services. In collaboration with local authorities, PMI deployed water tank trucks to deliver safe water directly to affected communities. The operation was conducted in two phases, distributing a total of 400,000 liters of clean water—120,000 liters in the first phase and 280,000 liters in the second. The photos (Figure 4) show residents, including children and elderly individuals, collecting water using jerry cans, buckets, and bottles—demonstrating the critical importance of this service in maintaining hygiene and preventing waterborne diseases in temporary shelters and flooded areas, as documented in the PMI Demak report *Emergency Response to the February–March 2024 Floods*.

As part of its early emergency response, the PMI Demak conducted rapid assessments across flood-affected areas to identify urgent needs and guide targeted interventions (Figure 5). Assessment activities were carried out in more than 10 sub-districts, including

Karangtengah, Sayung, Karanganyar, Mranggen, Wonosalam, Dempet, Mijen, Bonang, and Demak City. These locations encompassed dozens of villages, such as Wonorejo, Dombo, Kalicilik, and Karangmlati, where PMI teams gathered data on infrastructure damage, displacement conditions, and community vulnerabilities. Visual documentation shows PMI officers navigating flooded areas and interviewing affected residents, highlighting their active role in ensuring that humanitarian assistance was data-driven, timely, and responsive to actual field conditions.

Fig 5. Disaster assessment (Source: PMI Demak, internal documentation, 2024).

To support residents affected by the 2024 flood disaster, the PMI Demak provided widespread logistical distribution services across eight districts and one disaster response post. Relief operations covered numerous villages in sub-districts such as Karanganyar, Dempet, Sayung, Wonosalam, Wedung, Mijen, and even extended to neighboring Kudus. Key distribution points included the residence of the Demak Regent (pendopo) and various public locations across affected areas. The logistics support comprised essential supplies such as food items, hygiene kits, baby kits, cleaning tools, and sacks, ensuring that the basic needs of displaced and vulnerable populations were met quickly and efficiently during the emergency response phase (Figure 6).

Fig 6. Logistic support (Source: PMI Demak, internal documentation, 2024).

3.3.2 Health services

During the 2024 flood response, the PMI Demak provided health services at nine evacuation posts, delivering a total of 11 medical sessions. These services were carried out in various locations, including schools, religious facilities, and public offices across Karanganyar, Gajah, and Kudus. Medical teams served 358 patients who presented with common flood-related complaints such as coughs, colds, skin irritation, fatigue, and blood pressure concerns. Figure 7 shows PMI volunteers actively conducting health checks, distributing medications, and offering basic clinical consultations, demonstrating their vital role in maintaining community health and preventing illness outbreaks in emergency shelters.

Fig 7. Health Services (Source: PMI Demak, internal documentation, 2024).

As part of the post-disaster recovery efforts in Demak, the PMI carried out extensive environmental cleaning services in flood-affected areas starting from February 2024 until March 31, 2024 (5 weeks). These activities included well cleaning and disinfection, as well as the sanitization of public facilities such as schools, mosques, and village halls. The operations were implemented in multiple locations, including SDN Karanganyar 2, Balai Desa Karanganyar, Masjid Wonoketingal, SMAN 1 Karanganyar, and several surrounding villages. In total, PMI successfully cleaned 78 wells and 7 public facilities in the first phase, followed by 56 wells and 4 public facilities in the second phase. These services were crucial in restoring access to safe water and reducing the risk of disease outbreaks in the aftermath of the flood.

3.3.3 Psychosocial support services

As part of its psychosocial support services, PMI Demak provided targeted assistance to children in the Kedungwaru Lor public kitchen area, as described in the PMI Demak report *Emergency Response to the February–March 2024 Floods*. These interventions aimed to help young flood survivors cope with emotional stress and restore a sense of safety and normalcy. Activities involved interactive engagement and supportive communication in a safe, child-friendly environment. The images show PMI volunteers interacting warmly with children, emphasizing the importance of psychosocial care alongside physical aid during disaster response.

3.4 Resilience gap for disaster responders and psychosocial integration

During the 2024 Demak floods, the PMI deployed 123 volunteers across affected areas. However, available reports and media coverage did not indicate the implementation of structured psychosocial support for responders, as noted in the PMI Demak report *Emergency Response to the February–March 2024 Floods (2024)* and in the news coverage by Sorot

Indonesia (2025). Volunteers worked under stressful conditions, including prolonged deployments and exposure to traumatic scenes, without documented mechanisms for emotional recovery such as psychological first aid or peer support. This situation reflects the limited integration of responder well-being in disaster response systems.

In affected communities, particularly among children and older adults, displacement and disruption of daily routines created vulnerabilities to psychological distress. Studies highlight that stable and supportive environments are crucial for maintaining resilience in these groups [12].

One report detailed how rising floodwaters forced over 5,000 families (70,000 individuals) to evacuate from 30 villages in Demak, with many relocated to Kudus, as reported by Metro TV News (2025) in the article *"Ratusan Keluarga Terdampak Banjir di Demak Mengungsi ke Kudus"*. The disruption of housing, mobility, and basic routines significantly affected emotional well-being. A disaster volunteer explained:

> *"It kept spreading and rising, so hundreds of families had to be evacuated until this afternoon". (Fatoni, disaster volunteer, Metro TV News, 2025).*

Children were particularly vulnerable. In a psychosocial intervention report, officers noted that emotional stress and sadness were common among evacuated children. To reduce trauma and promote resilience, play therapy and motivational sessions were conducted:

> *"Specifically for the children's age group, we used play methods so they would not be carried away by sadness, and we also provided psychological counseling to residents affected by the flood". (Susilo, Jateng Daily, 2024).*

These findings reaffirm the critical importance of providing stable, emotionally supportive environments to buffer vulnerable groups against disaster-related stress, which aligns with evidence from coastal communities such as Timbulsloko, where cultural rituals and religious gatherings strengthen emotional resilience and place attachment [13].

Field documents also noted operational issues that may have affected psychosocial service delivery. These included spontaneous volunteer arrivals without structured assignment, fragmented coordination across local stakeholders, and a lack of real-time needs assessment. For example, during the second phase of clean water and sanitation efforts, coordination gaps were noted:

> *"The sub-district and the village didn't coordinate well during the second phase of clean water distribution and well-draining." (PMI Demak, 2024, Presentasi Banjir Demak, slide: Kendala yang Dihadapi Saat Layanan).*

Additional logistical constraints, such as limited access to flooded areas, inadequate transport equipment, and insufficient safety measures for volunteers, further complicated service deployment. These conditions likely constrained the precision and timeliness of psychosocial outreach during peak emergency periods. Further study is needed to understand how these gaps affect responder resilience and the integration of psychosocial components in future disaster response efforts in Demak.

4 Conclusion

In early 2024, Demak Regency, Central Java, faced two major flood events that affected 13 sub-districts, displaced over 130,000 residents, and caused extensive damage to housing, infrastructure, and livelihoods. The PMI Demak Branch mobilized 123 personnel to deliver a wide range of services, including evacuation, logistics, health care, and environmental sanitation. Using a retrospective qualitative approach, this study examined operational reports and media coverage to assess PMI's disaster response with a particular focus on psychosocial aspects.

The findings show that while physical relief was delivered through coordinated logistics and healthcare services, the psychosocial support—especially for responders—was not explicitly documented in PMI's operational reports. Available information suggests that children and older adults were among the groups receiving limited psychosocial activities through external actors, including PMI itself. However, there is insufficient data regarding systematic mental health interventions, particularly for PMI volunteers. The study also notes recurring operational issues, such as the arrival of spontaneous volunteers, limited coordination, and a lack of integrated data systems, which may have constrained the overall response.

This study highlights the importance of strengthening psychosocial integration and volunteer support mechanisms within disaster response systems. Further research is needed to explore the long-term psychosocial effects on both affected communities and disaster responders, and to inform future response planning in flood-prone areas like Demak.

Acknowledgements. The authors would like to express sincere gratitude to the Indonesian Red Cross (PMI) Demak Branch for granting access to operational reports and supporting the research process. The first author would like to express gratitude to the British Writing Academy for the training programs in writing qualitative research.

References

1. A. Guddo, Conceptualization of flooding disaster: A theoretical perspective of management. Disaster Adv. 17, 40–45 (2024). https://doi.org/10.25303/172da040045
2. V.T.T. Van, M.D. Nhat, L.N. Van, N.D. Tan, T.L.T. Hai, Designing an early flood warning system using SRF05 and S201 sensors. Proc. Int. Conf. Electr. Control Instrum. Eng. (ICECIE), Hanoi, Viet Nam (IEEE, 2024). https://doi.org/10.1109/ICECIE63774.2024.10815662
3. Badan Nasional Penanggulangan Bencana, Data dan Informasi Kebencanaan Bulanan Teraktual. Info Bencana 5, 1–19 (2024). https://bnpb.go.id/storage/app/media/Buletin Info Bencana/Buletin Info Bencana Januari 2024.pdf
4. A. Asrofi, S.R. Giyarsih, D.S. Hadmoko, The impact of tidal floods on poor households in the Sayung coast, Demak Regency, Indonesia. Indones. J. Geogr. 56, 456–465 (2024). https://doi.org/10.22146/ijg.94063
5. Pus. Data Inf. Komun. Kebencanaan BNPB, Buletin Info Bencana Februari 2024. Info Bencana 5, 2 (2024). https://bnpb.go.id/informasi-bencana/buletin-info-bencana-februari-2024
6. Palang Merah Indonesia, Posko TDB Banjir Kabupaten Demak (PMI, Demak, 2024).
7. G.A. Bowen, Document analysis as a qualitative research method. Qual. Res. J. 9, 27–40 (2009). https://doi.org/10.3316/QRJ0902027
8. Kabupaten Demak, Peraturan Bupati Demak Nomor 48 Tahun 2023 tentang Kajian Risiko Bencana Daerah Tahun 2023–2027 (Kabupaten Demak, 2023).
9. Badan Nasional Penanggulangan Bencana, Rencana Kegiatan Pendampingan Sosial Ekonomi dan Sumber Daya Alam dan Lingkungan di Kab. Bandung Barat & Kab. Demak (BNPB, Jakarta, 2025).
10. E. Junyanti, Dampak psikologis banjir terhadap siswa di SMAN 1 Karanganyar Demak: Sebuah studi analisis wacana kritis artikel Detiknews, "11 kecamatan di Demak banjir

imbas 6 tanggul jebol, 93 ribu jiwa terdampak" dengan pendekatan Norman Fairclough. Morfol. J. Ilmu Pendidik. Bahasa Sastra Budaya 2, 198–206 (2024).

11. Jateng Daily, Masyarakat terdampak banjir di Demak dan Kudus peroleh trauma healing. Jateng Daily, 1–2 (2024).
12. S.M. Southwick, G.A. Bonanno, A.S. Masten, C. Panter-Brick, R. Yehuda, Resilience definitions, theory, and challenges: Interdisciplinary perspectives. Eur. J. Psychotraumatol. 5, 25338 (2014). https://doi.org/10.3402/ejpt.v5.25338
13. M. Karmila, M.Y. Madrah, Culture and faith: Timbulsloko community adaptation and strategy in facing tidal flood disaster. IOP Conf. Ser. Earth Environ. Sci. 1524, 012008 (2025). https://doi.org/10.1088/1755-1315/1524/1/012008

Appendix. List of Primary Data Sources Analyzed

No	Document Title	Publisher/ Media	Year	Notes
1	Posko TDB Banjir Kabupaten Demak	Palang Merah Indonesia (PMI)	2024	Operational report
2	Laporan Tanggap Darurat Bencana Banjir Februari–Maret 2024	Palang Merah Indonesia (PMI)	2024	Operational report
3	Update Infografis Bencana Banjir Kabupaten Demak	Pemerintah Kabupaten Demak	2025	Government infographic
4	*Banjir Demak Terparah dalam 30 tahun terakhir – Apakah bencana ini berkaitan dengan Selat Muria?*	Antara News	2024	News report
5	*Banjir di Demak, Jawa Tengah, 05-02-2024*	Kementerian Kesehatan	2024	Situation update
6	*Ratusan Keluarga Terdampak Banjir di Demak Mengungsi ke Kudus*	Metro TV	2025	News report
7	*Banjir Demak: 88 Desa Terdampak, 12.982 Orang Mengungsi*	Kompas.com	2024	News report
8	*Banjir Demak: Apa penyebabnya dan sampai kapan hujan ekstrem melanda?*	Antara News	2024	News report
9	Keputusan Bupati Demak No. 360/56 Tahun 2024 tentang Perpanjangan Status Tanggap Darurat	Pemerintah Kabupaten Demak	2024	Government decree
10	Keputusan Bupati Demak No. 360.2/432 Tahun 2023 tentang Penetapan Status Siaga Darurat	Pemerintah Kabupaten Demak	2023	Government decree
11	Keputusan Bupati Demak No. 360/95 Tahun 2024 tentang Penetapan Status Transisi Darurat	Pemerintah Kabupaten Demak	2024	Government decree
12	*Masyarakat Terdampak Banjir di Demak dan Kudus Peroleh Trauma Healing*	Jateng Daily	2024	News report
13	*Dampak Psikologis Banjir Terhadap Siswa di SMAN 1 Karanganyar Demak*	Universitas Muhammadiyah / UGM	2024	Study/article
14	*PMI Siagakan Relawan dan Logistik Dampak Banjir di Demak*	Sorot Indonesia	2025	News report

Intra-hospital inventory optimization in disaster planning and response: An article reviews the tertiary care hospital perspective

Yaowapha Siripakarn[1,*], *Jing* Tang[1], *Natt* Leelawat[1]

[1]Master of Science Program in Risk and Disaster Management, Chulalongkorn University, Bangkok, Thailand

Abstract. Disasters, whether natural or anthropogenic, are significant challenges to hospital operations, especially in resource-constrained environments. Effective intra-hospital inventory optimization is pivotal for ensuring the continuity of care during surge events, including mass casualty incidents, pandemics, and infrastructure breakdowns. This narrative review synthesizes current evidence on hospital disaster workflows, inventory management strategies, and surge capacity frameworks. Emphasis is placed on operational resilience, allocation ethics, and the application of simulation and forecasting tools. Key models, including the 4S framework, just-in-time (JIT), and vital-essential-desirable (VED) classification, are discussed alongside their limitations during crises. The review highlights critical gaps in standard protocols and calls for the development of adaptive, ethically sound, and context-specific inventory management systems to improve disaster preparedness in hospitals.

1 Introduction

The current global disaster situation, both natural and man-made, is found to be increasing in frequency and severity each year [1]. Thailand is experiencing a similar trend, with a growing number of citizens affected by various disasters, largely due to rapid urbanization. These include floods, pandemics, chemical leaks, and fires. Disasters impact key societal capacities, including infrastructure, population well-being, economic stability, utilities, and healthcare. Healthcare systems are highlighted as a critical component in both the Sendai Framework and the Sustainable Development Goals (SDGs), as a crucial element for maintaining quality of life. The World Health Organization (WHO) emphasizes the importance of hospital disaster resilience and has developed the Hospital Safety Index [1], a tool for assessing and evaluating the safety and resilience of hospitals during emergencies. This index evaluates factors such as structural integrity, the availability of essential services, and communication systems to ensure the hospital's ability to function during crises. To preserve societal health during a disaster, healthcare systems must not only be capable of maintaining normal operations but also be able to manage an increase in medical demand [2]. The increase in medical demand during a disaster can result from injuries, such as those caused by drowning

* Corresponding author: yssiripakarn@gmail.com

or debris, infectious diseases resulting from poor sanitation, or pandemics that themselves emerge as disasters. However, disasters often lead to significant reductions in access to healthcare services and shortages of critical medical supplies, complicating both routine healthcare activities and acute disaster response efforts.

Lessons from past disasters, such as Hurricane Katrina in 2005 in the Gulf of Mexico, which resulted in more than 1800 fatalities, demonstrated that the loss of healthcare infrastructure affects not only disaster-related casualties but also individuals with chronic illnesses who require continuous care. The consequences may be more severe in low- and middle-income countries, where healthcare systems are ill-prepared to address medical needs and other critical medical demands during disasters.

Surge capacity is a key concept in disaster management, referring to "the ability of a health service to expand beyond its normal capacity to meet increased demand for clinical care" [3]. Its implementation is often complicated by unclear guidelines on the timing, methods, and extent of capacity surges, which can lead to both resource shortages and excesses. Effective surge capacity relies on a well-functioning incident management system and an adequate supply of resources, including space, medical supplies, personnel, and special considerations. Mutual aid agreements between healthcare facilities (public and private), emergency services, and other sectors (e.g., fire, police, military) are vital to prevent overwhelming a single facility or area. Despite the importance of surge capacity, significant challenges remain in the allocation of resources during disasters. For instance, during the COVID-19 pandemic [4-7], the WHO declared COVID-19 a Public Health Emergency of International Concern, resulting in a sharp increase in the demand for medical supplies and personnel. Numerous reports from hospitals worldwide highlighted shortages in critical resources, such as ventilators, personal protective equipment (PPE), and medical staff, illustrating the urgent need for effective resource distribution during disasters.

In tertiary hospitals, which typically operate at or near full capacity under normal conditions, the allocation of resources becomes increasingly complex during disasters due to the involvement of multiple departments. The demand for resources may fluctuate throughout the disaster response period, depending on patient flow and the treatments required. For instance, an initial surge in patient volume typically occurs in the emergency department, followed by patient distribution to other areas of the hospital.[6]· The increased demand for critical care during disasters, particularly from patients with infectious diseases, necessitates the expansion of hospital capabilities, including staffing, medical equipment, and physical infrastructure.

Resource distribution and communication failures during disasters can result in inefficiencies, such as the misallocation of supplies and personnel. For example, uncoordinated inventory management often results in shortages in some departments while other areas experience surpluses [4]. Effective planning, including the identification and prioritization of critical healthcare functions, is essential to ensure continuity of care during disasters [3]. Studies have examined optimized hospital inventory management models for both normal operations and disaster response [8-12]. In addition to ensuring the availability of sufficient medical supplies, inventory management must also prioritize the safety, cost-effectiveness, and expiration dates of these supplies. Previous approaches, such as JIT inventory management, stockpiling, strategic reserves, and flexible surge capacity models, have been widely discussed in the literature. However, the lack of standardized guidelines and crisis needs assessment for resource allocation further complicates disaster response efforts. This gap highlights the need for a framework that follows the principles of medical disaster management, particularly the priority of saving lives under conditions of limited resources. Therefore, it is crucial to establish clear guidelines and systems for resource allocation within hospitals to ensure that healthcare systems can meet the increased demand during disasters. Improper resource allocation can lead to disparities in care, with some areas over-resourced while others remain underserved, adversely affecting patient outcomes. This review identifies critical gaps in existing protocols and inventory management systems, underscoring the need for strengthened approaches to enhance hospital disaster preparedness.

2 Frameworks for Hospital Disaster Response and Inventory Management Models

2.1 Hospital workflow and hospital incident command system response during disaster and mass casualty incidents

Hospitals play a critical role in responding to disasters and public health emergencies. The level of hospitals, typically based on available resources and infrastructure, substantially influences response capacity, resulting in considerable variability in disaster preparedness and capabilities across institutions [12]. Although several studies have examined hospital surge capacity in different settings [4,5,8]. A systematic review noted that most previous research has compared hospitals based on urban versus rural location, rather than level of care [6]. This systematic review found that while urban and rural hospitals differ in available resources, no study has explicitly analyzed differences in surge capacity by hospital level. Furthermore, due to the lack of specific location data, no conclusive inferences could be made regarding the disparity in surge preparedness between rural and urban hospitals.

2.2 Demand-surge time hospital workflow

The processes involved in patient care are interconnected with various parts of the hospital, including those related to direct treatment processes, such as nurses, physicians, operating rooms, and supportive functions such as medical records, finance and treasury, and public utility service. In emergencies or during patient surges, the characteristics of patients' flow remain closely related to multiple areas, but the volume tends to increase (Figure 1).

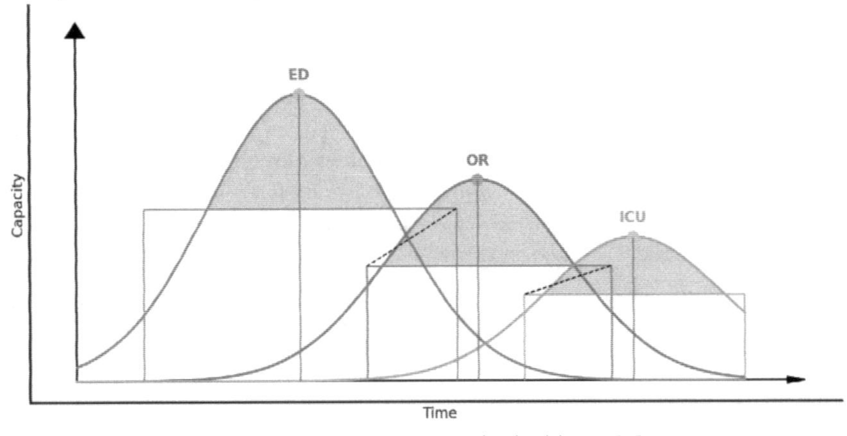

Fig. 1. Hospital patients and resource peak during casualty incidence [5].

Krintina L. et.al., through a comprehensive mass casualty response simulation, found that although the complexity of hospital patient flow remains during disaster response, the surge in patient volume and resource demands shift over time, from the point of entry or emergency department to other areas of the hospital providing definitive care [6].

2.3 Hospital inventory management strategies

Inventory management is the process of overseeing the flow of goods and materials by monitoring stock levels, coordinating procurement and storage, and ensuring efficient utilization of resources. It involves planning, demand forecasting, order management, replenishment, and tracking to maintain adequate supply while minimizing costs. In hospitals, inventory management extends to the availability, storage, distribution, and utilization of medical supplies, equipment, and pharmaceuticals. Unlike typical cost-driven models, hospital systems must prioritize patient service levels, making it essential to ensure the availability of high-quality healthcare items in an uncertain and fluctuating environment. The traditional strength of healthcare systems lies in delivering high-quality care through skilled physicians, nurses, and staff supported by robust infrastructure, advanced technology, and reliable medical supplies. During a disaster, the focus has shifted toward operational resilience-ensuring that healthcare delivery remains functional under all circumstances, including during disasters and crises. This includes capacity planning, resource allocation, inventory management systems, demand forecasting, scheduling, and other operational activities [11].

Hospital inventory management adopts the concept of industrial management, including ABC inventory management, Just-in-Time (JIT), and Just-in-Case (JIC) approaches, to optimize the availability of medical supplies. The ABC inventory management categorizes items based on their value and importance, allowing for prioritized monitoring and control of high-value items. The ABC classification for prioritization of inventory management may not be well-suited for medical use [11]. This is because certain medications or equipment may be essential for treatment but may not fall under the high-value category of Group A. Therefore, the vital, essential, and desirable (VED) principle is used, which considers the necessity of items over their cost and physical volume. JIT, by contrast, aims to focus on maintaining inventory levels with real-time demand, reducing the need for excessive stockpiling. A study reported that using the JIT approach in hospital inventory management can maximize cost-effectiveness and minimize wastage and overstocking [8]. However, JIT may not be appropriately used during a casualty incident or disaster due to its tardiness and inability to respond to timely and uncertain increases in medical demand. In such contexts, the JIC approach provides a complementary strategy, ensuring the pre-positioning of critical resources to maintain operational continuity during emergencies.

Therefore, the appropriateness of using these concepts in hospital inventory management is still questioned due to the complexity of patient flow and treatment procedures. Hospital operations involve diverse departments and specialized care pathways. Some procedures are lifesaving and require immediate action, but they may involve low-value items and occur infrequently during normal circumstances. Particularly during demand surges in disasters, the varying needs of patients and unpredictable demand patterns limited the effectiveness of the aforementioned inventory approaches. Furthermore, measuring the losses and benefits in medical procedures is more challenging compared to industrial aspects, as the outcome directly affects individuals' lives and disabilities. During the COVID-19 pandemic, it is considered a global supply chain crisis that has affected all parts of the world, including the medical and healthcare sectors. In response, studies have been conducted on the distribution and allocation of medical resources, resulting in various recommendations [3,7]. The findings of these studies during the COVID-19 period align with the previous research that the existing resource allocation models and inventory management may not be suitable for medical resource allocation.

In a study to examine the suitability of different medical resource allocation models in both normal and disaster situations [7]. The study employed simulation software to evaluate three inventory strategies: reorder point policy, periodic order policy (POP), and JIT. The findings indicated that the Just-in-Time policy was the most cost-effective when there was no lockdown or only partial lockdown, as it prevented stockouts and overstocking. However, in the case of a complete lockdown, the periodic order policy was found to be a more suitable strategy.

2.4 Role of inventory management in hospital surge capacity, hospital inventory management during disasters

In situations with increased medical demands, such as disasters, mass casualty incidents, and mass gatherings, there is a process of resource augmentation for disaster management. This process is known as surge capacity. Surge capacity is "the ability to respond to a sudden increase in patient care demands" [3]. The concept of hospital surge capacity is the ability to expand capacity to accommodate and effectively manage a significant increase in patient volume to ensure that the entire hospital can maintain the quality of care and patient safety during scarce resources.

Barbisch and Koening (2006) defined surge capacity components consisting of '4 S'. HaSan et. al. also reviewed the activities in surging these four components [7]. The first component, staff, refers to the availability of healthcare professionals to meet increased demand, with strategies such as recalling off-duty staff, activating standby teams, mobilizing volunteers, recruiting retired professionals, and training non-specialists for trauma care. The second, stuff, including medical supplies, pharmaceuticals, and essential equipment, with strategies focused on reallocation and redistribution to meet urgent needs. The third component, structures, refers to physical space for patient care, which may expand to using non-clinical areas, increasing beds through safe early patients' discharge, reverse triage, and increasing interfacility transport capacity. Finally, systems involve the development and continuous updating of policies and procedures to guide effective surge planning and coordination.

2.5 Tools for surge capacity preparedness

There are some studies on how to estimate the resources needed during surge capacity [11,13], such as the Hospital Medical Surge Preparedness Index (HMSPI) and the Surge Simulation Tool (SST) or Web-based simulation tool. A study proposed a web-based simulation tool designed to enable emergency planners to proactively estimate hospital surge capacity [11]. Similarly, another study developed and tested an optimization model for early patient disposition through hospital surge capacity scenarios [13]. This model supports the development of effective preparedness and response plans during disasters and emergency surges. The HMSPI was developed by the U.S. Assistant Secretary for Preparedness and Response (ASPR) to quantify hospital-level preparedness for medical surge events, including disasters, pandemics, and mass casualty incidents (MCIs). The MSPI evaluates four key domains—staff, stuff, space, and systems—as well as operational resilience and continuity of operations planning (COOP). The tool is designed for self-assessment by hospitals to guide preparedness improvements. However, there is still an unclear point regarding surging capacity. The gaps revolve around determining the appropriate timing for surging capacity, the quantity of capacity that should be mobilized, and the timeframe within which it should occur.

Based on reported data, it was found that the time period before a disaster-related patient is transported to a hospital varied, depending on the type of disasters, the affected transportation systems, the population affected, regional healthcare facilities, and distances to healthcare facilities [14]. It has been reported that the impact may start within 90 minutes and reach its peak demand within 24 hours after the occurrence of a disaster. The report found that 95% of acute medical demand during disasters typically occurs within the first 24 hours.

This is in line with the recommendation of the Thai National Institute of Emergency Medicine, stating that all hospitals should be able to sustain their services independently for at least 24 hours before seeking external assistance.

Medical surge capacity typically involves the following steps: identifying the medical need, determining the resources required to address it in a timely manner, rapidly mobilizing those resources to the locations of patient demand, and managing them effectively to maximize their capacity and impact. To operationalize these steps, the U.S. Institute of Medicine and other frameworks on the allocation of scarce resources in disasters emphasize that healthcare delivery occurs along a continuum of care, with the level of care determined by resource availability [8]. This continuum includes conventional care, in which usual resources and standard level of care are maintained through the maximal use of the facilities existing beds, staff, and supplies, contingency care, where functionally equivalent care is delivered by adapting from usual practices; for example, boarding critical care patients in post anesthesia care areas; and crisis care, where inadequate resources are available to provide equivalent care, and treatment is provided at the best possible level given resource gaps. Crisis care is associated with increasing morbidity and mortality; these risks can be mitigated through implementing proactive resource use strategies.

The continuum of care framework guides decision-making on resource allocation and operational adjustments, allowing healthcare facilities to scale responses appropriately during mass-casualty incidents. Resource utilization strategies, such as prioritization protocols and modified standards of care, are critical for managing scarcity while maintaining life-saving interventions. Although the specific medical supplies requiring surge are not universally defined, the literature commonly refers to “critical resources,” including testing, personal protective equipment (PPE), intensive care unit (ICU) beds, ventilators, therapeutics, and vaccines, as those warranting prioritization. Emanuel et al. [9] provided key recommendations for resource allocation during the COVID-19 pandemic, emphasizing the maximization of benefit by saving the most lives and years of life, prioritizing frontline healthcare workers for critical interventions, and ensuring equality through random allocation among patients with similar prognoses. They further stressed that prioritization should vary by intervention and adapt to emerging scientific evidence, that individuals participating in research should be recognized in allocation decisions, and that scarce resources should be allocated equitably across patients with COVID-19 and those with other medical conditions.

The COVID-19 pandemic caused global shortages of essential medical equipment and supplies, including PPE, ventilators, medications, and oxygen, with significant impacts on healthcare personnel [15]. These constraints necessitated resource allocation strategies that considered cost-effectiveness, lifespan, patient age, and comorbidities. Ethical principles—beneficence, non-maleficence, autonomy, and justice—guided decision-making, alongside utilitarian approaches emphasizing the greatest good for the greatest number. In 2021, the Ministry of Public Health issued Thailand’s guideline for critical-care resource allocation, which diverged from conventional severity-based prioritization by emphasizing survival likelihood, patient characteristics, and disease prognosis [15]. According to previous studies, decisions regarding the allocation of intensive care resources in the context of the COVID-19 pandemic, as recommended by the German interdisciplinary Association for Intensive and Emergency Medicine, working group on ethics in medicine, German Society for Palliative Medicine, should be based on the best available information. This includes the following: (1) information on the patient’s current clinical condition; (2) Information about the patient’s wishes (current/declared in advance/ previously orally expressed or presumed); (3) Medical history/clinical assessment of comorbidities; (4) Medical history and clinical assessment of the general health status (including frailty, e.g., according to the Clinical Frailty Scale); (5) Laboratory parameters regarding points 1 and 3, if available; (6) Scores with prognostic relevance (e.g., sepsis-related organ failure assessment [SOFA] score)

Some studies have proposed strategies for optimizing hospital inventory and surge capacity, including the application of machine learning in simulation settings and predictive analytics [14]. These approaches aim to create a more flexible, context-sensitive approach to enhance disaster response and facilitate resource sharing between responding departments within hospitals. However, there remains a lack of standardized protocols or models specifically designed for adjusting inventory management during hospital disaster response.

3 Gaps and future directions

Despite increasing recognition of the importance of inventory optimization in disaster contexts, several critical gaps remain. Firstly, there is an absence of standardized hospital-level protocols specifically tailored for inventory planning during disasters, which limits consistency and preparedness across institutions. Secondly, the integration of predictive analytics and real-time inventory tracking systems into hospital operations remains inadequate, undermining the ability to anticipate and respond efficiently to supply chain disruptions. Thirdly, coordination across hospital departments is often deficient, leading to redundant stockpiling in some areas and critical shortages in others. Lastly, many institutions lack adequate training programs focused on surge modeling, resource triage, and the ethical dimensions of inventory management, further weakening their operational readiness in times of crisis.

Future directions should prioritize the development of context-sensitive, simulation-based inventory models that can adapt to the unique challenges of different healthcare settings during disasters. In addition, hospitals should invest in digital inventory platforms that offer real-time visibility, enabling timely decision-making and efficient resource allocation. Strengthening cross-sector partnerships is also essential to facilitate mutual aid agreements and coordinated resource pooling, enhancing overall system resilience during large-scale emergencies.

4 Conclusion

Intra-hospital inventory optimization is fundamental to effective disaster response. As health systems face increasingly frequent and complex threats, it is critical to transition from cost-minimization models to resilience-oriented strategies. By integrating ethical principles, predictive tools, and coordinated response systems, hospitals can better safeguard their capacity to deliver care during disasters.

References

1. World Health Organization; Pan American Health Organization, Hospital Safety Index: Guide for Evaluators, 2nd ed. (WHO, Geneva, 2015).
2. N. Masbi, N.S.A. Malek, A.S.A. Razali, M.I. Ismail, Challenges of providing special care services in hospitals during emergencies and disasters: a scoping review. BMC Emerg. Med. 24, 238 (2024). https://doi.org/10.1186/s12873-024-01160-1
3. S. Mehrotra, H. Rahimian, M. Barah, F. Luo, K. Schantz, A model of supply-chain decisions for resource sharing with an application to ventilator allocation to combat COVID-19. Manuf. Serv. Oper. Manag. 23, 4, 888–908 (2020). https://doi.org/10.1287/msom.2020.0922
4. K.L. Montán, P. Örtenwall, M. Blimark, C. Montán, S. Lennquist, A method for detailed determination of hospital surge capacity: a prerequisite for optimal preparedness for

mass-casualty incidents. Eur. J. Trauma Emerg. Surg. 49, 2, 619–632 (2023). https://doi.org/10.1007/s00068-022-02081-z

5. Association for Health Care Resource & Materials Management, Supply Chain Disaster Preparedness Manual (AHRMM, Chicago, 2020).
6. M.K. Hasan, S.M. Nasrullah, A. Quattrocchi, P. Arcos González, R. Castro-Delgado, Hospital surge capacity preparedness in disasters and emergencies: a systematic review. Public Health 225, 12–21 (2023). https://doi.org/10.1016/j.puhe.2023.03.009
7. S. Mehrotra, H. Rahimian, M. Barah, F. Luo, K. Schantz, A model of supply-chain decisions for resource sharing with an application to ventilator allocation to combat COVID-19. Manuf. Serv. Oper. Manag. 23, 4, 888–908 (2020). https://doi.org/10.1287/msom.2020.0922
8. J.L. Hick, D. Hanfling, S.V. Cantrill, Allocating scarce resources in disasters: emergency department principles. Ann. Emerg. Med. 59, 3, 177–187 (2012).
9. E.J. Emanuel, R. Upshur, B. Thome, M. Parker, A. Glickman, G. Persad, C. Zhang, C. Boyle, M. Smith, J.P. Phillips, Fair allocation of scarce medical resources in the time of COVID-19. N. Engl. J. Med. 382, 21, 2049–2055 (2020).
10. D.E. Marcozzi, R. Pietrobon, J.V. Lawler, M.T. French, C. Mecher, J.E. Peffer, et al., Development of a Hospital Medical Surge Preparedness Index using a national hospital survey. Health Serv. Outcomes Res. Methodol. 20, 1, 60–83 (2020). https://doi.org/10.1007/s10742-020-00208-6
11. E. Saha, P.K. Ray, Modelling and analysis of inventory management systems in healthcare: a review and reflections. Comput. Ind. Eng. 139, 106561 (2020). https://doi.org/10.1016/j.cie.2019.106561
12. B. Balkhi, A. Alshahrani, A. Khan, Just-in-Time approach in healthcare inventory management: Does it really work? Saudi Pharm. J. 30, 1830–1835 (2022). https://doi.org/10.1016/j.jsps.2022.10.013
13. M.K. Hasan, S.M. Nasrullah, A. Quattrocchi, P. Arcos González, R. Castro-Delgado, Hospital surge capacity preparedness in disasters and emergencies: a systematic review. Public Health 225, 12–21 (2023). https://doi.org/10.1016/j.puhe.2023.09.017
14. A. Khorram-Manesh, Developing a conceptual framework for flexible surge capacity based on complexity and collaborative theoretical frameworks. Public Health 208, 46–51 (2022). https://doi.org/10.1016/j.puhe.2022.04.012
15. Thailand Society of Critical Care Medicine; Ministry of Public Health Thailand, Thailand's Guideline for the Allocation of Critical-Care Resources in the Context of the COVID-19 Pandemic (Thailand Society of Critical Care Medicine, Bangkok, 2021).

Psychosocial Support and Disaster Preparedness in Early Childhood Education: Strategies for Coping with Volcanic and Environmental Hazards in Indonesia

Dewi Mufidatul Ummah[1,2,*], *Koentjoro* Soeparno[2], *Muh Aris* Marfa'i[3], *Putri* Praditya Pertiwi[2], and *Muna* Faiza Amatullah[4]

[1]Faculty of Medicine and Health Sciences, Khairun University, Indonesia
[2]Faculty of Psychology, Gadjah Mada University, Yogyakarta, Indonesia
[3]Faculty of Geography, Gadjah Mada University, Yogyakarta, Indonesia
[4]Faculty of Sharia, IAIN Ternate, Indonesia

Abstract. Psychological preparedness is essential for early childhood educators, who play a central role in supporting children's emotional stability during disasters intrinsically linked to environmental hazards. However, this dimension remains underexplored, particularly in vulnerable and geographically isolated ecosystems such as volcanic islands in Indonesia. This study aims to examine the psychological preparedness of early childhood teachers on Ternate Island, focusing on how they develop adaptive strategies in the face of recurrent volcanic hazards. The research was conducted in a community routinely exposed to eruptions from Mount Gamalama and the cascading environmental risks they generate. Employing a qualitative sequential multimethod design, data were gathered through open-ended questionnaires completed by 62 experienced teachers, followed by a focus group discussion with seven participants who had direct disaster exposure. Thematic analysis identified five interrelated dimensions of psychological preparedness: personal psychological readiness, cognitive knowledge and experience, emotional-spiritual coping, social-communal support, and environmental control strategies. The focus group further underscored the need for sustained psychosocial support throughout all phases of disaster management and improved geospatial literacy among educators to better interpret and respond to environmental signals. The findings demonstrate that repeated disaster exposure enhances psychological preparedness and adaptive capacity for environmental change. This study contributes to disaster and environmental education by offering a context-sensitive framework for strengthening preparedness in early childhood education, integrating emotional, spiritual, and practical strategies tailored to local ecological conditions.

1 Introduction

Indonesia, an expansive archipelagic nation situated along the Pacific Ring of Fire, is among the most disaster-prone regions in the world due to its high levels of tectonic and volcanic activity. The country is home to more than 127 active volcanoes, many of which are located

* Corresponding author: dewimufidatulummah1984@mail.ugm.ac.id

near densely populated areas, creating substantial risks for communities and critical infrastructure [1]. Past eruptions, such as those of Mount Merapi in 2010 and Mount Marapi in 2023, resulted in fatalities, large-scale displacement, and severe economic losses [2]. These events serve as crucial reminders of the dangers posed by living close to active volcanoes. There is a critical need for comprehensive disaster preparedness to reduce the impact of disasters and increase community resilience in the face of environmental hazards.

As the effects of climate change intensify, the frequency and severity of volcanic eruptions, along with cascading environmental risks such as ashfall and floods, are expected to increase, further endangering vulnerable populations [3]. One particularly distinctive case is Mount Gamalama, a prominent Type A stratovolcano located on the small yet historically significant volcanic island of Ternate in North Maluku, Indonesia (Figure 1). This enchanting island, characterized by its lush landscapes and rich biodiversity, has been significantly shaped by Gamalama's explosive nature. Despite its limited geographic size, Gamalama has recorded more than 70 eruptions since the 16th century, many of which occurred with little or no warning [3-7]. The volcano's location atop a densely populated island intensifies the risks associated with its eruptions. The catastrophic lahar in 2011 serves as a reminder of the volcano's destructive potential, causing significant damage to housing and infrastructure. The risks are further intensified by the island's dense population, where more than 63% of residents are classified as highly exposed to volcanic hazards according to the national disaster vulnerability index [3]. This convergence of frequent eruptions, high exposure, and limited warning time makes Ternate one of the most vulnerable small-island environments in Indonesia.

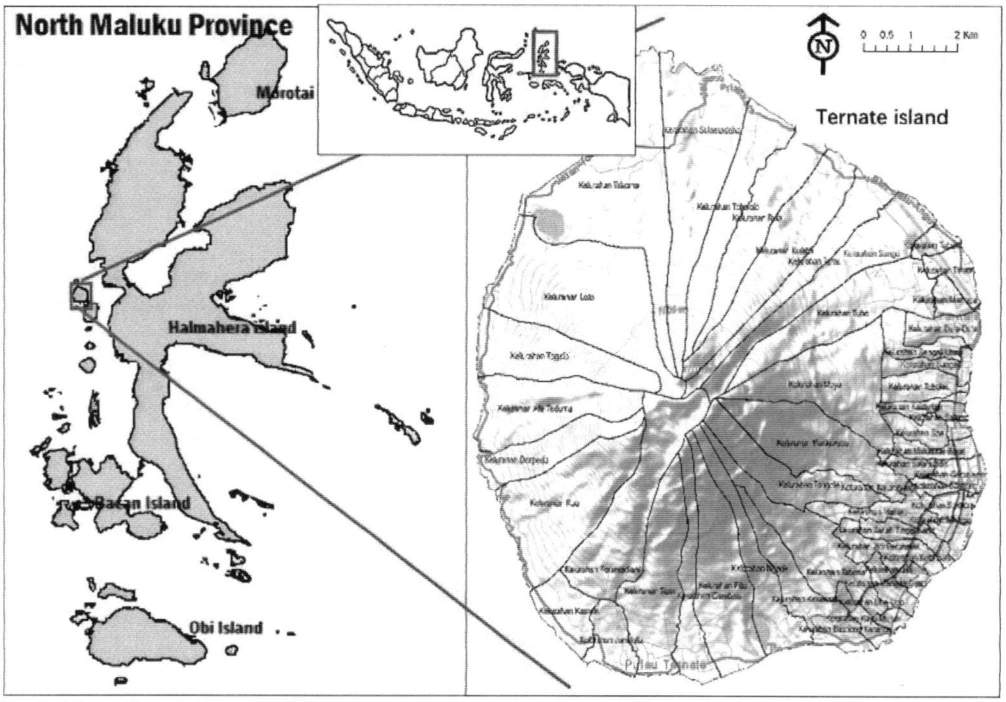

Fig. 1. Map of study location [3].

Despite the high level of volcanic risk, preparedness among communities in Ternate remains inadequate. Studies have shown that disaster awareness is generally low, with unsafe housing construction and limited knowledge of evacuation procedures contributing to this lack of awareness [3]. A cross-sectional survey revealed that more than 70% of families had little understanding of disaster management, reflecting weak household-level readiness [6]. Furthermore, shortcomings in early warning communication and the absence of clear evacuation routes during the 2011 and 2015 eruptions hindered timely responses and increased exposure to harm [7]. This situation underscores a critical gap between Ternate's high disaster vulnerability and the community's limited preparedness capacity.

Among the most vulnerable groups in volcanic disasters are young children, whose emotional and cognitive development can be severely disrupted by sudden evacuations, ashfall, or prolonged school closures [8]. Episodes of fear, anxiety, and confusion are commonly reported, particularly when children are separated from caregivers or receive insufficient guidance during emergencies. In this context, early childhood educators hold a central role in providing psychological support and maintaining stability in learning environments. However, literature research indicates that preschool teachers often lack adequate training, resources, and coping strategies to address children's psychological needs during volcanic events [9]. his gap underscores the importance of strengthening psychological preparedness in early childhood education as a vital component of disaster risk reduction and environmental education on the island.

Young children are among the most at-risk groups during volcanic disasters, particularly in areas with frequent eruptions, such as Ternate Island. The proximity to Mount Gamalama exposes early childhood populations to hazards such as ashfall, lahar floods, and sudden evacuations. These events can disrupt routines and affect children's emotional and cognitive development [10]. Local studies have shown that many preschool-aged children experience anxiety and confusion during eruptions, especially when separated from caregivers or lacking clear guidance. Early childhood educators in Ternate often lack the training and resources to provide psychological support in such situations. In rural areas of the island, limited access to safe learning spaces and health services further increases the risk. These challenges highlight the urgent need for child-focused disaster risk strategies, including age-appropriate education, teacher preparedness, and family engagement. Strengthening these components is essential for building psychological resilience in children living near active volcanoes.

Research in other volcanic settings has shown that psychological preparedness varies considerably across age groups. For example, studies in Stromboli, Italy, reported that adolescents demonstrated higher adaptive responses to volcanic hazards compared to younger children, mainly due to greater cognitive maturity and access to hazard-related information [11]. Similarly, research conducted in the aftermath of the Mount Unzen eruption in Japan indicated that adults were better able to cope with evacuation and post-disaster stress through established social support networks, whereas younger populations exhibited higher levels of anxiety and dependency [12]. In the case of Mount St. Helens in the United States, long-term follow-up studies found that while adults eventually developed effective coping strategies, children continued to display prolonged symptoms of fear and confusion, underscoring their heightened vulnerability. These comparisons highlight that, although all populations living in volcanic environments face significant psychological challenges, early childhood groups require distinct forms of preparedness and targeted educational interventions.

Despite the growing body of research on disaster preparedness in Indonesia, the psychological dimension within early childhood education remains underexplored, particularly in small-island contexts exposed to frequent volcanic hazards. Addressing this gap, this study aims to examine the psychological preparedness of early childhood teachers on Ternate Island in relation to volcanic hazards, with a focus on their coping strategies, knowledge, and support systems. Through this approach, the study aims to provide a context-sensitive framework for incorporating psychological preparedness into disaster management and environmental education, thereby improving the safety and wellbeing of young children in high-risk environments.

2 Methods

2.1 Study design

This research employs a qualitative multi-method sequential design, involving the collection of multiple qualitative data sets to understand the study's problems [13]. This method uses a level of integration through a joint display and reporting approach to present the integration of all data in one matrix. This study is divided into open-ended questionnaires and a focus group discussion (FGD). The procedural diagram of the compound research in shown in Figure 2.

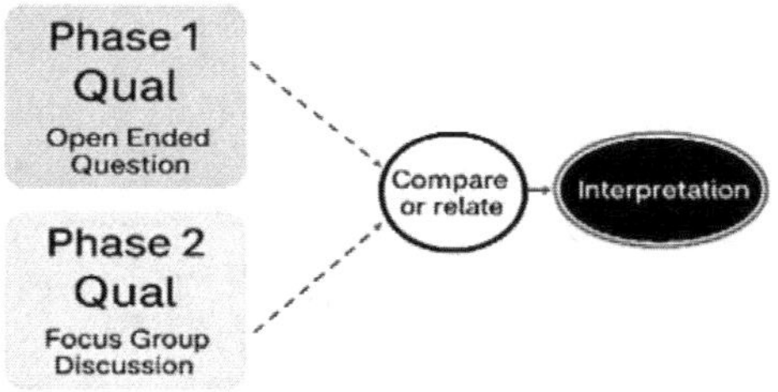

Fig. 2. Procedural diagram of research.

In this study, phase 1 was collected through open-ended questions posed to respondents, focusing on their understanding of psychological preparedness for the Gamalama eruption disaster. The second qualitative phase of data collection uses the FGD. Participants conducted a group discussion about their experience in school based on the psychological aspects of disaster preparedness developed by Gadjah Mada University. Then, data were systematically organized following the seven-stage framework established by Braun and Clarke, which includes: (1) transcription, (2) initial reading, (3) coding, (4) theme searching, (5) theme review, (6) theme definition and naming, and (7) report writing. The qualitative data flow is illustrated in the accompanying Figure 3.

Fig. 3. Stages of qualitative data analysis.

This research utilised MAXQDA 2024 software for the first four phases of their study, while the later phases were carried out manually through collaborative discussions within the research team. MAXQDA is a software tool designed to facilitate the analysis of data through methods such as coding, categorisation, and the assessment of result validity. Identifying themes in stages four to six was performed using the thematic analysis framework established by Moran-Ellis et al. Integration in multimethod research is carried out at the research design, interpretation, and reporting levels. A joint display approach will be used in the interpretation and reporting section, which presents the data from the qualitative study into a matrix. This aims to gain new knowledge beyond the results of qualitative and quantitative studies conducted separately. The joint display is both a method and a cognitive framework for integrating data, based on a deliberate process and rational thinking.

2.2 Ethics and Consent

Ethical approval for this research was obtained from the Research Ethics Committee of the Faculty of Psychology, Gadjah Mada University, Number: 9748/UN1/FPSi

1.3/SD/PT.01.04/2024. Participants completed the consent form either online via the form or by signing the printed version and returning it to the researcher. Participation is voluntary, and consent may be withdrawn at any time. Participant identification information is used solely for survey administration. Pseudonyms are used to protect participant identity during data analysis and reporting.

2.3 Participants

The sample of this study was parents who live on Ternate Island, North Maluku. The study's qualitative open-ended question sample consisted of 62 preschool teachers with the inclusion criteria: 1) teaching in preschools located in Ternate, and 2) residing on the island of Ternate. Participants were given a questionnaire in Google Form format distributed through WhatsApp social media. Participants of the second phase are preschool teachers in disaster-prone locations of the Mount Gamalama eruption. Researchers collected data at Tubo, Northern Ternate, which was one of the nearest locations to the top of Mount Gamalama. This exploratory study of seven teachers was conducted with the criteria of participants: 1) having experienced a volcanic disaster, and 2) teaching in preschools located in the most vulnerable areas. Participation was voluntary, and all participants read the form, giving consent to participate in this study, which is attached to the questionnaire. Respondents in FGD (N=7) also consented to recording with a recorder in the data collection process. The demographic data of the study participants can be seen in Table 1 below.

Table 1. Demographics of participants.

Characteristics	Category	Frequency	Percentage
Gender	Female	62	100
	Male	0	0
Education	Undergraduate	54	86.89
	Masters	3	4.92
	High school	5	8.2
Duration of domicile	0-10 years	12	20
	11-20 years	16	25
	21-30 years	22	35
	31-40 years	6	10
	41-50 years	6	10
Tribes	Ternate	32	50.82
	Jawa	9	14.76
	Makian	6	9.84
	Maluku	4	6.56
	Bugis	4	6.56
	Tidore	3	4.92
	Tobelo	2	3.28
	Others Tribe	3	4.92
Age	21-30 years old	20	32.79
	31-40 years old	21	34.43
	41-50 years old	18	29.51
	51-60 years old	2	3.28
Region	North Ternate	25	40.98
	Centre Ternate	17	27.87
	Temate Selatan	15	24.59
	Temate Island	4	6.56
Total		100	

Based on Table 1, it is known that a total of 62 early childhood education teachers participated in this study. The demographic composition reflects that all of the respondents are women, indicating the gender bias in early childhood education in the study area. In terms of age distribution, most of the participants (34.43%) were in the 31–40 years age group, suggesting a relatively young but professionally active teaching population. This was followed by 32.79% aged 21–30 years, 29.51% aged 41–50 years, and 3.2% aged 51 years and above. This age distribution suggests that the majority of respondents are in their optimal working age group and can use their physical capabilities in combination with their potential knowledge and skills to help achieve effective disaster preparedness.

In terms of their level of education, the majority, 86.9% held an undergraduate degree, 5.2% had a Master's degree, and 8.2% had a Senior High School education. This heterogeneity indicates a broad spectrum of formal educational backgrounds in early childhood teachers, which could contribute to taking up and interpreting disaster preparedness knowledge. Regarding years of tribes, the majority of the participants (50.8%) reported from Ternate, 14.76% Javanese, 9.8% from Makian, and 24.6% other. These numbers indicate that the sample represents various participants.

Finally, 62.9% of respondents had experienced volcanic eruptions that made them appropriate informants for this study. Their direct exposure to disaster is an enriching aspect of the data available, and enables a rich understanding of experiences of psychological readiness and practical responses in Early Childhood Education as influenced by direct experiences of disaster events.

3 Results

3.1 Open-ended questions analysis

The researcher conducted a thematic analysis of the results of the respondents' open-ended questions and narrowed them into several themes that describe respondents' psychological preparedness for Gamalama's eruption disaster. The themes are shown in Figure 4.

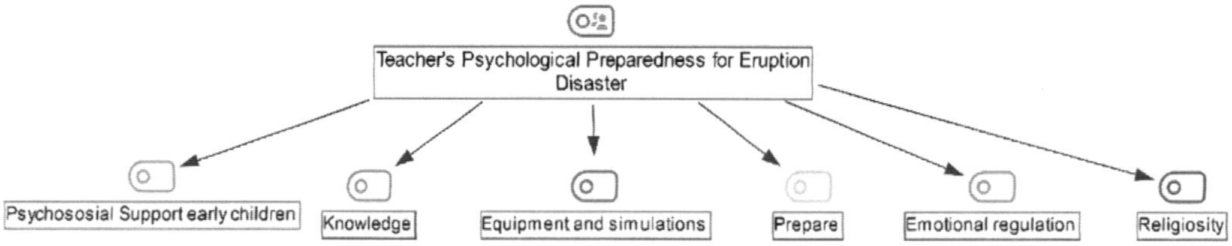

Fig. 4. Open-ended question themes.

After going through transcription and recognition reading, the researcher finds at least 339 codes based on the respondents' answers. Then the researcher classified them into six themes: psychosocial support for early childhood, knowledge, equipment and simulations, preparation, emotional regulation, and religiosity. Six themes were reviewed and defined as follows:

1. Psychosocial support for early childhood: the importance of the provision of child-focused psychosocial support for young children in Gamalama Volcano eruption preparedness. It focuses on intervention and prevention approaches offered by families, schools, or community entities to promote the psychological health, emotional security, and strength of children before and in the wake of times of disaster.

2. Knowledge: respondents' awareness and comprehension of volcanic hazards, such as identification of eruption indicators, disaster risks, evacuation procedures, risk mitigation methods, and overarching disaster management frameworks.
3. Equipment and simulations: the essentiality of enough emergency equipment, infrastructure, and routine simulation exercises or drills to improve readiness for volcanic eruptions. It emphasises the need for physical resources and practical training activities that enhance the psychological preparedness and capability of people and communities to respond successfully to eruption-related catastrophes.
4. Preparation: signifies the practical readiness actions implemented by people and communities before volcanic catastrophes. This encompasses logistical arrangements, such as the procurement of food supplies and pharmaceuticals, the delineation of evacuation routes, the establishment of emergency shelters, and synchronised community readiness initiatives.
5. Emotional regulation: responders' competencies and methodologies for regulating emotional responses such as stress, anxiety, dread, and panic that frequently occur during volcanic catastrophes. It includes coping strategies, stress management methods, and emotional self-regulation abilities that aid in preserving psychological balance and emotional stability amid crises.
6. Religiosity: the role of religious beliefs and spiritual practices in influencing individuals' psychological preparedness and coping mechanisms during volcanic eruptions. It highlights respondents' reliance on faith-based practices, rituals, prayer, and support from religious institutions, which collectively contribute to emotional comfort, acceptance of disaster situations, and enhanced resilience in the face of volcanic disasters.

3.2 Focus group discussion analysis

Researchers conducted a thematic analysis of the respondents' answers and narrowed them down to several themes. The synthesis process of the emerging themes is shown in Figure 5.

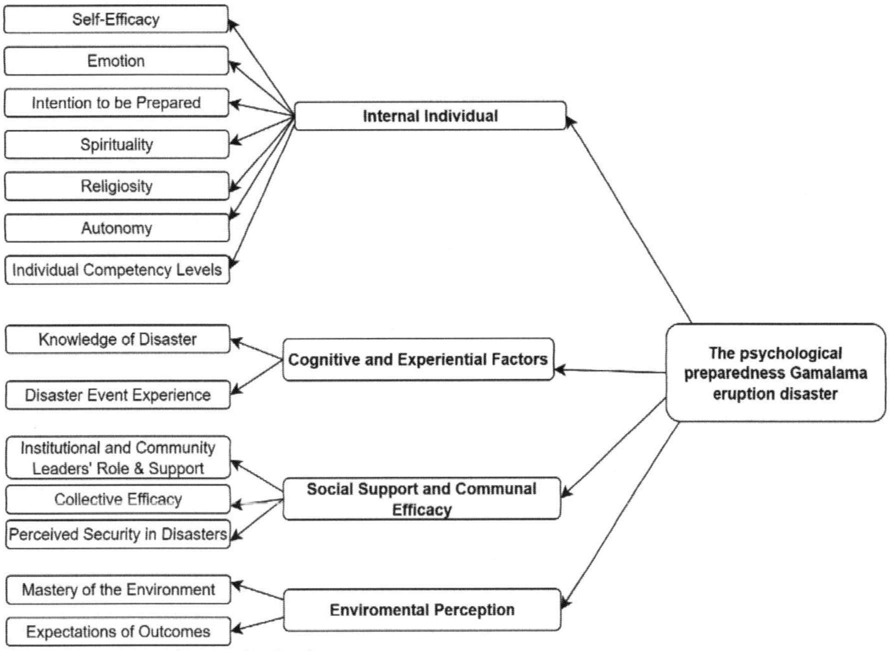

Fig. 5. Focus group discussion's themes.

This research has found four themes of psychological preparedness of early childhood educators for the Gamalama volcanic eruption: internal individual factors, cognitive and experiential factors, social support and communal efficacy, and environmental perception. These themes correspond to the multidimensional context of psychological preparedness, which extends beyond technical preparedness to encompass personal, social, and contextual dimensions.

Internal individual factors comprise basic psychological constructs, such as self-efficacy, emotional regulation, spiritual-religious identity, autonomy, intentions to prepare, and levels of individual skills. These resources serve as psychological building blocks that support teachers' resilience and ability to work effectively under stress.

"Preparations have already been made, using that bag, putting documents into the bag, and immediately evacuating to a safer place." (Miss D).

"If I am sure, yes, because through prayer I always ask for protection from Allah SWT, so I am sure that it can ward off calamities." (Miss A).

"In my opinion, the disaster occurred because of human actions". (Miss H).

"In the Kololikie way, they descend the mountain. That's the language of Kololikie and Kololigam, which is for cleaning... They are just villagers, they walk around wearing white clothes, reciting dhikr and so on." (Miss D).

Cognitive and experiential variables (e.g., prior disaster experiences; knowledge about disaster management). More than anything else, it does so by capturing the importance of antecedents (i.e., experience and information) in influencing risk perception and response decisions during emergencies.

"It was raining from the mountains. It rained for a long time, and it was very heavy. So, in terms of potential, yes, in my opinion, the biggest floods are in Tubo." (Miss A).

"Animals that appear or come out into residential areas can be an indication". (Miss F).

Based on the previous discussion, social support and community efficacy reflect the impact of external resources, institutional support, community leadership, confidence, and safety in the school and neighbourhood. These results confirm the previous evidence indicating that social cohesion and institutional responsiveness are essential factors in decreasing psychological distress during crises.

"It's getting dark on the mountain, the weather is changing, they are doing community service, picking up trash from the gutters and carrying sand." (Miss F).

"When it rains, for example, when it rains too hard, it means that the head of the neighbourhood association in Tubo goes to the flooded area to check on the situation". (Miss Y).

Finally, environmental perception, control over one's environment, and expectancy toward an outcome represent educators' perception of control and belief that they can respond effectively when faced with emergencies.

"So, at that point, I had to start stimulating myself that the more I panicked, no one could help me, it was just me. That's why I had to suggest to myself that I had to be able to calm myself down." (Miss F).

When combined, these four domains create an overarching model of psychological preparedness that recognises the dynamic relationship between personal agency, cognitive insight, social connection, and environmental engagement. The conceptual model reported in this study provides a comprehensive perspective for both understanding and improving young children's psychological preparedness in disaster-prone settings with implications for early childhood education practice.

3.3 Joint display data integration

The joint display table describes the integration of results from the research (Table 2). The synthesis of the FGDs and the open-ended responses led to five key themes of psychological preparedness of early childhood teachers for volcanic disasters. These integrating themes include: personal psychological readiness, cognitive knowledge and experience, emotional-spiritual coping, social-communal support, and environmental control strategies. Each theme represents an interplay of personal, cognitive, emotional, social, and contextual factors that underpin teachers' ability to respond appropriately in a crisis event.

Table 2. Joint display integrating data.

Integrative theme	Findings from Study 1	Findings from study 2	Meta-inference
Personal psychological readiness	Prepare	Self-efficacy, autonomy, intention to prepare	Teachers emphasize the importance of being mentally and personally prepared. The intention to be ready arises from their responsibility as early childhood educators.
Cognitive Knowledge and Experience	Knowledge	Knowledge of disasters, disaster experience	Knowledge and past experiences shape cognitive preparedness. Access to accurate information encourages proactive behaviour.
Emotional-spiritual coping	Emotional regulation, religiosity	Emotion, spirituality, religiosity	Teachers rely on emotional strategies and religious values to manage stress and provide emotional stability for children during disasters.
Social-communal support	Psychosocial support for early childhood	Institutional support, collective efficacy	Teachers not only need social and institutional support but also play a critical role in providing psychosocial care and calm for young children.
Environmental control strategies	Equipment and simulations	Mastery of environment, outcome expectation	Teachers feel more confident when they have access to evacuation tools and participate in simulations, which enhances their sense of control.

The first theme, personal psychological readiness, includes subthemes of self-efficacy, autonomy, spiritual grounding, and intention to be ready. The teachers believed that they were competent in their preparedness to be proactive and exercise good judgment in the face of emergencies, largely due to a strong consciousness of professional responsibility and internal motivation. Their preparedness was reinforced through a sense of purpose, reflecting the cultural relevance of spiritual and religious coping mechanisms for dealing with ambiguity.

The second, cognitive knowledge and experience, reflects the influence of the knowledge of disasters and prior disaster-related experiences as predictors of preparedness. Teachers with prior experience with disasters had more structured, logical thinking and a higher level of awareness regarding appropriate responses. Such experiences served as mental models that made teachers more confident in handling similar situations in the future.

The third, emotional-spiritual coping, highlights the coping strategies employed by teachers to regulate stress, anxiety, and fear. Teachers frequently employed emotional regulation strategies to regain calm and composure, which also served to soothe the children. References were frequently made to the strength of religiosity in terms of psychological strength and moral clarity, suggesting that personal belief systems underpinned emotional stability.

The fourth theme, social-communal support, relates to the external structures, e.g., institutional support, support from community leaders, and collective efficacy. Teachers expressed the importance of feeling supported by their schools and local communities for feeling secure and self-assured. These communities fostered collective responsibility and strengthened psychological resilience.

The fifth theme, environmental control strategies, involves teachers' sense of their ability to comprehend and manage the environment immediately around them. This encompassed perceived control over physical spaces, availability of resources, and the anticipated effectiveness of preparedness behaviours. Staff's level of environmental mastery was positively related to an additional sense of psychological readiness for disaster (feelings of confidence in being mentally able to protect children during an emergency).

Taken together, these themes offer a comprehensive view of psychological preparedness by emphasising the importance of attending to the emotional, cognitive, spiritual, social, and environmental dimensions of early childhood educators' preparation for disaster response.

4 Discussion

This study has provided a holistic view of the psychological preparedness of early childhood teachers in Ternate, a volcanic island setting where the risks of eruption are both sudden and recurrent. The five interconnected dimensions, including personal psychological readiness, cognitive knowledge and experience, emotional and spiritual coping, social communal support, and environmental control strategies, offer an integrated framework for understanding how teachers prepare themselves to support young children in contexts of environmental hazard. These dimensions also interact to shape a comprehensive preparedness profile that is especially relevant in geographically constrained and high-risk environments such as small volcanic islands.

One of the most significant contributions of this study lies in highlighting the dual role of early childhood educators. First, as individuals who must sustain their own emotional preparedness, and also as caregivers responsible for the psychological safety of children. Teachers described themselves as emotional stabilisers, emphasising that their composure and routines directly influence children's sense of security during disaster. This confirms earlier research that positions educators as key actors in children's emotional recovery following disasters [14]. Yet the present findings extend this understanding by situating the role within an environmentally hazardous island context. In Ternate, where external psychosocial services are often delayed due to geographic isolation, the teacher's role as both caregiver and emotional first responder becomes even more crucial.

The findings also reinforce that psychological preparedness is not solely emotional but equally cognitive and practical. Teachers who possessed disaster-related knowledge and hands-on experience through drills or simulations reported feeling more confident and capable of fulfilling their roles during emergencies. This aligns with cognitive behavioural disaster preparation models, which emphasise that preparedness requires both knowledge and skill competencies [15]. In the context of volcanic hazards, this dimension extends beyond generic preparedness to include environmental literacy, such as interpreting hazard maps, recognising geophysical warning signs, and understanding evacuation routes. Bridging the gap between abstract knowledge and concrete action through regular drills and contextualised training is therefore critical to ensuring that teachers can respond effectively when Gamalama erupts.

Preparedness intention further emerged as a key antecedent of teachers' willingness to engage in proactive readiness activities. This intention, shaped by a sense of professional responsibility and perceived personal control, resonates with the Theory of Planned Behaviour [30], which highlights intention as the immediate predictor of behaviour. The data suggest that when teachers feel autonomous and motivated, they are more likely to initiate classroom-based preparedness practices such as storytelling, mock evacuations, and hazard-related discussions with children. Strengthening this internal motivation is essential, as it transforms preparedness from a compliance-based activity into a proactive and sustained commitment, especially in environments where volcanic hazards are recurrent and unpredictable.

Beyond cognitive knowledge, the study highlights internal coping mechanisms (emotion regulation and religiosity) as essential resources that sustain teachers' functionality during eruptions. Emotion regulation practices are not only for teachers' stability but also as a guide for children through uncertainty. Religiosity, meanwhile, surfaced as a culturally embedded strategy that provided psychological meaning, hope, and moral direction. This finding echoes Pargament's religious coping theory, which underscores the protective qualities of spirituality in times of crisis. In the Indonesian context, religious practices serve both personal and communal functions, reinforcing teachers' ability to remain resilient within shared belief systems while also transmitting a sense of safety to their students.

Equally significant is the role of teachers as providers of psychosocial support in classroom settings during and after disasters. Participants emphasised their obligation to maintain safety, stability, and emotional comfort for children, often suspending routine academic activities to focus on recovery-oriented practices such as storytelling, drawing, or singing. These strategies align with global evidence that educator-led, school-based psychosocial interventions can effectively reduce post-traumatic stress symptoms and facilitate emotional recovery in children. In the small-island context of Ternate, where access to external psychosocial professionals may be limited during crises, teachers' capacity to deliver trauma-informed care becomes indispensable. This underlines the importance of integrating psychosocial competencies into teacher training and disaster education frameworks, ensuring that educators are equipped to function as both protectors and healers in times of environmental hazard.

Taken together, these findings illustrate that psychological preparedness among early childhood educators on Ternate Island is shaped by a dynamic interplay of cognitive, emotional, social, and cultural factors. The prominence of religiosity as a coping mechanism reflects a culturally embedded contribution of this study, underscoring the importance of aligning preparedness strategies with local belief systems and communal practices. At the policy level, the results point to the urgent need to integrate psychosocial skills and disaster preparedness modules into teacher education programs, complemented by institutional support for regular drills and context-specific training. For disaster risk reduction in small-island environments, strengthening teachers' dual role as both caregivers and community-based psychosocial actors is vital to safeguard children's wellbeing during volcanic crises. Future research should expand this framework by exploring how culturally grounded coping strategies can be systematically incorporated into formal disaster education and by examining the transferability of these findings to other volcanic and environmentally vulnerable regions.

Several limitations should be considered. First, the qualitative design provided rich descriptions but limited the extent to which the findings can be generalised beyond Ternate. Future studies could adopt longitudinal or mixed-methods approaches to capture how preparedness evolves. Second, the reliance on self-reported accounts introduces possible bias. Observational methods or psychometric tools would help triangulate the evidence. Third, the focus on early childhood teachers excluded other educator groups whose preparedness may differ. Comparative studies across education levels and across other volcanic regions such as Stromboli in Italy, Unzen in Japan, or Mount St. Helens in the United States would strengthen the framework's broader relevance. Finally, while religiosity was prominent here, the interplay between spiritual and secular coping strategies remains an area for further exploration. Addressing these gaps could inform the development of more comprehensive and culturally sensitive disaster education strategies.

5 Conclusion

These five themes interact with each other to highlight the intertwined nature of psychological preparedness. It is not a single trait but rather an interdisciplinary concept anchored in cognitive comprehension, emotive regulation, spiritual orientation, behavioural action, pragmatic preparedness, and societal duty. This comprehensive approach represents a significant addition to current models of preparedness for disaster, specifically as it applies to early childhood educators, for whom the psychological demands are particularly intense.

The findings of this study hold several important practical implications for disaster preparedness initiatives, particularly within the context of early childhood education in disaster-prone regions. The study highlights the need for targeted professional development programmes that go beyond technical preparedness and focus on psychosocial competencies. Training modules should integrate emotional regulation strategies, trauma-informed teaching approaches, and culturally grounded coping mechanisms such as religious and spiritual resilience. This would enable early childhood educators not only to manage their stress but also to provide emotional stability for young children during and after disaster events.

Furthermore, findings underscore the importance of embedding disaster preparedness into early childhood education policy. Ministries of education and local authorities should consider formalising disaster response protocols within school operations, including regular simulation drills, classroom-based preparedness activities, and access to child-centred emergency kits. These measures not only enhance environmental control but also foster a sense of safety and predictability among both teachers and children. The result suggests that community-based collaboration is critical, as schools cannot act in isolation. Partnerships with local disaster management agencies, religious institutions, parent associations, and neighbourhood leaders can strengthen collective efficacy. Such collaborations may involve joint evacuation planning, psychosocial outreach, or the development of culturally sensitive preparedness materials. These partnerships also reflect the social-communal ethos identified as central to teachers' psychological preparedness.

The conclusions of this research call for a human-centred approach to disaster preparedness, one that values the emotional, cognitive, spiritual, and social dimensions of preparedness. Strengthening these areas is critical not only for safeguarding children's well-being but also for advancing disaster education in other small-island and hazard-prone contexts. By recognising and responding to the lived realities of early childhood educators, we can develop more compassionate and effective disaster education strategies that protect and empower both teachers and the children in their care.

Acknowledgements. We gratefully acknowledge 'Pembina 7 Early Childhood Education' in Ternate Island for the knowledge and assistance provided during the data collection process.

References

1. A. Hidayat, D. S. Hadmoko, M. A. Marfai, and B. W. Mutaqin, "Volcanic hazard knowledge and preparedness of small island community on the flank of Gamalama volcano Ternate Island–Indonesia," *GeoJournal*, vol. 88, no. 2, pp. 1251–1263, 2023, doi: 10.1007/s10708-022-10682-9.

2. S. D. Andreastuti, E. T. Paripurno, S. Subandriyo, D. K. Syahbana, and A. S. Prayoga, "Volcano disaster risk management during crisis: implementation of risk communication in Indonesia," *J. Appl. Volcanol.*, vol. 12, no. 1, pp. 1–20, 2023, doi: 10.1186/s13617-023-00129-2.
3. M. R. Lessy, J. Lassa, and K. K. Zander, "Understanding Multi-Hazard Interactions and Impacts on Small-Island Communities: Insights from the Active Volcano Island of Ternate, Indonesia," *Sustain.*, vol. 16, no. 16, 2024, doi: 10.3390/su16166894.
4. B. G. Dewanto, D. S. Hadmoko, N. F. Ramadhani, and A. M. Julius, "Multitemporal satellite images for monitoring the volcanic activities and geothermal potential of Ternate Island's Gamalama Volcano, Indonesia's densest active volcanic island," *Remote Sens. Appl. Soc. Environ.*, vol. 38, no. April, p. 101555, 2025, doi: 10.1016/j.rsase.2025.101555.
5. A. Hidayat, M. A. Marfai, and D. S. Hadmoko, "The 2015 eruption of Gamalama volcano (Ternate Island–Indonesia): precursor, crisis management, and community response," *GeoJournal*, vol. 87, no. 1, pp. 1–20, 2022, doi: 10.1007/s10708-020-10237-w.
6. F. A. Djafar, D. L. Setyowati, and E. Suharini, "Ternate s People Readiness in Facing the Eruption of," vol. 8, no. 5, pp. 139–152, 2020.
7. A. Faral *et al.*, "A 22,000-year tephrostratigraphy record of unidentified volcanic eruptions from Ternate and Tidore islands (North Maluku, Indonesia)," *J. Volcanol. Geotherm. Res.*, vol. 423, p. 107474, 2022, doi: 10.1016/j.jvolgeores.2022.107474.
8. K. R. Ronan and D. M. Johnston, "Behaviourally-based interventions for children following volcanic eruptions: An evaluation of effectiveness," *Disaster Prev. Manag. An Int. J.*, vol. 8, no. 3, pp. 169–176, 1999, doi: 10.1108/09653569910275364.
9. R. Salam, E. Marthinu, R. Achmad, P. A. Rohmah, and W. Lestari, "Disaster Preparedness for Children : Systematic Literature Review Study," vol. 10, no. 2, pp. 148–156, 2025.
10. M. P. Pacheco, S. M. Pereira, and I. E. Rego, "Seismic preparedness of families with children: Measures and dynamics," *Int. J. Psychol.*, vol. 56, no. 1, pp. 30–39, 2021, doi: 10.1002/ijop.12694.
11. A. Turchi, R. Lumino, D. Gambardella, and M. F. Leone, "Coping Capacity, Adaptive Capacity, and Transformative Capacity Preliminary Characterization in a 'Multi-Hazard' Resilience Perspective: The Soccavo District Case Study (City of Naples, Italy)," *Sustain.*, vol. 15, no. 14, 2023, doi: 10.3390/su151410877.
12. M. Horii, K. Yamaoka, H. Y. Kim, S. Takewaki, and T. Kunitomo, "Comparative Study of Literacy Enhancement on Volcanic Disaster Reduction for the Residents and Visitors in Mt. Ontakesan and Other Volcanic Areas," *J. Disaster Res.*, vol. 19, no. 1, pp. 159–172, 2024, doi: 10.20965/jdr.2024.p0159.
13. John W. Cresswell, *Designing and Conducting Mixed Methods Research*, 3rd editio. California: Sage Publications, Inc, 2018.
14. Y. R. Lee, S. N. Park, M. R. Lee, and E. Nam, "Influencing factors of early childhood teachers' disaster preparedness," *Front. Public Heal.*, vol. 11, no. November, 2023, doi: 10.3389/fpubh.2023.1249736.
15. D. Paton, M. Millar, and D. Johnston, "Community resilience to volcanic hazard consequences," *Nat. Hazards*, vol. 24, no. 2, pp. 157–169, 2001, doi: 10.1023/A:1011882106373.

Comparing the accuracy of fracture diagnosis using minimal-resource vs. standard-resource plain radiographs

Safrizal Rahman[1], *Muhammad* Ariq Ersan[2], *Iskandar* Zakaria[3,*]

[1]Department of Surgery, Zainoel Abidin General Hospital/Faculty of Medicine, Universitas Syiah Kuala, Banda Aceh, Indonesia
[2]Faculty of Medicine, Universitas Syiah Kuala, Banda Aceh, Indonesia
[3]Departement of Radiology, Zainoel Abidin General Hospital/Faculty of Medicine, Universitas Syiah Kuala, Banda Aceh, Indonesia

Abstract. Plain radiographs remain a fundamental diagnostic tool for identifying long bone fractures. In resource-limited settings, adaptation of radiographic methods is essential to ensure diagnostic reliability. This study aimed to compare the accuracy of plain radiograph interpretations performed with minimal resources against those using standard resources, thereby optimizing diagnostic approaches in constrained environments. An observational analytic study with a cross-sectional design was conducted using medical records of extremity fracture patients at Dr. Zainoel Abidin Regional Public Hospital (RSUDZA). Radiographs obtained with minimal resources, either anteroposterior (AP) or lateral (Lat) views, were compared with standard-resource images that combined both views (AP + Lat). Diagnostic test parameters, including accuracy, sensitivity, and specificity, were calculated. Among 55 samples, AP view readings achieved 89.1% accuracy, 95.3% sensitivity, and 80% specificity. Lateral view readings showed 87.3% accuracy, 91.1% sensitivity, and 70% specificity. Standard-resource interpretation provided the highest accuracy at 96.3%. These findings indicate that even in minimal-resource conditions, plain radiographs can yield reasonably reliable diagnostic results, particularly due to their high sensitivity. The study underscores the importance of adapting radiographic protocols in low-resource environments and suggests that simplified methods may serve as effective alternatives. Further research is recommended to validate these findings in larger populations.

1 Introduction

During disasters, X-ray imaging is one of the most urgently needed diagnostic modalities for suspected fractures, as it is the fastest and most accessible method for identifying bone injuries [1,2]. Plain radiographs are widely applied due to their low radiation dose, affordability, and diagnostic reliability, particularly in assessing fracture characteristics, displacement, and alignment using standard two-projection views [3–6]. Traumatic fractures

* Corresponding author: iskandarzakaria@usk.ac.id

frequently arise from traffic accidents, workplace injuries, and violence, but also occur commonly in disaster settings, where rapid diagnosis is vital for timely treatment [7,8].

Yet in disaster situations, radiological imaging often faces severe resource constraints. Equipment, imaging films, trained operators, and power supply may be limited or unavailable, making standard diagnostic practices difficult to maintain [9,10]. These limitations highlight the need for adaptable radiographic strategies that preserve diagnostic accuracy even under minimal-resource conditions. Addressing this gap is directly tied to disaster preparedness and the resilience of health systems.

The Sendai Framework for Disaster Risk Reduction (2015–2030) emphasizes strengthening disaster preparedness, enhancing health system resilience, and ensuring continuity of essential medical services during crises [10]. Within this framework, the capacity to diagnose fractures efficiently despite constrained resources represents both a clinical and public health priority. Therefore, this study aims to compare the diagnostic accuracy of plain radiographs obtained with minimal resources against those produced under standard conditions, with the goal of informing strategies that integrate radiological services into resilient health systems and future disaster preparedness efforts.

2 Methods

This cross-sectional diagnostic accuracy study used secondary data from medical records and radiology archives at RSUDZA Banda Aceh, Indonesia. Plain radiograph interpretations obtained under minimal-resource conditions were compared with those using standard resources, with the reference standard defined as the official reports issued by board-certified radiologists. The study population included patients clinically suspected of extremity fractures who underwent plain radiography between January and April 2024. Inclusion criteria were complete medical records, availability of anteroposterior and lateral projections, and a definitive radiologist's interpretation; exclusion criteria were incomplete records, inadequate image quality, or absence of a radiologist's report. Total sampling was applied, yielding 55 eligible radiographs.

Ethical clearance was obtained from the RSUDZA Ethics Committee, and all data were anonymized to maintain confidentiality. The study design was explicitly contextualized within disaster risk reduction, recognizing that resource limitations often occur in disaster-affected areas. By evaluating diagnostic accuracy under constrained versus standard conditions, this study seeks to inform strategies for resilient health systems and ensure continuity of essential radiological services during disasters.

3 Results

A total of 55 patients met the inclusion and exclusion criteria. The sample was dominated by older adults, particularly those aged 60–69 years, who accounted for over 30% of the total. This skewed age distribution suggests that fractures disproportionately affect elderly patients, an important consideration for both diagnostic accuracy and resource planning. For clarity, patients were grouped into broader age ranges (Table 1). Men were more frequently affected than women (67.3% vs. 32.7%). Most cases (81.8%) were confirmed fractures, with the femur being the most common site.

Table 1. Characteristics of respondents (n=55)

Characteristics	Category	Frequency (n)	Percentage (%)
Age	<30	14	25.5
	30–49	20	36.4
	50–69	21	38.2

	≥70	0	0
Gender	Male	37	67.3
	Female	18	32.7
Fracture status	Fracture confirmed	45	81.8
	No fracture	10	18.2
Imaging location	Brachii	8	14.5
	Antebrachii	15	27.3
	Femoris	17	30.9
	Cruris	15	27.3

When comparing radiological expertise against the standard resource reference, diagnostic accuracy was high (96.3%), with sensitivity 97.8% (95% CI: 88.5–99.9) and specificity 90.0% (95% CI: 55.5–99.7). The PPV was 97.8% (95% CI: 88.5–99.9), while the NPV was 90.0% (95% CI: 55.5–99.7) (Table 2).

Table 2. Diagnostic performance of radiology expertise compared with standard resources (n=55)

Radiology expertise	Reference: Fracture confirmed	Reference: No fracture	Total
Positive	44	1	45
Negative	1	9	10
Total	45	10	55

Sensitivity : 97.8% (95% CI: 88.5–99.9)
Specificity : 90.0% (95% CI: 55.5–99.7)
Accuracy : 96.3% (95% CI: 87.3–99.5)
PPV : 97.8% (95% CI: 88.5–99.9)
NPV : 90.0% (95% CI: 55.5–99.7)

With minimal AP-view resources, diagnostic performance decreased (accuracy 89.1%). Sensitivity remained high at 95.3% (95% CI: 84.2–99.4), but specificity fell to 80.0% (95% CI: 44.4–97.5). The PPV was 91.1% (95% CI: 78.8–97.5), while the NPV dropped notably to 66.7% (95% CI: 35.4–88.7). The results of the diagnostic test, utilizing radiological examination expertise with minimal resources in the AP view, showed an accuracy of 89.1%, with sensitivity and specificity values of 95.3% and 80%, respectively. The positive predictive value and negative predictive value were 91.1% and 66.7%, respectively (Table 3).

Table 3. Diagnostic performance of minimal AP-view imaging compared with radiology expertise (n=55)

Minimal AP-view	Reference: Fracture confirmed	Reference: No fracture	Total
Positive	41	2	43
Negative	4	8	12
Total	45	10	55

Sensitivity : 95.3% (95% CI: 84.2–99.4)
Specificity : 80.0% (95% CI: 44.4–97.5)
Accuracy : 89.1% (95% CI: 77.4–95.9)
PPV : 91.1% (95% CI: 78.8–97.5)
NPV : 66.7% (95% CI: 35.4–88.7)

The diagnostic test using minimal lateral-view imaging achieved an accuracy of 87.3%, with a sensitivity of 91.1% and specificity of 70.0%. While the sensitivity was acceptable, the relatively low specificity and negative predictive value (63.6%) suggest a higher risk of misclassifying patients without fractures as fracture cases and, more importantly, missing true fractures. Compared with both standard-resource imaging (accuracy 96.3%, specificity

90%) and AP-only imaging (accuracy 89.1%, specificity 80%), the lateral view performed worse overall. This weakness is clinically significant in disaster contexts, as reliance on lateral-only films could increase false reassurance and lead to untreated injuries (Table 4).

Table 4. Diagnostic performance of minimal Lateral-view imaging compared with radiology expertise (n=55)

Minimal lateral-view	Reference: Fracture confirmed	Reference: No fracture	Total
Positive	41	3	44
Negative	4	7	11
Total	45	10	55

Sensitivity : 91.1% (95% CI: 78.8–97.5)
Specificity : 70.0% (95% CI: 34.8–93.3)
Accuracy : 87.3% (95% CI: 75.5–94.7)
PPV : 93.2% (95% CI: 81.3–98.6)
NPV : 63.6% (95% CI: 30.8–89.1)

4 Discussion

The present study found that the highest incidence of fractures occurred in the 60–64 age group, consistent with global evidence that fracture risk increases substantially with age due to osteoporosis and other degenerative changes. While this aligns with Bergh (2021) and related studies, the relevance of this finding within the Indonesian context lies in the growing aging population, which underscores the importance of strengthening fracture management strategies locally. Similarly, the observed male predominance (67.3%) is in agreement with several international reports, yet the discussion should emphasize that this pattern reflects sociocultural and occupational exposures unique to the study setting, such as traffic-related trauma and high-risk labor.

An important contribution of this study lies in the diagnostic accuracy evaluation of minimal-resource imaging, particularly the AP view. The AP-only approach demonstrated reasonably high sensitivity (95.3%) and accuracy (89.1%), suggesting that it may serve as a practical substitute when full radiographic series are not feasible. However, the specificity (80%) and negative predictive value (66.7%) were relatively low, raising concern about false negatives and the risk of missed fractures. The lateral-only approach performed even worse, with reduced specificity and predictive values. These findings need to be explicitly contrasted with the standard two-view protocol, which provided the highest accuracy (96.3%) and should remain the gold standard when resources allow.

This distinction is critical in the context of disaster risk reduction, where imaging resources may be severely limited. In mass-casualty or rural settings, reliance on single-view imaging may expedite triage and facilitate treatment initiation. Nevertheless, the lower specificity and negative predictive value indicate that AP-only imaging should be applied cautiously, ideally supported by clinical judgment and, where possible, follow-up imaging. This application-oriented perspective is more valuable than restating prior studies and highlights the novelty of this study as one of the few to empirically assess minimal-resource fracture imaging in an Indonesian tertiary hospital setting.

Despite these contributions, the study has several limitations. First, the sample size was relatively small (n=55), which restricts the precision of accuracy estimates. Second, the retrospective design and reliance on secondary medical records may introduce reporting bias and limit the ability to control for confounding variables such as mechanism of injury or comorbidities. Third, the absence of inter-rater reliability testing reduces confidence in the generalizability of the diagnostic accuracy findings. Future research should adopt prospective designs, larger sample sizes, and incorporate disaster risk reduction simulation contexts to validate whether AP-only imaging can safely support triage and fracture diagnosis in disaster-prone and resource-constrained environments.

5 Conclusion

This study demonstrates that minimal-resource AP radiographs can achieve reasonably high diagnostic accuracy (89.1%) and sensitivity (95.3%), but with lower specificity (80%) and negative predictive value, raising concern for potential missed fractures. Lateral-only views performed less favorably, confirming that standard two-view imaging remains the most reliable approach.

The findings highlight both the potential and the limitations of minimal-resource imaging in disaster or resource-constrained settings. While AP-only radiographs may serve as a pragmatic triage tool when resources are severely limited, they cannot fully substitute for standard protocols. Clinical judgment and follow-up imaging are essential to mitigate false negatives.

The study's small sample size, retrospective design, and single-center setting limit the generalizability of its results. Future prospective, multi-center studies in disaster-prone regions are recommended to further evaluate the role of minimal-resource radiography in building resilient and adaptive health systems aligned with disaster risk reduction priorities. Hospitals in disaster-prone areas should consider incorporating minimal-resource AP radiography as a rapid triage option, while ensuring pathways for confirmatory imaging when resources allow.

Acknowledgements. We would like to express our deepest gratitude to the researcher's parents for their unwavering support, to the supervisors and examiners for their patience and guidance, and to the Rector, Dean, Program Director, and Thesis Coordinator of the Faculty of Medicine at Syiah Kuala University. Our sincere thanks also extend to all parties who contributed to the completion of this thesis.

References

1. B. Yunus, K. Bandu, The effect of x-ray radiation on children. Makassar Dent. J., **8**, (2019).
2. M. Akhadi, Sinar-X menjawab masalah kesehatan, 1st ed., Ayu (ed.) (Deepublish Publisher, Yogyakarta, 2020).
3. N.I.A. Halim, F. Mohd Zaki, H.A. Manan, Z. Mohamed, An evaluation of the quality of plain radiograph interpretations by radiology trainees: A single institution experience. Diagnostics, **12**, 1885 (2022). https://doi.org/10.3390/diagnostics12081885
4. T. York, C. Franklin, K. Reynolds, G. Munro, H. Jenney, W. Harland, et al., Reporting errors in plain radiographs for lower limb trauma: a systematic review and meta-analysis. Skeletal Radiol., (2021). https://doi.org/10.1007/s00256-021-03821-9
5. K.H. Miao, J.H. Miao, Radiological diagnosis and imaging of femoral shaft fractures. Anatomia, **2**, 282–299 (2023). https://doi.org/10.3390/anatomia2030023
6. R. Kraus, K. Dresing, Rational usage of fracture imaging in children and adolescents. Diagnostics **13**, 2334 (2023). https://doi.org/10.3390/diagnostics13142334

7. Z. Jin, H. Lv, M. Li, Z. Hou, X. Lian, W. Chen, et al., Epidemiological investigation of hospitalized patients with traumatic fractures: a cross-sectional study. J. Int. Med. Res., **49**, 1–12 (2021). https://doi.org/10.1177/0300060520988565
8. S. Aydin, O. Kazci, B. Ece, M. Kantarci, Earthquakes from a radiological perspective: what is demanded from the radiologists, and what can we do? A pictorial review. Diagn. Interv. Radiol., **29**, 1–8 (2023). https://doi.org/10.5152/dir.2023.23094
9. A. Karera, P. Engel-Hills, F. Davidson, Radiology image interpretation services in a low-resource setting: medical doctors' experiences and the potential role of radiographers. Radiography, **30**, 560–566 (2024). https://doi.org/10.1016/j.radi.2024.01.010
10. A. Karera, P. Engel-Hills, F. Davidson, Radiographers' experiences of image interpretation training in a low-resource setting. Radiography, **29**, 590–596 (2023). https://doi.org/10.1016/j.radi.2023.02.002

Evolving Disaster Resilience in Indonesia: Insights from Implementation of Local Wisdom through Science, Technology, Engineering, and Mathematics-Based Learning in Schools

Syahrul Ridha[1,2,3,4,*], *Susilawati* Susilawati[4,5], *Hizir* Sofyan[4,6], *Yopi* Ilhamsyah[4,7], and *Puspita Annaba* Kamil[1]

[1]Deparment of Geography Education, Universitas Syiah Kuala
[2]Tsunami and Disaster Mitigation Research Center (TDMRC), Universitas Syiah Kuala
[3]Research Center for Cultural and Social Studies, Universitas Syiah Kuala
[4]STEM Research Center, Universitas Syiah Kuala
[5]Department of Physic Education, Universitas Syiah Kuala
[6]Deparment of Mathematic, Universitas Syiah Kuala
[7]Department of Marine Sciences, Universitas Syiah Kuala

Abstract. Since disasters can happen at any time, having strong disaster resilience is crucial. Thus, it is critical to consider how local wisdom will endure. This research question is about how to integrate local wisdom and STEM-based disaster resilience for disaster mitigation. Therefore, this research aims to reveal the importance of this integration. Data collection was carried out by distributing questionnaires to teachers in Aceh province. A total of 144 teachers participated in this research. The results of this research are that local wisdom values must be integrated into the learning process to improve post-disaster recovery through STEM. In this way, it is hoped that local wisdom can be applied to learning at school through STEM. Therefore, stakeholders, society, and government must pay attention to this situation to create disaster-resistant students and communities.

1 Introduction

Indonesia is a region that is vulnerable to various natural hazards. Starting from geological, environmental, and hydrometeorological hazards, Indonesia has a lot of experience with disasters. In this case, to create a disaster-prepared society, efforts need to be made to reduce disaster risk [1,2]. One way is through strengthening disaster education in formal institutions, for example, schools. In schools, disaster learning should be achieved well, namely by increasing students' knowledge and participation in disaster risk reduction [3]. However, many schools in Indonesia do not implement STEM-based education or introduction to local

[*]Corresponding author: syahrul.ridha@usk.ac.id

wisdom to improve disaster mitigation in their curriculum. Even though STEM is seen as one of the main drivers of sustainable development worldwide, STEM is very important in increasing global competitiveness and creating innovation [4].

Nowadays, conversations on lowering the risk of disasters by using local knowledge and educating people about the dangers of natural hazards are becoming more and more significant globally [5]. This is to create a society that is ready and resilient in facing disasters. Indonesia is a country with a region that has a high level of disaster vulnerability. Disasters are events or series of events that threaten and disrupt people's lives and livelihoods caused by either natural or non-natural factors or human factors, resulting in human casualties, environmental damage, property loss, and psychological impacts. Natural hazards continue to occur in Indonesia, including geological, environmental, hydrometeorological, and social disasters. These dangers require mitigation and adaptation carried out in an integrative manner by the government, academics, and society. The government can prepare regulations that reduce the risk of disasters in Indonesia. Meanwhile, academics can make disaster knowledge a curriculum achievement and goal at both primary, secondary, and higher education levels. Apart from that, the community plays an active role in protecting the environment, such as managing waste properly and using sustainable natural resources [6].

Disasters cannot be avoided, but the negative impacts or risks of disasters can be reduced. Disaster risk must be reduced through disaster risk management. The concept of disaster risk management has experienced a paradigm shift from a traditional approach to a holistic (comprehensive) approach. Traditionally, disasters are events that cannot be avoided and victims must immediately receive help. Several factors can determine the number of victims in a disaster [7]: (a) lack of understanding of the characteristics (hazards) of disasters, (b) attitudes and behavior that endanger the quality of resources (vulnerability), (c) lack of early warning information causing a lack of preparation, (d) helplessness or inability to face disasters.

Schools are one of the institutions that can play a role in DRR activities and have the task of brightening the life of the nation and improving community welfare through the application of science and technology in society. Therefore, schools play an important role in improving community skills for disaster prevention and risk reduction, including student communities. Disaster education for the community is an effort to provide awareness and knowledge to the community regarding disasters and the actions that need to be taken before, during a disaster and after a disaster [7]. The application of disaster education has been used in several countries as a first step in reducing disaster risk. The research question is what are teachers' perceptions of STEM-based integration of local wisdom to improve disaster mitigation in schools? To answer this question, this research examines teachers' perspectives on integrating STEM-based local wisdom in schools. So this study explains an alternative sustainability role for STEM in integrating local wisdom for disaster mitigation.

2 Literature Review

2.1 Local Wisdom

Local wisdom is local knowledge used by local people to survive in an environment that is integrated with belief systems, norms, and culture, and is expressed in traditions and myths that have been adhered to for a long time [8]. Local wisdom has a broader scope than just traditional knowledge [9]. This coverage has become an embodiment and implementation of articulation in the form of traditional knowledge that is understood by humans or communities who interact with the natural surroundings. Thus, local wisdom is cultural knowledge possessed by a certain group of people, including models of sustainable environmental management, including maintaining its relationship with nature through wise and wise use and preservation of the environment.

One of the famous local cultures in Aceh after the Indian Ocean tsunami was *Smong*. The story of *Smong*, which means tsunami or high tide, is traditional knowledge from the people of Simeulue, Aceh, Indonesia. The history of *Smong* has been ingrained in society through social interactions since 1907 [10]. Knowledge about this history which has been passed down from generation to generation succeeded in increasing the awareness of the Simeulue community regarding the tsunami disaster in the Indian Ocean in 2004. The history of the *Smong* community was preserved through traditional oral channels and developed through various cultural channels after the 2004 tsunami. Community resilience can be increased when the community local communities actively incorporate learning from the history of the *Smong* people into their current recovery efforts. Thus the *Smong* story shows an important relationship between local wisdom and disaster risk reduction.

In the context of education, local knowledge needs to be considered, such as: (a) to increase disaster resilience, the independent curriculum must include local wisdom values, (b) the Pancasila Profile Strengthening Project in the Merdeka Curriculum can help integrate local cultural values and disaster resilience character, and (c) the Disaster Safe Education Unit program plays an important role in reducing disaster risk. Therefore, the government, community, and stakeholders must pay attention to these conditions so that students and communities become disaster resilient.

2.2 Disaster Resilience and Mitigation

The ability of a person, community, organization, or system to face, mitigate the impact of, and recover from disasters such as earthquakes, floods, fires, or other natural or human-induced disasters is called disaster resilience [11]. Some important components of disaster resilience are as follows: (a) Prevention and mitigation are efforts to reduce the risk and impact of disasters before they occur, such as by building strong infrastructure and good spatial planning. (b) Preparedness is thorough preparation, such as training, simulations, and providing emergency equipment. (c) Response is quick, and effective action is taken when a disaster occurs, such as evacuation and rescue. (d) Recovery encompasses reconstruction and rehabilitation, restoring conditions to their original state or even better after a disaster.

An action carried out to reduce disaster risk is called disaster mitigation. Disaster risk is the potential loss due to a disaster that occurs in an area and within a certain period, which can be in the form of death, injury, disease, risk of life, loss of security, displacement, damage, or loss of property, and disruption to the environment. Natural disaster risk is measured or evaluated using the natural disaster risk assessment method. This method is done by knowing the level of risk and creating a disaster risk map. In general, the assessment takes place in three stages, namely assessing the level of danger or threat of disaster, assessing vulnerability and capacity, and assessing disaster risk. In general, the risk of natural disasters is managed by paying attention to factors that can reduce the risk of natural disasters in a planned, coordinated, integrated, and comprehensive manner through prevention and mitigation, preparedness, and early warning efforts. These risks are dangers or threats that cause disasters and vulnerabilities related to the elements affected by these dangers (people and assets), including the ability to deal with hazards.

Mitigation efforts are efforts to reduce disaster risk and its impact on communities in disaster-prone areas. Apart from that, mitigation efforts are also carried out to avoid disasters. Mitigation activities are carried out through physical development as well as increasing awareness and increasing capacity to overcome disaster threats. As a preventive action/effort that aims to minimize the negative impact of a disaster that is predicted to occur in the future in a particular area, mitigation is a long-term investment for the welfare of all levels of society. Disaster mitigation activities are carried out through planning and implementing spatial planning based on disaster risk analysis; regulations in the fields of planning, infrastructure development, and construction layout; and providing education, training, and guidance, both traditional and modern.

Mitigation can be classified into two types, namely structural mitigation, and non-structural mitigation. Structural mitigation is an engineering effort, both natural and man-made, that can prevent or reduce the possibility of disasters and their impacts. This form of mitigation is any physical construction intended to reduce or avoid potential impacts caused by the threat of danger or the application of engineering techniques to create the strength and resilience of a structure or system. Non-structural mitigation is a non-technical effort that involves adjusting and adapting human activities to structural mitigation and other efforts. These mitigation actions are all actions that do not involve physical construction and use knowledge, practices, or agreements to reduce risks and impacts, including through policies and legislation, awareness, training, and community education.

2.3 STEM for Disaster Resilience and Mitigation

STEM is the abbreviation for Science, Technology, Engineering, and Mathematics. STEM, which was initiated by the United States, is an approach that combines these four sciences in an integrated manner in problem-based learning methods and everyday contextual events. STEM-based learning while applying knowledge and skills to solve a case. This approach is presented as a 21st-century learning approach to produce human resources equipped with quality cognitive, psychomotor, and emotional skills.

STEM learning includes four main components: science, technology, engineering, and integrated mathematics. In science, students must be able to utilize the natural knowledge they obtain to solve problems in everyday life. Regarding technology, students must be able to collaborate by using technology to convey information and process the data they find. In engineering, students collaborate on their findings to find the right solution or even create a product. In mathematics, students can use their mathematical knowledge when processing data. Therefore, students will be trained in critical and creative thinking skills, apart from that, students will be trained in how to solve problems in everyday life. In another context, the STEM field is seen as important for global economic development and societal welfare.

The STEM approach systematically combines the four aspects of science in a problem-based learning approach to understanding disaster mitigation. Study STEM while applying knowledge and skills to solve disaster problems holistically, not just from one perspective. The multidisciplinary combination of science, technology, engineering, and mathematics is very suitable for the analysis of disaster problems. STEM can shape the character of students who can understand a concept or knowledge (science) and combine this knowledge with skills that have been mastered and developed (technology) so that they can create or construct a method (engineering) with appropriate analysis and calculations (mathematics) [12]. In this case, the main concept of the STEM method focuses on solving complex cases related to the characteristics of a disaster hazard.

3 Material and methods

3.1 Research Design

This research design uses descriptive quantitative research. The researcher used this design to describe the stem-based integration of local wisdom to increase disaster resilience in schools. Apart from that, this research also uses field survey techniques to survey schools affected by disasters during emergency conditions.

3.2 Collection and Analysis Data

Data was collected by giving questionnaires to teachers in Aceh Province Indonesia. A total of 144 teachers were respondents in this research. Questionnaire data totaling 11 questions were analyzed using descriptive statistics to see the percentage of teachers' responses to STEM-based integration of local wisdom to improve disaster mitigation. In this research. To find out teachers' perspectives on STEM-based integration of local wisdom in improving disaster mitigation, in this research there were 11 questions distributed to teachers via Google Form. These 11 questions are representative of generalizing teacher perceptions (Table 1).

Table 1. Questions in teacher perception questionnaires.

No	Response Items
1	Familiarity and understanding of STEM
2	Effective STEM implementation strategies for disaster mitigation
3	The urgency of STEM in school learning for disaster mitigation
4	Application of STEM for disaster mitigation
5	Barriers to the application of STEM for disaster mitigation
6	The importance of integrating STEM-based local wisdom for disaster mitigation
7	The reason integration of local wisdom is needed in disaster mitigation
8	Local wisdom integration strategies in STEM-based disaster mitigation
9	Methods for integrating local wisdom for disaster mitigation
10	Obstacles in integrating disaster mitigation in STEM-based learning
11	Teachers need to be provided with training to develop STEM for disaster mitigation

4 Results and Discussion

Teachers need to know about STEM in this 21st-century education. Integrating a STEM approach into the application requires understanding and analysis on the part of educators. Integrating these four components into learning is a must and this is a difficulty for teachers. Table 2 below shows teachers' perceptions of the existence of STEM in learning at school.

Table 2. Teachers' Perspectives on STEM-Based Integration of Local Wisdom.

No	Response Items	*f* (%)
Familiarity and understanding of STEM		
1	Have heard of it but are still confused about its application	40,3%
2	Often heard but not yet maximally understanding the concept of its application	36,8%
3	Often hear and understand implementation strategies	1,4%
4	I often hear about it and have already applied it, even though I don't yet fully understand its characteristics	9,7%
5	Often hear, understand the strategy, and have implemented it	0,7%
6	Never heard	11,1%
Effective STEM implementation strategies for disaster mitigation		
1	Integrating extra curricular activities	24,3%
2	Integrate in co-curricular	22,2%
3	Integrating intra-curricular activities	29,9%
4	Developing STEM-based teaching tools	59,7%
5	Developing STEM-based media	52,1%

6	Never implemented it	17,4%
7	Integration with entrepreneurship	0,7%
The urgency of STEM in school learning for disaster mitigation		
1	Useless	0,7%
2	Less useful because it is difficult to implement	4,9%
3	Helping students produce products	54,2%
4	Helping students solve problems	62,5%
5	Help students understand the concept	60,4%
6	Not implemented yet	9,1%
Application of STEM for disaster mitigation		
1	Never done	59,7%
2	I have, although I don't fully understand the concept of its application	36,8%
3	Have experience and understand of the implementation strategy	2,1%
4	Have applied several times	1,4%
Barriers to the application of STEM for disaster mitigation		
1	The characteristics of STEM applications are not yet understood in depth	70,8%
2	Facilities not yet supported	43,1%
3	Suitable learning tools are not yet available	28,5%
4	Inappropriate time allocation	9%
5	The characteristics of STEM applications are not yet understood in depth	7%
The importance of integrating STEM-based local wisdom for disaster mitigation		
1	Very important	50,7%
2	Important	48,6%
3	Less Important	0,7%
4	Not important	0%
The reason integration of local wisdom is needed in disaster mitigation		
1	Preserving the local wisdom	79,7%
2	Introducing culture to the younger generation	79%
3	Presenting contextual learning	46,2%
Local wisdom integration strategies in STEM-based disaster mitigation		
1	Integrate into teaching materials	66,7%
2	Packaging IT-based cultural content	61,8%
3	Developing teaching tools based on local wisdom	52,8%
4	Developing media based on local wisdom	42,4%
Methods for integrating local wisdom for disaster mitigation		
1	Implement a knowledge management system	57,6%
2	Demonstration of local wisdom at school	75%
3	Local wisdom FGD between schools and indigenous communities	45,1%
Obstacles in integrating disaster mitigation in STEM-based learning		
1	There is a lack of teacher understanding regarding disaster mitigation	67,4%
2	Lack of socialization of disaster mitigation in schools	75,7%
3	Lack of availability of teaching tools that integrate disaster mitigation in learning	55,6%
Teachers need to be provided with training to develop STEM for disaster mitigation		
1	Strongly agree	68,1%
2	Agree	31,3%
3	less agree	0,7%
4	Don't agree	0%

4.1 Familiarity with STEM Terms

STEM learning is a form of innovative learning that can develop various skills needed by students to face the challenges of the 4.0 era. STEM can help students develop thinking skills, gain comprehensive knowledge, and master various daily life problems such as disaster problems. In STEM learning, students are directly involved in the process of exploring the concepts they have learned. In this case, teachers already know or are familiar with STEM but are still confused about its application (40.3%). Apart from that, teachers also often hear but do not yet fully understand the concept of applying STEM in learning (36.8%). Another thing that was also conveyed was that most teachers still have difficulty in implementing the STEM approach. Among the four STEM components, the engineering component is the most difficult to implement in designing a STEM approach to learning. Therefore, regarding disasters, it is necessary to develop interior and teacher competency in understanding STEM to integrate into various fields, such as integrating local culture for disaster mitigation. The familiarity of STEM for teachers in learning (Figure 1).

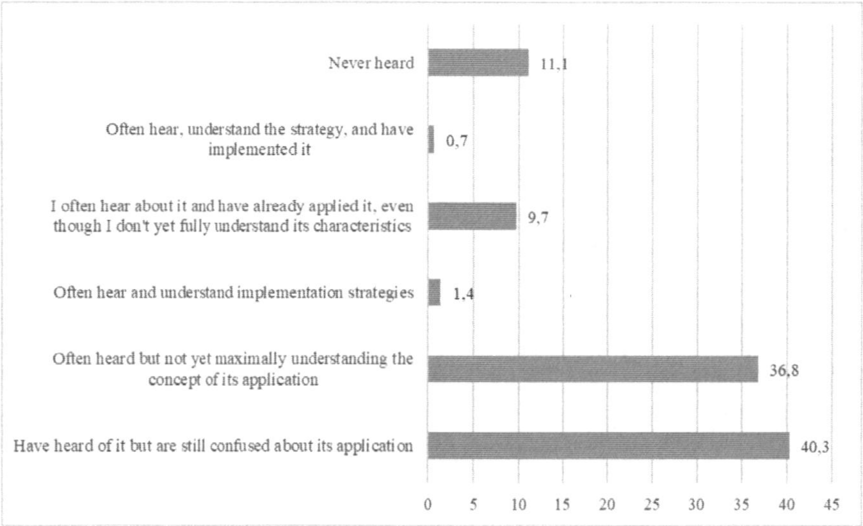

Fig. 1. STEM familiarization for teachers in learning.

4.2 Strategy and Urgency for Implementing STEM

What is very urgent in integrating STEM-based local culture for disaster mitigation in schools is developing STEM-based learning tools and media in learning. Developing learning tools is an effort to plan to learn so that it can be carried out effectively and efficiently, and learning objectives can be achieved. In addition, because disasters are not known when they occur, knowledge of disasters can be increased through contextually developed learning media to understand disaster mitigation. In this way, students can think critically and creatively in solving environmental and disaster problems. The percentage of strategies and urgency for implementing STEM in learning (Figure 2 a & b).

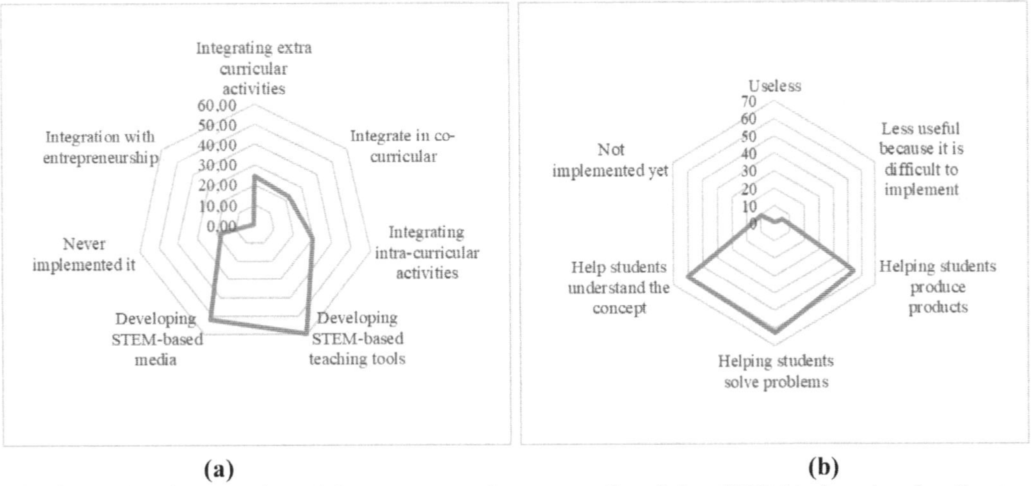

Fig. 2. Teachers' perception of the strategy and urgency of applying STEM in learning for disaster mitigation.

4.3 Application of STEM and its Obstacles

Most STEM applications for disaster mitigation have never been carried out (59.7%). The STEM approach to learning should be well understood by teachers and students so that the four STEM components can be implemented well. This research found several obstacles that cause the application of STEM to be hampered, namely: the characteristics of the application of STEM are not yet understood in depth (70.8%), the facilities or means and infrastructure are not yet supportive (43.1%), the appropriate learning tools are not yet available (28.5%), and the allocation of learning time is not appropriate (9%). The conditions for applying STEM for disaster mitigation and the obstacles (Figure 3 a & b).

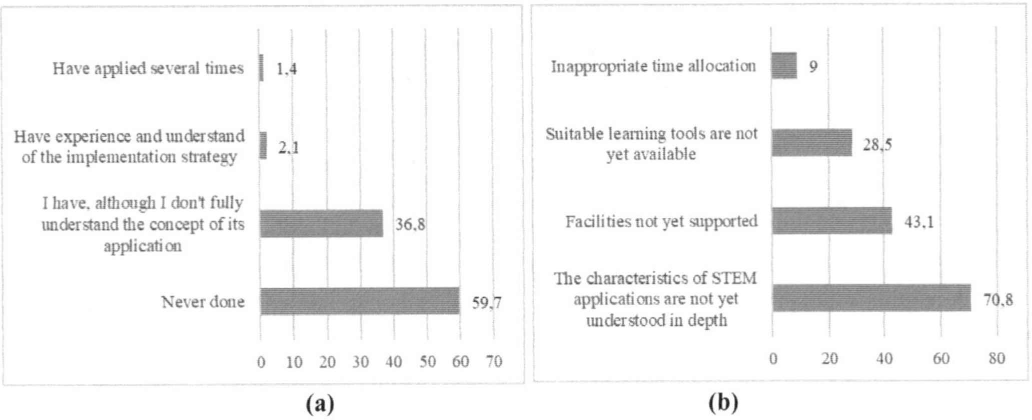

Fig. 3. Conditions of STEM application for disaster mitigation and obstacles.

4.4 The Importance of STEM-Based Local Wisdom Integration

Local wisdom can be integrated from various aspects. These aspects are all forms of knowledge, beliefs, insights, and ethics that guide human behavior in life as an ecological community. Apart from that, local wisdom also concerns knowledge, understanding, and customs regarding humans and nature, and understanding the relationships between all residents of an ecological community. Thus, local wisdom cannot be separated from science, technology, engineering, and mathematics (STEM) to create a sustainable life. Therefore, local wisdom can be applicable and pragmatic with a philosophical basis that can be applied in school learning. It starts by talking about the dry season and how to get water using the *Senggot* well. A well where the water is drawn is not with a pulley but with a long stick. This is in line with the findings in this research, namely 51% of teachers consider that the integration of STEM-based local wisdom for disaster mitigation is very important, and 48% consider it important. This assumption is based on the reason that integrating STEM-based local wisdom into learning can: (a) preserve local wisdom, (b) introduce culture to the younger generation, and (c) provide contextual learning. The Importance of STEM-Based Local Wisdom Integration for Disaster Mitigation (Figure 4 a & b).

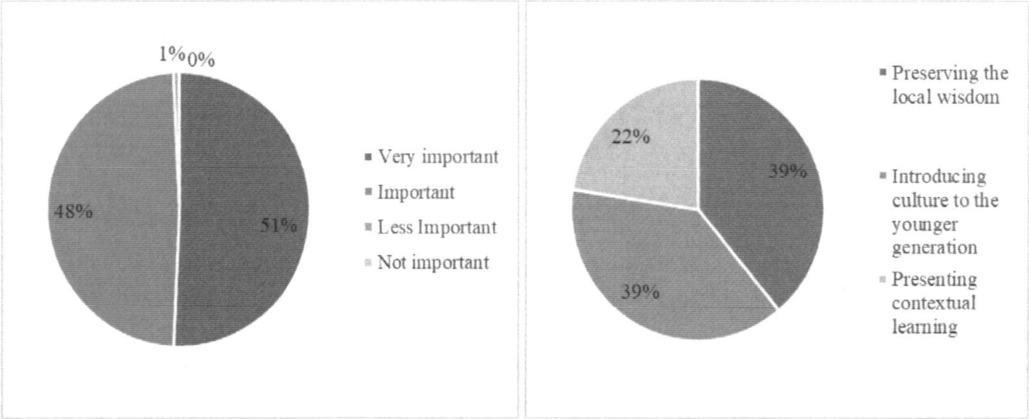

(a) **(b)**

Fig. 4. The importance of STEM-based local wisdom integration for disaster mitigation.

4.5 STEM-Based Local Cultural Integration Strategies and Methods

Now that we know that integrating local wisdom to improve STEM-based disaster mitigation is very important, how are integration strategies and methods carried out? Integration strategies can be carried out by (a) integrating STEM-based local wisdom into teaching materials (66.7%), (b) packaging IT-based cultural content (61.8%), (c) developing learning tools (52.8%), (d) developing culture-based learning media (42.4%). To implement this strategy through methods: demonstration of local culture at school (75%), implementing of a knowledge management system (57.6%), and local culture FGD between schools and indigenous communities (45.1%). The strategy and method for integrating local culture for STEM-based disaster mitigation (Figure 5 a & b).

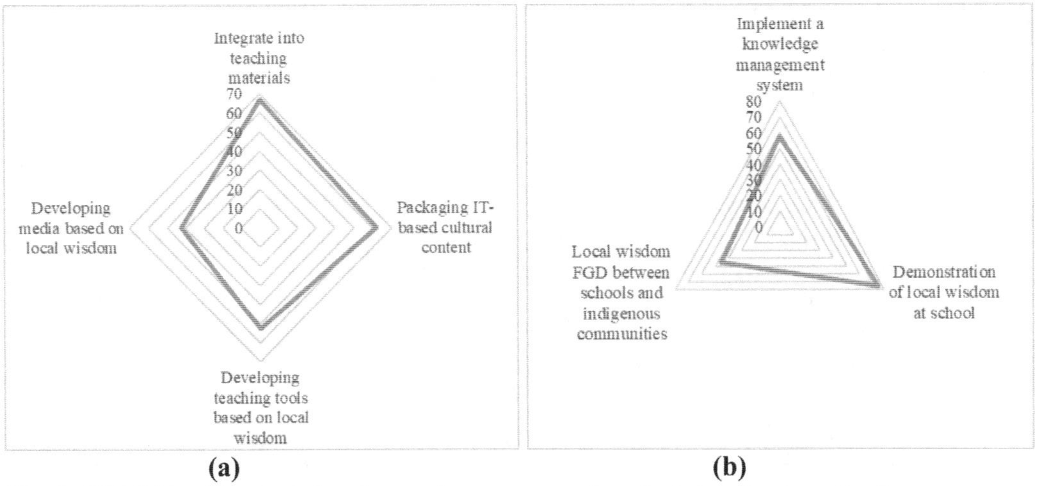

Fig. 5. STEM-based local wisdom integration strategy and methods for disaster mitigation.

4.6 Obstacles in Integrating STEM-Based Local Wisdom for Disaster Mitigation

The obstacles faced in integrating STEM-based local wisdom to improve disaster mitigation are weak teacher understanding regarding disaster mitigation (34%), lack of socialization of disaster mitigation in schools (38%), and lack of availability of learning tools that integrate disaster mitigation in learning (28%). Thus, this obstacle must be overcome through special training carried out by the government, schools, and institutions with an interest in education and disasters. This can be proven by teacher perceptions that 68% of teachers strongly agree that special training is to develop STEM for disaster mitigation. The obstacles to integrating STEM-based local wisdom and teachers' perceptions of training that must be carried out by the government, schools, and stakeholders in education (Figure 6 a & b).

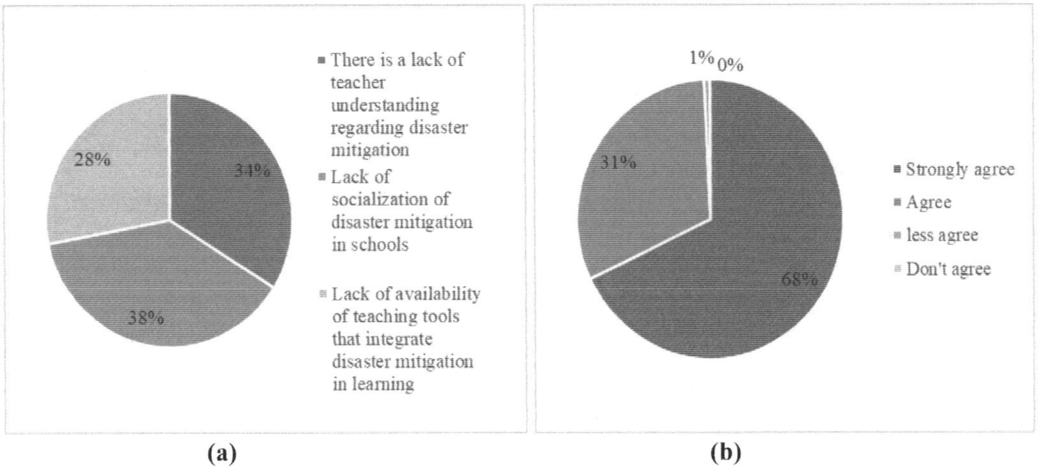

Fig. 6. Obstacles to Integrating STEM-based local wisdom.

Apart from the training solutions above, based on the literature review, there is a government program relating to disasters, namely the Disaster Safe Education Unit Program (SPAB) which is one of the efforts to prevent and overcome the impact of natural disasters on educational units in Indonesia. To reduce disaster risk among the community, such as educational institutions, an organization is needed that can sustainably manage disasters. In this case, the Indonesian government through the Ministry of Education and Culture has formed a Disaster Safe Education Unit and issued regulations regarding the implementation of the Disaster Safe Education Unit program abbreviated as SPAB [13]. Apart from that, currently, the Indonesian Government through the Secretary General of the Ministry of Education, Culture, Research, and Technology has issued technical guidance regulations for implementing disaster-safe education unit programs. Education units can be educational service groups that provide education through formal, non-formal, and informal channels at all levels and types of education. Education units can develop disaster risk reduction programs based on disaster conditions, especially before a disaster, during an emergency, and after a disaster. Pre-disaster is a normal situation, without disaster or risk of disaster. A disaster emergency is a situation caused by a disaster and determined by the government for a certain period, including emergency preparedness, emergency response, and emergency response transition recovery based on recommendations from disaster response agencies. Post-disaster is a condition set by the government to carry out recovery, reconstruction, and restoration efforts.

According to the findings, teachers encounter a variety of difficulties, such as integrating STEM with local knowledge, practical difficulties, and their comprehension of STEM. The results show that most teachers do not know how to relate STEM education to local contextual values. STEM-based teaching resources are thought to be the most successful method for incorporating culture into education [14]. In addition, other research states that there is an increase in students' scientific literacy skills as a result of implementing STEM-based student worksheets. Figure 7 shows the integration of STEM in local wisdom in learning at school. A powerful strategy to make education more contextual, culturally relevant, and engaging for students is to combine STEM (Science, Technology, Engineering, and Mathematics) with local wisdom in school learning. This approach not only increases students' understanding of STEM but also promotes their appreciation of cultural heritage, critical thinking, and problem-solving skills based on real-world and local experiences [15].

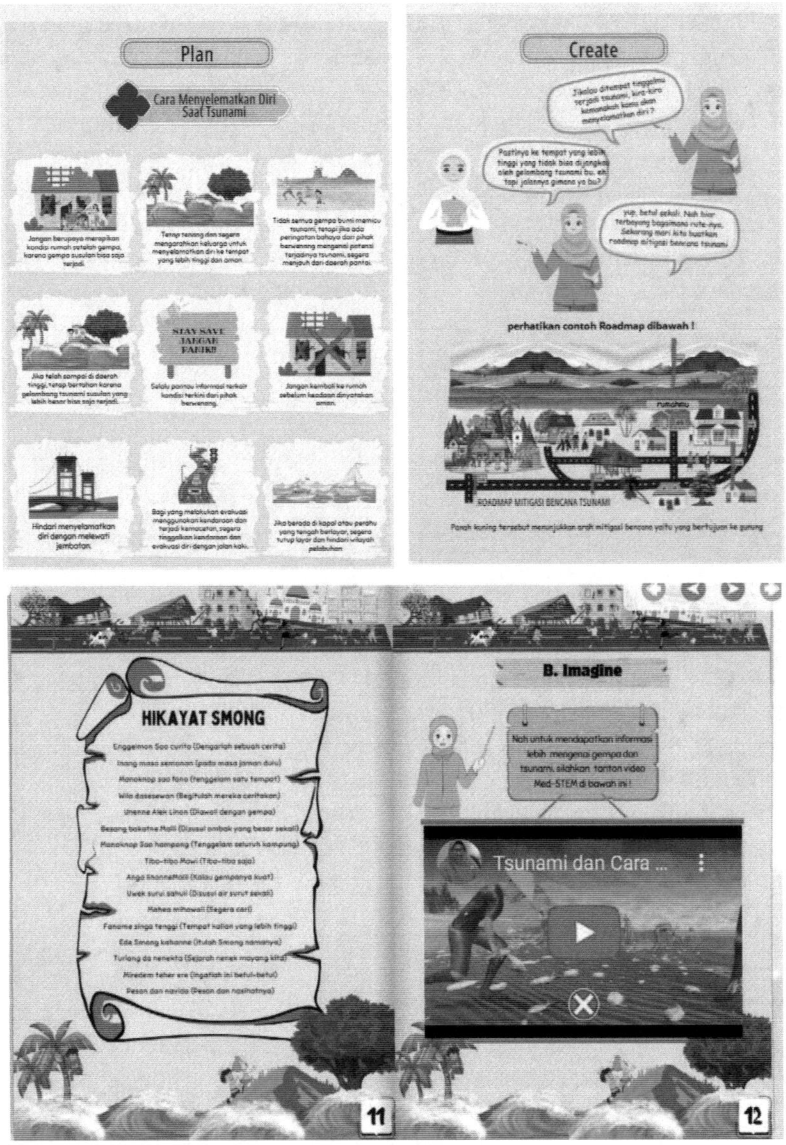

Fig. 7. STEM-Based Integration in Local Wisdom in Learning at School

5 Conclusion

Based on the results and discussion above, this research concludes that teachers already know or are familiar with STEM but are still confused about its application (40.3%). What is very urgent in integrating STEM-based local culture for disaster mitigation in schools is developing STEM-based learning tools and media in learning. Preparing learning tools is a learning plan so that it can be carried out effectively and efficiently, and learning objectives can be achieved. However, most STEM applications for disaster mitigation have never been carried out because the characteristics of STEM applications have not been understood in depth. Apart from that, local wisdom cannot be separated from science, technology, engineering, and mathematics (STEM) to create a sustainable life. Therefore, local wisdom can be applicable and pragmatic with a philosophical basis that can be applied in school learning. Thus, integration strategies that can be carried out include integrating STEM-based local wisdom into teaching materials, packaging IT-based cultural content, developing learning tools, and developing culture-based learning media. This strategy is carried out through local culture demonstration methods in schools, implementing a knowledge management system in schools, and local culture FGDs between schools and indigenous communities. The obstacles faced in integrating STEM-based local wisdom include teachers' weak understanding regarding disaster mitigation, lack of socialization of disaster mitigation in schools, and lack of availability of learning tools that integrate disaster mitigation in learning. Thus, this obstacle must be overcome through special training carried out by the government, schools, and institutions with an interest in education and disasters. Apart from that, there is a government program relating to disasters, namely the Disaster Safe Education Unit Program (SPAB) which is one of the efforts to prevent and manage the impact of disasters on educational units in Indonesia.

Acknowledgements. We would like to thank the Republic of Indonesia's Ministry of Education, Culture, Research, and Technology for providing financing for this study through the Regular Fundamental Scheme (633/UN11.2.1/PT.01.03/DPRM/2023).

References

1. R. S. Oktari, K. Munadi, R. Idroes, and H. Sofyan, International Journal of Disaster Risk Reduction **51**, 101881 (2020)
2. Syamsidik, R. S. Oktari, A. Nugroho, M. Fahmi, A. Suppasri, K. Munadi, and R. Amra, International Journal of Disaster Risk Reduction **54**, 102052 (2021)
3. P. E. Suarmika, I. B. Putu Arnyana, I. W. Suastra, and I. G. Margunayasa, International Journal of Disaster Risk Reduction **72**, 102874 (2022)
4. O. Olaitan and N. Mavuso, Perspectives in Education **4**, 19 (2022)
5. L. Mönter and K.-H. Otto, Journal of Geography in Higher Education **42**, 205 (2018)
6. P. A. Kamil, S. Utaya, Sumarmi, and D. H. Utomo, International Journal of Disaster Risk Reduction **43**, 101411 (2020)
7. W. Adiyoso, *Manajemen Bencana: Pengantar Dan Isu-Isu Strategis* (Bumi Aksara, Jakarta, 2018)
8. N. Nopriyasman, G. Asnan, A. Fauzi, I. P. Hastuti, A. H. Ritonga, V. Kurniawan, and R. Mairiska, International Journal of Disaster Risk Reduction **107**, 104497 (2024)
9. K. Palma, D. Jofré, S. Huenul Colicoy, S. Inzunza, N. Guerrero, V. Cabello, C. Martínez Reyes, and J. Castañeda González, International Journal of Disaster Risk Reduction **119**, 105281 (2025)
10. A. Rahman, A. Sakurai, and K. Munadi, International Journal of Disaster Risk Reduction **29**, 13 (2018)
11. T. Terblanche, L. O. De Sousa, and D. Van Niekerk, Jàmbá Journal of Disaster Risk Studies **14**, 1 (2022)
12. I. Zulfiya, Sumarmi, S. Wagistina, and F. Rosyida, Jurnal Integrasi Dan Harmoni Inovatif Ilmu-Ilmu Sosial **3**, 828 (2023)
13. Permendikbud No. 33, (2019)
14. S. Susilawati, H. Sofyan, S. Ridha, S. Wahyuni, and Y. Ilhamsyah, in (2024), pp. 2035–2043
15. M. Ogunniyi and P. N. Iwuanyanwu, African Journal of Research in Mathematics, Science and Technology Education **28**, 319 (2024)

Women in the Context of Disasters: A Bibliometric Analysis of Global Research Over Three Decades

Ikhwan Amri[1,2,*], *Ismiatul Ramadhian* Nur[1,3], and *Rizka* Puspitasari[1]

[1]Tsunami and Disaster Mitigation Research Center, Universitas Syiah Kuala, Banda Aceh, Indonesia
[2]Faculty of Teacher Training and Education, Universitas Samudra, Langsa, Indonesia
[3]Faculty of Da'wah and Communication, UIN Ar-Raniry, Banda Aceh, Indonesia

Abstract. As natural hazard risks escalate globally, growing attention has been directed toward understanding how disasters affect women. This study applies a bibliometric analysis of publications between 2005 and 2025 retrieved from Scopus, focusing on gender and natural hazards. Using VOSviewer and RStudio, we assessed publication growth, prolific contributors, and thematic clusters. Results reveal a strong upward trend, with the United States emerging as the most influential country. Seven thematic clusters highlight issues such as gendered impacts of climate change, women's empowerment, and vulnerability to psychological disorders and gender-based violence. This study provides a valuable reference for scholars and policymakers seeking to strengthen gender-sensitive disaster risk frameworks.

1 Introduction

In recent years, there has been growing scholarly and policy-level interest in the intersection between gender and disaster. Disasters do not merely cause physical destruction and economic loss; they also tend to exacerbate existing social inequalities, including gender disparities. Numerous studies have shown that women often face disproportionate impacts compared to men, not due to biological differences alone, but mainly because of entrenched social structures that limit their access to resources, information, and decision-making processes [1, 2].

Nonetheless, portraying women solely as victims of disaster oversimplifies their roles and underestimates their agency. In many contexts, women act as key agents of resilience, sustaining households, organizing community-based relief efforts, and initiating grassroots recovery strategies. These contributions are often overlooked in official disaster narratives, despite playing a significant role in ensuring effective emergency responses and long-term recovery [3]. This shift in understanding aligns with a broader paradigm transition in disaster studies (i.e., from a reactive, emergency-focused approach to one that emphasizes risk reduction, preparedness, and resilience). Within this framework, gender is increasingly

* Corresponding author: ikhwan.amri@unsam.ac.id

recognized as an essential analytical lens for understanding the social dimensions of disaster impacts and for designing inclusive and equitable risk reduction strategies [4].

Although research on gender and disaster has expanded substantially, a systematic synthesis that maps the global development of this field remains lacking. The existing literature is scattered across various disciplines and regions, making it challenging to identify overarching patterns, such as the evolution of research trends, key contributors, and thematic priorities.

To address this gap, the present study adopts a bibliometric approach, a quantitative method for analyzing scientific publications to uncover the structural dynamics and knowledge development within a particular domain. By examining publication data from 2005 to 2025, this study aims to provide a comprehensive overview of how gender and disaster issues have evolved in academic discourse over time, while identifying thematic strengths and research gaps that remain underexplored. The findings of this study are expected to inform researchers, practitioners, and policymakers seeking to strengthen gender-sensitive disaster risk reduction efforts. Moreover, the bibliometric insights may serve as a foundation for developing a more strategic and responsive future research agenda that aligns with global resilience goals and community needs.

2 Research methods

This study employed the Preferred Reporting Items for Systematic Reviews and Meta-Analyses (PRISMA) protocol in collecting publication metadata related to women and disasters (Figure 1). The first stage involved identifying relevant literature from a scientific database. Scopus was selected as the primary search engine due to its strength in providing high-quality publication records along with relatively comprehensive metadata [5]. The literature search was conducted on May 19, 2025, using the terms "woman", "women", "female", "girl*", and "gender" in the Title, combined with the Boolean operator AND with "natural hazard*", "disaster*", "earthquake*", "tsunami*", "volcan*", "landslide*", "hurricane*", "typhoon*", "cyclone*", "drought*", "flood*", "wildfire*", and "forest fire*" within the Title, Abstract, and Keywords. The extracted metadata was exported as a comma-separated values (.csv) file. During the identification phase, a total of 2,916 publications were retrieved from Scopus; however, 14 duplicate records were identified and subsequently removed.

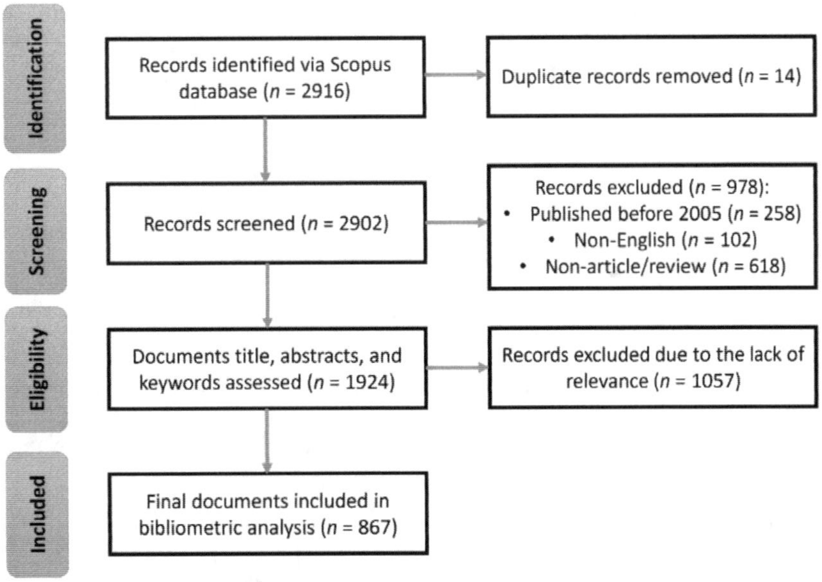

Fig. 1. PRISMA protocol for document selection.

In the subsequent phase, additional documents were excluded based on specific criteria. The study required documents to be published between 2005 and 2025 and written in the English language. Furthermore, only articles and review papers were included in the analysis. In total, 978 records were excluded after further screening of titles, abstracts, and keywords to assess their relevance to the research topic. Papers that focused on non-natural hazards or primarily addressed climate change without a disaster-related context were also excluded. Finally, 867 documents were selected for the bibliometric analysis.

Data analysis was conducted using key software tools for bibliometric review, specifically BiblioShiny version 4.0 and VOSviewer version 1.6.20. Microsoft Excel was also used as a supporting tool to generate frequency graphs for specific metrics. The analysis covered the following aspects: (1) the yearly trend of publications, (2) the most prolific journals, authors, and countries, (3) the most cited documents, and (4) author keyword mapping.

3 Results and discussion

3.1 Annual publication trends

Over the past two decades, scholarly attention to women and disaster has significantly increased, reflecting the growing recognition of the need for gender-sensitive research in this interdisciplinary field. Figure 2 illustrates the annual number of publications on women and disasters from 2005 to 2025, revealing key trends. In the early years (2005–2010), research output was relatively low, with fewer than 20 publications annually. This modest number likely reflects the initial phase of integrating gender considerations into disaster studies, characterized by limited awareness of the specific impacts that disasters have on women.

However, a noticeable upward trend began around 2011, coinciding with the emergence of global policy frameworks such as the Hyogo Framework for Action and the Sendai Framework for Disaster Risk Reduction, which emphasized the integration of gender perspectives into disaster planning. During this period, the number of annual publications steadily increased, surpassing 40 per year by 2015. The post-2015 period saw a rapid expansion in research output, with the number of publications more than doubling over the next decade. By 2020, annual publications exceeded 80, and the growth continued through to 2025. This surge can be attributed to factors such as enhanced international collaboration, increased funding for gender and disaster research, and the growing connection between disaster studies and other global agendas like public health, climate change, and sustainable development.

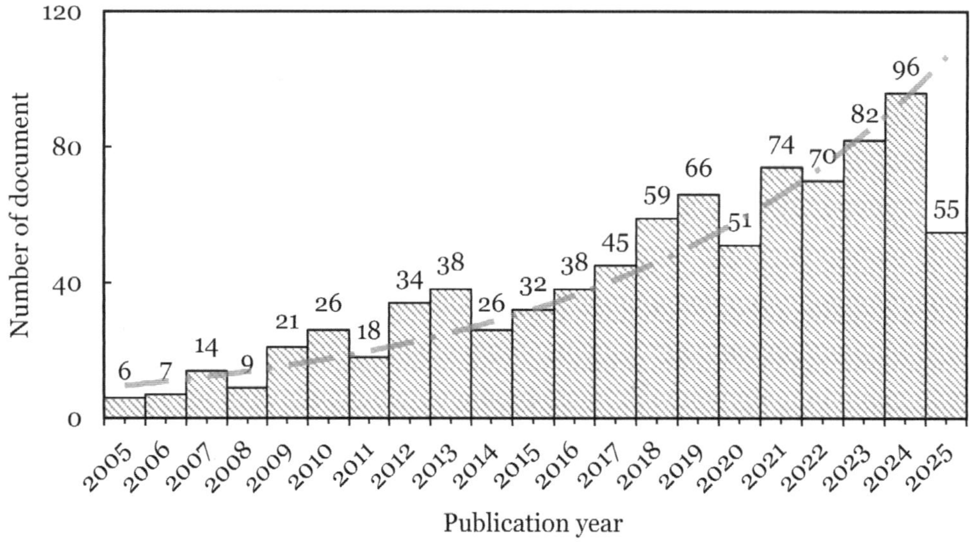

Fig. 2. Annual publication trends in women and disaster research (2005–2025).

3.2 Sources

The documents included in this analysis were distributed across 468 sources. Publications on women and disasters have been published in journals with varied scopes, covering fields such as disaster science, development studies, gender studies, health sciences, psychology, geography, economics, social sciences, and multidisciplinary outlets. Table 1 depicts the 10 most productive journals in disaster research focusing on women. These journals were generally regarded as highly reputable, with most indexed in Q1 or Q2 based on the 2024 SCImago Journal Rank (SJR). The majority were published by major academic houses known for their high-quality outputs, such as Elsevier, John Wiley & Sons, Springer Nature, Taylor & Francis, and MDPI. In general, the top sources were dominated by journals specializing in disaster-related research. *International Journal of Disaster Risk Reduction* ranked first with 59 documents. Additionally, it represented the most impactful source, based on citation count (1,008 citations), underscoring its role as a key research outlet in this field. *International Journal of Environmental Research and Public Health* ranked second with 20 documents, while *Disasters* contributed 18 documents, making it the third most influential journal in disaster research related to women.

Figure 3 illustrates the journals with the highest H-index, indicating their local impact within the dataset. *International Journal of Disaster Risk Reduction* once again stood out as the most influential journal, achieving an H-index of 18, meaning that 18 of its published articles were cited at least 18 times each. It was followed by *Disasters*, *International Journal of Environmental Research and Public Health*, and *Journal of Affective Disorders*, which recorded H-index values of 13, 11, and 9, respectively. Further down the list, five journals shared an H-index of 7, including *Asian Journal of Women's Studies*, *Disaster Prevention and Management: An International Journal*, *Gender and Development*, *Natural Hazards*, and *Sustainability (Switzerland)*.

Table 1. Most productive journals.

Journal Title	Publisher	Number of Documents	Number of Citations
International Journal of Disaster Risk Reduction	Elsevier	59	1008
International Journal of Environmental Research and Public Health	Multidisciplinary Digital Publishing Institute (MDPI)	20	401
Disasters	John Wiley & Sons	18	512
Disaster Medicine and Public Health Preparedness	Cambridge University Press	16	200
Australian Journal of Emergency Management	Australian Institute for Disaster Resilience	15	68
Natural Hazards	Springer Nature	13	270
Disaster Prevention and Management: An International Journal	Emerald Publishing	11	180
World Development	Elsevier	10	407
Gender and Development	Taylor & Francis	10	297
Sustainability (Switzerland)	Multidisciplinary Digital Publishing Institute (MDPI)	10	150

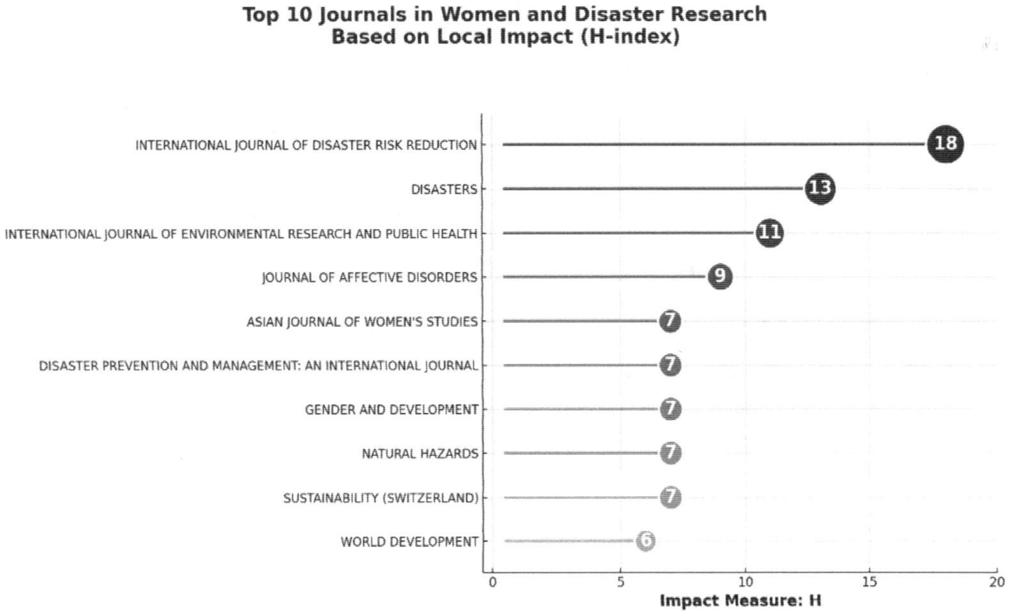

Fig. 3. Sources' local impact by H-index.

3.3 Authors

The bibliometric analysis of the 867 documents included in this study reveals the most productive authors in the field of women and disaster research. Table 2 presents the top 10 authors based on the number of publications. E.W. Harville ranks as the most prolific author, with 12 publications and 379 citations. This is followed by S. Sohrabizadeh, with 10 publications and 149 citations, and S. Dema Moreno, with 7 publications and 46 citations additionally, five authors: M. Alston, J.E. Rhodes, C. Carmassi, L. Dell'Osso, and D. Parkinson each contributed 6 documents to the field.

Table 2. Most productive authors.

Author Name	Number of Documents	Number of Citations
E.W. Harville	12	379
S. Sohrabizadeh	10	149
S. Dema Moreno	7	46
M. Alston	6	307
J.E. Rhodes	6	251
C. Carmassi	6	196
L. Dell'Osso	6	196
D. Parkinson	6	21
A. Rossi	5	246
P. Buekens	5	226

3.4 Co-authorship by countries

Figure 4 displays the geographical distribution of scientific production related to women and disasters. Darker shades of blue indicate a higher number of publications, while lighter shades represent lower publication output. Countries shown in gray did not have any relevant publications recorded in the dataset. The United States emerged as the most active country (n = 681). The following most significant contributors were Japan (n = 298), Australia (n = 237), China (n = 187), Italy (n = 133), and India (n = 131). Representation from African countries remains very limited, highlighting the need for enhanced cross-national collaboration and capacity building in research.

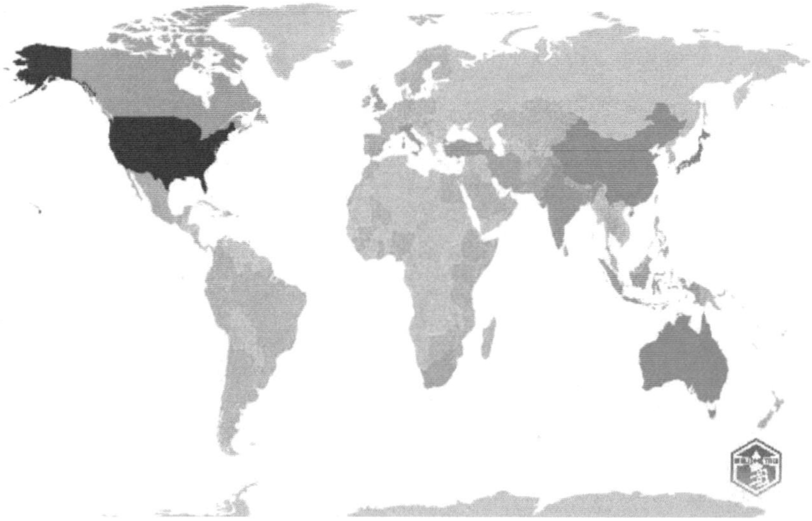

Fig. 4. Country scientific production.

In terms of citations, the United States also leads with a total of 3496 citations, underscoring its dominant role in publishing and influencing disaster studies. The United Kingdom follows with 1747 citations, highlighting its contribution to global research where gender and disaster studies intersect. Australia, with 1071 citations, particularly in the context of disaster risk reduction and gender studies in the Pacific and Southeast Asia regions. China's research output has surged to 955 citations, reflecting the country's growing influence in the field, driven by its increasing focus on integrating gender perspectives into disaster management strategies. Other countries, such as Italy and Canada, contributed 867 and 654 citations, respectively.

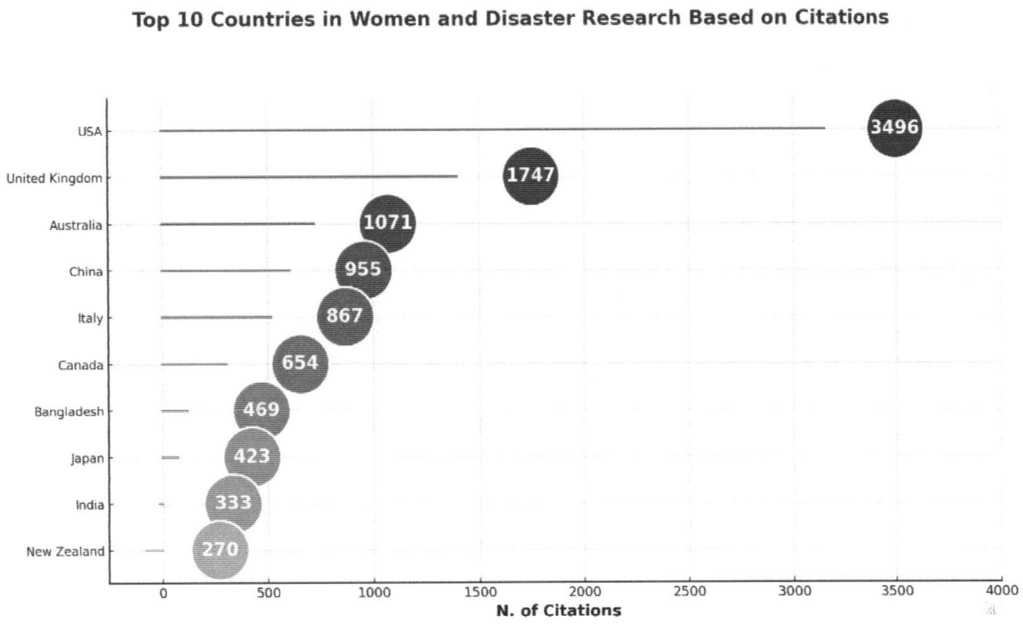

Fig. 5. Most cited countries.

3.5 Most cited documents

The impact of research within the field of women and disaster is reflected not only in publication counts but also in how frequently individual studies are cited by the scientific community. Table 3 presents the ten most cited documents in this domain, based on citation data extracted from the Scopus database. The most highly cited article, authored by Neumayer and Plümper (2007), examines disasters that disproportionately affect women due to social and economic inequalities, rather than biological differences [6]. This work, with 769 citations, examines how disaster vulnerability and women's socioeconomic status impact the gender gap in life expectancy, revealing that higher female vulnerability is associated with lower socioeconomic status. The second most cited article by Ajibade et al. (2013), with 198 citations, examines how women in Lagos, Nigeria, experience and are affected by flash floods, particularly focusing on the July 2011 flood, using a gendered perspective [7]. It aims to understand how socioeconomic status, class, and location shape women's vulnerability, recovery, and adaptive capacity, and to highlight the need for targeted interventions that support low-income women in coping with climate-related disasters. Other highly influential papers focus on diverse issues, including the relationship between climate change and gender equality in developing states [8]. The purpose of this study is to examine how climate change negatively affects gender equality and women's rights, particularly in developing countries.

It investigates whether the unequal burdens women face from climate impacts, such as loss of assets, increased domestic responsibilities, and reduced access to resources, are reflected in broader social and institutional inequalities.

Table 3. Most cited documents.

Document Title	Author(s)	Year	Number of Citations	Source Title
The gendered nature of natural disasters: The impact of catastrophic events on the gender gap in life Expectancy, 1981-2002	Neumayer E; Plümper T.	2007	769	*Annals of the Association of American Geographers*
Urban flooding in Lagos, Nigeria: Patterns of vulnerability and resilience among women	Ajibade I; McBean G.; Bezner-Kerr R.	2013	198	*Global Environmental Change*
Climate change and gender equality in developing states	Eastin J.	2018	188	*World Development*
Gender, age and circumstances analysis of flood and landslide fatalities in Italy	Salvati P; Petrucci O.; Rossi M.; Bianchi C.; Pasqua A.A.; Guzzetti F.	2018	166	*Science of the Total Environment*
Social support mobilization and deterioration after Mexico's 1999 flood: Effects of context, gender, and time	Norris F.H; Baker CK; Murphy A.D.; Kaniasty K.	2005	156	*American Journal of Community Psychology*
Full and partial PTSD among young adult survivors 10 months after the L'Aquila 2009 earthquake: Gender differences	Dell'Osso L; Carmassi C.; Massimetti G.; Daneluzzo E.; Di Tommaso S.; Rossi A.	2011	137	*Journal of Affective Disorders*
PTSD symptoms in children and adolescents 28 months after a flood: Age and gender differences	Bokszczanin A.	2007	132	*Journal of Traumatic Stress*
Critical Mass of Women on BODs, Multiple Identities, and Corporate Philanthropic Disaster Response: Evidence from Privately Owned Chinese Firms	Jia M.; Zhang Z.	2013	132	*Journal of Business Ethics*
Vulnerability of women to climate change in arid and semi-arid regions: The case of India and South Asia	Yadav S.S; Lal R.	2018	130	*Journal of Arid Environments*
Violence against women and natural disasters: Findings from post-tsunami Sri Lanka	Fisher S.	2010	129	*Violence Against Women*

Most of the ten papers explore how disasters affect women differently from men. They highlight that women often face greater risks during events like floods, earthquakes, and landslides, not because of biological differences, but due to social and economic inequalities. Several studies look at how gender and age influence survival rates, mental health issues, and access to support during and after disasters. In low-income or developing areas, women tend to recover more slowly and face more challenges due to limited resources and traditional gender roles. Overall, these papers highlight the importance of understanding the gendered impacts of disasters and advocate for more inclusive policies that take into account women's specific needs and vulnerabilities.

3.6 Keyword co-occurrence analysis

The keyword co-occurrence analysis served to identify terms that were frequently used within the dataset. By examining these commonly occurring terms, it was possible to estimate the main themes within the research field. To determine the most relevant keywords, a minimum occurrence threshold was set. Only keywords that appeared at least five times were included in the analysis. After applying this criterion, a total of 99 keywords met the threshold and were used for further visualization.

Figure 6 illustrates the most commonly used keywords, with "gender" (n = 180), "women" (n = 127), and "disaster" (n = 88) appearing as the top-ranking terms. Keywords such as "earthquake", "drought", and "flood(s)" were also among the most frequently mentioned, likely indicating the specific types of natural hazards most often addressed in the literature. The prominence of "climate change" among the top five keywords underscores the growing recognition of its connection to disaster risk and its disproportionate impact on vulnerable populations, particularly women.

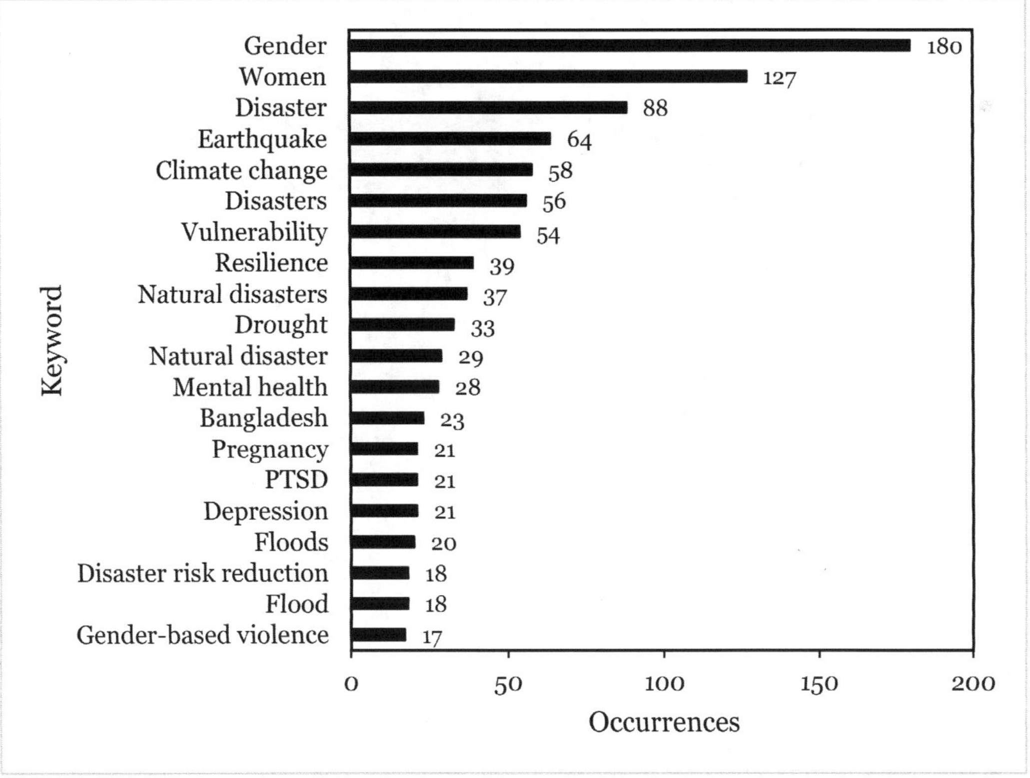

Fig. 6. Most commonly occurring author keywords in the bibliometric dataset.

The terms "vulnerability" and "resilience," while conceptually contrasting, appeared in close frequency, suggesting that existing research frequently explores both key dimensions of disaster risk management: on one hand, women's social vulnerabilities, and on the other, their adaptive capacities in the face of disaster. Psychosocial-related terms such as "mental health", "PTSD", and "depression" were also standard, reflecting growing awareness of the gendered differences in mental health outcomes, as women are often found to experience different rates of psychological distress during and after disasters compared to men [9]. The inclusion of "pregnancy" as a frequently cited keyword indicates a focused interest in a particularly vulnerable subgroup, acknowledging that pregnant women face unique risks and needs in disaster contexts [10]. Additionally, the presence of the term "gender-based violence" underscores evidence that disaster situations can exacerbate violence against women [11].

Furthermore, a network visualization of author keywords was constructed using VOSviewer software (Figure 7). The size of each circle represents the frequency of keyword occurrences, while the color indicates the cluster to which each keyword belongs. Keywords that appear closer to one another in the visualization are considered to have a higher degree of relatedness [12].

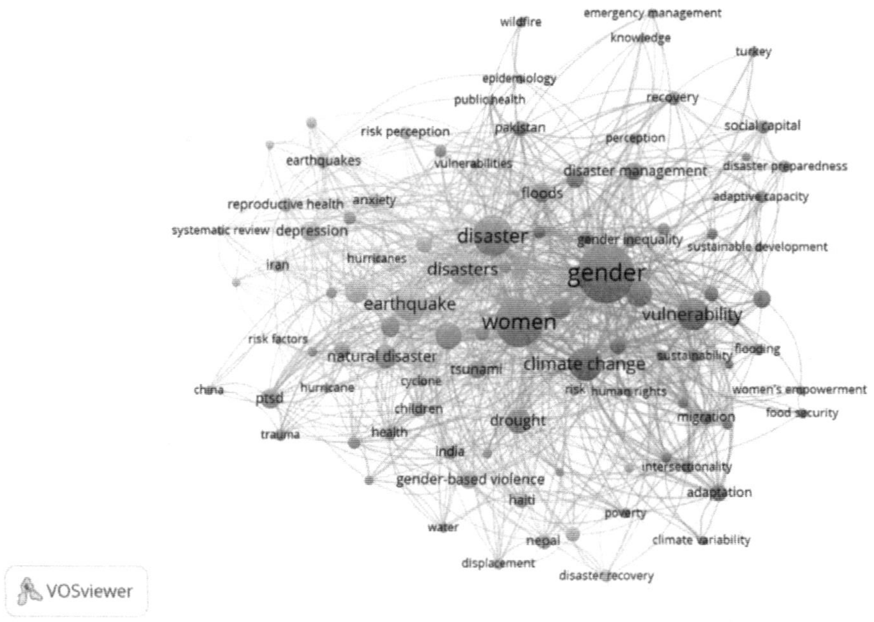

Fig. 7. Network visualization of author keywords (weights: occurrences).

A total of seven distinct clusters emerged from the network visualization of author keywords, with a total link strength of 1,686 and 868 connections among the terms. The first cluster (red) centers on themes related to the gendered impacts of climate change and the integration of gender perspectives into disaster risk reduction strategies [13]. This cluster includes key terms such as "climate change" (n = 58), "vulnerability" (n = 54), "disaster risk reduction" (n = 18), "adaptation" (n = 16), "gender equality" (n = 14), and "gender mainstreaming" (n = 11). The second cluster (green) reflects issues surrounding mental health and disaster-related psychological impacts, represented by keywords such as "disaster" (n = 88), "natural disasters" (n = 37), "PTSD" (n = 21), "stress" (n = 12), "trauma" (n = 8), and "post-traumatic stress disorder" (n = 8). This cluster also includes "pregnancy" (n = 21) and "pregnant women" (n = 14), indicating a subset of research that focuses on the psychological impacts of disasters on pregnant women [14]. The third cluster (dark blue) is associated with the empowerment and agency of women in disaster management. It features keywords such as "gender" (n = 180), "resilience" (n = 39), "disaster management" (n = 17), "social capital" (n = 10), "adaptive capacity" (n = 9), and "women's empowerment" (n = 5), highlighting research that emphasizes the active roles women play in building community resilience [15]. The fourth cluster (yellow) shares thematic similarities with the second cluster due to the presence of psychological terms such as "mental health" (n = 28), "depression" (n = 21), "anxiety" (n = 11), and "posttraumatic stress disorder" (n = 7). However, it is more focused on earthquake-related disasters, as indicated by keywords like "earthquake" (n = 64) and "earthquakes" (n = 12).

The fifth cluster (purple) indicates a regional focus on South Asia, as reflected by keywords such as "Bangladesh" (n = 23), "India" (n = 10), and "South Asia" (n = 5). These are closely linked with hydrometeorological hazards, including "drought" (n = 33), "floods" (n = 20), and "cyclone" (n = 6). The sixth cluster (light blue) highlights the issue of violence against women, represented by the keyword "gender-based violence" (n = 17). Other terms within this cluster include "Nepal" (n = 14), "Haiti" (n = 12), and "disaster recovery" (n = 9). Finally, the seventh cluster (orange) contains keywords such as "perception" (n = 8), "emergency management" (n = 6), and "knowledge" (n = 5). This smaller group points to a growing interest in how individuals and communities (particularly women) perceive, understand, and respond to disaster risks.

4 Conclusion

This study evaluated the current developments in women-focused disaster research using a bibliometric approach spanning 2005–2025. Publications have steadily increased, with the International Journal of Disaster Risk Reduction emerging as the most influential journal, and E.W. Harville as the most productive author. The United States leads in both output and citation impact. Co-occurrence analysis identified multi-dimensional themes, including climate change impacts, women's empowerment, and vulnerabilities such as mental health and gender-based violence. These findings highlight the importance of integrating gender-sensitive approaches into disaster risk reduction frameworks.

Acknowledgements. The submission fee for this article was funded by the Dana PTNBH UPT Mitigasi Bencana USK for the 2025 fiscal year.

References

1. S. Bradshaw, Gender, development and disasters (Edward Elgar Publishing, Cheltenham, 2015).
2. F.T. Zahra, M.G.C. Rodil, M.F. Khan, Gender and disaster resilience: A systematic review. Int. J. Disaster Risk Reduct. **70**, 102799 (2022). https://doi.org/10.1016/j.ijdrr.2021.102799
3. J. D'Cunha, V. Le Masson, L. Savage, Women's leadership in disaster risk reduction: Evidence from South and Southeast Asia, (UN Women, New York, 2021).
4. J.C. Gaillard, Vulnerability, capacity and resilience: Perspectives for climate and development policy. J. Int. Dev. **22**, 218–232 (2010). https://doi.org/10.1002/jid.1675
5. A.H. Achadi, I. Amri, D. Ruslanjadi, R. Tanaka, Two decades of bibliometric exploration on leadership in disaster management. Disaster Adv. **17**, 24–32 (2024). https://doi.org/10.25303/176da024032
6. E. Neumayer, T. Plümper, The gendered nature of natural disasters: The impact of catastrophic events on the gender gap in life expectancy, 1981–2002. Ann. Assoc. Am. Geogr. **97**, 551–566 (2007). https://doi.org/10.1111/j.1467-8306.2007.00563.x

7. I. Ajibade, G. McBean, R. Bezner-Kerr, Urban flooding in Lagos, Nigeria: Patterns of vulnerability and resilience among women. Glob. Environ. Change **23**, 1714–1725 (2013). https://doi.org/10.1016/j.gloenvcha.2013.08.009
8. J. Eastin, Climate change and gender equality in developing states. World Dev. **107**, 289–305 (2018). https://doi.org/10.1016/j.worlddev.2018.02.021
9. C. Carmassi, H.S. Akiskal, D. Bessonov, G. Massimetti, E. Calderani, P. Stratta, A. Rossi, L. Dell, Gender differences in DSM-5 versus DSM-IV-TR PTSD prevalence and criteria comparison among 512 survivors to the L'Aquila earthquake. J. Affect. Disord. **160**, 55–61 (2014). https://doi.org/10.1016/j.jad.2014.02.028
10. E. İbici Akça, N. Gökbulut, A. Şenoğlu, Pregnant women's depression and posttraumatic stress levels after the large-scale Turkey earthquakes: a cross-sectional study. Women Health **64**, 736–744 (2024). https://doi.org/10.1080/03630242.2024.2402789
11. A.M. Thurston, H. Stöckl, M. Ranganathan, Natural hazards, disasters and violence against women and girls: a global mixed-methods systematic review. BMJ Glob. Health **6**, e004377 (2021). https://doi.org/10.1136/bmjgh-2020-004377
12. Wibowo, A., Amri, I., Surahmat, A., Rusdah, R., Leveraging artificial intelligence in disaster management: A comprehensive bibliometric review. Jàmbá J. Disaster Risk Stud. **17**, a1776 (2025). https://doi.org/10.4102/jamba.v17i1.1776
13. N.K. Bharwani, T. Hodges, D. Lepadatu, D.X. Rodriguez, Just leader? No, lideresa! Experiences of female leaders working in climate change disaster risk reduction and environmental sustainability in the global south. Environ. Res.: Climate **3**, 045008 (2024). https://doi.org/10.1088/2752-5295/ad6a7a
14. Z. Qu, D. Tian, Q. Zhang, X. Wang, H. He, X. Zhang, L. Huang, F. Xu, The impact of the catastrophic earthquake in China's Sichuan province on the mental health of pregnant women. J. Affect. Disord. **136**, 117–123 (2012). https://doi.org/10.1016/j.jad.2011.08.021
15. H.K. Tran, N.K. Downes, Narratives of women's resilience to flood risks in Ho Chi Minh City, Vietnam. Environ. Urban. Asia **14**, 90–103 (2023). https://doi.org/10.1177/09754253231168693

E3S Web of Conferences **651**, 02011 (2025) https://doi.org/10.1051/e3sconf/202565102011
AIWEST-DR 2025

Tsunami pedestrian evacuation simulation for Camaná, Peru: Perspectives for improving evacuation performance

Jheyder Perez[1,*], *Luis* Moya[1], *Julio* Ramirez[1], *Edgard* Gonzales[2], *Erick* Mas[3], *Bruno* Adriano[3], and *Shunichi* Koshimura[3]

[1]GERDIS Research Group, Department of Engineering, Pontificia Universidad Católica del Perú, Av. Universitaria 1801, San Miguel, 15088, Lima, Peru
[2]Faculty of Geology, Geophysics and Mines, Universidad Nacional de San Agustín de Arequipa, Santa Catalina Nro. 117, Arequipa, Perú
[3]International Research Institute of Disaster Science, Tohoku University, Aoba 468-1 E401, Aramaki, Aoba-ku, Sendai, 980-8572, Miyagi, Japan

Abstract. Optimizing pedestrian evacuation in the face of a tsunami remains a critical challenge for safeguarding human lives. Agent-based models combined with reinforcement learning techniques offer a powerful framework to simulate complex evacuation scenarios, where agents learn to make decisions and identify safe routes in real time. This study focuses on improving evacuation efficiency along the coast of Camaná, Arequipa, Peru. We propose the use of numerical simulations to model pedestrian movement under the guidance of a reinforcement learning-based system. Under current transportation network conditions, only 16.6% of the population is able to reach a safe area in a tsunami scenario similar to the 2001 event. To address this, several modifications to the transportation network were proposed, including the addition of new evacuation paths and the construction of vertical evacuation structures. With the incorporation of 12 new paths and 6 vertical evacuation structures, the percentage of the population reaching safety increases to 73%. These findings provide a scientific basis for planning and implementing improvements to evacuation infrastructure in tsunami-prone areas.

1 Introduction

The readiness of various stakeholders in developing cities for natural disasters remains limited [1], including communities, businesses, governments, and scientific institutions. Over recent decades, earthquakes have caused considerable loss of life and highlighted deficiencies in emergency response efforts, especially in Asia and the Americas [2]. However, countries with strong preventive cultures and advanced tools like early warning systems have managed to lessen their damages and losses. This suggests that the impact of disasters on cities is inversely related to levels of preparedness, capacity, behavior, and the efficiency of early

* Corresponding author: jheyder.perez@pucp.edu.pe

responses by communities and institutions. Therefore, exploring solutions related to human behavior in risk scenarios—such as understanding how people interact within city and street networks after a disaster—is crucial.

Currently, various innovative and technological techniques, including artificial intelligence, are being studied to develop solutions. Along with Agent-Based Modeling (ABM) simulations, these approaches enable better assessments of evacuation processes. One of these methods in computer science is Machine Learning, which includes Reinforcement Learning (RL)—where an agent learns to perform actions in an environment to maximize reward. To save lives and improve community resilience to tsunamis in coastal areas, we propose adopting the work of [3, 4]. The goal is to design an evacuation plan to create an intelligent Evacuation Guidance System (EGS) for the coast of Camaná beach resort, Arequipa, Peru.

This work aims to improve tsunami evacuation modelling and advance our understanding of human–city interactions. Prior studies [4] often encode an agent's state as bare (x,y) coordinates—omitting contextual cues such as nearby evacuees—or as a full image of the environment at each time step, a representation that is computationally heavy and scales poorly to city-scale domains. In contrast, we define the state as a compact set of environment-derived features that guide agent decisions and learn policies via reinforcement learning; specifically, we train an on-policy SARSA agent. Our reward function promotes actions that increase evacuation success while accounting for heterogeneous departure times and time-varying street congestion throughout the simulation. In controlled experiments, the learned SARSA policy outperforms a Dijkstra's shortest-path baseline—delivering faster egress and reduced network congestion—indicating improved collective pedestrian behavior [4].

The EGS will play a vital role in early disaster response by supporting early warning systems and guiding evacuations during emergencies. It will also assist in identifying the most effective evacuation routes in various scenarios, thereby increasing the likelihood of evacuating more individuals within the flood zone. This study proposes creating a model within an EGS to simulate agent-based evacuations for the Camaná beach resort in Arequipa. Chosen for its high tsunami inundation risk, the goal is to boost the Expected Survival Rate (ESR). The implementation will require calibrating the EGS using local factors such as 1) the street network layout, 2) population density, and 3) designated evacuation zones.

2 Study area and Methods

2.1 Camaná, Peru

The Camaná beach resort area, located in the province of Camaná, includes beaches in the districts of Camaná, Samuel Pastor, and Quilca. These regions are connected by a street network that shares similar vulnerabilities, as shown in Fig. 1. Human settlements with the same names as the beaches are situated nearby (Figs. 1b and 1c). Figure 1b illustrates the street network within the beach area, while Figure 1c highlights the limited pedestrian access to designated refuge sites. Three factors cause this restricted access: 1) topographical variations, 2) insufficient signage, and 3) high-risk zones that cannot be effectively mitigated.

The population census within the ten communities in the beach circuit of the Camaná resort study area, focusing on those in flood-prone zones, recorded a total of 655 individuals. This data was carefully gathered in the censuses of 2007 and 2017 for the study area. However, for research purposes, this number does not represent the entire population of the Camaná resort area, as it omits certain groups not usually included in standard counts. These groups include the periodic population and the transient population, which only reside there during the summer. As a result, the population is roughly estimated at four thousand using fewer comprehensive estimates, and around twenty thousand based on more detailed estimates [5, 6, 7].

Camaná, a resort town, is highly vulnerable to tsunamis. Its current evacuation plans are static, showing transportation routes with arrows on maps, but they do not consider how pedestrians interact with the existing transport network. The Peruvian government has not yet explored solutions like alternative routes, vertical evacuation, or additional horizontal evacuation options from flood-prone areas. The map, created in 2003 by the National Institute of Civil Defense (INDECI) [8], highlights the primary evacuation route as the road connecting the beach to the Pan-American Highway South. It also marks zones for gathering, shelter, and temporary stay, but lacks designated access routes between these zones—crucial for the effectiveness of vertical evacuation and supplementary exits.

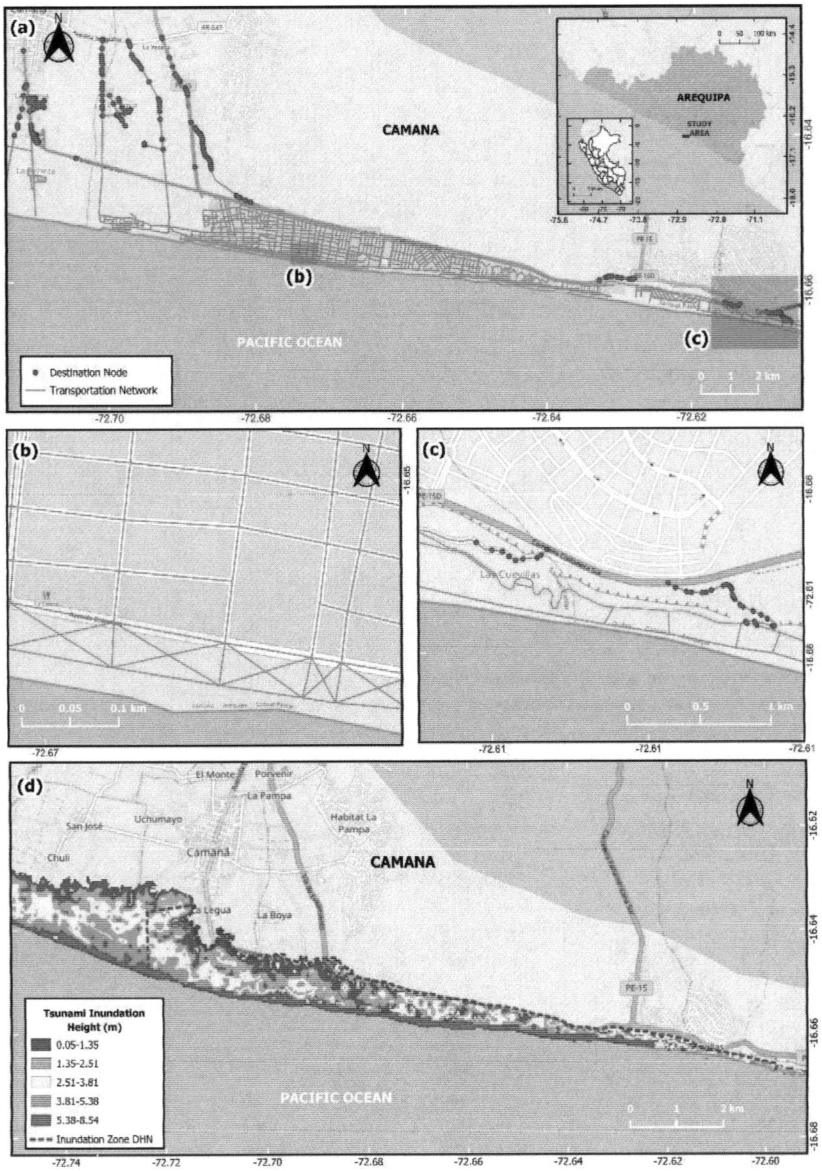

Fig. 1. (a) Location of the study area. The yellow lines constitute the graph representing the transportation network, the red marks are the location of destination nodes, and the inset shows the location of the study area in Arequipa, Peru. (b) and (c) Closer look at two zones of the study areas. (d) Tsunami inundation height based on numerical simulation. The dashed red polygon denotes the inundated area provided by the Office of DHN [9].

2.2 Tsunami Simulation at Camaná

The numerical simulation aligns with Bruno Adriano's proposed model. The employed equations, action mechanisms, and detailed procedures are documented in [10]. Fig. 1d illustrates the tsunami-induced flooding in Camaná. Results show the flood affects all blocks in the resort town, extending roughly 1.96 km perpendicular and 70 km parallel to the coast, with maximum heights reaching 8.54 meters. Most streets in Camaná experience floodwaters up to about 6 meters high. Such inundation can completely submerge 84.9% of single-story and 14.2% of two-story homes, totalling 99% of residences fully flooded. Meanwhile, 1% of homes (7 out of 790), which have three or more floors, may remain partially dry, providing safety for their occupants.

2.3 Intelligent Evacuation Guidance System

We utilize the approach suggested by [10] for tsunami evacuation simulation, combining reinforcement learning with an agent-based framework to enhance evacuation efficiency. The simulation depicts the movement of pedestrians, starting from within the inundation zone to safe areas. To this end, the research follows the EGS process, as shown in Fig. 2. The street network shown in the figure is obtained from an open source from OpenStreetMap, the tsunami simulation from [10], and the population from an official source. For calibration, as mentioned above, we use RL as the basis for MBAs. Furthermore, for the performance evaluation process, we use training-level data, where we have the number of evacuees at each point in time. The results presented correspond to the computational model in its 7,000th learning simulation. The selected simulation number corresponds to 100% learning by exploration, which provides us with the optimal pedestrian evacuation route. In addition, the results in terms of the total number of agents evacuated remain constant. Each simulation was carried out for 40 minutes, and a 5-minute Rayleigh mean was considered. Further details on the calibration process can be found in [4]. We assess the evacuation in two scenarios based on population size: one with about 4,500 agents, representing a time when only residents are present, and another with 20,300 agents, reflecting the summer period when beach occupancy is significantly higher. With a quantity of 4 and 18 agents per node within the street network for each experiment, respectively.

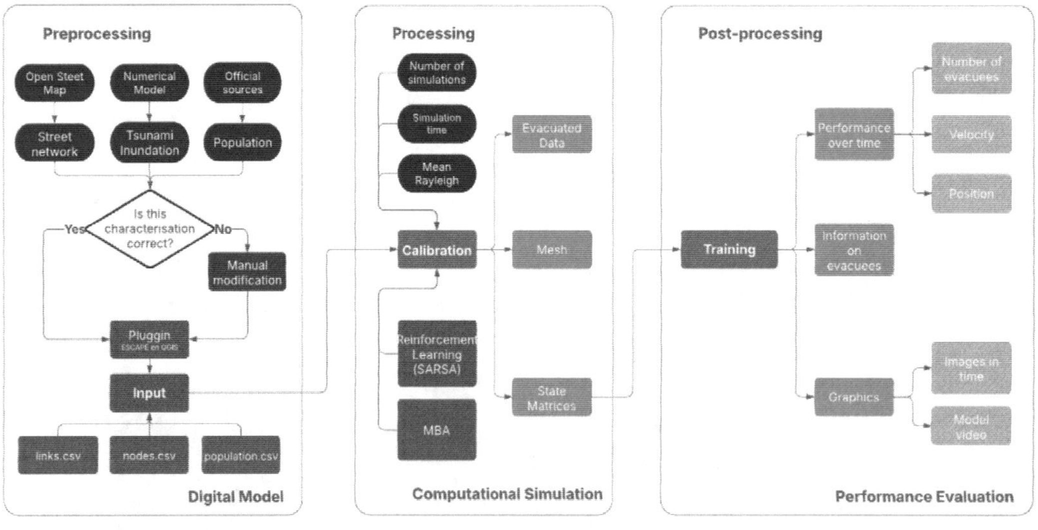

Fig. 2. Tsunami evacuation model. Preprocessing: Represents the process of digitizing the study zone. Processing: The application of the Computational Simulation with the RL Technique based on MBA. Post-processing: Includes analysis of system Performance Evaluation.

3 Results

This chapter assesses the performance of the computational model created for the street network and agents of Camaná resort. Four experiments with sixteen scenarios are presented to guide the development of three pedestrian evacuation plans for a tsunami.

3.1 Actual state of the transportation network

Figures 3 and 4 depict the number of pedestrian agents reaching destination nodes over time. In the 4,500-agent scenario, only 747 agents (17%) evacuated successfully within 14 minutes, matching the estimated tsunami arrival time from numerical simulations. By 23 minutes—the observed arrival during the 2001 tsunami—1,967 agents (44%) had reached safety. With 20,300 agents, 2,368 (12%) managed evacuation within 14 minutes, and 4,551 (22%) did so within 23 minutes. The lower survival rates in larger populations are mainly due to slower evacuation caused by higher pedestrian density. Increased congestion significantly hampers movement. An important factor is the placement of destination nodes at the resort's edges, forcing pedestrians to cover large distances to reach safety. These results indicate that the current transportation network requires major improvements to boost evacuation efficiency.

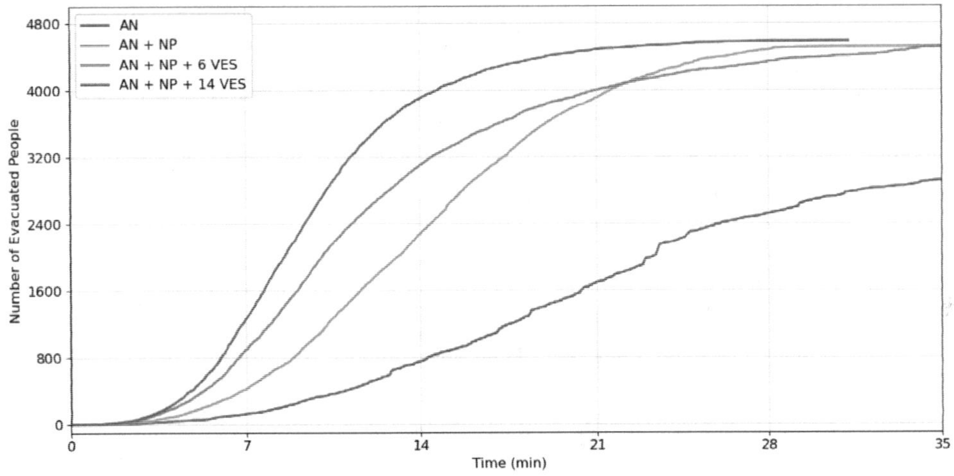

Fig. 3. Number of evacuated people vs time of a scenario with 4,500 pedestrian-agents. The red line denotes the result from the actual transportation network, the orange line represents the results of adopting additional pathways to the transportation network, and the green and blue lines are the results of using, in addition to the additional pathways, 6 and 14 vertical evacuation structures, respectively. AN: Actual network, NP: New pathways, VES: Vertical evacuation structures.

3.2 Transportation network with additional destination points

As previously discussed, several measures can be adopted to improve the tsunami evacuation process. This study emphasizes adding more destination nodes to boost accessibility and evacuation efficiency. Specifically, we suggest connecting the evacuation zone recommended by INDECI with additional pathways. Visual inspection identified 12 potential locations for constructing paths to this zone (Figs. 5 and 6). These modifications significantly enhanced evacuation results. For 4,500 pedestrian agents, 2,270 (50%) reached safety in 14 minutes, and 4,156 (92%) evacuated within 23 minutes. With 20,300 pedestrians, 6,902 (34%) reached destination nodes in the first 14 minutes, and 13,558 (67%) evacuated within 23 minutes. These improvements considerably increased evacuation rates compared to the current transportation network. The most notable improvement was in the 4,500-agent scenario, where safety within 23 minutes rose from 44% to 92%. The 20,300-agent scenario showed the least progress, with evacuees within 14 minutes increasing from 13% to 34%. While these enhancements are significant, further measures are necessary to ensure satisfactory evacuation outcomes, especially at higher population densities.

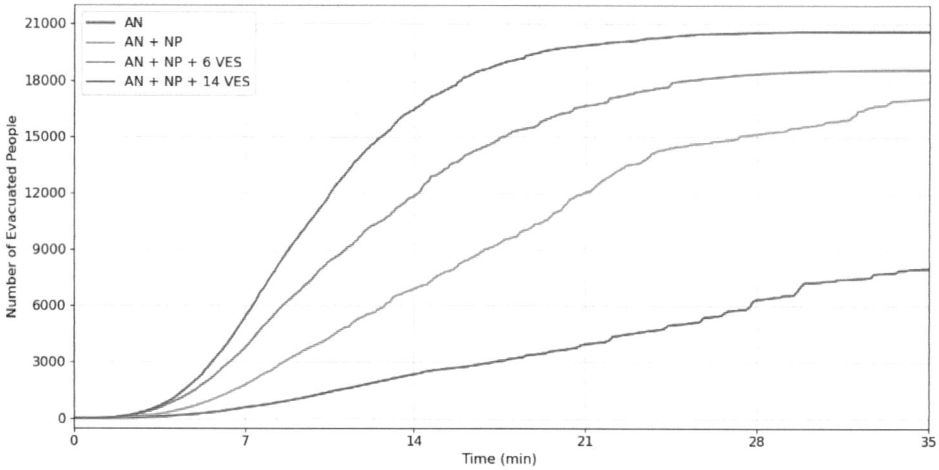

Fig. 4. Number of evacuated people vs time of a scenario with 20,300 pedestrian-agents. The red line denotes the result from the actual transportation network, the orange line represents the results of adopting additional pathways to the transportation network, and the green and blue lines are the results using, in addition to the additional pathways, 6 and 14 vertical evacuation structures, respectively. AN: Actual network, NP: New pathways, VES: Vertical evacuation structures.

Fig. 5. Photos of the potential location to connect additional pathways with the current transportation network. Their location is shown in Fig 6.

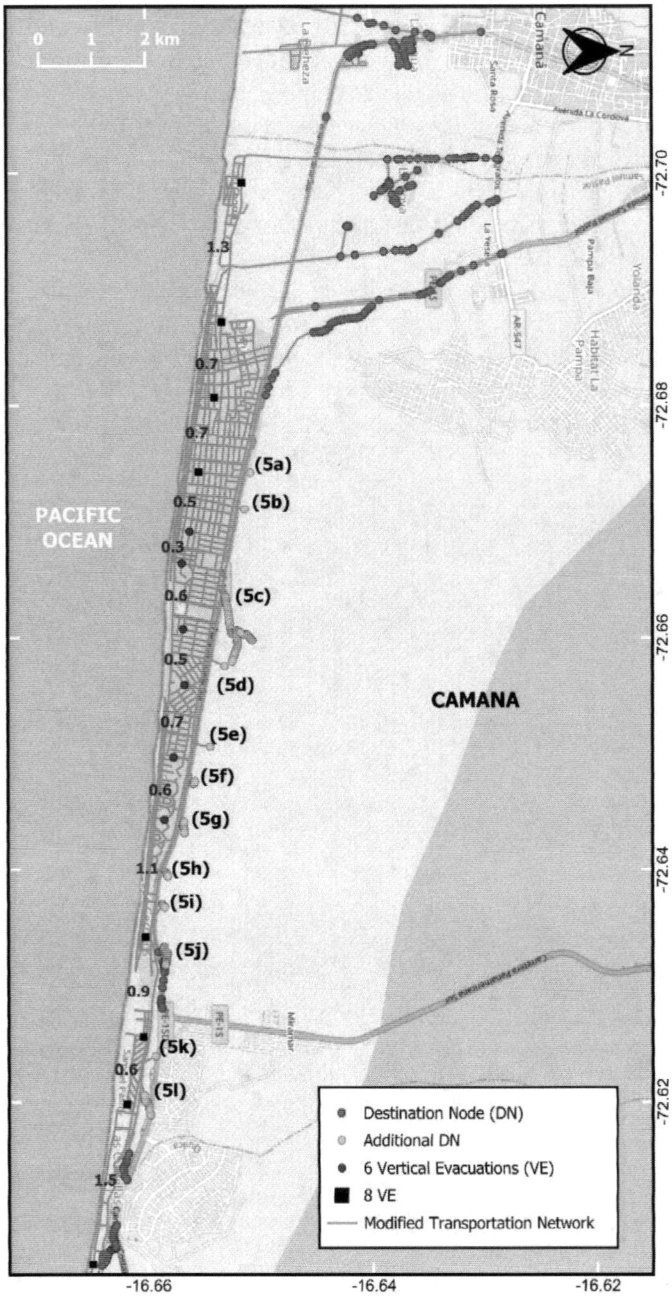

Fig. 6. Modification of the transportation network (See original network in Fig. 1a). The green circle marks denote new destination nodes representing new pathways that evacuate the inundation zone (See Fig. 5 for photos of their potential locations). The blue circle and black square marks are new destination nodes representing vertical evacuation structures. The blue number is the spacing distance between vertical evacuation structures.

Alongside expanding pathways within the evacuation zone, we evaluated how placing destination nodes—identified as vertical evacuation structures—along the coast affects evacuation. These points were examined in various setups, with the number of destination nodes varying and evenly spaced along the shoreline. Our goal was to determine the minimum number of vertical evacuation structures needed to boost evacuation efficiency significantly. As shown in Fig. 7, results reveal notable trends in the percentage of pedestrians evacuated with different counts of vertical evacuation nodes. Initially, the survivor rate increases linearly with up to four nodes. However, adding more nodes—such as six—produces a more substantial improvement, surpassing what the linear trend would suggest. A similar pattern appears with 14 vertical evacuation nodes.

Among 4,500 exposed pedestrians, 68% evacuate successfully within 14 minutes when six vertical evacuation nodes are installed. This figure rises to 84% with 14 nodes. However, during a 23-minute evacuation period, adding vertical evacuation structures does not produce significant improvements, indicating that most evacuees would likely have already reached safety within this time.

As the total exposed population increases to 20,300 pedestrians, six vertical evacuation nodes enable 58% of individuals to reach safety within 14 minutes and 85% within 23 minutes. Increasing the number of nodes to 14 raises these percentages to 79% and 97%, respectively, within the same durations. These results highlight the significant potential of vertical evacuation nodes to enhance safety, particularly in densely populated areas.

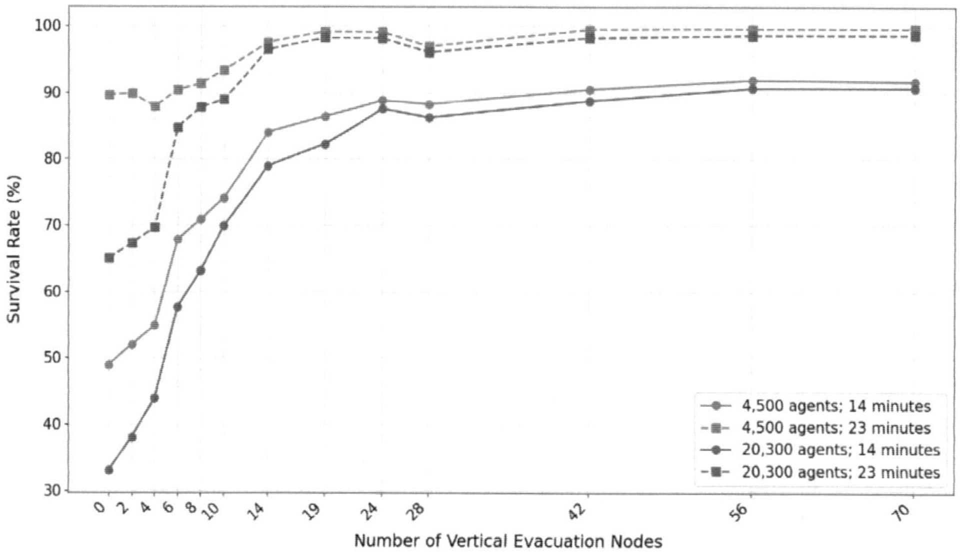

Fig. 7. Percentage of evacuated pedestrians under different numbers of vertical evacuation structures.

4 Discussion

The present study aims to evaluate the design of various pedestrian evacuation plans in response to tsunami inundation hazards by implementing a Reinforcement Learning algorithm within an agent-based model framework. The computational model, established as an intelligent guidance system for pedestrian evacuation processes, comprises between 1,124 and 1,159 street intersection nodes, with either 4 or 18 agents per node, to simulate the estimated population. Subsequently, four experiments or scenarios are proposed to reflect the designs outlined in this research, assessing the impact of adding destination nodes and increasing the number of vertical evacuation structures. These include a current street network and an improved network with specific modifications. In defining the locations of the evacuation nodes, several key considerations were identified and analysed, such as access timeliness, topography, availability, congestion in the current scenario, and the primary junctions between roads. A total of sixteen scenarios has been conducted, with the primary objective of enhancing the performance of the expected survival rate.

5 Conclusions

The results obtained from the current or baseline condition indicate a significant risk in the study area, characterized by latent danger and high vulnerability. Only 22% of the population is able to evacuate, at best, 23 minutes after the arrival of the wave in the scenario with the highest population density. This demonstrates that the existing network necessitates substantial modifications to enhance its survival rate in the face of tsunamis. Conversely, the first and second evacuation plans, in both low and high population density scenarios, demonstrate an improvement in performance. There has been an increase of up to 2,189 (46.1%) and 9,007 (42.6%) additional evacuated agents over the actual network within 14 and 23 minutes, respectively. Moreover, the most notable performance enhancements are observed with the increase in the number of vertical structures within the street network, accompanied by additional exits, as exemplified in the third proposed plan. The number of evacuated individuals increases by 826 (18.9%) and 1,623 (35.1%) within the low population density scenario, relative to the transportation network with supplementary destinations. In the high population density scenario, there is an additional evacuation of 4,946 (24.7%) and 9,565 (45.9%) individuals. These two optimal results correspond to the first 14 minutes following the tsunami wave's arrival, with 6 and 14 vertical evacuations, respectively.

This study was funded by PROCIENCIA Applied Research Programs (PE501078853-2022).

References

1. C. Cordova-Arias, J. W. Pérez Aguinaga, Quantitative Indicators for Multi-Hazard Community Resilience Assessment of Informal Settlements. Proceedings of the 18th World Conference on Earthquake Engineering (WCEE2024) (June 2024).
2. L. Ceferino, Y. Merino, S. Pizarro, L. Moya, B. Ozturk, Placing engineering in the earthquake response and the survival chain. Nat. Commun. 15, 4298 (2024). https://doi.org/10.1038/s41467-024-48624-3
3. Mas, E., Moya, L., Koshimura, S., Tsunami evacuation guidance using reinforcement learning algorithm, 17th World Conference on Earthquake Engineering, Japan (2020)
4. E. Mas, L. Moya, E. Gonzales, S. Koshimura, Reinforcement learning-based tsunami evacuation guidance system. Int. J. Disaster Risk Reduct. 115, 105023 (2024). https://doi.org/10.1016/j.ijdrr.2024.105023
5. PCM, Información territorial del departamento Arequipa: información departamental, tasa de dependencia de la población. Viceministerio de Gobernanza Territorial, Presidencia del Consejo de Ministros (2017). https://cdn.www.gob.pe/uploads/document/file/1870459/Arequipa_Información%20Territorial%20Completo.pdf.pdf
6. Redacción RPP, Camaná: evacuarán más de 20 mil veraneantes por simulacro de tsunami. RPP (2012). https://rpp.pe/peru/actualidad/camana-evacuaran-mas-de-20-mil-veraneantes-por-simulacro-de-tsunami-noticia-454247
7. PREDES, Video evacuación ante tsunami en La Punta, Camaná por Centro de Estudios y Prevención de Desastres (PREDES). YouTube (2012). https://www.youtube.com/watch?v=jBK-b1sLyP8
8. M. Toledo Gonzales-Polar, G. Núñez Monar, J. Aspilcueta Barbachán, Plan de prevención ante desastres: usos de suelo y medidas de mitigación, ciudad de Camaná y zonas aledañas. Proyecto INDECI-PNUD PER/02/05. Ciudades sostenibles. Instituto Nacional de Defensa Civil – INDECI (2003).

9. Directorate of Hydrography and Navigation, Flood maps (in Spanish). Directorate of Hydrography and Navigation (2024). https://www.dhn.mil.pe/cnat/cartas-inundacion
10. B. Adriano, E. Mas, S. Koshimura, Y. Fujii, S. Yauri, C. Jimenez, H. Yanagisawa, Tsunami inundation mapping in Lima, for two tsunami source scenarios. J. Disaster Res. 8(2), 274–284 (2013). https://doi.org/10.20965/jdr.2013.p0274

E3S Web of Conferences **651**, 02012 (2025) https://doi.org/10.1051/e3sconf/202565102012
AIWEST-DR 2025

Tsunami Readiness in Kudat, Sabah: Examining Community Trust, Risk Perception, and Preparedness

Noor Diyana Fazan Ahmad[1,*], *Mohd Muhaimin* Ridwan Wong[2], *Haliza* Zahari[2], and *Noorul Syifa* Mohd Isa [3]

[1]Centre for Research and Innovation Management, National Defence University of Malaysia, Kuala Lumpur, Malaysia
[2]Humanitarian Assistance and Disaster Relief Research Centre, National Defence University of Malaysia, Kuala Lumpur, Malaysia
[3]Community and Nation Building, Yayasan Telekom Malaysia, Level 6, TM Annexe 2, Jalan Pantai Jaya, 59200 Kuala Lumpur, Federal Territory Kuala Lumpur, Malaysia

Abstract. One of the emerging critical concerns on disaster risk reduction is tsunami preparedness as its impact is substantial to damaging infrastructure and to cause loss of lives. Kudat located at the tip of Sabah is an area that is prone to a tsunami disaster originating from either the Brunei Slide, Sulu Sea, or from the Marina Trench. Although there have been efforts to increase their awareness and preparedness, there remains a lack of localised studies exploring how community level factors affect disaster readiness. Thus, this study aims to address the gap by examining socio-economic factors, trust in agencies, risk perception, intention to prepare and preparedness of the local community in Kudat. The study employed a quantitative approach through self-administered questionnaires among local leaders and members of the community. Data obtained was then analysed using Statistical Package for Social Sciences (SPSS) to identify their perceptions and patterns. Preliminary findings indicate that there is a positive perception towards the agencies, intention to prepare and have taken necessary steps to prepare for a tsunami disaster. However, financial preparedness is expected to be impeded due to their average household income. The findings suggest that the community can be prepared with the right intervention by the local government and agencies however their financial capacity could affect their recovery. This study offers practical recommendations for policy and capacity building efforts that can be adapted in other tsunami prone areas. The importance of how locally driven strategies is underlined in study to enhance tsunami preparedness and community resilience.

1 Introduction

Tsunamis are one of the most devastating disasters to occur as it impacts the livelihood of the community due to its catastrophic nature. Large scale natural disasters such as tsunamis greatly affect developing countries as they often lack the resources and governance to cope with the disaster [1]. Various efforts globally have been made to strengthen disaster risk reduction and resilience especially is coastal regions that are vulnerable to natural disasters. These efforts include installing early warning systems, training the local responding agencies and preparing the local authority and community through awareness campaigns. Historically, Malaysia is mostly affected by floods and landslides however 2004 is the first time Malaysia was impacted by the 2004 Indian Ocean tsunami which resulted in 64 lost lives in Malaysia.

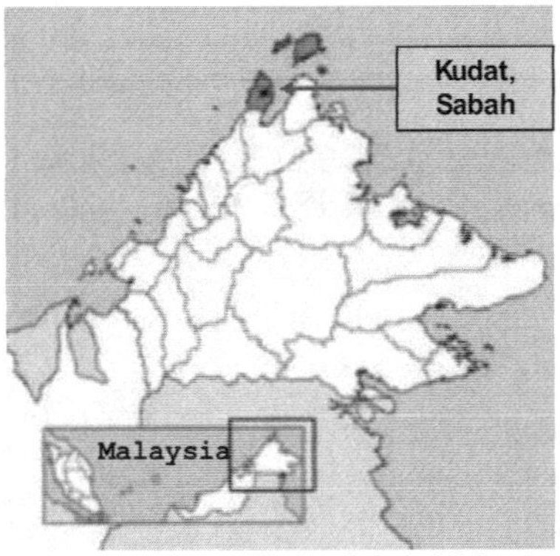

Fig 1. Location of Kudat, Sabah, Malaysia

To have a beter prepared toward the tsunami, the Malaysian government has undertaken necessary action such as installing the Malaysian National Tsunami Early Warning System and educated the local authority and community based

*Corresponding author: n.diyana@upnm.edu.my

on the Sendai Framework for Disaster Risk Reduction (2015-2030) that integrates both top-down and bottom-up approach. Based on a past study, Kudat, a district in Sabah located in East Malaysia had the earliest record of a tsunami disaster which was triggered by the Zamboanga earthquake in the Philippine. Over the last decade, several studies have found that Kudat is vulnerable to a tsunami disaster due to an underwater landslide in the Baram Delta as the potential source [2]. Even though the source is a rare occurrence, it is one of the most powerful natural disasters as it results in a higher run-up and more challenging to warn [3].

Despite the high risk of tsunami in Kudat (see Fig. 1), limited studies have been conducted to examine the local community level factors that influence disaster readiness as it is critical to design effective and sustainable interventions. Thus, this study seeks to address the gaps by examining socio-economic factors as well as trust in government agencies, risk perception, preparedness intention and the preparation for a tsunami disaster among the community in Kudat, Sabah.

Trust in government agencies have been found to have an influence on the community to participate in community preparedness activities and is impacted by their political ideology [4]. However, another study found that trust does not significantly impact intention and preparedness. On the other hand, the same study findings show a positive influence of risk perception and awareness on intention to prepare and actual preparedness [5]. It is imperative to understand the role of trust in the context of Malaysia as people are more reliant on authorities especially in unfamiliar disasters who is seen to have knowledge on the disaster. As for risk perception, one study found that by increasing perceived severity or risk perception, understanding has the strongest indirect influence on intention to prepare [6]. In a more disaster-prone country like Japan, experiencing and exposure to disasters effects their perception of risk and more inclined to take proactive measures [7]. Since Kudat has never experienced a tsunami, it is imperative that this study examines their level of risk perception as it has been found that those who have never been affected have lower levels of risk perception compared to those who have [8]. Intention to prepare and preparedness is widely used in previous studies [5 – 7] to examine the level of preparedness to tsunami as it provides how inclined they are to prepare for a disaster and whether they have taken the precautionary measures to reduce the impact of the disaster to their livelihoods. Besides that, intention is a better predictor of preparedness as it is less subjected to bias to previous experiences to disasters, strategies for public education and culture [9].

Through examining these factors, this study aims to provide insights that are context-specific for a more effective and community focused disaster risk reduction strategies. This study also aims to contribute by demonstrating how locally driven approaches that when supported by effective multi-level partnerships can enhance preparedness in vulnerable communities.

2 Methods

This study employed a quantitative approach by applying the multi-stage sampling method to select the respondents in tsunami prone areas in Kudat, Sabah. Several areas were selected based on their geographic location comprising of 4,985 households was then stratified into villages based on their vulnerability to tsunamis. A total of four (4) villages were selected namely Kg. Pantai Bahagia, Kg. Pengaraban, Kg. Tanjong Kapor and Kg. Landong Ayang and households from each village was selected systematically with the assistance of the head of villages and were represented by either their head of household or member of the household aged 18 and over. Prior to the data collection, permission from the district office was obtained and was granted to ensure that the local authority is informed of the on-going study. The head of villages were briefed regarding the study to ensure that the household is represented accurately. Ethical approval was obtained through the University's Ethics Committee before the data collection process to ensure that the respondents are treated with respect and their rights are protected as well as to ensure the researchers obtain voluntary and informed consent via a consent form before distributing the survey.

A total of 166 self-administered questionnaires were obtained from four villages during the data collection process which was carried out in February 2025. The total number of questionnaires obtained is statistically significant using Slovin's formula at a 95% confidence level and a 7.5% margin of error which is within the acceptable range for representing the community. Before handing out the survey, a brief explanation of the study and how to answer the form was done to ensure the respondents understand and give consent to participate in the study. The self-administered survey was then passed to representatives of the household, and a token of appreciation was given to those that returned the completed survey and a consent form (see Fig. 2).

Fig 2. Local community filling up survey form

The questionnaire consisted of five sections namely socio-economic background, risk perception, trust in agencies, intention to prepared and preparedness. The first section includes data such as age, education level, household income level, number of household members, and working status. A total of eight (8) statements were included in the risk perception, five (5) statements in the trust in agencies section and four (4) statements in the intention to prepare section using a Likert-type scale ranging from 1 (strongly disagree) to 5 (strongly agree). The preparedness was measured using 18 'Yes' and 'No' statements. Statistical Package for Social Science (SPSS) was used to analyse the data through descriptive analysis to determine the levels of risk perception, trust in agencies, intention to prepare and preparedness for a tsunami disaster. The levels of each variable were determined using SPSS by calculating the total score for each variable and categorizing them based on percentile values. Scores from the minimum value up to the 25th percentile were classified as low level. Scores between the 25th and 50th percentiles were classified as moderate level, while scores above the 50th percentile up to the maximum attainable score were classified as high level. These levels will differ based on the total score calculated for each variable in the study.

3 Results and discussion

3.1 Socio-demographic background

Based on the data collected in the study, a total of 166 questionnaires were obtained from a total of four villages which were Kg. Pantai Bahagia (n=36), Kg. Pengaraban (n=26), Kg. Tanjong Kapor (n=57), and Kg. Landong Ayang (n=47). Based on the descriptive analysis of the data, half of the respondents were Male (51.8%) and had an average age of 41 years old. As for education level, half of the respondents completed high school (57%) and 25% did not attend school. The average household income of the respondents was RM941 and had an average 6 people per household. As for their working status, most of them were reported to be unemployed while only 27% have their own businesses and 17% have subsistence works.

These findings suggest that the respondents' socio-economic background fall far below the average household income for the district of Sabah which is reported at RM3,797 [10]. This can be attributed to a higher percentage of respondents who are unemployed. Besides that, the highest education level reported is completed high school as education as enrollment to primary and secondary schools have improved over the last decade. Teacher development as well as better access in remote areas have provided more educational opportunities for the younger generation in rural areas. This improvement could provide a better understanding of the impact of tsunami in the area among the youth which could influence the older generation to have better preparedness.

Table 1. Socio-demographic background of respondents

Background	**N=166**	**Average**	**Percentage (%)**
Gender			
Male	86		52
Female	80		48
Age		41	
Highest education level			
Did not attend school	42		25
Primary School	26		16
Secondary School	94		57
Tertiary level	3		2

No. of household members		6	
Household income		RM940.00	
Working status			
Government Sector	8		5
Private Sector	3		2
Entrepreneur	44		27
Subsistence work	28		17
Not working	80		48
Retired	3		1

3.2 Risk perception

Based on the findings, the average score for risk perception was found to be 3.4 while the level of risk perception was reported at a moderate level (46%). The item with the highest mean was tsunami is threat to their personal safety (4.50) and a tsunami can affect their daily activities such as work and free time (4.38). The respondents' perception towards the threat of a tsunami disaster is relatively high as they perceive that it can disrupt their daily lives and pose a risk to their personal and community's safety. They also strongly believe that if a tsunami disaster occurs in the future, they will be affected by it. These findings suggest that the respondents are aware of the risk of tsunami in their village which is critical as the respondents are those living in a tsunami prone area. These findings indicate that the respondents recognise the potential severity of a tsunami disaster, and they would be directly affected by it reflecting a high situational awareness to their vulnerability. Awareness in important especially in disaster prone areas as previous studies have found it to be one of the key drivers of preparedness behaviours [5-6].

Table 2. Risk perception level

Level of risk perception	Frequency (N)	Percentage (%)
Low	48	29
Moderate	77	46
High	41	25

As for the trust in agencies, the respondents reported a mean score of 4.3 which shows a high sense of trust in the agencies providing rescue and relief services in the event of a disaster. The respondents highly agree that they are comfortable to depend on the information provided by the relevant government agencies (4.40) and are confident that these agencies are doing a good job in developing policies to protect the people (4.40). Besides that, the respondents also agree that they can depend on the agencies to deliver news on tsunami related issues and incidents (4.37) as well as using the guidelines provided by the agencies if they were to face a tsunami risk (4.34). The high level of trust in agencies is supported by previous studies which found that in an unfamiliar situation, people will rely on government agencies as their main source of information and depend on them to protect them through developing policies. One study emphasised the central role of government agencies in risk communication and public compliance during disasters [11] while another found that trust influences the willingness of an individual to follow instructions and engage in preparedness activities and behaviours [12]. On the other hand, preparedness strategies should consider trust as a bottom-up approach to involve the community in disaster management processes as they are more dependent on agencies due to the unfamiliarity with the disaster.

Table 3. Trust in agencies level

Level of trust	Frequency (N)	Percentage (%)
Low	49	30
Moderate	43	26
High	74	44

The mean score for intention to prepare is 4.2 and the level of intention was found to be at a high level (46%). The respondents have a high intention to prepare a plan for evacuation in the event of a tsunami disaster (4.29) and gather information on tsunami risk (4.28). Respondents also intent to involve themselves with the community to discuss methods to reduce the impact of a tsunami (4.09) which suggests a collective nature of the disaster preparedness. Overall, these findings suggest that the respondents living in potentially high tsunami risk areas have the intent to prepare for a tsunami disaster through designing an evacuation plan and seeking out relevant information from trusted sources. This is inline with findings from a previous study which found that individuals were more likely to engage in disaster preparedness actions when they have the access to relevant resources and felt being empowered to act [13].

Table 4. Intention to prepare level

Level of trust	Frequency (N)	Percentage (%)
Low	44	26
Moderate	46	28
High	76	46

The level of preparedness among the respondents was found to be at a moderate level (55%) and the respondents were mostly prepared by keeping a flashlight with them (95%) and storing important documents in an easily accessible location (92%). The respondents have also prepared necessary items in a bag that can be grabbed during a tsunami disaster (81%) and are aware of the tsunami safe zones (76%) and routes (71%). As for health preparedness, 64 percent of respondents' family members have learnt first aid and 72 percent have received vaccines. The least prepared by respondents are financial preparation such as purchasing life insurance (21%), additional car insurance for protection against natural disasters (21%) and home protection insurance (16%) which exposes a critical gap for long-term recovery planning. This increases the dependency of the respondents on the government for financial support for recovery. However, almost half of the respondents were reported to have savings worth three months of income (45%) suggesting a capacity to manage short-term recovery after a disaster. These findings are in line with previous studies where individuals are more likely to prepare for physical preparedness compared to financial preparedness which is also critical for recovery. Past research have highlighted the importance of a comprehensive preparedness strategy that integrates both practical readiness such as preparing supplies and financial resilience that can influence recovery post disaster [14].

Table 5. Preparedness level

Level of preparedness	Frequency (N)	Percentage (%)
Low	43	26
Moderate	91	55
High	32	19

4 Conclusion

Overall, this paper has examined the preparedness among the communities at risk of a tsunami disaster in Kudat, Sabah which focuses on socio-economic status, risk perception, trust in government agencies, intention to prepare, and actual preparedness levels. The findings from this study revealed that the risk perception was moderate however, the respondents are very concerned about the potential impact and disruption of a tsunami disaster on their daily lives and personal safety which indicates a high situational awareness. Findings show that trust in agencies was reported at a high level as respondents show a strong confidence in the agencies' ability to provide relevant and reliable information as well as implementation of effective policies and procedures that can protect them. Similarly, the intention to prepare for tsunami among respondents was also found to be high, particularly in relation to evacuation planning and seeking information from trusted sources.

Despite positive awareness, trust in agencies and intention to prepare, the actual preparedness reported was not on attitudes and intentions. The actual preparedness was uneven. While most of the respondents had taken practical readiness measures such as preparing grab bags, storing important documents in a safe place, knowledge of evacuation routes and safe zones, a significant gap was found in financial preparedness. A very small number of the respondents were reported to have life, vehicle and home insurance which is critical during the recovery process to continue their livelihoods. Although the respondents reported having three months' worth income in savings, it may not be sufficient to sustain their expenses in the event of a major disaster such as a tsunami. In combination with the low income of households reported, it is expected that a top-bottom approach will be necessary to support those affected by the tsunami.

Given the socio-economic constraints and risk faced by coastal communities in Kudat, Sabah, there are several recommended targeted interventions by the government to minimise the gap. To overcome the financial constraints faced by the community, accessible financial tools should be included as one of the tool such as savings plan backed by the government with incentives or micro insurance that specifically caters low income and high-risk communities. Through these kinds of initiatives, the government should also leverage the high trust in agencies for more effective risk communication and behavioural influence to ensure the local community are fully prepared to face the impact of the disaster and recover quickly. Additionally, culturally and localised programs should be carried out to ensure that awareness programs collaborate with local authorities and leaders as a form of two-way knowledge transfer. Collaborative efforts by various agencies and private sector are essential to ensure a more dynamic and holistic community resilience activities. The role of academia could also be a potential opportunity to strengthen global partnerships especially from countries with vast experiences in tsunami disaster such as Japan and Indonesia. This collaboration is inline with the Sendai Framework for Disaster Risk Reduction 2015 – 2030 which affirms the importance of a bottom-up approach to empower the local community to be more resilient towards disasters. Overall, this study supports the need for a locally driven disaster strategies supported by multi-level partnerships, integrating social trust, economic resilience, and culturally relevant community engagement to enhance tsunami preparedness and foster long-term disaster resilience.

Acknowledgements. The authors of this study want to express our gratitude to Yayasan TM for providing the funds through the 'GLC Demi Rakyat dan Negara' Grant to conduct this project (Project code: K0011-UPNM/2024/KTP/YTM/1). We also appreciate the support from the local authority and community leaders in Kudat, Sabah for their assistance throughout the data collection process. All the support from NADMA is also highly valued for their contribution while carrying out this project. We would also like to acknowledge all individuals and organisations who have contributed to the successful completion of this project either directly or indirectly.

References

1. J. Rosas-Muñoz, M. Baquedano-Rodríguez, What can we learn about the natural disasters and multidimensional poverty link for artisanal fishery in Chile?. Marine Policy. **144**, 105225 (2022)
2. L. H. Koh, W. K. Tan, S. Y. Teh, Chai, M. F., Simulation of Potentially Catastrophic Landslide Tsunami in North West Borneo Trough. International Journal of Environmental Science and Development. **7**(12), 889 (2016)
3. United Nations International Strategy for Disaster Reduction. Tsunami Disaster Risk 2016: Past impacts and projections. Centre for Research on the Epidemiology of Disasters; United Nations International Strategy for Disaster Reduction (2016).
4. P. Repetto, P. Cordón, N. Bronfman, The relationship between political ideology and current earthquake and tsunami preparedness. Global Environmental Change. **72**, 102444 (2022)
5. P. C. Cisternas, L. A. Cifuentes, N. C. Bronfman, P. B. Repetto, The influence of risk awareness and government trust on risk perception and preparedness for natural hazards. Risk Analysis. **44**(2), 333-34 (2024).
6. A. K. S. Ong, Y. T. Prasetyo, P. Kusonwattana, N. Yuduang, S. F. Persada, R. Nadlifatin, M. M. L. Cahigas, T. Chuenyindee, K. Thana, Factors affecting the intention to prepare for tsunami in Thailand. Ocean & Coastal Management. **233**, 106464 (2023)
7. M. Fukurai, L. G. Ludwig, Risk perception to preparedness: 2011 pre-and post-great Tōhoku Japan earthquake. International Journal of Disaster Risk Reduction. **116**, 105074 (2025)
8. A. Cerase, L. Cugliari, Something still remains: factors affecting tsunami risk perception on the coasts hit by the Reggio Calabria-Messina 1908 event (Italy). Sustainability. **15**(3), 2787 (2023)
9. L. Jang, J. J. Wang, D. Paton, N. Y. Tsai, Cross-cultural comparisons between the earthquake preparedness models of Taiwan and New Zealand. Disasters. **40**(2), 327–345 (2016)
10. Department of Statistics Malaysia (DOSM), 2020. https://open.dosm.gov.my/dashboard/kawasanku/Sabah/parlimen/P.167%20Kudat.
11. M. Siegrist, G. Cvetkovich, Perception of hazards: The role of social trust and knowledge. Risk analysis : an official publication of the Society for Risk Analysis. **20**(5), 713–719 (2000)
12. D. Paton, Risk communication and natural hazard mitigation: How trust influences its effectiveness. International Journal of Global Environmental Issues. **8**(1/2), 2 (2008)
13. M. K. Lindell, R. W. Perry, Communicating environmental risk in multiethnic communities: Sage Publications (2003)
14. N. Kapucu, Collaborative emergency management: better community organising, better public preparedness and response. Disasters. **32**(2), 239-262 (2008)

Building Malaysia's First Tsunami-Ready Community in Kudat, Sabah

Mohd Muhaimin Ridwan Wong[1,*], *Haliza* Mohd Zahari[1], Chai *Mui Fatt*[2], *Noor Diyana* Fazan Ahmad[3], *Noorul Syifa* Mohd Isa[4], *and Badrul Hisham* Abdul Samad[1]

[1]Humanitarian Assistance and Disaster Relief Research Centre, Strategic and Security Studies Institute, National Defence University of Malaysia, Kem Perdana Sungai Besi, 57000 Kuala Lumpur, Federal Territory Kuala Lumpur, Malaysia
[2]Sabah Meteorological Office, 7th Floor, Wisma Dang Bandang, Beg Berkunci No. 54, 88995 Kota Kinabalu, Sabah, Malaysia
[3]Centre of Research and Innovation Management, National Defence University of Malaysia, Kem Perdana Sungai Besi, 57000 Kuala Lumpur, Federal Territory Kuala Lumpur, Malaysia
[4]Community and Nation Building, Yayasan Telekom Malaysia, Level 6, TM Annexe 2, Jalan Pantai Jaya, 59200 Kuala Lumpur, Federal Territory Kuala Lumpur, Malaysia

Abstract. Tsunami Ready Recognition Programme (TRRP) was introduced as a performance-based approach to strengthen tsunami preparedness among coastal communities. While communities across 31 countries have been recognised as "Tsunami Ready" (TR) globally, coastal communities in Malaysia have yet to be certified even though past experience and studies provided irrefutable evidence that coastal cities and towns in the country are exposed to tsunami risks. This paper reports on a two-phase project implemented in Kudat district in northern Sabah, Borneo Malaysia which operationalised the UNESCO/IOC's guidelines on TRRP as a roadmap to develop the first candidate for a TR community in Malaysia. A table-top exercise conducted with local stakeholders provided input necessary to improve the local tsunami emergency response plan. Post-program surveys also showed a significant increase in the understanding of tsunami risks and response plans, better awareness on safe tsunami evacuation, higher confidence in responding to tsunami warnings, and greater willingness to participate in future risk reduction programs. Additionally, this paper maps Kudat communities' progress in each TR indicator, identifies persistent gaps for effective tsunami risk reduction locally, as well as recommendations to cultivate a culture of tsunami readiness at all levels and move Malaysian coastal communities towards the UNESCO/IOC certification.

1 Introduction

As disasters become more frequent globally in recent decades, effective disaster risk management is essential at national, local, and individual levels. In 2015, the Sendai

* Corresponding Author: m.muhaimin@upnm.edu.my

Framework for Disaster Risk Reduction (SFDRR) 2015 – 2030 was adopted at the 3rd United Nations Conference on Disaster Risk Reduction with the aim to improve the effectiveness of disaster risk management (DRM) across United Nations member states. Building upon the emphasis on disaster preparedness in the preceding Hyogo Framework for Action 2005 – 2015, this new framework calls for the prevention of new, and reduction of existing risks by addressing vulnerability and exposure as two key factors to effectively reduce disaster risks [1]. SFDRR recommended the pursuit of these DRM goals through implementation of comprehensive measures encompassing structural, environmental, economic, education, political, and social aspects, all while putting communities at the heart of DRM initiatives.

In a community-centric DRM approach, improving knowledge and capacities of communities to ensure that they will be able to anticipate, respond and recover from disaster events effectively are necessary to build preparedness and reduce disaster risks [2]. One of the critical actions to build preparedness is contingency planning, whereby disaster risks are analysed, and action plans are made in advance to enable efficient and coordinated responses by all stakeholders during disaster events. This includes identifying institutional roles and responsibilities, resources, communication plans, and operational arrangements for all relevant stakeholders in times of need. Contingency planning also requires stakeholders to undertake training programs and simulation exercises to anticipate, discuss and solve potential issues that may be faced during disasters such as alternative source of supplies, incapacitation of emergency response agencies, or the compounding impacts caused by natech disaster events. As a crucial component of disaster risk management, failure to develop sufficient preparedness would often lead to increased casualties, exacerbated losses, and overall cause setbacks to the affected population's development.

The impact of inadequate preparedness was clearly documented in one of most extreme disaster of the 21st century, the 2004 Indian Ocean tsunami. The event on 26th of December 2004 caused more than 200,000 casualties and displaced about 1.7 million others across 14 countries surrounding the Indian Ocean [3]. In Aceh, the westernmost province of Indonesia and one of the closest coastal areas to the epicentre of the triggering M9.1 earthquake, nearly 130,000 deaths were recorded [4]. In comparison to this extremely high number of casualties, on nearby Simeulue island located only 40 km from the epicentre, only seven deaths were directly caused by the tsunami among its 80,000 inhabitants despite widespread destruction of physical infrastructures on the small island [5]. Studies into the phenomenon highlighted how disaster preparedness integrated in the local's culture allowed them to recognise signs of impending tsunami and contributed to their swift evacuation to higher ground [6]. Increased risk due to the lack of preparedness against tsunami had its impact in Malaysia as well, despite the country farther from the source. The tsunami travelled more than 700 km to reach Langkawi Island to the northwest of Peninsular Malaysia where it claimed the life of a disabled, elderly woman, before travelling further south and caused more casualties. In total, around 70 Malaysian lives were lost in the country, and 30 others were lost overseas including in Sri Lanka, Indonesia, and Thailand [7]. Like Aceh, the casualties in Malaysia were linked to the absence of an early warning system and low preparedness among the population [7]. The 2004 event was the first tsunami disaster experience for the Malaysian government and communities, leading to heightened tsunami awareness and marked improvement in tsunami disaster risk management strategies.

2 Background of the Project

One of the positive impacts of the disaster was the widespread increase in tsunami awareness, its risks, and the importance of building preparedness within communities exposed to the hazard. Inspired by the TsunamiReady program that was established in the United States as early as 2001, the international community was introduced to the community-based readiness program with a series of pilot projects between 2011 and 2014. Recognitions in Anguilla and the British Virgin Islands proved its value and the Intergovernmental Oceanographic Commission (IOC) endorsed a drafted guideline in 2015, which allowed for further refinement of the program between 2017 and 2022 based on its implementation in sixteen countries across the Caribbean, Pacific, Indian Ocean and Mediterranean. Their collective experience led to the publication of the Standard Guidelines for the Tsunami Ready Recognition Programme (TRRP) in April 2022 by UNESCO and the Intergovernmental Oceanographic Commission with refined readiness indicators categorised into three groups: assessment (ASSESS), preparedness (PREP), and response (RESP). As of July 2025, the program has recognised 100 communities across 39 countries as Tsunami Ready [8]. Indonesia remains the only country with such recognition among Southeast Asian nations, with 21 communities awarded the TRRP recognition.

While Malaysia has yet to obtain recognition under this program, efforts to improve tsunami preparedness among its coastal communities are well underway. Between 2006 and 2018, over 90 tsunami awareness and exercise programs were conducted with most of them involving communities affected by the 2004 tsunami in Peninsular Malaysia, and several others for coastal communities in East Malaysia [9]. Installation of early warning systems, disaster alert systems, setting up of disaster relief funds, and development of national standard operating procedures were among the measures implemented by the Malaysian government throughout the years [10]. Despite these measures, the initiatives remained largely top-down, and progress toward the community-driven, bottom-up DRM envisioned in the Sendai Framework has been limited. Preparedness action through contingency planning has also gained interest in Malaysia. Two community-based tsunami emergency response plans (TERPs) were developed in 2015 for communities in Balik Pulau and Batu Ferringhi on Penang Island [11, 12], while the third TERP was developed for communities in Kudat, Sabah in 2016 [13]. To the researchers' knowledge, no new TERP has been developed since then for any at-risk communities in the country.

After two decades have passed since the 2004 tsunami disaster, initiatives to enhance tsunami preparedness have been noticeably dwindling. Awareness campaigns decreased significantly with programs being conducted only once a year between 2018 and 2023, and the COVID-19 pandemic restrictions meant that none were held between 2020 to 2022. To regain the lost momentum of years prior, a collaborative project was proposed to Yayasan Telekom Malaysia, a local foundation whose focus areas include community and nation building, to revitalise tsunami readiness among coastal communities in Malaysia. For this purpose, Kudat district in East Malaysia was selected as the project area due to two main reasons. Firstly, its 2016 TERP was developed with limited involvement from local community leaders and was subsequently not communicated widely to community members; and (2) located between the South China Sea and Sulu Sea, Kudat is one of the few areas in East Malaysia with very high tsunami risk. Past studies have shown that Kudat may be affected by tsunamis originating from subduction zones around the region [14, 15]. In particular, tsunami originating from the Sabah Trough is of serious concern due to its triggering mechanism being a submarine mass failure, which can cause destructive tsunami that can reach Kudat within a short span of time [15]. Additionally, given that no community in Malaysia has pursued the Tsunami Ready recognition, the initiative also presented an opportunity to develop the first Tsunami Ready community in the country. As such, this paper presents the progress, achievements and findings from the two-phase project to build tsunami preparedness for communities in Kudat, Sabah, Malaysia.

3 Methods

The project was proposed with four specific objectives:

1. To update the Tsunami Emergency Response Plan (TERP) for the communities of Kudat, Sabah.
2. To raise community awareness in Kudat about tsunami hazard and associated risks.
3. To assess the current level of disaster preparedness among Kudat residents.
4. To strengthen local capacity for tsunami response through a community-based disaster risk management (CBDRM) approach.

To achieve the above objectives, the project is divided into two linked phases to align technical revision with public adoption of the local emergency response plan. The first phase convened representatives of the district disaster management committee to revise the older tsunami emergency response plan for the district. Following this revision and update, the second phase which consisted of a four-day community engagement program was conducted to disseminate and operationalise the plan. Sequencing these activities into two phases avoided circulation of provisional guidance and allowed for local stakeholders' feedback to be incorporated into the response plan. More details on the activities during these two phases are provided below.

3.1 Phase 1: Review of the 2016 Tsunami Emergency Response Plan

Considering nearly ten years have passed since the first TERP for Kudat was developed, the first course of action was to revise the plan based on input from Kudat's locals. For this purpose, a two-day workshop was conducted bringing together 57 representatives from the local disaster management committee, NGOs, educational institutions, and hotel operators. Activities were conducted with the steps described below.

1. Hazard briefing by the Sabah state Meteorological Office covering tsunami generation mechanism, warning systems, and local warning arrangements. The National Disaster Management Agency also delivered a briefing on the national directive that needs to be adhered to by government agencies for DRM.
2. Open discussion to review the 2016 TERP to identify obsolete or missing information, as well as suggestions for improvement from the locals' perspective.
3. Participatory mapping in which local stakeholders analysed and redrew evacuation maps for different zones in the district.
4. Ground truthing field visit to verify the practicality and safety of proposed safe evacuation routes and zones/shelters.
5. Table-top exercise (TTX) using the worst-case scenario to assess understanding of roles of each stakeholder for different tsunami warning levels.

Feedback received from participants during the workshop, including from the TTX, were analysed by comparing each suggestion to the 2016 TERP. Changes were accepted only if they complied with the national NADMA Directive No. 1, which governs the roles of specific agencies in managing disaster risks, verified in the field during the ground truthing visit, and unanimously agreed upon by the project team.

3.2 Phase 2: Community roadshows

After the TERP revision, the team conducted four one-day roadshows to engage local communities on tsunami risk and communicated the localised emergency response plan. Four venues were selected based on space availability and practicality: the Institute of Islamic Studies and Dakwah Sabah (IPDAS) Kudat branch in Sikuati, the Ministry of Youth and Sports Arena in Kudat town, St Peter Primary School, and Lok Yuk Batu 1 (Chinese) Primary

School. A total of 713 participants were the direct beneficiaries from the roadshows, with the program itself following a standard sequence:

1. concise talk on tsunami hazards and Malaysia's warning system;
2. walkthrough of the updated Kudat TERP;
3. earthquake "drop-cover-hold" drill;
4. group planning of neighbourhood evacuation actions using colour-coded maps;
5. practical sessions for basic lifesaving and first-aid skills; and,
6. hand-over of printed TERP copies, tsunami evacuation posters, and emergency/ trauma kits to local representatives.

Parallel to these roadshows, fourteen permanent tsunami information and evacuation signs were installed at key roadside locations approved by local public works department and the district council board.

3.3 Data collection and analysis

Two instruments were used to assess participants' learning outcomes:

1. 1Pre- and post-workshop surveys during the Phase 1 workshop were used to gauge participants' understanding of tsunami hazard, local disaster response mechanism, and confidence to take actions during emergencies. A five-point Likert scale was used in these surveys.
2. Brief exit survey at each roadshow to understanding, subjective response readiness, knowledge of safe evacuation areas, and overall satisfaction with the engagement program. The exit survey was limited to adults (≥18 years) as the roadshows were conducted publicly with transient attendance, making it impractical to obtain parental consent on site. The instrument was also written for adult literacy levels with 5-point Likert scale, further supporting the decision to restrict sampling for adults alone. Therefore, minors were excluded to meet safeguarding and ethics requirement.

For both of these surveys, they were practice-oriented, and analyses were descriptive by design with no confidence internals planned a priori. Relative changes in scores are reported to indicate magnitude of improvement. A more comprehensive community survey and analyses were also conducted following the roadshow to collect data pertaining to the third objective of the project, though this is covered in a separate research article.

4 Findings and Discussion

4.1 TERP revision

For the 2016 TERP review, aside from several rewording of the sub-headings, the components of the TERP remained the same as listed in Table 1. However, substantial changes were made content-wise in terms of information presentation, data updates, map updates, and more elaborate descriptions. Given that Kudat district consists of a more concentrated town area, with smaller villages scattered along its coastline, a section to delineate the geographic scope of the current TERP was included to avoid confusion among locals. It also serves as a point for future improvement that can be undertaken by the local communities themselves. Additionally, an emphasis statement was included for local stakeholders that the TERP is a living document requiring revision, updates, and improvement from time to time to reflect current state of knowledge, lessons learned, and advances in disaster management practices. In the new edition, the District Officer, who is the Chairperson of the District Disaster Management Committee, was named as the person responsible to ensure that this process is implemented with the support of members of the committee and community.

4.2 Phase 1 workshop outcomes

4.2.1 Knowledge gains

Comparison between pre- and post-workshop responses showed that the average understanding of the TERP rose from 2.9 (out of a maximum of 5.0) before the workshop to 4.4, indicating a shift from "fair" to "good" comprehension. Similar improvement was seen in their familiarity with the concept of "safe evacuation" and identifying its routes whereby the share of participants (out of 57 persons) who could identify correct procedures rose from 51% to 97% after the workshop. This large gain suggests that the mapping exercise, field verification and scenario discussion during TTX were effective in translating plan details into practical knowledge.

Table 1. Comparison of headings between the 2016 and 2025 TERPs.

No.	2016 TERP	2025 TERP
1.	Introduction to the TERP	Introduction
2.	Objectives of the TERP	Objectives of the TERP
3.	Overview of the Kudat community	Kudat District
4.	Geographic map of Kudat	Geographic scope (of the TERP)
5.	Major tsunami threats to Kudat District	Tsunami threats to Kudat
6.	Tsunami alert and warning system	Tsunami alert and warning system
7.	Disaster emergency declaration	Natural tsunami warning signs
8.	Emergency response groups	Local warning system
9.	Activation of the Kudat TERP	Community warning process
10.	Evacuation zones (description)	Emergency response groups
11.	Emergency centres (description)	Safe evacuation (process)
12.	Dissemination of the TERP to the community	Tsunami signages
13.	Tsunami signages	Dissemination of the TERP to the community
14.	Tsunami emergency response checklist (based on section 8)	Kudat District Disaster Management Committee
15.	Tsunami safe evacuation maps	Tsunami disaster management flow chart
16.	Tsunami disaster management flow chart	Tsunami safe evacuation maps

4.2.2 Readiness

About 97% of participants reported feeling "confident" or "very confident" in applying what they have learned from the workshop during an actual tsunami warning situation. This confidence in responding to emergencies is critical to be built during preparedness activities to supports the project's aim of fostering a proactive safety culture. Observation during the TTX also confirms this evaluation as all response agencies are aware of their roles, readily collaborates with their fellow emergency responders to ensure local communities' safe and timely evacuation, as well as capable of finding alternative solutions to challenges arising from a worst-case tsunami scenario.

Feedback on the workshop itself was overwhelmingly positive with 97% expressing satisfaction with the program, 95% rated the content and activities as suitable to meet the program's objectives, and 92% felt that their expectations were fully met. Comments collected during the plenary debrief mentioned the hands-on evacuation-route mapping ad verification exercise, as well as the TTX as particularly helpful for clarifying roles among stakeholders and boosting their readiness to respond during tsunami emergencies.

4.3 Phase 2 community roadshow outcomes

4.3.1 Community reach

The four-day program, held at four different venues, was able to reach a total of 713 community members covering 11 coastal villages, three primary schools, one secondary school, one vocational college, and 1 institute of higher education (IHE). These communities are tabulated in Table 2 below for reference. Rates of attendance ranged from 46 participants on the first day, to 383 participants (highest attendance) on the third day. The breadth of participants coverage achieved during this roadshow ensured that adult decision-makers, youths, and younger children received consistent preparedness message that can be cultivated in households across the district.

Table 2. List of communities reached during the four one-day roadshows.

Type	Communities reached	
Coastal villages	1. *Kampung Korina* 2. *Kampung Pantai Bahagia* 3. *Kampung Muhibbah* 4. *Kampung Pengaraban* 5. *Kampung Tanjung Kapor Lama* 6. *Kampung Tanjung Kapor Tengah*	7. *Kampung Landung Ayang Laut* 8. *Kampung Landung Ayang Bukit* 9. *Kampung Landung Ayang Darat* 10. *Kampung Gumpa* 11. *Kampung Ayer*
Primary schools	1. *SK St Peter* 2. *SJK(C) Lok Yuk Batu 1* 3. *SK Lok Yuk*	
Secondary school	*SMK Sikuati 1*	
Vocational college	*Kolej Vokasional* Kudat	
IHE	Institute of Islamic Studies and *Dakwah* Sabah, Kudat branch	

4.3.2 Gains in knowledge and self-efficacy

A brief exit questionnaire completed by a total of 114 adult participants at the end of each roadshow showed that 93% of respondents rated their understanding of local tsunami risk as "good" or "very good" (score 4–5 on a five-point scale). Their confidence in executing protective actions also rose with 87% feeling ready to evacuate themselves and their families upon warning, and the same percentage knowing the nearest designated safe evacuation areas. In contrast, a smaller share of 73% expressed confidence to deliver first aid. This presents a notable base for future skills training, suggesting the need for longer training period and addition of practical modules integrating disaster response scenarios. Workshop logistics and content were well received with virtually all attendees declaring themselves satisfied with the programme, and 97% showed interest to participate in similar events in the future. Open-ended feedback highlighted their needs for more hands-on drills and community exercises, a longer roadshow calendar to reach coastal villages further away from the main town, and a stronger media engagement to amplify key messages on tsunami risk management.

Suggestions received either verbally or through the survey point towards clear priorities for future projects, which are to extend the preparedness program into practical exercises, and to institutionalise it as an annual, district-wide preparedness building activity.

4.4 Progress towards Tsunami Ready recognition

The indicators for the Tsunami Ready Recognition Program (TRRP) listed by UNESCO/IOC were compared to activities undertaken in Phase 1 and 2 of the project, as well as to findings from the participant surveys. Based on this comparison, Kudat communities have now met 10 out of 12 global indicators required for TRRP. The two unfulfilled indicators are ASSESS-2 (the number of people at risk in the tsunami hazard zone is estimated), and ASSESS-3 (economic, infrastructural, political, and social resources are identified). Due to these, a roadmap to full compliance is proposed to the National Disaster Management Agency (NADMA) as follows:

1. To establish a governing structure for TRRP. A National Tsunami Ready Board (NTRB) needs to be established with NADMA as the lead to serve as the official accrediting body in Malaysia. This milestone is proposed to be achieved by the 3rd quarter of 2025, followed by the formation of an official, local Tsunami Ready Liaison Committee (TRLC) in Kudat by 4th quarter 2025. The formation of these two structures would expedite the collection of data required to fulfil the remaining indicators in Kudat. Additionally, these structures would also encourage the development of Tsunami Ready communities at other localities in the country.
2. To develop a tsunami exposure and capacity inventory for Kudat. This action is proposed specifically to fulfil ASSESS-2 indicator whereby district government (as the TRLC) needs to be mobilised to compile all necessary data that is already available at individual agencies and departments. The TRLC is also advised to partner with local universities to expedite this data compilation process.
3. To integrate state and district DRM plans into the new TERP. This action is recommended to achieve ASSESS-3 more efficiently. The TRLC should align development plans pertaining to Kudat district, as well as available resources, budget, and assets, as further improvement to the current TERP. This will not only expedite the fulfilment process but also allow for all stakeholders to be aware of available resources that can contribute towards their tsunami disaster resilience.
4. To maintain PREP and RESP related commitments from 2026 onwards. The TRLC, through support from NTRB, is recommended to conduct at least three tsunami awareness campaigns annually and one emergency response exercise biannually. Each activity should also be documented rigorously to build the evidence base for TRRP application and future recognitions.

In spite of the recommendations above, some challenges that are common to DRM or CBDRM can be anticipated based on experiences implementing the project in Kudat. Firstly, bureaucratic lead times to form the NTRB and TRLC, and securing support from relevant entities is expected to delay its fulfilment by several months at the very least. Without this formal structure and mandate, progress for the other actions cannot be realistically achieved. Sustained funding is also an anticipated challenge, as local stakeholders are still expecting continuous financial support from the federal or state governments for any DRM activities. This dependency will undoubtedly reduce the sustainability of current progresses. Turnover in local leadership also presents a huge threat, as erosion of institutional memory was identified as the main reason many current stakeholders were unfamiliar with the 2016 TERP. Subsequent to delays caused by these challenges, community fatigue and disinterest may also be expected. Without routine engagement, enthusiasm within local communities towards building tsunami preparedness will wane, a situation further exacerbated by complacency and a false sense of safety that are reinforced by infrequency of tsunami events.

5 Conclusion

This paper documents the activities, impacts and actionable plan towards building the first Tsunami Ready community in Malaysia. The findings provided evidence that a well-structure program, while short in duration can broaden community knowledge regarding tsunami risks, while also delivering measurable progress towards tsunami readiness for coastal communities in Kudat, a relatively small district in East Malaysia. Phase 1 of the project delivered three main outcomes:

1. Improved understanding on tsunami risks: stakeholders reported clearer understanding of local tsunami hazards, warning levels, roles of each stakeholder, and safe evacuation zones.
2. Enhanced confidence to act on tsunami warnings: stakeholders expressed greater confidence to act upon noticing tsunami signs or receiving tsunami warnings as specific actions are clearly outlined in the TERP, and evacuation routes have been verified.
3. Operational product: stakeholder-endorsed and field-verified TERP has successfully been produced for reference and implementation by locals in Kudat.

The subsequent community roadshows engaged diverse range of communities, reinforced correct tsunami knowledge, cultivated interest on tsunami among the younger generation, and left tangible assets in the form of updated TERP, trauma kits, and fourteen permanent tsunami-evacuation signages. These outputs satisfied ten of the twelve global Tsunami Ready indicators, bringing Kudat closer than any other Malaysian communities to formal TRRP recognition. An actionable plan was also proposed to the national and local authorities to guide the fulfilment of the two remaining indicators. Nonetheless, fulfilling these indicators and sustaining tsunami preparedness requires overcoming bureaucratic constraints, maintaining political will, and securing financial support for tsunami risk management activities. Despite these challenges, it is believed that Kudat is well-positioned to become not only Malaysia's first Tsunami Ready community but also a replicable model for other vulnerable districts and communities in the country.

Acknowledgements. The authors expressed their gratitude to the National Disaster Management Agency, Kudat District Office, Kudat Native Court, and the Natural Disaster Research Centre, Universiti Malaysia Sabah for all their support. The project would not be as effective without their cooperation and assistance. This project was funded by *Yayasan Telekom* Malaysia through the GLC *Demi Rakyat dan Negara* (GDRN) grant 2024. The views and opinions expressed in this article are those of the authors and do not necessarily reflect the official policy or position of any affiliated agency of the authors.

References

1. United Nations Office for Disaster Risk Reduction (UNDRR), Sendai framework for disaster risk reduction (2015), accessed online on 28 June 2025 at: https://www.undrr.org/publication/sendai-framework-disaster-risk-reduction-2015-2030
2. United Nations Office for Disaster Risk Reduction (UNDRR), Report of the open-ended intergovernmental expert working group on indicators and terminology relating to disaster risk reduction (2016), accessed online on 5 July 2025 at: https://www.undrr.org/publication/report-open-ended-intergovernmental-expert-working-group-indicators-and-terminology
3. D. Arcas, V. Titov, Sumatra tsunami: lessons from modelling. Surv. Geophys. **27**, 679–705 (2006).
4. United Nations Information Management Service (UNIMS) in collaboration with the Badan Rehabilitasi dan Rekonstruksi (BRR), Tsunami recovery report (UNIMS/BRR, Banda Aceh, 2005).
5. Badan Rehabilitasi dan Rekonstruksi (BRR) and the World Bank, Rebuilding a better Aceh and Nias: Stocktaking of the reconstruction effort (BRR/World Bank, Banda Aceh, 2005).
6. B.G. McAdoo, L. Dengler, G. Prasetya, V. Titov, Smong: how an oral history saved thousands on Indonesia's Simeulue Island during the December 2004 and March 2005 tsunamis. Earthq. Spectra **22**, S661–S669 (2006).
7. F.-R. Ahmadun, M.M.R. Wong, A.M. Said, Consequences of the 2004 Indian Ocean tsunami in Malaysia. Safety Sci. **121**, 619–631 (2020).
8. UNESCO/IOC, Tsunami Ready viewer (2025), accessed online on 7 July 2025 at: https://tsunamireadyviewer.ioc-tsunami.org/
9. Malaysian Administrative Modernisation and Management Planning Unit (MAMPU), Tsunami exercise and public awareness campaign program (2021), accessed online on 8 July 2025 at: https://archive.data.gov.my/data/ms_MY/dataset/program-latih-amal-tsunami-dan-kempen-kesedaran-awam
10. A.M. Said, F.-R. Ahmadun, A.R. Mahmud, F. Abas, Community preparedness for tsunami disaster: a case study. Disaster Prev. Manag. **20(3)**, 266–280 (2011).
11. A.M. Said, F.-R. Ahmadun, E.A. Bakar, N. Sulaiman, Tsunami emergency response plan Balik Pulau, Penang (UPM, Selangor, 2015).
12. A.M. Said, F.-R. Ahmadun, A.R. Mahmud, T.A. Mohamed, Tsunami emergency response plan Batu Ferringhi, Penang (UPM, Selangor, 2015).
13. Malaysia Meteorological Department (METMalaysia), Tsunami emergency response plan Kudat (Ministry of Science, Technology and Innovation, Putrajaya, 2016).
14. M.H. Dao, P. Tkalich, E.S. Chan, K. Megawati, Tsunami propagation scenarios in the South China Sea. J. Asian Earth Sci. **36**, 67–73 (2009).
15. W.K. Tan, S.Y. Teh, H.L. Koh, Tsunami run-up and inundation along the coast of Sabah and Sarawak, Malaysia due to a potential Brunei submarine mass failure. Environ. Sci. Pollut. Res. Int. **24**, 15976–15994 (2017).

Why Aren't There More Cooperative Housing Construction Projects? An Analysis of Responsibility Framing in Post-Disaster Housing Recovery in Türkiye

Fatma Özdoğan[1,*], *Binali* Tercan[2], *Neşe* Bozkurt[3], and *Hülya* Koç[4]

[1]Université de Montréal, School of Architecture, H3S 2C2 Montréal (Québec), Canada
[2]Bolu Abant İzzet Baysal University, Urban Planning Department, 14030 Bolu, Türkiye
[3]Isparta University of Applied Sciences, Faculty of Technology, 32200 Isparta, Türkiye
[4]Dokuz Eylül University (formerly), Urban Planning Department, 35390 İzmir, Türkiye

Abstract. As Türkiye continues to recover from the devastating 2023 Southeastern Anatolia Earthquakes, housing shortages and limitations of conventional post-disaster housing models have become increasingly evident. Lessons from previous disasters point out the value of alternative solutions like cooperative housing. Yet, despite acknowledged benefits, their implementation remains minimal. Here, we examine why cooperative housing remains underrepresented in Türkiye's disaster recovery by analysing discursive framings of responsibility articulated by various actors and their interactions with existing legal and institutional contexts. Employing critical discourse analysis, we conducted 25 semi-structured interviews with representatives of cooperative housing groups, professional chambers, and local authorities, along with a corpus of 135 policy documents, professional reports, academic articles, and press materials covering the period from the 1999 Marmara Earthquake to the second anniversary of the 2023 Southeastern Anatolia Earthquakes in 2025. Findings reveal that dominant narratives emphasising rapid, centralised, and standardised state-led reconstruction significantly restrict the emergence and institutional acceptance of cooperative housing alternatives. We highlight persistent institutional barriers and tensions between official disaster governance discourses and local experiences. Theoretically, this analysis demonstrates how discursive framings shape governance outcomes, illustrating the ways hegemonic narratives limit the acknowledgement of alternative practices. Practically, we advocate targeted regulatory reforms and strengthened institutional support mechanisms to transform abstract calls for participation into actionable cooperative housing policies and practices.

*Corresponding author: fatma.ozdogan@umontreal.ca

1 Introduction: Housing in the Times of Crisis

Hazard-related disasters have significantly increased in recent decades, drawing global attention to critical challenges in housing provision and reconstruction. Climate change further intensifies these challenges by exacerbating extreme weather events such as floods, heat waves, storms, and wildfires. These disasters disproportionately impact communities already struggling with inadequate housing. At the same time, a global housing crisis persists, characterised by widespread shortages and poor living conditions.

Current disaster recovery approaches commonly repeat problematic patterns that reduce the effectiveness of housing reconstruction. These include relocating communities to peripheral urban areas, neglecting necessary infrastructure and community services, and failing to address socio-economic and cultural needs. Centralised decision-making often ignores informal economies and imposes new construction technologies unfamiliar to local populations, creating difficulties in both community involvement and long-term maintenance. Addressing these persistent issues requires decentralised governance, community participation, and housing strategies that respond directly to local conditions and promote social justice.

The devastating 2023 Southeastern Anatolia Earthquakes vividly illustrate these complexities. The earthquakes severely damaged or destroyed approximately 280,000 buildings across 11 provinces, leaving around 1.5 million people homeless in Türkiye alone [1]. The Turkish government responded swiftly by initiating a massive, centralised reconstruction program, aiming to build approximately 650,000 new housing units [2].

This centralised reconstruction strategy aligns closely with broader housing policies under President Recep Tayyip Erdoğan's government. Such policies, emphasising rapid, large-scale construction, have historically prioritised quantity over quality and hindered efforts to address underlying social and environmental injustices. To accelerate reconstruction following the 2023 earthquakes, the Turkish government reinforced this centralised approach by granting increased land expropriation powers to the Ministry of Environment, Urbanisation, and Climate Change (MEUPCC). Extensive land expropriations continue in various areas of Hatay province and other affected regions.

However, this approach has encountered significant resistance, particularly regarding expropriations of agricultural lands and centuries-old olive groves. In Hatay, for example, along with many regions, residents have actively resisted, confronting excavation teams uprooting agricultural lands, and olive and citrus trees, leading to intense disputes. Legal challenges and sustained local protests eventually prompted administrative courts to temporarily halt some expropriation orders in late 2023. Civil society groups, including the professional chambers, have also intervened, criticising the choice to expropriate fertile agricultural lands rather than utilising available public land.

Given these tensions, alternative housing models emphasising democratic and participatory processes, such as cooperative housing, present viable solutions. Research highlights cooperatives' capacity to increase resident satisfaction, reinforce community ties, and foster solutions adapted to local social, cultural, and economic realities, all of which represent significant advantages in disaster-affected areas. Moreover, such models promote greater transparency, accountability, and active resident participation in decision-making, thus counteracting many pitfalls associated with centralised recovery. Despite these documented advantages, such as preserving pre-disaster social ties, facilitating community participation, and promoting local economic continuity, cooperative approaches have largely been overlooked within Türkiye's previous and ongoing reconstruction initiatives.

Despite considerable criticism of centralised housing production, land expropriations, and standardized building designs imposed by TOKİ (Türkiye's Mass Housing Administration under the MEUPCC), many earthquake survivors continue to view housing reconstruction primarily as the state's responsibility. This paradox clearly emerged during field research conducted by the main author in the summer of 2024 in Antakya's central Akevler neighbourhood. When discussing alternative solutions such as cooperative housing, residents frequently expressed scepticism or reluctance, highlighting their expectation that the government should rebuild their homes. One resident captured this sentiment clearly: *"Why should I build my own house? I am a victim of the earthquake; providing housing is the state's responsibility."* Similarly, during our interview in May 2024, representatives from the Chamber of Urban Planners in Hatay acknowledged this tension. Despite their criticism of centralised policies and rapid land expropriations, they argued that only centralised reconstruction could realistically manage recovery from such extensive destruction.

These varied perspectives on housing provision and responsibility form the core motivation of this article. They prompt us to critically examine how responsibility for post-disaster housing is defined, contested, and negotiated by different actors. We explore the broader political, institutional, and social factors shaping these competing definitions of responsibility. Our analysis reveals how centralised governance structures and dominant disaster narratives limit the development of alternative approaches, such as cooperative housing. Ultimately, this article seeks to deepen the understanding of post-disaster housing responsibility and emphasise the possibilities of more inclusive, participatory, and community-responsive approaches.

2 Theoretical Framework: Discursive Agency and the Three Spheres of Responsibility

We employ the concept of discursive agency, drawing on the work of Leipold and Winkel (2017), to examine how actors construct, negotiate, and challenge responsibility in post-disaster housing recovery [3]. Inspired by their conceptualisation, we describe discursive agency as the strategic capacity of actors to shape meanings within established institutional structures and dominant narratives. By engaging in deliberate framing and positioning, actors influence how responsibility is articulated, who is held accountable, and which responses are recognised as legitimate. In post-disaster contexts, responsibility becomes highly contested due to competing priorities and urgent demands for effective action. Various stakeholders, including state institutions, professional chambers, civil society groups, and affected communities, actively mobilise discursive strategies to assert authority, advocate for inclusion, and redistribute accountability. However, their ability to influence these discourses depends significantly upon their institutional authority, access to symbolic resources, and political legitimacy. Consequently, discursive agency highlights how power relations shape both the definition of responsibility and perceptions regarding the legitimacy of recovery efforts.

To enhance our analysis, we integrate Karen O'Brien's (2018) framework of the three spheres of transformation: political, practical, and personal (see Fig. 1). Each sphere corresponds to a distinct dimension of responsibility, encompassing institutional obligations, implementation practices, and normative values. These spheres function interdependently, collectively structuring how responsibility emerges in post-disaster recovery processes.

The political sphere involves institutional and legal responsibility, encompassing governance frameworks, policy mechanisms, and administrative structures that assign mandates and formal obligations. Actors actively engage in this sphere by affirming or contesting institutional legitimacy, invoking legal norms, or advocating policy change. Additionally, political institutions mediate societal demands and transform these into enforceable responsibilities, thereby determining what forms of accountability are institutionally recognised and practically achievable. Thus, institutional arrangements significantly influence the parameters within which practical recovery activities occur.

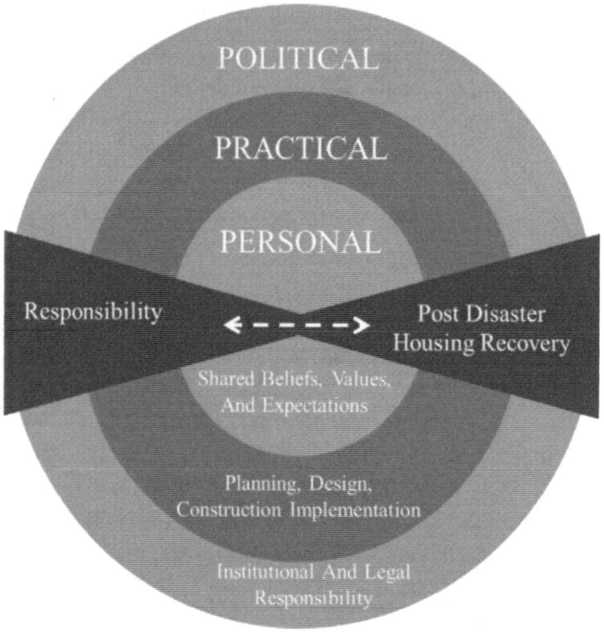

Fig. 1. Three Spheres of Responsibility for Post-Disaster Housing Recovery adapted from O'Brien (2018) [4]

Linked closely to institutional frameworks, the practical sphere addresses responsibility through the implementation of recovery activities, such as planning, design, construction, and distribution. Within this sphere, responsibility is evaluated according to operational performance criteria, including efficiency, compliance with standards, and measurable outcomes. Actors emphasise their technical expertise and track record to legitimise their authority. Nevertheless, practical implementation always reflects underlying political decisions and broader societal expectations. Therefore, successful implementation requires not only technical proficiency but also adequate institutional support and alignment with prevailing social values.

The personal sphere encompasses normative considerations, including shared beliefs, values, and expectations. This sphere informs who society believes should act, for whom actions should be taken, and upon which grounds these actions are justified. Actors invoke housing recovery narratives such as justice, solidarity, or entitlement to reinforce or challenge prevailing assumptions. These normative discourses strongly influence institutional priorities and shape perceptions regarding the fairness and legitimacy of practical outcomes.

These three spheres operate through mutual reinforcement and continuous interaction. Institutional decisions establish conditions that facilitate or constrain practical activities. Practical outcomes, in turn, reinforce or undermine institutional credibility. Normative expectations shape both institutional reforms and implementation priorities. Thus, responsibility emerges dynamically through interactions among institutional frameworks, implementation practices, and societal values.

Integrating discursive agency with these interconnected spheres offers a comprehensive analytical framework for examining responsibility in post-disaster housing recovery. This approach highlights how actors actively shape narratives, how power circulates across political, practical, and personal domains, and the conditions under which transformative shifts in responsibility occur.

3 Methodology: Capturing Responsibility in Post-Disaster Housing Narratives

We employ Critical Discourse Analysis (CDA) to examine how responsibility for post-disaster housing reconstruction is framed by various actors in Türkiye, stressing cooperative housing [5]. CDA enables the exploration of how language constructs and communicates responsibilities across critical dimensions of post-disaster housing reconstruction.

The empirical dataset comprises 135 documents covering the period from the 1999 Marmara Earthquake to the second anniversary of the 2023 Southeastern Anatolia Earthquakes, recognised as the modern era of disaster management in Türkiye. Documents include policy documents, professional reports, academic publications, and press releases collected through targeted keyword searches in Turkish via Google Scholar and Google News. Keywords included "*konut kooperatifi*" (housing cooperative), "*yeniden inşa*" (reconstruction), "*afet sonrası konut*" (post-disaster housing), and "*katılım*" (participation). Specific legal documents analysed include Law No. 7269 (1959) (Law on Measures and Assistance to Be Taken Due to Disasters Affecting Public Life [*Umumî Hayata Müessir Afetler Dolayısıyla Alınacak Tedbirlerle Yapılacak Yardımlara Dair Kanun*]), Law No. 6306 (2012) (Law on Transformation of Areas under Disaster Risk [*Afet Riski Altındaki Alanların Dönüştürülmesi Hakkında Kanun*]), Law No. 4123 (1995) (Law on Services Regarding Damages and Destruction Due to Natural Disasters [*Tabii Afet Nedeniyle Meydana Gelen Hasar ve Tahribata İlişkin Hizmetlerin Yürütülmesine Dair Kanun*]), Law No. 3194 (1985) (Zoning Law [İmar Kanunu]), Law No. 1163 (1969) (Cooperatives Law [*Kooperatifler Kanunu*]), the Site Selection Circular (2013) for Post-Disaster Housing (*Yer Seçimi Genelgesi*), and the Regulation on Determination of Rights Holders Due to Disaster (2013) (*Afet Sebebiyle Hak Sahibi Olanların Tespiti Hakkındaki Yönetmelik*).

To systematically analyse these documents, we developed an inductive coding framework aligned with the three spheres of responsibility as shown in Table 1. The framework categorises narratives into main codes reflecting political (institutional/legal), practical (implementation), and personal (normative) dimensions of responsibility. Each main code is subdivided into specific sub-codes representing concrete responsibilities, such as governance coordination, site determination, construction oversight, and community engagement. This approach allows us to identify how discursive strategies operate across institutional obligations, practical implementation, and normative justifications.

Complementing the document analysis, we conducted 25 semi-structured interviews between March 2023 and June 2025. We selected interviewees based on their direct involvement, expertise, or experience related to cooperative housing initiatives, disaster housing recovery processes, or relevant governance roles, including representatives from cooperative housing initiatives (e.g., *İzmir* Public House, *Kalkınma Atölyesi*, *Borkop*-Cooperatives, *Yerküre* Local Studies Cooperative), professional chambers (urban planners and architects in *İzmir*, *Elazığ*, and *Hatay*), and city officials in *Kastamonu* and *Samandağ*, *Hatay*. Interviews explored participants' perceptions of how responsibility was framed, contested, or reinforced within political frameworks, practical execution, and personal expectations and definitions of housing recovery responsibility.

To ensure validity, analytical consistency, and depth, we triangulated findings across government texts, legal documents, academic literature, media sources, and interview transcripts. Comparative analysis facilitated the identification of converging, diverging, or shifting narratives across institutional contexts. We maintained reflexivity throughout the analysis by acknowledging the inherent political nature of discourses and our positionalities within critical urban studies.

Table 1. Coding Framework for Responsibility Framing in Post-Disaster Housing Reconstruction

Sphere of Responsibility	Main Codes	Sub-Codes	Actors Involved
Political (Institutional/Legal Responsibility)	Governance frameworks and structures	Central governance authority, Local authority roles, Interagency coordination	Central government, local authorities, professional chambers, NGOs
	Legal and policy mechanisms	Legal mandates, Policy development, Regulatory frameworks	Central government, professional chambers, local authorities
Practical (Implementation Responsibility)	Implementation processes	Site selection, Land acquisition, Infrastructure provision, Housing design and construction oversights	Central government, local authorities, private contractors, professional chambers
	Performance and outcomes	Quality assurance, Efficiency indicators, Compliance standards	Central government (TOKİ), third-party inspectors, local authorities
Personal (Normative Responsibility)	Normative justifications	Values, expectations, Solidarity narratives, Justice and equity principles	Affected communities, civil society organizations, NGOs
	Community engagement and participation	Stakeholder involvement, Participatory practices, Consensus-building, Conflict resolution	Local communities, professional chambers, local authorities, NGOs

4 Findings

In Türkiye, all cooperative models, including those in housing, agriculture, trade, and services, are regulated under a single legal framework: the Cooperatives Law No. 1163, enacted in 1969. This law defines cooperatives as member-owned entities established to meet common economic, social, and cultural needs through democratic decision-making and collective ownership. This law defines cooperatives as member-owned entities established to meet common economic, social, and cultural needs through democratic decision-making and collective ownership. The origins of this model go back to the *Bahçelievler* Cooperative (see Fig. 2), founded in Ankara in 1934 to provide affordable housing for civil servants [6].

Cooperative housing gained momentum, especially between 1945 and 1980, supported by strong state institutions. During this period, membership rose from approximately 5,400 to 100,000, and SSK financed over 200,000 cooperative housing units [6].

A turning point came with the enactment of the Mass Housing Law No. 2487 in 1981, which integrated cooperative housing into the national mass housing strategy, enabling cooperatives to reach about 30% of licensed housing production by the late 1980s. However, the 1990s brought increasing regulatory instability and diminishing financial support, reducing their market share to around 20% by the end of the decade. The early 2000s marked a sharp decline: the termination of the Housing Fund in 2001 and the rise of the TOKİ restructured the housing sector around a centralised, state-led model. By 2020, cooperatives accounted for only 1% of licensed housing (ibid).

Fig. 2. Current condition of one of the remaining examples from the *Bahçelievler* Cooperative project (source: main author)

The centralisation that weakened housing cooperatives in Türkiye also shaped post-disaster housing production [7]. Since the 1999 Marmara Earthquake, reconstruction has followed a consistent model: rapid, standardised, centrally managed, and often located on urban peripheries. The Law on Measures to be Taken After Disasters (Law No. 7269), enacted in 1959, provided the legal foundation for state-led recovery efforts.

Fig. 3. Post-Disaster Housing Project Examples from Zafran Neighbourhood after the 2020 *Elazığ* Earthquake (top) and *Bayraklı* Neighbourhood after the 2020 İzmir Earthquake (bottom) (source: main author)

This top-down framework was reinforced by the establishment of the Disaster and Emergency Management Authority (AFAD) in 2009 and further expanded through the Law on the Transformation of Areas Under Disaster Risk (Law No. 6306), adopted in 2012. This law granted the state authority to redevelop areas not officially designated as disaster zones, framing the process as risk mitigation while enabling large-scale urban transformation with limited public oversight. Although certain policies promote participatory planning, they remain insufficient in practice. For example, the Circular on Procedures and Principles Regarding the Preparation of Zoning Plans After Disasters (Circular No. 3003), issued in 2013, encourages stakeholder involvement in land-use decisions for post-disaster housing. However, it does not provide mechanisms for meaningful influence by affected populations or other actors. From the 1999 Marmara Earthquake to the 2023 Southeastern Anatolia Earthquakes, post-disaster housing production in Türkiye has followed a clear pattern (examples shown in Fig. 3). Regardless of regional differences, the housing tends to be uniform in design, implemented by the TOKİ, and disconnected from the existing urban and social fabric.

4.1 Cooperatives as a Post-Disaster Housing Model: *Düzce* Hope Houses and Public House Projects

While centralised frameworks have come to dominate both affordable and post-disaster housing production in Türkiye, two recent initiatives have challenged this trend by offering alternative models: the *Düzce* Hope Homes (*Düzce Umut Evleri*) and İzmir Public House (İzmir Halk Konut) projects illustrate how cooperative housing can offer inclusive and locally grounded alternatives to standardised, top-down approaches.

The *Düzce* Hope Homes initiative began following the devastating earthquakes in August and November 1999, significantly affecting *Düzce* province and the broader Marmara region. These earthquakes resulted in approximately 18,000 deaths, displaced over 250,000 people, and severely damaged or destroyed around 113,000 buildings. Despite comprehensive governmental reconstruction programs, tenants, who made up a significant proportion of the affected population, were excluded from housing support provided under Law No. 7269, which favoured property owners. Facing this exclusion, tenants founded the *Düzce* Homeless Earthquake Victims Housing Cooperative in 2003 (see Fig. 4). The government, through TOKİ, facilitated this process by offering the land at a subsidised rate and allowing payments in instalments over ten years. However, requests for direct state funding or long-term, low-interest construction credits from TOKİ were denied, compelling the cooperative to seek alternative funding.

Between 2013 and 2016, the cooperative members actively engaged in the participatory workshops, collaborating with architects, engineers, planners, sociologists, journalists, academics, and students. This collaborative process ensured the residents' input shaped their future homes. As a cooperative member expressed in an interview conducted by Hazal Gümüş, "*Initially, we knew nothing about design, but this process was genuinely enjoyable and empowering. We learned how and what to do. Now, no one can say they didn't want something a certain way, because they were consulted at every step*" [8].

Construction began in 2016 with significant direct involvement by cooperative members, who contributed financially through membership fees and regular contributions, and substantial *sweat equity* by participating in construction work. Additionally, volunteers, including architects and engineers from various national and international universities, offered technical assistance. The cooperative also organised fundraising campaigns, such as "#harcımızaUMUTkat," to solicit small donations and material support from the broader community [9].

By 2020, the project delivered 234 homes, turning former tenants into empowered homeowners. Nevertheless, the cooperative faced substantial obstacles, including lengthy bureaucratic procedures that delayed permits and significant financial constraints due to limited external funding and the rejection of bank loans and governmental credits. Despite these hurdles, the initiative significantly reduced construction costs through innovative, resource-efficient strategies and extensive voluntary labour.

Fig. 4. Participants of Düzce Hope Homes in front of the project sign [9]

Today, *Düzce* Hope Homes represents a compelling model of cooperative housing in post-disaster contexts, demonstrating its potential for fostering social cohesion, resilience, and sustainable urban recovery through collective action, participatory governance, and strategic financial management.

A comparable bottom-up initiative emerged in the aftermath of the 2020 İzmir earthquake. Unlike the *Düzce* case, the central government did not declare the affected area a disaster zone. Consequently, Disaster Law No. 7269 was not fully implemented, and homeowners of moderately damaged buildings were excluded from formal reconstruction programs [10]. The limited insurance compensation provided proved insufficient for repairs or rebuilding, leaving many households without viable recovery options.

To address this institutional gap, affected residents established the İzmir Public House (*İzmir Halk Konut*) cooperative in 2021, aiming to develop inclusive and comprehensive housing solutions. İzmir Metropolitan Municipality played a crucial facilitating role by supporting the creation of limited liability cooperatives, providing technical assistance, and implementing zoning amendments to increase the floor area ratio, thus significantly improving the feasibility of reconstruction. Additionally, the municipality reduced bureaucratic barriers and ensured municipal enterprises provided construction services at minimal profit margins. These supportive measures considerably strengthened the project's viability and enabled cooperative members to actively participate in shaping their built environment.

Beyond tangible reconstruction outcomes, the cooperative model fostered a strong sense of solidarity and collective agency among residents. Illustrating this impact, a member of the first completed apartment as part of the project (Fig. 5) stated during our interview, "We didn't know our neighbours before this. Now, we design and build together and want to turn this into a space for workshops on civil rights, disasters, and climate change." This experience of community-driven reconstruction inspired other local initiatives, encouraging people to actively communicate, organise, and take ownership of their recovery process.

Despite these positive developments, the cooperative initiative encountered significant bureaucratic challenges. Due to the historical decline in housing cooperatives, officials were unfamiliar with the required legal processes and administrative requirements. This lack of institutional knowledge caused prolonged delays as residents struggled to navigate outdated regulations and ambiguous procedural steps.

In addition, granting additional floors intended to reduce unit costs and enhance affordability became another point of debate. While financially beneficial, this planning adjustment drew criticism regarding its compatibility with existing urban planning regulations. *İlker Kahraman*, President of the İzmir Chamber of Architects, articulated this concern in our interview, stating, "Granting additional floors to reduce the cost burden on cooperative members is understandable, but this approach does not align with existing urban development plans."

Fig. 5. Dilber Apartment, İzmir Public House: 3D model (left, image courtesy of *Halk Konut*) and construction photo (right, photograph by the main author)

In addition to administrative obstacles, financial accessibility remained a critical issue, particularly for retirees and lower-income households with limited savings or access to credit. Public distrust toward housing cooperatives further complicated the initiative. Due to past experiences with failed or incomplete projects, many approached new cooperative efforts with caution. As noted by a founding member of *Yerküre* Local Studies Cooperative, a scientific research and development cooperative, during our interview, negative perceptions associated specifically with housing cooperatives extend broadly, causing hesitation and caution toward other types of cooperatives as well, even those with well-functioning and successful models.

Despite these challenges, *İzmir Halk Konut* currently includes approximately 1,000 homeowner members and plans to construct around 800 housing units. The cooperative underlines transparency, participatory design, and collective decision-making throughout the process. Much like the *Düzce* Hope Homes, this case illustrates the potential of cooperative housing to serve as an equitable and community-driven alternative in post-disaster recovery. It also points out the importance of legal clarity, institutional capacity, and public trust in advancing such models.

Both cooperative housing cases demonstrate the potential of cooperative models to deliver equitable, participatory, and empowering housing solutions. At the same time, they underscore common struggles such as bureaucratic inefficiencies, financial barriers, and internal community dynamics that must be acknowledged and addressed in future cooperative housing initiatives. Reflecting on these challenges, a founding board member of *Güney Çankaya Vişnelik* Housing Cooperative in Ankara, who also advocates for legal improvements in cooperative systems in Türkiye through the *Borkop* Cooperative initiative, highlighted during our interview that the perception of cooperatives as inherently slow is misleading. They discussed their experience following the 2023 earthquakes in Adana province, explaining that despite strong community support and concerted efforts to propose cooperative housing as a viable alternative for reconstruction, negative perceptions of cooperatives, legal hurdles, and authorities' preference for seemingly faster and more established models such as those provided by TOKİ prevented the initiative's success.

4.2 Responsibility in Narratives of Post-Disaster Housing

Responsibility narratives significantly shape the landscape of post-disaster housing recovery in Türkiye, influencing the allocation of roles, decision-making processes, and overall governance outcomes.

In the political sphere, post-disaster housing recovery in Türkiye is predominantly framed by a centralised, state-oriented discourse. Government narratives consistently position the state as the primary authority responsible for swift and comprehensive housing provision following disasters. This framing is reflected in official campaigns such as the "Reconstruction of the Century," launched after the devastating 2023 *Kahramanmaraş* earthquakes. President Recep Tayyip Erdoğan explicitly stated, "Rest assured, just as we overcame the massive earthquake in Van, as we overcame those in *Bingöl*, Malatya, and Elazığ, and just as we swiftly rebuilt the buildings after the disaster in İzmir and delivered them to their owners, likewise, have no doubt we will rapidly do the same in these 10 provinces and hand over the homes to the owners" [11]. Similarly, following the 1999 Marmara earthquake, President at the time Süleyman Demirel reassured the public under the compelling slogan *"Father's Promise!" (Baba Sözü!)* that "the state will construct 26,000 temporary houses, delivering a total of 34,000 homes by the end of November 1999, with the goal that no one remains in tents by that time" [12]. These narratives consolidate the central government's role as the sole effective entity capable of rapidly addressing post-disaster housing needs, marginalising community-led or decentralised alternatives, and justifying expansions of governmental power, including significantly increased expropriation rights, to facilitate rapid recovery processes.

This centralised approach is further institutionalised within Türkiye's legal and regulatory frameworks, reinforcing state-centric control over post-disaster housing decisions. Key legislation, including Disaster Law No. 7269 and Urban Transformation Law No. 6306, explicitly consolidates authority within governmental bodies such as the Ministry of Environment, Urbanisation, and Climate Change, and provincial governors. Law No. 7269 assigns significant decision-making power regarding disaster-affected areas and reconstruction to central institutions (Law No. 7269, Article 2). Similarly, TMMOB explicitly criticises Law No. 6306, stating, "Local administrations are sidelined, all implementations are carried out by the central government, and local people are excluded from decision-making mechanisms" [13, p. 69]. Additionally, Circular No. 3003 reinforces centralisation through detailed procedures for site selection managed predominantly by committees coordinated by AFAD provincial offices, lacking clear mechanisms for integrating divergent local perspectives, there by marginalising meaningful participatory opportunities [10].

At the practical level, opaque procurement processes and centralised planning significantly constrain transparency and community participation in post-disaster housing recovery implementation. Reconstruction projects managed by central authorities, notably TOKİ, frequently attract criticism for favouritism and insufficient accountability. Investigative reports after the 2023 earthquakes highlighted, "In the billions-worth housing tenders organised for earthquake victims by the Ministry, contractors known for their closeness to the government were directly negotiated with" [14]. Urban planners further emphasise that decisions regarding site selection, housing design, and relocations "Unfortunately, scientific and ethical principles were ignored; all authority was transferred to the central government and TOKİ, and community-oriented participatory processes were disregarded" [15].

In the personal sphere, dominant narratives significantly influence public expectations and perceptions regarding responsibility for housing reconstruction. Many citizens expect comprehensive state intervention following disasters, reflecting historical patterns of centralised governance. Municipal officials in *Samandağ* highlighted during our interviews,

"Local municipalities lack resources and capacities to independently manage large-scale post-disaster housing projects; thus, central governmental leadership is essential". Likewise, representatives from the Chamber of Urban Planners in Hatay stated, "Local participation is crucial, yet considering the scale of post-disaster reconstruction, central government leadership is necessary, along with strengthened local participation." Besides, historical scepticism toward cooperative models further complicates the acceptance of community-led alternatives, often influenced by past experiences of inefficiency and perceived corruption. TMMOB explicitly notes these concerns: "One of the most significant problems in cooperatives arises from membership conditions and share transfers. Given our country's housing problems, high costs, high rents, and the lengthy duration of cooperatives making costs unpredictable, this diverts cooperatives away from their goal of producing social housing for low-income groups."

Despite these reservations, urban planners, scholars, and civil society organisations, including TMMOB, advocate persistently for more inclusive, transparent, and participatory recovery processes. TMMOB Hatay Urban Planners representatives specifically emphasise during the anniversary of the 2023 earthquakes, "We call upon all authorities, before any further irreparable mistakes are made, to implement a reconstruction approach that is planned, programmed, participatory, in the public interest, and transparent and holistic in line with sound urban planning principles" [15].

Overall, responsibility narratives across political, practical, and personal spheres reveal a predominantly centralised recovery framework in Türkiye. This centralisation systematically limits cooperative housing solutions and reinforces existing power dynamics, creating barriers to inclusive, participatory approaches. Addressing these challenges requires comprehensive reforms in regulatory frameworks, institutional practices, and societal attitudes, fostering more democratic and responsive reconstruction practices.

5 Discussion

Our analysis identifies significant barriers to the broader implementation of cooperative housing within Türkiye's post-disaster recovery efforts, highlighting how these barriers are deeply embedded within dominant narratives and institutional frameworks. Centralised governance emphasises a state-driven discourse prioritising rapid reconstruction, efficiency, and standardised solutions. This dominant narrative, continually reinforced through official communication and public discourse, profoundly shapes public expectations and perceptions of effective disaster recovery. As a result, communities often align themselves with immediate state-led interventions, despite recognising inherent shortcomings, due to the persuasive framing of rapid and uniform housing provision as responsible and necessary.

5.1 The Influence of Outdated Regulatory Frameworks and Lack of Event-Based Recovery Strategy

Türkiye's existing regulatory frameworks structurally reinforce and perpetuate the dominant discourse of centralised post-disaster housing. Key legislative instruments, including Disaster Law No. 7269 (1959), Law No. 4123 on Services Regarding Damages and Destruction Due to Natural Disasters (1995), Zoning Law No. 3194 (1985), and the Regulation on Determination of Rights Holders Due to Disaster (2013), collectively institutionalise rigid, centralised decision-making processes. These regulations grant significant discretionary powers to state authorities, enabling rapid land expropriation and mass housing production while significantly limiting meaningful community engagement. The strong central control makes it hard to adjust rebuilding efforts to fit local needs, sidelining community-led housing options and keeping a dependence on one-size-fits-all, centrally run solutions.

Moreover, current legislation lacks clear provisions for tenants, thereby excluding significant portions of the affected population from accessing equitable housing recovery solutions. Issues regarding accurate damage assessments, equitable definition of rights holders, and inclusion of tenants underscore critical gaps in the existing legal framework.

Additionally, the Cooperatives Law No. 1163 (1969), designed to regulate all types of cooperatives under a single legislative framework, poses significant procedural and bureaucratic obstacles. The absence of housing-specific regulatory provisions complicates the establishment, management, and sustainability of housing cooperatives, limiting their practical implementation and effectiveness.

There is a critical need for effective implementation of disaster-specific strategic planning frameworks, such as the "Sakınım Plans (Risk Mitigation Plans)" advocated by Murat Balamir, which have already been integrated into the Spatial Plans Preparation Regulation (Mekânsal Planlar Yapım Yönetmeliği). These frameworks provide tailored, event-based recovery strategies through detailed damage assessments and region-specific planning, significantly enhancing local adaptability and fostering greater resilience. Adopting these comprehensive, city-scale reconstruction strategies based on localised urban conditions would substantially improve recovery efforts, facilitate community participation, and encourage context-sensitive housing solutions.

5.2 From Abstract Participation to Concrete Models

Participation has understandably become a recurring demand in post-disaster housing debates, yet it's meaning often remains ambiguous. Too frequently, it is practiced with consultation exercises or symbolic inclusion, rather than understood as a process of genuine co-production in which affected people share decision-making power. Clarifying what participation entails, and more importantly, identifying who holds ultimate authority, is essential for transforming participatory claims into substantive outcomes.

The cases of *Düzce* Hope Homes and İzmir Public House provide valuable insights into what such transformation requires. In *Düzce*, structured design workshops and collective resource mobilisation enabled residents to act as decision makers rather than households reduced to the role of beneficiaries. In İzmir, municipal facilitation through technical assistance, transparent budgeting, and regulatory adjustments ensured that participatory processes were institutionally anchored and more durable. Both cases demonstrate that when affected households are situated as co-producers, and when professional and institutional support is available, participation can generate tangible benefits: stronger social cohesion, improved accountability, and housing solutions more closely aligned with local realities.

At the same time, these initiatives expose the limits of participation when it is not backed by sustained institutional and policy frameworks. Financial barriers, legal ambiguities, and reliance on exceptional local leadership or voluntary support constrain the scalability of otherwise promising models. Exploring both successful and incomplete examples, in Türkiye and abroad, can therefore provide useful insights into what participation should mean in practice. Analysing these cases makes it possible to distinguish between participation as rhetoric and participation as governance, and to draw lessons on the regulatory, financial, and organisational conditions that can move post-disaster housing beyond consultation toward genuine co-production.

5.3 Collaborative Pathways for Transformative Change

A transformative shift toward cooperative and community-led housing depends on collaboration among academia, disaster-affected communities, professional organisations, local authorities, and central authorities. Academic institutions can contribute by examining the conditions that support cooperative models and by documenting relevant practices and lessons. Sharing these insights with public institutions and the broader public may support more informed decision-making and facilitate wider knowledge exchange. Direct engagement of disaster-affected communities ensures that local knowledge, cultural contexts, and practical experiences shape reconstruction approaches, thereby increasing acceptance and effectiveness. Local authorities and professionals play crucial roles by providing technical and administrative expertise to ensure scalability, sustainability, and effective implementation of participatory models.

Finally, shifting toward a transformative discourse on housing requires recognising housing as a fundamental human right rather than merely a temporary or market-driven commodity. Comprehensive legal and institutional reforms must explicitly prioritise community participation, sustainability, and resilience. Integrating cooperative housing systematically within national recovery policies can significantly reduce vulnerabilities, promote equity, and establish more resilient and empowered communities capable of effectively addressing future disasters.

6 Conclusion

In this research, we have critically examined why cooperative housing initiatives remain underrepresented in Türkiye's post-disaster recovery, focusing particularly on the framing of responsibility by different actors and its interaction with centralised governance and institutional frameworks. Using critical discourse analysis, we explored narratives promoted by state actors, professional chambers, local authorities, and affected communities. We analysed a comprehensive corpus of 135 documents and conducted 25 semi-structured interviews, revealing that dominant disaster narratives emphasising rapid, standardised, and centrally managed reconstruction systematically marginalise cooperative and community-based alternatives.

Our findings show that the deep-rooted narratives and centralised practices, strongly embedded in legal and institutional frameworks such as Disaster Law No. 7269 and the Urban Transformation Act (Law No. 6306), significantly shape public perceptions of housing responsibility. These structures frame the state as the primary, if not sole, actor responsible for post-disaster reconstruction, severely limiting community involvement and the acceptance of cooperative housing models. Despite clear evidence from successful cooperative initiatives like *Düzce* Hope Houses and İzmir Public House, which demonstrate substantial advantages in terms of social cohesion, resilience, and long-term sustainability, the broader adoption of these models remains minimal.

Addressing these barriers requires substantial reforms to Türkiye's regulatory and institutional structures. This involves legal revisions explicitly designed to facilitate greater community participation, flexibility, and the inclusion of cooperative housing within post-disaster recovery strategies. Furthermore, targeted financial and institutional support mechanisms specifically tailored to cooperative initiatives must be developed and implemented. Strategically documenting and widely disseminating successful cooperative housing cases can help shift perceptions and demonstrate their practical feasibility and benefits.

Future research could further explore comparative analyses and interdisciplinary approaches to better understand the specific conditions that enable cooperative housing initiatives to succeed in various disaster contexts. Systematically integrating cooperative housing within national policies and adopting a rights-based housing approach can reduce vulnerabilities, promote equity, and build resilient communities capable of effectively addressing future disasters. Realizing such transformative change requires collaborative efforts among academics, affected communities, professional organizations, and policymakers, fostering a socially just, inclusive, and effective post-disaster housing recovery framework.

References

1. Presidency of Strategy and Budget, 2023 Kahramanmaraş and Hatay earthquakes report. Presidency of the Republic of Türkiye, Ankara (2023).
2. Anadolu Ajansı, Cumhurbaşkanı Erdoğan: Deprem bölgesinde 650 bin konut yapacağız. Anadolu Ajansı (2023). https://www.aa.com.tr/tr/gundem/cumhurbaskani-erdogan-deprem-bolgesinde-650-bin-konut-yapacagiz/2896206
3. S. Leipold, G. Winkel, Discursive agency: (re-)conceptualizing actors and practices in the analysis of discursive policymaking. Policy Stud. J. 45(3), 510–534 (2017). https://doi.org/10.1111/psj.12172
4. K. O'Brien, Is the 1.5 °C target possible? Exploring the three spheres of transformation. Curr. Opin. Environ. Sustain. 31, 153–160 (2018). https://doi.org/10.1016/j.cosust.2017.10.004
5. N. Fairclough, Critical discourse analysis: The critical study of language. Routledge (2013). https://doi.org/10.4324/9781315834368
6. H. Koç, Kentsel konut sunumunda giderek unutulan bir organizasyon: Konut yapı kooperatifleri. İDEALKENT 13(37), 2002–2026 (2022). https://doi.org/10.31198/idealkent.1117042
7. E. Altınok, Türkiye'de konut kooperatifçiliği: Sınırlar, imkânlar, öneriler. Mekanda Adalet Derneği, İstanbul (2025).
8. H. Gümüş, Bir müştereklеştirme deneyimi olarak Düzce Umut Evleri. beyond.istanbul (2021). https://beyond.istanbul/2021/11/12/bir-mustereklestirme-deneyimi-olarak-duzce-umut-evleri1-2/
9. Düzce Umut Atölyesi, Düzce Umut Evleri dünyanın en önemli 10 konut projesi arasında (2017). https://duzceumutatolyesi.wordpress.com/2017/12/13/duzce-umut-evleri-dunyanin-en-onemli-10-konut-projesi-arasinda/
10. F. Özdoğan, G. Lizarralde, B. Herazo, The politics of land management after disasters: The case of post-earthquake reconstruction in Türkiye. Disaster Prev. Manag. (2024). https://doi.org/10.1108/DPM-12-2023-0320
11. AFAD, Cumhurbaşkanımız Sn. Erdoğan: Deprem anından itibaren devlet olarak tüm kurumlarımızla sahadayız. AFAD, Gaziantep (2023). https://www.afad.gov.tr/cumhurbaskanimiz-sn-erdogan-deprem-anindan-itibaren-devlet-olarak-tum-kurumlarimizla-sahadayiz-23-merkezicerik
12. Hürriyet, Baba sözü: Kasım sonu çadırda kimse kalmayacak. Hürriyet Gazetesi (1999). https://www.hurriyet.com.tr/gundem/baba-sozu-kasim-sonu-cadirda-kimse-kalmayacak-39108189
13. TMMOB İnşaat Mühendisleri Odası, Kentsel dönüşüm nedir? Sorular, sorunlar, çözümler. TMMOB İnşaat Mühendisleri Odası, Ankara, İMO/21/03 (2021). https://www.imo.org.tr/Eklenti/7388%2C185884028pdf.pdf?0=
14. Ç. Toker, Deprem ihaleleri 50'yi geçti. T24 (2023). https://t24.com.tr/yazarlar/cigdem-toker/deprem-ihaleleri-50-yi-gecti%2C39081
15. TMMOB-Şehir Plancıları Odası Hatay İl Temsilciliği, TMMOB Şehir Plancıları Odası Hatay İl Temsilciliğinden 6 Şubat açıklaması. Atayurt Gazetesi, Hatay (2025).

https://doi.org/10.1051/e3sconf/202565103001

https://atayurtgazetesi.com.tr/tmmob-sehir-plancilari-odasi-hatay-il-temsilciliginden-6-subat-aciklamasi/

Improving Competence in the Seismic Code Implementation for Housing in Indonesia using Visualization and Practical Exercises

Setya Winarno[1,*], *Catur* Singgih[1], and *Julius Ayodeji* Fapohunda[2]

[1]Civil Engineering Department, Universitas Islam Indonesia, Yogyakarta, Indonesia
[2]Built Environment Department, Cape Peninsula University of Technology, Bellville, South Africa

Abstract. Recent data on earthquake tragedies reveals that the seismic code implementation for residential houses is remarkably rare in Indonesia. Improvement of construction actors' competence for this situation is important. This paper examined the application of video-based visualization and practical exercises as educational tools for seismic code application to improve the competence of construction actors in the context of non-engineered residential houses in Indonesia. Video-based visualizations were compiled from several social media sources, and practical exercise materials were developed utilizing different essential aspects of quality building materials. There were 19 masons and carpenters involved in this study through a focus group discussion to evaluate the improvement of competencies before and after seeing the video and participating in practical exercises in the Bantul District, Indonesia. The results have demonstrated an enhancement in competence of around 29.01% among participants before and after seeing the video and participating in practice. A significant correlation (r=0.602) was observed between the participants' educational background and their improvement in the application of seismic codes. The age group of 42-43 years has had the most significant enhancement in competency compared to others. Work experience and age show no significant correlation (r=0.129 and r=-0.216 respectively) with the increase of competence.

1 Introduction

Historically, the impact of earthquakes has resulted in horrific disasters globally, regarded as one of nature's most terrifying and devastating phenomena. According to the Indonesian Seismic Zonation, about 60% of cities and urban centers in Indonesia, around 290 out of 481, are situated in regions classified as reasonably high to very high seismic zones [1]. The earthquake in Yogyakarta, Indonesia, on May 27, 2006, resulted in around 6,000 fatalities and the destruction of over 155,000 houses. The earthquakes in Palu and Cianjur, Indonesia, in 2018 and 2022 caused the serious destruction or collapse of 20,129 and 21,465 houses, respectively [2, 3]. These earthquakes underscore the widespread vulnerability of Indonesian

* Corresponding author: winarno@uii.ac.id

housing due to insufficient earthquake-resistant construction. The majority of the collapsed or severely damaged structures were non-engineered masonry buildings, composed of heavy materials, with or without reinforced concrete structures, particularly those owned by medium-low-income communities or as low-cost housing [4].

As traditional wooden buildings are hard to find and expensive, most homes in Indonesia are non-engineered structures, primarily consisting of heavy materials such as half-brick thick masonry strengthened with reinforced concrete frameworks. These constructions, referred to as "practical columns and beams," are often labelled as 'a new culture' [5], as noted by Boen. Moreover, many Indonesians have started to apply these "new culture" structures to raise their social status and cut down on maintenance costs. At the same time, these "new culture" buildings are often constructed without considering seismic performance, turning buildings into death traps. Boen has emphasized that if these "new culture" non-engineered houses are constructed using high-quality materials and competent workmanship, they are likely to withstand the most severe earthquakes predicted by the Indonesian seismic hazard map.

Characteristics of non-engineered structures have ignored basic earthquake-safety principles mentioned in Indonesian seismic codes (especially SNI 1726:2019) like confinement of masonry, ductile detailing of reinforcement, and continuous load path. Subsequent seismic incidents in Indonesia have revealed that most newly constructed cultural homes lack adequate earthquake resilience, with just a small number built according to recommended seismic resistance standards. While some reconstruction efforts aimed at providing safer housing, many homes still showed discrepancies with local characteristics and did not fully address principles of earthquake-resilient construction [6]. Although some homeowners have adopted reinforced concrete frames for added strength, many modifications have not been aligned with proper seismic-resistant building principles [7]. Numerous deficiencies in construction management include lack of detail in design and planning, absence of building permits, and failure to adhere to technical specifications for earthquake resilience. Despite ongoing efforts to improve post-earthquake reconstruction and promote earthquake-resilient housing in Indonesia, significant gaps remain in aligning local construction practices with seismic-resistance standards [8]. In particular, there is limited emphasis on systematic training and capacity building for grassroots construction actors - such as masons and carpenters - who are central to the delivery of non-engineered housing. This gap highlights the urgent need for approaches that not only strengthen technical competencies in seismic-safe construction but also preserve local cultural traditions in housing design.

Research undertaken prior to the Yogyakarta earthquake indicated that 93.5% of residential housing in Yogyakarta City included non-engineered constructions [9] in which 84.8% of non-engineered houses were constructed from heavy materials, specifically masonry buildings with one or half brick thickness. Among these, 1% were extremely traditional houses without reinforcement and adequate maintenance, while 83.8% represented what appeared to be 'the new culture' housing. There were 8.7% of houses constructed using lightweight materials, including teakwood in traditional and historic structures and were non-engineered. Some of them were among the poorest of the poor, constructed with extremely lightweight materials that may withstand significant seismic activity and are less deadly in the event of a collapse. The last category comprised just 6.5% of the engineered houses, which were ultimately owned by wealthy people.

The issue of sociocultural dynamics has become increasingly prominent during the implementation of seismic measures in non-engineered houses, which involve more than just physical intervention, since this construction practice is frequently influenced by local wisdom of social cohesion and community spirit. Small to medium-sized contractors, foremen, masons, and carpenters are the primary actors in non-engineered building processes, typically employing vernacular methods, local resources, and labor-intensive techniques, often relying on unwritten rules and without official contracts or arrangements. In certain instances, homeowners may have complete control over the whole construction process, leaving contractors with no bargaining power to propose alternative methods, if applicable. They all frequently possess inferior education, inadequate training, limited access to knowledge, and less attention from modern science and technology. They acquire building skills organically via hands-on experience and observation of local practices. Ultimately, they are the primary factors in the implementation of seismic elements in the building of non-engineered structures. Consequently, it is essential to reconcile the gaps between knowledge and action to prevent the generation of new risks by permanently altering building practices in accordance with traditional cultural knowledge and local social cohesiveness. An exemplary training-based intervention has been conducted in Pacitan, Java Island, Indonesia, targeting construction workers (contractors, foremen, masons) to enhance their competence and understanding of earthquake catastrophe preparedness [10]. However, there remains a critical research gap in systematically measuring the extent of knowledge improvement or increment gained through such training and capacity-building programs for grassroots construction actors in the context of non-engineered housing.

The role and challenges of social media and visual tools have emerged following a workshop on 'seismic risk management of non-engineered housings' held in 2006 in Yogyakarta [9] that seismic codes must be articulated in simple language that is socially and culturally acceptable, easily implementable, and cost-effective. Effective information transmission and knowledge transfer among construction actors is recognized as the primary approach to address poor communication [11], hence enhancing the effective utilization of resources within the non-engineered construction sector. Also, grass root construction actors exhibit a preference for visual content over written material in the dissemination of these codes. Due to the visual content, social media, through visual information, offers a range of interactive modules, educational videos as a dynamic distribution method, and learning materials that are readily accessible to the general audience, enabling anyone to independently learn how to integrate seismic characteristics. The utilization of visual information via social media has the potential to enhance public education for evaluating community resilience [12]. Similarly, this can complement and improve current knowledge of seismic code implementation associated with disaster preventive strategies. This has been recognized as a valuable means of distributing information and offering guidance to the public [13, 14]. In addition, inferior building materials typically contribute to the collapse of housings during earthquakes [5]. Consequently, visualization aimed at enhancing the competence of construction actors must incorporate both simple video-based visualizations sourced from various social media platforms and practical exercises focused on identifying quality building materials. Materials such as river sand, gravel, cement, steel bars, bricks, and lightweight concrete blocks need to be tested visually to distinguish which materials are either good or bad.

Despite the constantly evolving nature of social media, there is a lack of systematic research regarding its impact on the knowledge improvement of construction actors. This paper aims to assess the improvement of knowledge or competency regarding seismic characteristics among construction actors in the context of non-engineered housing, comparing the period before and after watching the video-based visualizations and the visual evaluation of quality materials through a focus group discussion.

2 Research Methodology

2.1 Focus Group Discussion

A house in Trirenggo Village, Bantul District area, Special Province of Yogyakarta that survived the 2006 earthquake was chosen as the site of the focus group discussion (FGD), presenting about how seismic-resistant homes can be used as a practical classroom, motivational symbol, and community proof-of-concept. This helps the FGD participants learn more about the topic and become more emotionally committed to it. The selection of the survived house, indeed, facilitated sense of place and sense of control to the FGD's mind. Nineteen local construction practitioners, including masons and carpenters through a purposive sampling, participated in the FGD. Before participation, ethical approval was secured, and informed consent was acquired from all participants by the signing of a letter of agreement during the session. Innovative initiatives and new synergies were easily absorbed over this process. These were easily integrated into the preexisting community structures and recognized the importance of each individual's participation. They were categorized into their educational background, work experiences, and ages, as mentioned later in Table 2.

The videos used for learning visual media come from various social media platforms, such as Facebook, TikTok, and Instagram. This visualization included site, design, seismic features, good construction materials, and other characteristics. Figure 1 and 2 presents examples of anchor for tie beam and steel bar arrangements. Participants in the FGD were presented with locally obtained samples of high-quality and substandard materials, including river sand, gravel, cement, steel bars, and bricks, for comparative analysis in determining the best material. The Indonesian standard (SNI 2847-2019) was utilized as benchmark while assessing the quality of the material. The presentation of those several construction materials was provided both via visualized videos and actual samples.

During the FGD, the participants examined their level of competence both before and after watching the developed visualization through a previously piloted questionnaire. The definition of competence was the demonstrated ability among masons and carpenters to apply knowledge (understanding seismic risks), skills (evaluating material quality), and attitudes (awareness of earthquake-resistant construction importance) to perform tasks effectively in the application of seismic features in common houses. Assessment has been aligned with Indonesian National Work Competency Standards for construction workers. A questionnaire pilot test was conducted via a focused discussion with four personnel from the Bantul District Disaster Office. The participants were purposively selected based on their significant involvement in disaster training programs for general masons and carpenters, ensuring that their experience was relevant to the study context. The draft questionnaire was systematically reviewed during the session to evaluate the clarity, relevance, and comprehensiveness of each item, with particular attention to the relevance of wording, terminology, and sequencing. Feedback obtained from the discussion was carefully documented and utilized to refine the instrument, thereby strengthening its content and construct validity, in addition to its practical applicability. The final version of the updated questionnaire was then made ready for use in the next round of data collecting.

Furthermore, participants were tested to distinguish which materials were either good or bad both before and after watching the visualization and evaluating the actual material samples. Specifically for measuring the quality of concrete block material, there was a drop test on light weight concrete block on hard soil (Figure 3) and if there was no broken block after the test, this concrete block was an acceptable quality. Their competence improvement was elaborated further, and a correlation between the competence improvement and participants' characteristic was then analyzed. SPSS program was used in this computation.

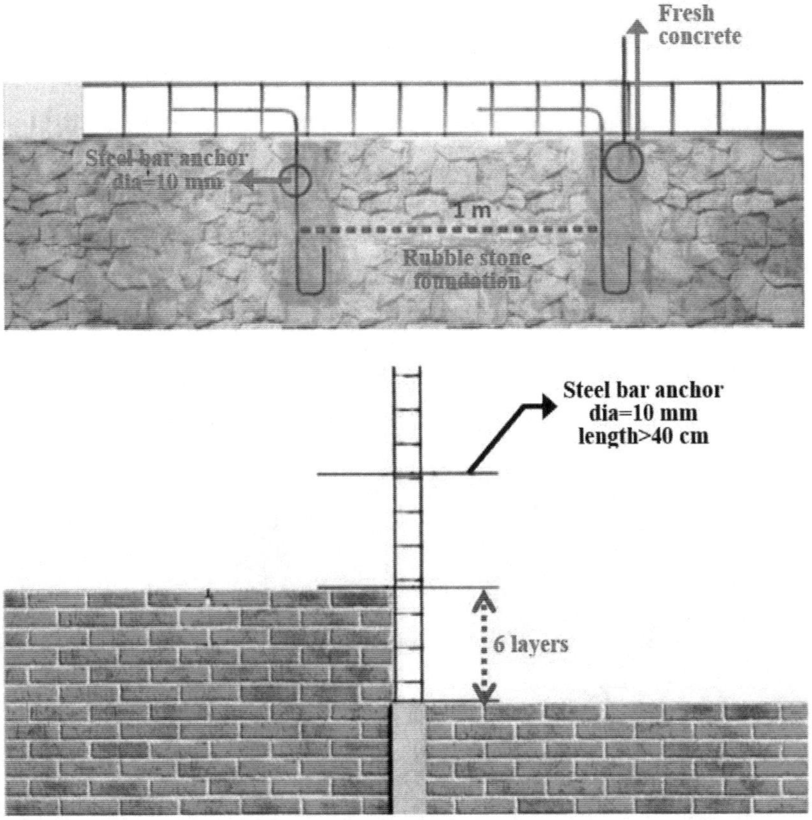

Fig. 1. Examples of the video contents: anchor for tie beam and wall (edited from Permen PUPR No. 5/PRT/M/2016).

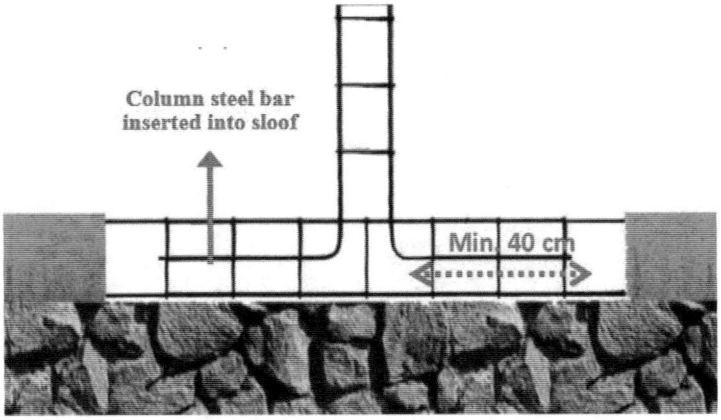

Fig. 2. Examples of the video contents: steel reinforced connection at bottom (edited from Permen PUPR No. 5/PRT/M/2016).

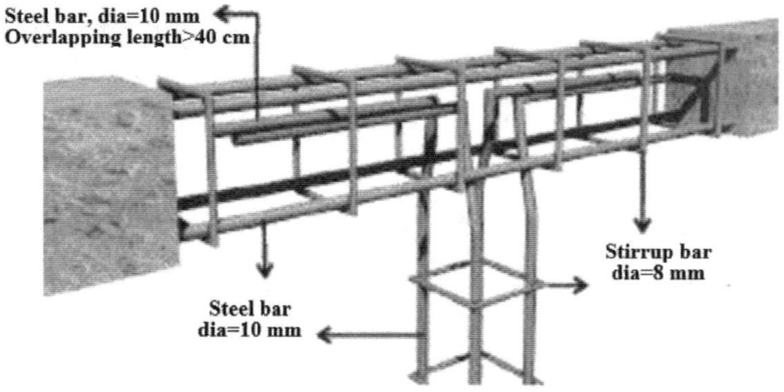

Fig. 3. Examples of the video contents: steel reinforced connection at top (edited from Permen PUPR No. 5/PRT/M/2016).

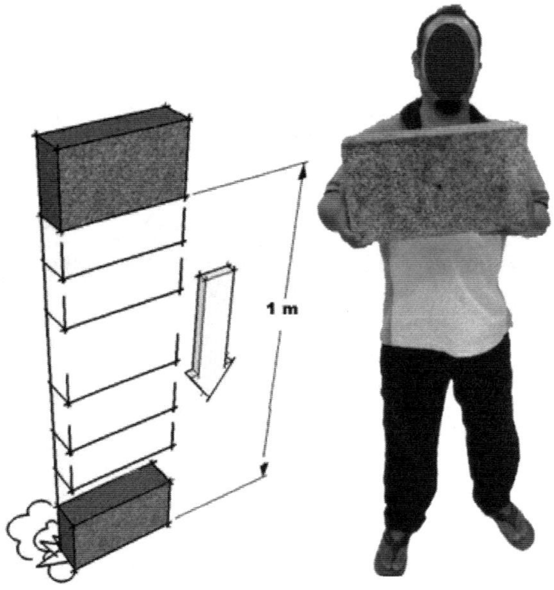

Fig. 4. Drop test on light weight concrete block.

2.2 Correlation Coefficients

The correlation coefficient is used to find how strong a relationship is between two variables. There are different types of correlation values, but Pearson's is the most popular. A correlation measure that is often used in linear regression is Pearson correlation, which is also written as Pearson's R. The methods give back a number between -1 and 1. A value of 1 means there is a strong positive relationship, a value of -1 means there is a strong negative relationship, and a value of 0 means there is no relationship at all. There is a critical value from a table (taken Pearson's Correlation Table) for two-tails tests. If the calculated Pearson's correlation coefficient is greater than the critical value from the table, then the relationship is statistically significant.

2.3 Questionnaires

A piloted questionnaire was administered to assess participants' competence both prior to and following exposure to instructional videos and practical exercises. The instrument focused on evaluating knowledge related to the implementation of seismic codes in residential housing, with its content developed based on established literature [9]. As shown in Table 1, the questionnaire comprises 12 items designed to measure respondents' understanding of seismic code application in residential buildings, with each correct response awarded a score of 4.

Table 1. Twelve question indicators for a good implementation of seismic features [9].

Criteria	Question indicators	Optional answers with a randomly selected correct answer
Steel reinforcement	1. Good steel reinforcement detail: stirrup spacing <15 cm	(1) stirrup spacing > 20 cm; (2) stirrup spacing 15-20 cm; (3) stirrup spacing <15 cm; (4) stirrup spacing <15 cm and hook 135°
	2. Good steel reinforcement detail: beam-column connection	(1) lapping longitudinal > 40 cm; (2) lapping longitudinal 30-40 cm; (3) lapping longitudinal 20-30 cm; (4) lapping longitudinal 10-20 cm
Lateral load path	3. Existence of lateral load path: plinth band	(1) plinth band; (2) brick layered band
	4. Existence of lateral load path: vertical reinforcement in all wall corners	(1) concrete practical column; (2) wood column; (3) bamboo concrete column
	5. Existence of lateral load path: roof-gable band	(1) roof band; (2) simple anchors to roof structure
	6. Existence of lateral load path: lintel band	(1) lintel band; (2) brick band
Anchors	7. Existence of foundation anchors to wall and wall anchors to vertical reinforcement	(1) foundation anchors; (2) wall anchors; (3) foundation and wall anchors
Building material	8. Good quality of cement	(1) grey colors; (2) loose material; (3) lump cement
	9. Good quality of sand and gravel	(1) dirty sand-gravel; (2) homogenous size of sand-gravel; (3) coarse sand; (4) solid gravel; (5) heavy gravel
	10. Good quality of steel reinforcement	(1) dirty steel; (2) new steel; (3) used steel; (4) diameter >10 mm
	11. Good quality of concrete block	(1) fragile concrete block; (3) lightweight material; (3) heavy material; (4) strong material
Workmanship	12. Good quality of workmanship	(1) junior mason; (2) senior mason; (3) agile mason; (4) high productivity mason

3 Results and discussion

A Focus Group Discussion (FGD) with 19 participants from the house construction sector was held. FGD participants were classified into three categories: academic background, age, and job experience. Participants were first briefed on the aim and objectives of the FGD, followed by a pre-test administered via a questionnaire assessing their current knowledge on the implementation of earthquake-resistant housing practices they have undertaken thus far. A visualization video and practical exercise demonstrating effective procedures in the execution of earthquake-resistant building requirements were then presented to enhance their understanding. Subsequent to watching visualization and practical activity, participants were requested to complete an identical questionnaire to assess the extent of their enhanced comprehension and knowledge. Table 2 presents the test scores of FGD participants before to (pre-test) and subsequent to (post-test) viewing the visualization and engaging in practice.

Table 2. The test scores of FGD participants: pre-test and post-test.

Participant number	Ages	Educational background	Work experiences, years	Pre-test score*	Post-test score *	Knowledge Increment
(1)	(2)	(3)	(4)	(5)	(6)	(7)=(6)-(5)
#1	32	High school (SMA)	15	21	28	7
#2	41	High school (SMA)	2	24	30	6
					Average	6.5
#3	42	Secondary school (SMP)	25	16	25	9
#4	53	Secondary school (SMP)	35	22	28	6
#5	56	Secondary school (SMP)	20	16	20	4
#6	56	Secondary school (SMP)	35	15	22	7
#7	57	Secondary school (SMP)	40	19	24	5
					Average	6.2
#8	43	Elementary school (SD)	23	15	23	8
#9	38	Elementary school (SD)	18	15	20	5
#10	48	Elementary school (SD)	25	14	17	3
#11	49	Elementary school (SD)	25	18	21	3
#12	50	Elementary school (SD)	33	16	23	7
#13	50	Elementary school (SD)	29	17	22	5
#14	53	Elementary school (SD)	11	17	20	3
#15	53	Elementary school (SD)	32	23	25	2
#16	60	Elementary school (SD)	40	16	21	5
#17	66	Elementary school (SD)	50	21	24	3
					Average	4.4
#18	50	Incomplete elementary school (<SD)	27	12	15	3
#19	43	Incomplete elementary school (<SD)	22	7	10	3
					Average	3
			Average	17.05	22	4.94
			Percentage of knowledge increment			29.01%

*) Maximum score is 48

Table 2 indicates that grassroot construction actors in the context of non-engineered residential houses in Indonesia may enhance their understanding of implementing seismic elements in buildings through dynamic visualization and practical use as an effective tool. Similar findings have revealed that the educational intervention effectively enhanced participants' knowledge and practices about earthquake preparedness [13]. Other study has presented on how education level affects competence of construction workers in Indonesia [14].

The maximum knowledge increment score is 6.5 for the FGD participant possessing a high school (SMA/SMK) educational background. Conversely, the minimum increment is 3 for respondents with an educational background of incomplete primary school (<SD). The knowledge improvement among respondents with educational backgrounds of SD and SMP is 4.4 and 6.2, respectively. The mean enhancement of competence or knowledge before and after viewing and practicing the visualization is 29.01%. Similar study provided 33% of knowledge increments [10]. The respondents aged approximately 42-43 have demonstrated the most substantial increase in their knowledge or competence.

These findings indicate that visualization-based training can effectively enhance knowledge across diverse educational backgrounds, with greater improvement observed among participants with higher baseline education levels, aligning with previous research highlighting the role of education in facilitating knowledge acquisition [15]. Moreover, higher levels of prior education significantly provide a foundational understanding and better cognitive abilities for learning, though programs must be tailored to participants' existing knowledge and context to ensure efficacy. Participants with more formal education may find it easier to process new information and apply it in practical contexts, while less-educated participants may require more support and context-specific guidance for effective learning transfer.

According to Table 2, Figure 5 illustrates the characteristics of FGD participants about their educational background and the increase in knowledge seen between the pre-test and post-test.

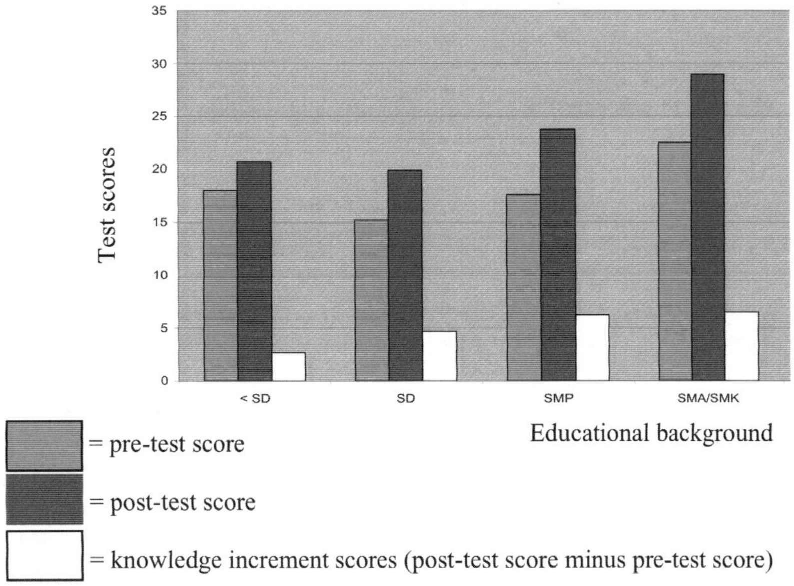

Fig. 5. The histogram of the educational background, test scores, and knowledge gains of FGD participants

A correlation between educational background and knowledge enhancement is arranged to supplement the aforementioned conclusion. The correlation coefficient (r_{result}) is 0.602, whereas the critical value (r_{table}) is 0.456 at a 95% significance level. This result indicates that the correlation is beyond the critical threshold, signifying a positive significant association between educational background and knowledge enhancement. The relationship between work experience, age, and knowledge enhancement is not statistically significant. These indicate a result of $r = 0.129$ for work experience and $r = -0.216$ for ages below the crucial value, with $r_{table} = 0.456$. Figure 6 illustrates the comprehensive correlation results.

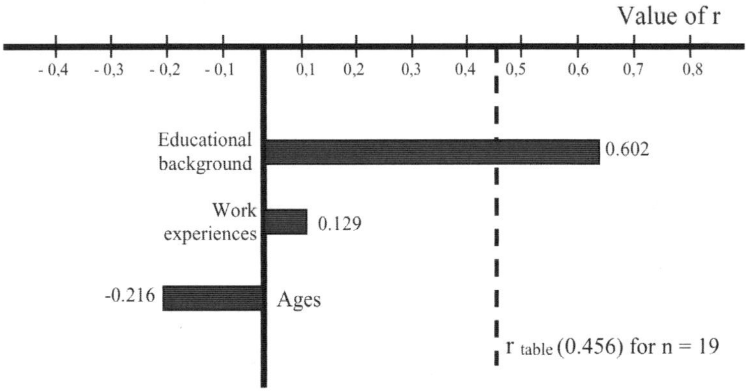

Fig. 6. The correlation between educational background, work experience, age, and enhancement of knowledge.

This study demonstrates that the learning strategy for grassroots construction actors using dynamic visualization and practical exercises has had positive outcomes. This study analyzed the changes in knowledge among 19 respondents from house construction sectors before and after viewing and practicing a dynamic visualization of seismic code execution in residential structures. Table 2 presents that all respondents clearly indicated an enhancement in their knowledge after viewing and practicing the visualization, however the degree of knowledge change varied among them. Three attributes - educational background, age, and job experience - have been analyzed for their link with the knowledge changes of respondents before and after viewing and practicing the visualization. The results indicate a robust positive correlation of 0.602 between educational background and knowledge growth. An enhancement in educational background signifies an augmentation in the worth of knowledge.

This study clearly indicates that dynamic visualization is essential for construction professionals to enhance their understanding. Moreover, higher education enhances knowledge acquisition more effectively. Construction actors engaged in the frontline of projects must get rigorous training to ensure the effective implementation of seismic codes, hence facilitating the development of more earthquake-resistant structures universally. The government, responsible for public safety against seismic events, must communicate the advantages of dynamic visualization to all stakeholders in construction. A multitude of parties should be convened and consolidated to undertake and share this effort and responsibility.

4 Conclusions

This study has successfully evaluated the improvement of knowledge and competency regarding seismic characteristics among construction actors in the context of non-engineered housing. By comparing the period before and after video-based visualization and the visual evaluation of quality materials through a focus group discussion, it was revealed that construction actors can substantially enhance their understanding of implementing seismic features in buildings. The mean improvement of 29.01% demonstrates the effectiveness of visualization-based training as an educational tool. Moreover, knowledge enhancement was obvious across different educational backgrounds, with participants of higher baseline education showing greater improvement, as provided in the moderate correlation ($r = 0.602$). In contrast, age and work experience showed no significant correlation with knowledge increment, suggesting that structured learning interventions may play a more critical role than experiential learning alone in improving earthquake-resilient construction practices.

However, these findings should be interpreted with caution due to certain limitations. The participants in this study were exclusively from Trirenggo Village, which may not represent the diversity of grassroots construction actors across Indonesia. Future research should therefore broaden the scope to encompass various regions, cultural contexts, and construction practices. Additionally, while video materials on seismic features are widely available through various social media platforms, their quality, content, and focus are not always the same. To build a stronger culture of prevention, there is a need for systematic effort of standardized training videos supported by government incentives, policies, and institutional discretion. Such an approach would not only strengthen training and capacity building among grassroots construction actors but also combine local traditions with seismic-resilience standards to reduce long-term housing vulnerability in Indonesia.

Acknowledgements. Some works contributed to this paper were funded by DPPM UII based on Contract No: 002/Dir/DPPM/70/Pen.PengembanganInstitusi/IV/2025. The authors express gratitude to the management of the Faculty of Civil Engineering and Planning at Universitas Islam Indonesia for providing the research facilities for this work. In particular, the author would like to thank many technicians in the Innovation Center of UII during the production of video and materials.

References

1. M. Irsyam, *et al.*, Development of new seismic hazard maps of Indonesia 2017, in *Proceedings of the 19th International Conference on Soil Mechanics and Geotechnical Engineering*, Seoul: ISSMGE Asian Technical Committee 7, 525–1528, (2017)
2. UNITAR, Indonesia: Comprehensive satellite detected building damage assessment overview as of 19 October 2018, Geneva, Switzerland, (2018)
3. Kemenpupr, *Cianjur Earthquake Response: 2022*. Jakarta: Satuan Tugas Penanggulangan Bencana PUPR, (2022)
4. N. Giordano, *et al.*, Seismic fragility models for typical non-engineered URM residential buildings in Malawi, *Structures*, **32**, 2266–2278, (2021). doi: 10.1016/j.istruc.2021.03.118
5. T. Boen, H. Imai, S.E. Suryanto, Earthquake resistant non-engineered construction in developing countries utilizing appropriate technology, *Eur. J. Adv. Eng. Technol.*, **7**, 8, 49–63, (2020)
6. A. Suryandono, P. Wong, Locally based approach for prefabricated housing – Case study: Indonesia, in *Proceeding of ACSA Fall Conference*, Philadelphia, 195–199, (2012)
7. N.C. Idham, Earthquake failures on buildings and the role of architect on building safety, *J. Archit. Built Environ.*, **45**, 2, 153–164, (2018). doi: 10.9744/dimensi.45.2.153-164
8. K.S. Pribadi, D. Kusumastuti, S.A.H. Sagala, R. Wimbardana, Post-disaster housing reconstruction in Indonesia: Review and lessons from Aceh, Yogyakarta, West Java and West Sumatera Earthquakes, in *Disaster Recovery*, Springer Nature Link, Ed., Tokyo: Springer Nature Link, 197–223, (2013)
9. S. Winarno, A. Griffith, P. Stephenson, Reducing earthquake risk to non-engineered buildings: A study of design and construction practices in Indonesia, *Int. J. Constr. Manag.*, **10**, 1, (2010). doi: 10.1080/15623599.2010.10773139
10. E.W. Pradana, *et al.*, Improving construction workforce competence to raise earthquake disaster awareness and preparedness, *J. Masy. Mandiri*, **6,** 6, 4689–4699, (2022). doi: https://doi.org/10.31764/jmm.v6i6.11075

11. J.A. Fapohunda, Evaluation of site managers' hindrances towards optimal utilisation of construction resources, *J. Eng. Des. Technol.*, **12**, 3, 348–363, (2014). doi: 10.1108/JEDT-03-2012-0013
12. B. Rachunok, J. Bennett, R. Flage, R. Nateghi, A path forward for leveraging social media to improve the study of community resilience, *Int. J. Disaster Risk Reduct.*, **59**, 102236, (2021). doi: 10.1016/j.ijdrr.2021.102236
13. A.K.C. Bhandari, M. Rahman, O. Takahashi, Enhancing earthquake preparedness knowledge and practice among Nepalese immigrants residing in Japan, *Sci. Rep.*, **13**, 4468, (2023). doi: 10.1038/s41598-023-31729-y
14. I. Amala, A. Utiarahman, F. Achmad, Analysis of the influence of education level on the competency of building construction workers in Gorontalo Province, *Compos. J.*, **4**, 1, 11–18, (2024)
15. L. Fergusson, Learning by… Knowledge and skills acquisition through work-based learning and research, *J. Work. Manag.*, **14**, 2, 184–199, (2022). doi: 10.1108/JWAM-12-2021-0065

Assessing Public Transport Vulnerability to Urban Flooding: A GIS Approach Incorporating Network Risk and Population Exposure

Orfeas Karountzos[1], *Marcos Julien* Alexopoulos[2,*], *Theano* Iliopoulou[2], *Christina* Iliopoulou[3] and *Denis* Istrati[2,4]

[1]Laboratory of Transportation Engineering, National Technical University of Athens
[2]Dept. of Water Resources and Environmental Engineering, National Technical University of Athens
[3]Department of Civil Engineering, University of Patras
[4]Leichtweiß Institute for Hydraulic Engineering, Technical University of Braunschweig, Germany

Abstract. Urban flooding increasingly disrupts public transportation, critical for vulnerable populations including elderly and low-income groups. Most research on flood resilience in transportation focuses on private vehicles, overlooking vulnerable users' reliance on public transport. This study addresses that gap by assessing flood-related disruption across the surface public-transport network of a large municipality in Athens, Greece. Using a two-dimensional coupled hydrological–hydraulic model, we simulate a 100-year flood and estimate water depths across the road network. Public transport and road infrastructure data are integrated in a Geographical Information System to identify segments where depths exceed 0.3 m, an operational safety threshold used for conservative detour screening. Critical links, affected bus stops and disrupted routes are systematically mapped, while the affected population is estimated through isochrone curves that delineate 15- and 30-minute accessibility. This approach pinpoints vulnerable areas, nodes and corridors in the network, supporting disaster preparedness and operational planning. The findings facilitate flood-risk maps tailored to public transport, enabling transit operators and municipal authorities to pre-emptively design alternative service plans.

1 Introduction

Urban flooding constitutes a paramount environmental hazard in densely built metropolitan areas, where impervious surfaces and aging drainage networks amplify runoff generation. This risk has been magnified by rapid urbanization and inadequate planning. Two recent European catastrophes illustrate the severity of such events: in May 2023, torrential storms forced the overflow of 23 rivers in Italy's Emilia-Romagna region—resulting in 17 fatalities, the evacuation of over 1000 residents, and € 8.5 billion in damages; in October 2024, unprecedented flash floods in Spain's Valencia region claimed more than 200 lives, imposed a –0.2 percentage-point drag on fourth-quarter GDP, and inflicted over € 10 billion in business losses. Flooding has generated approximately $85 billion in global economic losses and $13 billion in insured losses in 2023, with a protection gap of 78 percent, according to Aon's 2024 Climate and Catastrophe Insight report. Such events not only threaten critical infrastructure and economic stability, but also foreshadow the vulnerability of regions like

* Corresponding author: mjalexopoulos@gmail.com

Attica, Greece, where extensive impervious cover and antiquated stormwater systems exacerbate flood risk.

Modeling practices are an established tool in managing flood risk. In contexts where gauged flows are unavailable, the rain-on-grid (RoG) methodology directly applies spatially distributed precipitation onto two-dimensional hydrodynamic meshes, obviating synthetic hydrograph derivation and enabling robust flood simulation in ungauged basins. Immediately thereafter, comparative analyses demonstrate that both Hydrologic Engineering Center's River Analysis System (HEC-RAS) and Iber solvers accommodate RoG inputs effectively, though solver selection hinges on terrain steepness and computational expediency; moreover, RoG frameworks achieve spatial extents commensurate with integrated 1D/2D configurations while streamlining setup complexity. Building on solver fidelity, balanced watershed decomposition techniques refine sub-catchment delineation, attenuating timing biases and enhancing hydrograph realism. Judicious depression management preserves natural storage, forestalling spurious connectivity and unrealistic flow patterns. Along those lines, terrain-preprocessing advances dovetail with DEM resolution considerations: resolutions of 1–5 m affords optimal hydrographic fidelity (KGE $\approx$ 0.91) balanced against computational demand, whereas coarser grids induce slope smoothing and inundation overestimation.

Urban-scale flood modeling demands precise emulation of the interplay between precipitation, runoff processes, and built infrastructure to accurately forecast inundation patterns. Central to this challenge is the faithful incorporation of buildings, which obstruct flow, curtail infiltration, and reconfigure surface routing, yet are frequently oversimplified or omitted, yielding substantial errors in urban flood extents and depths. To rectify this, a suite of techniques has been advanced: the Building Holes approach excises building footprints from the computational mesh [1], whereas the Building Blocks (BB) and Stubby Building (SB) methods augment the digital elevation model by elevating structure footprints to represent physical obstructions [2]. In contrast, the Building Resistance strategy preserves topography but embeds hydraulic impedance directly into land-cover layers via adjusted Manning's coefficients, thereby simulating urban drag without modifying elevation [3].

Flooding poses a growing threat to urban transportation systems, with the potential to simultaneously disrupt multiple components of a network and severely compromise mobility. Public transport systems are particularly vulnerable, as flooded roads, bus stops, and terminals can lead to widespread service interruptions, passenger delays, and economic losses [4]. Despite this, much of the existing research on transport vulnerability focuses on the removal of individual network components, often overlooking the compounded effects of simultaneous failures—an outcome frequently associated with natural disasters like floods [5]. Understanding how flood events impact transport infrastructure requires spatially informed methods capable of capturing environmental exposure. In this context, integrating Geographic Information Systems (GIS) with flood modeling offers a tangible approach to identify flood-prone links and stops, assess their criticality, and inform contingency planning. This study adopts such an integrated framework to evaluate the flood-related vulnerability of public transport networks, aiming to support the development of resilient and adaptive transit systems in the face of climate-driven hazards. The study site selected is the municipality of Chalandri.

Recent work highlights that transport disruption under extreme events is systemic, with cascading effects from localized infrastructure failures to network-wide accessibility losses [6], implying that resilience assessments should link exposure, operational thresholds, and equity impacts within a single mapping workflow. Coupling hydrodynamic inundation with transit accessibility helps identify stops that cross safety thresholds and quantify the knock-on effects on reachable destinations, providing a concise decision layer for agencies that must design detours and communicate disruptions under uncertainty [7]. Against this backdrop, the present analysis is intentionally parsimonious: it combines standard hydrodynamic practice with operational criteria and pedestrian catchments to support Disaster Risk Reduction (DRR)-aligned contingency planning without requiring additional behavior or traffic-simulation modules [8].

This study quantifies surface public-transport vulnerability to urban flooding by integrating two-dimensional hydrodynamic outputs with GIS overlays of stops, network segments, and pedestrian catchments, and by estimating the population whose practical access is reduced under a conservative operational threshold of 0.3 m water depth. The aim is to assess how inundation depths from a design flood translate into (i) stop-level service vulnerability and (ii) spatially distributed losses of short-horizon transit access for residents, and what are the immediate planning implications under a minimum-information operational framework. Within disaster-risk-reduction practice, the focus is on operationally actionable mapping of vulnerable nodes and corridors and on transparent assumptions rather than on multi-scenario optimization, to enable contingency planning without expanding model scope.

2 Study Area and Data

2.1 Catchment overview

Chalandri is located within the Podoniftis basin, which encompasses 79.7 km² and exhibits a mean slope of 14%. For the present investigation, analysis is confined to a 66.3 km² sub-basin, omitting the diversion of the Penteli branch toward Rafina. Fig. 1 illustrates the delineated study area along with its elevation gradients. The sub-basin is markedly urbanized, with impervious structures occupying approximately 60% of its surface. Its flood regime has been punctuated by significant inundation events, most prominently the catastrophic flood of 21 October 1994, when overtopping of the Podoniftis stream resulted in 17 fatalities and extensive damage to residential, commercial, and infrastructural assets within the municipalities of Nea Ionia, Nea Filadelfia, and Perissos [9]. Chalandri is selected as a representative urban municipality within the Podoniftis basin because its drainage setting, documented flood history, and reliance on surface public transport combine to create operational exposure that is informative for contingency planning under design storms. The choice allows for an assessment of bus and trolleybus vulnerability and access impacts in a dense, transit-served environment where inundation can propagate service effects beyond the flood footprint.

2.2 Hydrologic & hydraulic input data

2.2.1 Design Rainfall

In the canon of hydrological design and flood-risk quantification, design rainfall curves constitute the foundational construct. Adhering to the schema promulgated by the Hellenic Ministry of Energy and Environment's official flood-risk platform (https://floods.ypeka.gr/sdkp-lap/omvries-2round/), the curves employed herein map rainfall intensity x (mm h^{-1}), time scale k (h)—often mis denominated as "duration"—and return period T (years) onto a probabilistic framework that yields locale-specific design events. The underlying stochastic model, as refined by Koutsoyiannis et al. (2024) [10], is rendered as:

$$x = \lambda_* \frac{(T/\beta_*)^{\xi} - 1}{(1 + k/\alpha)^{\eta_*}} \quad (1)$$

where $\alpha = 0.18$ h and $\xi = 0.18$ are invariant nationally, while λ_* (mm h^{-1}), β_* (years), and η_* vary spatially on a 5 km grid. A basin-representative point curve arises from area-weighted averaging of these parameters in a GIS environment, yielding for the study area:

$$x = 67.84\frac{(T/0.020)^{0.18} - 1}{(1 + k/0.18)^{0.72}} \qquad (2)$$

Thereafter, the Alternating Blocks method partitions the 100-year total depth into discrete segments, positioning the maximum increment at mid-duration and symmetrically arranging lesser depths outward. A 12 h hyetograph—exceeding the basin's ≈3 h concentration time—is thus synthesized. To convert point intensities to the spatial averages required for hydrodynamic simulation, an areal reduction factor φ is applied, defined for 1 min–25 d and 1–30 000 km^2:

$$\varphi = \max\left\{1 - \frac{0.048A^{0.36-0.01 lnA}}{k^{0.35}}, 0.25\right\} \qquad (3)$$

where A (km^2) and k (h) are as previously defined.

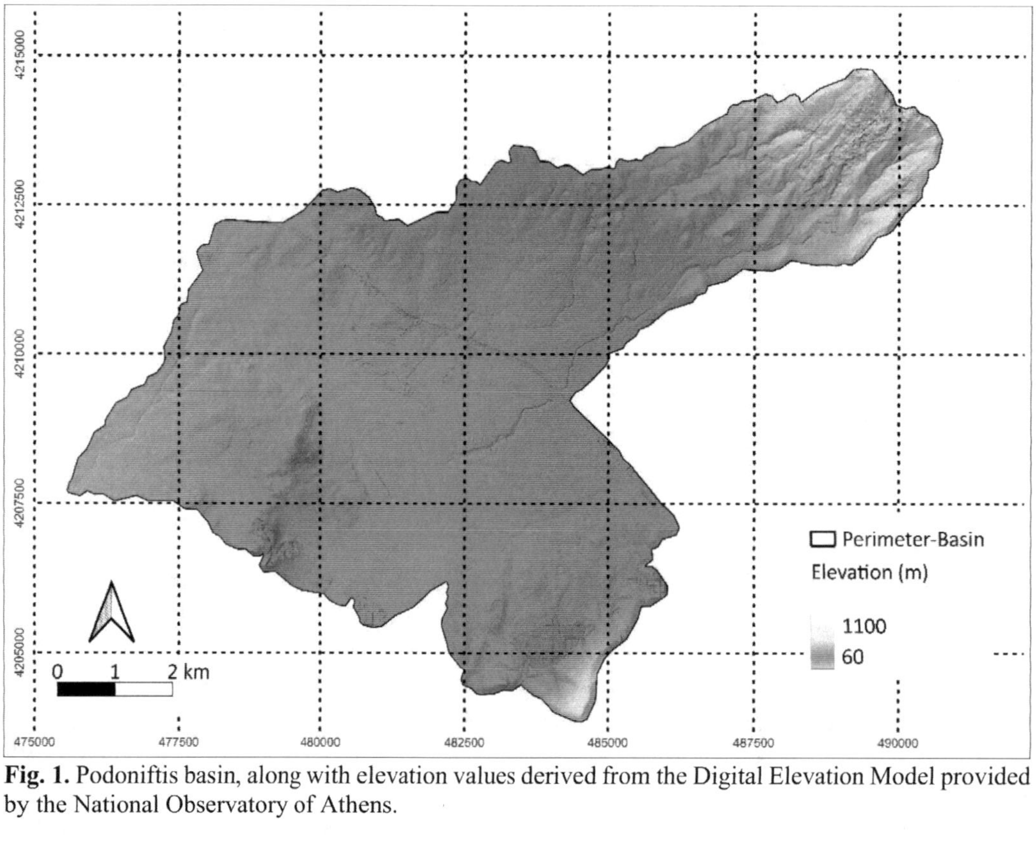

Fig. 1. Podoniftis basin, along with elevation values derived from the Digital Elevation Model provided by the National Observatory of Athens.

2.2.2 Losses

The foregoing protocol defines the gross hyetograph, yet only the soil's infiltration capacity governs excess (active) rainfall. Utilizing the USDA Soil Conservation Service approach, we assume an initial abstraction period τ_0 during which all precipitation h_0 is absorbed ($h_o = 0.1$ S, chosen over the canonical 0.2 S to reflect regional findings of 0.014–0.037. For total depth h, the excess h_e obeys:

$$h_e = \begin{cases} 0 & h \leq 0.1S \\ \dfrac{(h-0.1S)^2}{h+0.9S} & h > 0.1S \end{cases} \quad (4)$$

This formulation applies to both aggregate and sectional depths, such that the runoff deficit asymptotically approaches 1.1 S for large h. In the absence of measured runoff, S (mm) is inferred from the curve-number CN via:

$$S = 254\left(\frac{100}{\mathrm{CN}} - 1\right) \quad (5)$$

where CN (0–100), determined by soil type, land use, and antecedent moisture, is mapped from land-cover typology.

Within the USDA-SCS framework (Eq. 5), spatially distributed curve-number (CN) values were derived from the European Space Agency's World Cover 2020 land-cover product (10 m resolution). Each class is reclassified to its corresponding CN using established lookup tables, thereby generating a CN raster that underpins the computation of spatially varying S and, hence, infiltration losses across the basin.

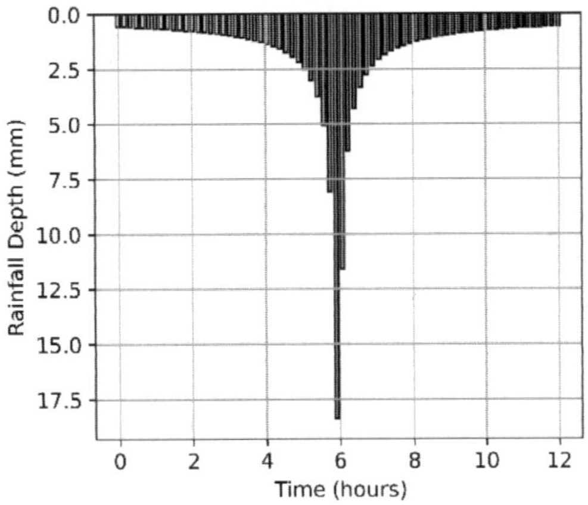

Fig. 2. Hyetograph for T_{100} rainfall intensity over a 12 h period.

2.2.3 Terrain

The digital elevation model (DEM) employed for both hydrologic and hydraulic analyses was sourced from the National Observatory of Athens and systematically modified to integrate the local hydraulic network—explicitly embedding bridge and culvert elevations to ensure accurate water routing beneath such structures. This terrain dataset was further conditioned (depression breaching, channel enforcement) to facilitate contiguous flow paths and is the DEM illustrated in Fig. 1.

2.3 Public-transport network data

Surface public transport data for the Municipality of Chalandri were retrieved and analyzed as part of the study. The area is served by a combination of municipal and Athens Mass Transit System (OASA)-operated services, including both buses and trolleybuses.

Specifically, Chalandri is serviced by 64 distinct OASA-operated bus and trolleybus lines, with a total of 192 public transport stops located within the municipality's administrative boundaries. Although metro stations are also situated within the study area, they were excluded from the analysis. This decision is based on the differing vulnerability of metro infrastructure to flooding events, as underground systems are subject to distinct safety thresholds and operational characteristics compared to surface transport services.

2.4 Socio-demographic data for exposure analysis

Population data for the Municipality of Chalandri were incorporated into the analysis to enhance the spatial accuracy of the study. Specifically, datasets from the most recent Greek Urban Spatial Analysis Units were employed. These units represent small-scale neighbourhood divisions and include comprehensive demographic data for the entire Region of Attica, within which the Municipality of Chalandri is located.

The population figures were subsequently spatially correlated with the 5-minute and 10-minute pedestrian catchment areas of the bus stops affected by potential flooding events. This allowed for an estimation of the population potentially impacted under such scenarios. Furthermore, in cases where the delineated catchment areas extended beyond the administrative boundaries of Chalandri, additional population data from adjacent Urban Spatial Analysis Units were also included, ensuring a more realistic and complete representation of the affected population.

3 Methodology

3.1 Hydrologic–Hydraulic Modelling

A uniform Cartesian mesh of 10 m resolution is used for the hydrologic/hydraulic simulation. The scale is sufficient to resolve building roof geometries while maintaining computational efficiency. It should be noted that HEC-RAS's RAS Mapper pre-processor extrapolates the high-resolution terrain data into each cell's elevation–volume and hydraulic property tables, so it retains sub-cell terrain detail regardless of the nominal grid spacing. In our simulation, we employ HEC-RAS's native RoG capability, which directly ingests spatial rainfall fields and numerically integrates the two-dimensional Saint-Venant (shallow-water) equations.

The conservation of mass is enforced via the continuity equation:

$$\frac{\partial h}{\partial t} + \nabla(\boldsymbol{V}h) = q \tag{6}$$

where h denotes water depth, V the two-dimensional velocity vector, and q external source/sink terms (precipitation, infiltration).

Momentum balance is captured by:

$$\frac{\partial V}{\partial t} + (\boldsymbol{V}\nabla)V = -g\nabla(h+z) + \mathrm{v_t}\nabla^2 V - \mathrm{c_f}V \tag{7}$$

where g is gravitational acceleration, z the bed elevation, Vt the horizontal eddy-viscosity coefficient, and cf a friction term reflecting surface roughness via locally assigned Manning coefficients.

Neglecting viscous diffusion and inertial transport yields the diffuse-wave approximation:

$$\frac{\partial V}{\partial t} + g\nabla(h+z) = gS_f \tag{8}$$

where Sf is the friction slope as computed from Manning's equation.

All simulations observe the Courant–Friedrichs–Lewy criterion to ensure numerical stability by constraining the time step relative to grid spacing and maximum flow velocity.

3.2 Building Representation (Stubby Building – SB)

For this study's urban-obstruction representation, building footprints were uniformly elevated by 0.5 m using the SB variant of the BB methodology to emulate the ingress delay imparted by raised entrances—commonly exceeding 0.3 m—above the ambient terrain [2]. Polygons delineating each structure were "burned" into the original DEM, thereby producing a modified surface that embeds the 0.5 m structural offset. The resultant elevation model, illustrating the SB modifications, is presented in Fig. 5.

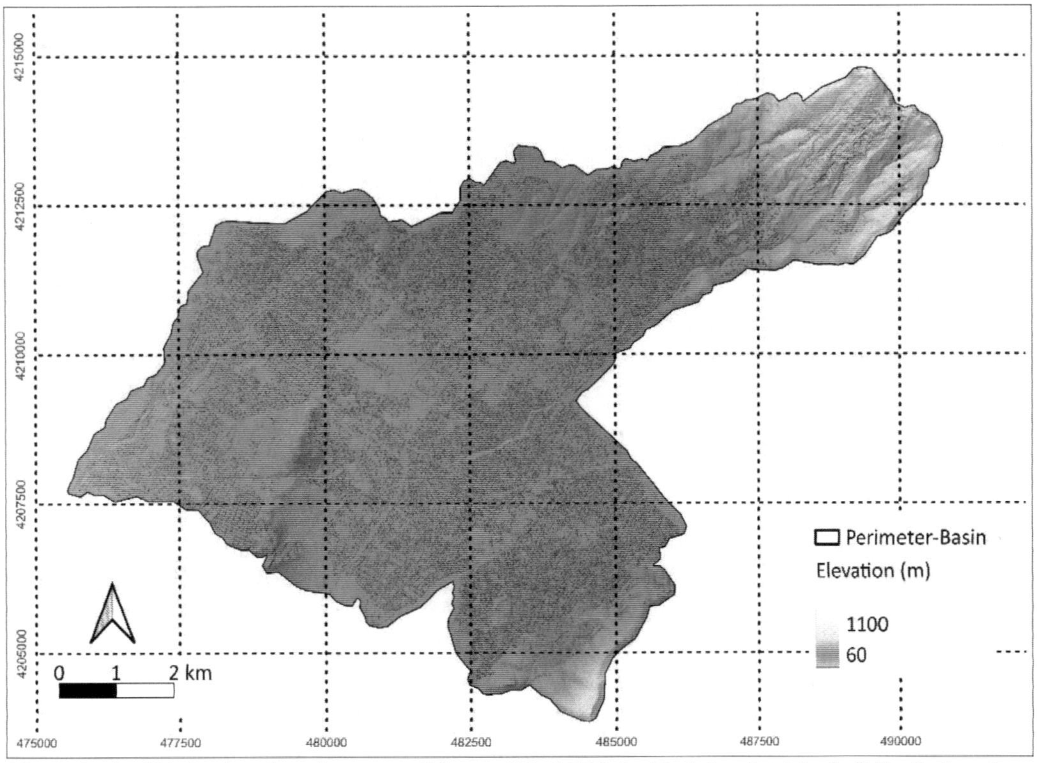

Fig. 3. Podoniftis basin, elevation map of the modified DEM with 0.5 m "stubby-building" elevations burned into building-footprint polygons.

3.3 Model validation and assumptions

Event-level service logs or remote-sensing-derived flood traces suitable for explicit calibration against past transport disruptions were not available for the study area; therefore, validation follows a best-practice, assumptions-transparent approach. Manning's n is assigned from ESA WorldCover land-cover classes [11], and curve numbers derive from established SCS lookups, with initial abstraction adjusted to regional ranges, consistent with widely adopted hydrodynamic workflows for urban basins. HEC-RAS 2D RoG is configured to satisfy Courant criteria with a 10 m mesh whose cell hydraulics are pre-processed from higher-resolution terrain, yielding inundation patterns consistent with known channel confinement and reported urban exceedances during design events in Podoniftis.

3.4 Critical-Infrastructure Overlay & Network Analysis

To evaluate which stops are affected, it is necessary to consider flood depth thresholds beyond which operations are unsafe or impractical. The safety of vehicles in floodwaters has been extensively studied. For light passenger vehicles, stability is typically compromised at water depths of around 0.3 meters. This threshold is based on multiple experimental and simulation-based studies and is often used to designate road closure conditions [12].

Even though bus operations exhibit a higher tolerance to floodwaters compared to passenger cars, hazardous conditions for buses also begin at water levels of approximately 12 inches (0.305 meters) at the stairwell level of the vehicle [13]. Typically, beyond this threshold, service modifications, including route suspensions or detours, are mandated by transportation agencies, with corresponding areas deemed unsafe for bus transit [14]. The 0.3 m cutoff is retained as a conservative operational screen because it aligns with stability limits for light vehicles and with agency practice for bus detours when water approaches the stairwell entry, while avoiding over-interpretation of vehicle-specific tolerances that vary by model and loading. This choice standardizes vulnerability mapping across stops and corridors and is not intended to substitute for agency-specific safety directives when detailed fleet and micro-siting information are available.

3.5 Accessibility & Population-Exposure Metrics

To operationalize these insights, a GIS-based flood impact analysis can overlay flood depth rasters with bus stop coordinates. Any stop located in an area where modelled or observed flood depth exceeds the 0.3-meter threshold can be flagged as potentially inaccessible or hazardous.

To assess the impact of flooding on accessibility, we complement the network vulnerability analysis with a population exposure assessment based on pedestrian accessibility. Specifically, we generate 5- and 10-minute walking catchments from each bus stop using network-based isochrone modelling. These walking thresholds represent realistic access distances for transit users under normal conditions. By overlaying these isochrones with flood extent maps, we identify which areas—and by extension, which populations—lose access to the transit network during flood events. Population data within the affected walking zones are then extracted to estimate the number of residents experiencing reduced or no access to public transport, thereby linking infrastructure vulnerability to social impact.

4 Results

4.1 Hydraulic Outputs

The inundation footprint extends across a vast portion of the Podoniftis basin, with a strong majority of cells exhibiting water depths at or below 0.3 m (Fig. 4). Obviously, pockets where depths exceed 0.3 m align closely with the primary stream channel.

However, exceedances of the 0.3 m threshold also manifest within the urban fabric. This applies for most coordinate quadrants that contain the basin. As an example, the reader may observe depths surpassing 0.3 m in the quadrant bounded by 482 500–485 000 E and 4 205 000–4 207 500 N and the quadrant from 477 500–480 000 E and 4 210 000–4 212 500 N.

 https://doi.org/10.1051/e3sconf/202565103003

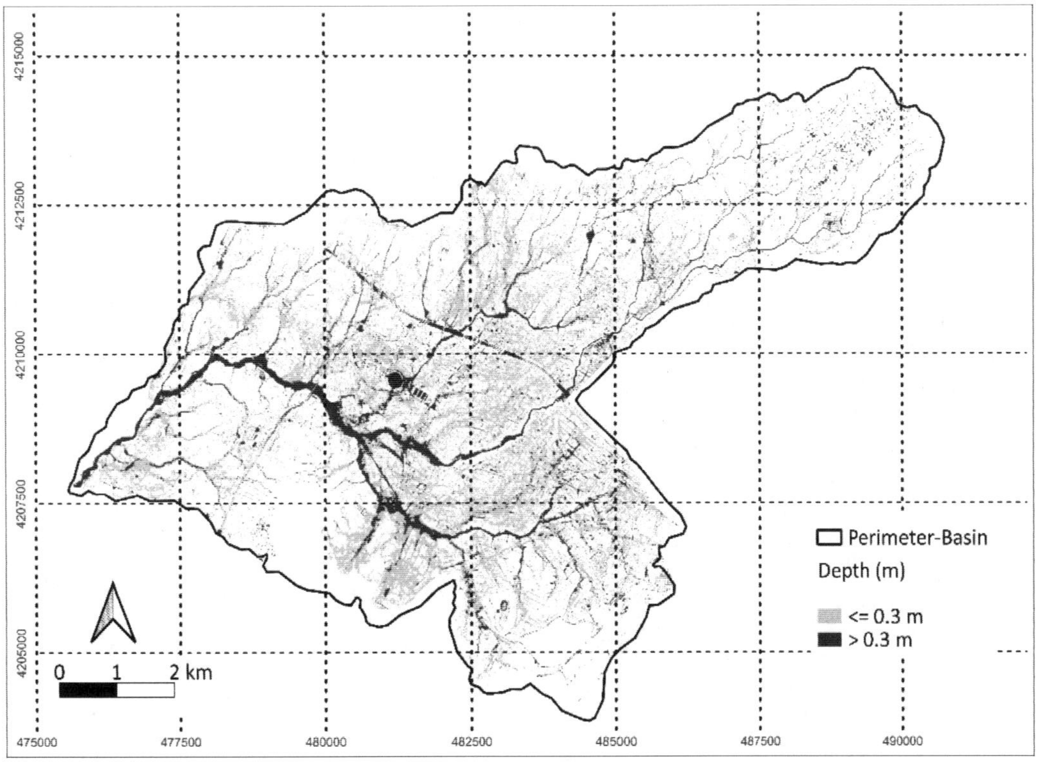

Fig. 4. Podoniftis basin inundation depths: areas ≤ 0.3 m in light blue and > 0.3 m in dark blue.

4.2 Disruption of Public-Transport Infrastructure

The analysis identified a total of 21 bus stops within the Municipality of Chalandri as vulnerable to severe flooding, based on the previously defined safety thresholds. Despite the limited number of affected stops, 62 out of the 64 OASA-operated bus and trolleybus lines serving the area were found to be impacted by potential flood-related disruptions. Fig. 5 illustrates the spatial distribution of the vulnerable bus stops, along with the corresponding bus and trolleybus lines projected to be affected (indicated in red) under severe flooding conditions.

While the disruption of a specific line segment does not necessarily result in the suspension of the entire route, it nonetheless poses significant operational challenges. Affected segments may lead to delays, partial route service, or increased headways, thereby reducing the reliability and efficiency of the public transport system. As such, the findings underscore the critical need for pre-defined, systematic rerouting strategies and contingency plans that can be promptly implemented to maintain network connectivity and ensure service continuity during flood events. These strategies should take into account the spatial distribution of affected stops, alternative road infrastructure, and passenger demand patterns, in order to optimize both operational feasibility and user accessibility under adverse conditions.

4.3 Accessibility & Population Impact

The results from the accessibility analysis are shown in Fig. 6 and Fig. 7, assuming 5-minute and 10-minute walking thresholds for bus stops.

Specifically, if all of the above bus stops were simultaneously disrupted by extreme flooding, a larger share of residents would be impacted. In the case of 5-minute catchment areas, 35,803 residents would be affected, while in the case of 10-minute catchment areas a total of 104,555 would be affected by such events, representing a 192% increase compared with the 5-minute catchment scenario.

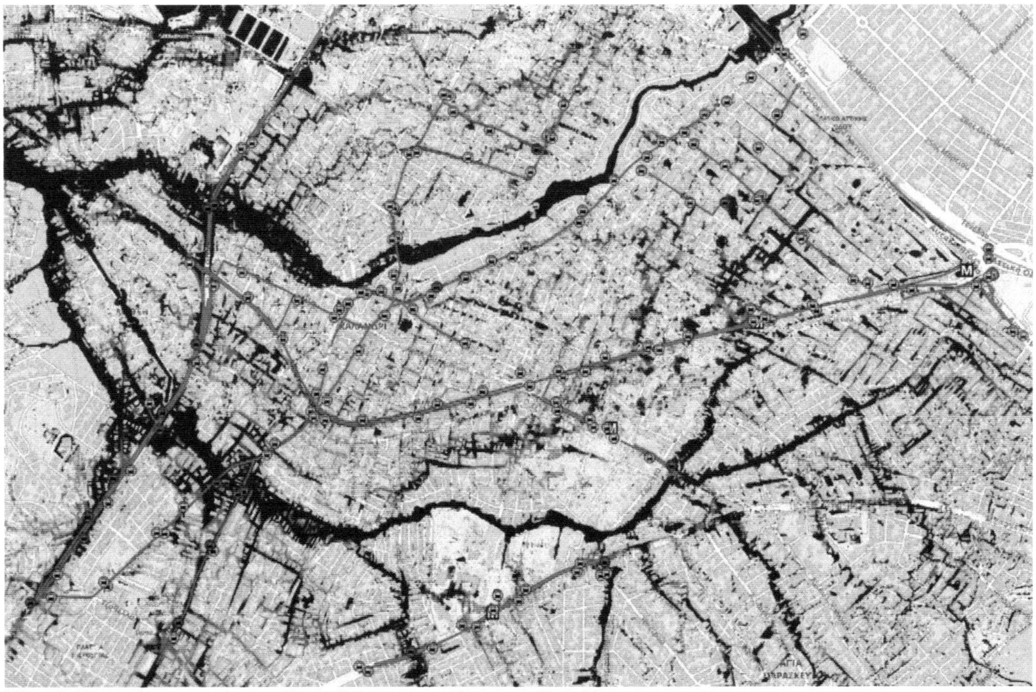

Fig. 5. Transport services affected in cases of severe flooding.

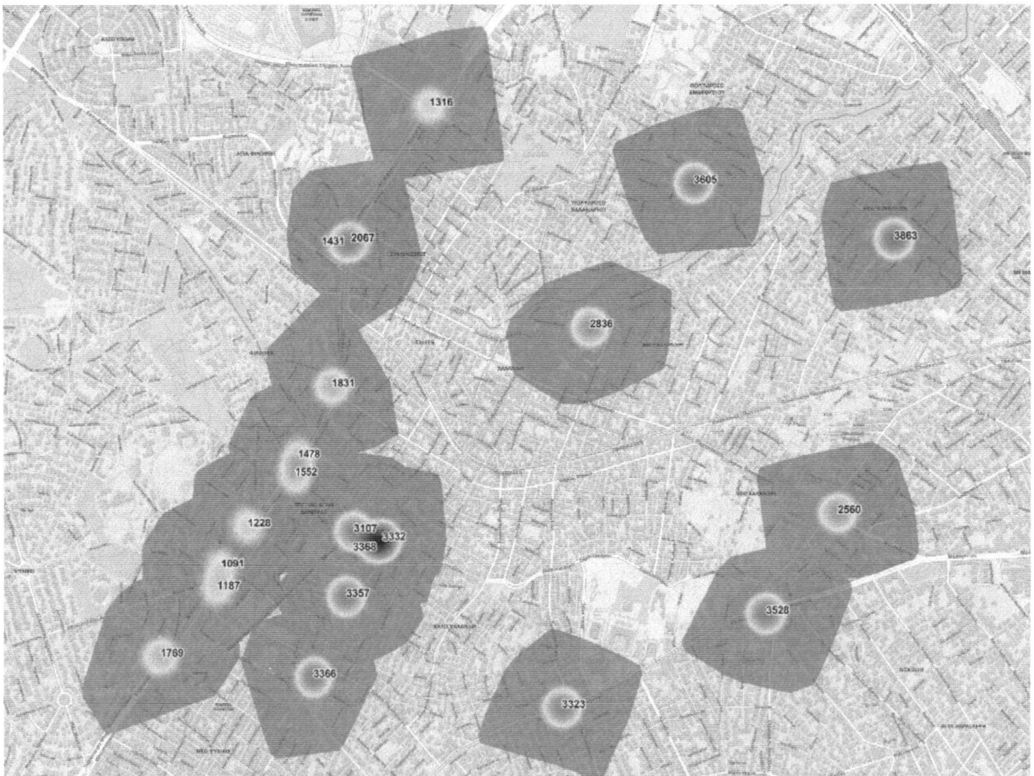

Fig. 6. Affected population per bus stop under 5-min walking threshold.

 https://doi.org/10.1051/e3sconf/202565103003

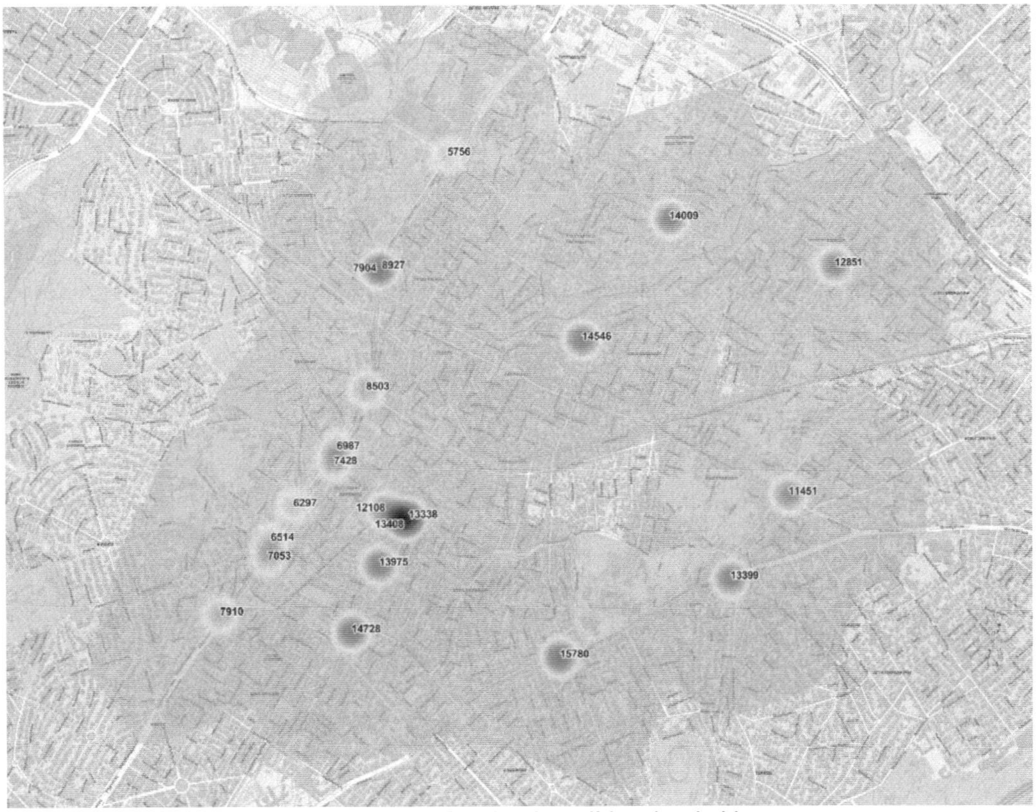

Fig. 7. Affected population per bus stop under 10-min walking threshold.

5 Discussion and Conclusions

Incorporating flood vulnerability into transport planning requires adaptive operations planning, fast information dissemination, and strategic infrastructure investment. Authorities can use transport modelling in conjunction with hydrodynamic models and historical flood records to assess network-wide vulnerabilities and prepare vulnerability mitigation strategies. The proposed approach enhances disaster preparedness in the context of vulnerability assessment of transit systems and critical component identification [5]. The framework allows operators to identify stops requiring elevation or relocation, design detour plans for high-risk zones and quantify potential service disruptions for network resilience analysis. Findings provide transferable insights for metropolitan areas in Greece and comparable international contexts where urban flooding threatens public transport continuity.

This study's limitations include uncertainties arising from the digital elevation model resolution, the completeness of the input dataset, and the uniform application of the 0.3-meter flood threshold, regardless of vehicle variance. Such factors may influence the precise extent of vulnerability estimates and showcase the need for cautious interpretation and further detailed validation with empirical flood and transit disruption data. Flood impact assessments focused on directly affected stops, yet the broader transit system experiences behavioural and operational responses, including altered passenger demand and route adjustments. While metro systems generally exhibit greater resilience due to underground infrastructure, their distinct operational profiles justify their exclusion from this surface-focused analysis. Disaster conditions typically reduce passenger demand and alter operational behaviour beyond directly inundated links; demand depression, temporary headway adjustments, and localized route short-turning can emerge even where stops remain nominally dry. These system-wide effects are not modelled here; instead, they are acknowledged as a distinct layer that agencies can evaluate using observed operations data once available, complementing the present exposure-based screening.

Future work should integrate multimodal resilience evaluations to comprehensively address urban mobility under flood scenarios. Still, the mapping framework is directly transferable to urban catchments with comparable morphology and surface-transport dependence, enabling operators to pre-position detours, stage stop elevations/relocations, and communicate access risk under design storms. By retaining conservative operational thresholds and standard hydrodynamic practice, the outputs are suited as a first-order DRR layer in data-limited environments and can be refined as event-level observations become available.

References

1. J.E. Schubert and B.F. Sanders, Building treatments for urban flood inundation models and implications for predictive skill and modeling efficiency. Adv. Water Resour. **41**, 49–64 (2012). https://doi.org/10.1016/J.ADVWATRES.2012.02.012
2. C. Iliadis, V. Glenis and C. Kilsby, Representing buildings and urban features in hydrodynamic flood models. J. Flood Risk Manag. **17**, e12950 (2024). https://doi.org/10.1111/jfr3.12950
3. R. Beretta, G. Ravazzani, C. Maiorano and M. Mancini, Simulating the influence of buildings on flood inundation in urban areas. Geosciences **8**, 77 (2018). https://doi.org/10.3390/GEOSCIENCES8020077
4. C. Iliopoulou, M.A. Makridis and A. Kouvelas, Improving transit network resilience against disruptions through path redundancy. Socioecon. Plann. Sci. **100**, 102228 (2025). https://doi.org/10.1016/J.SEPS.2025.102228
5. C. Iliopoulou and M.A. Makridis, Critical multi-link disruption identification for public transport networks: A multi-objective optimization framework. Physica A **626**, 129100 (2023). https://doi.org/10.1016/J.PHYSA.2023.129100
6. M. Vieira Passos, K. Barquet, J.C. Kan, G. Destouni and Z. Kalantari, Hydrometeorological resilience assessment of interconnected critical infrastructures. Sustain. Resil. Infrastruct. **10**(3), 267–283 (2025). https://doi.org/10.1080/23789689.2024.2446124
7. S. Gössling, C. Neger, R. Steiger et al., Weather, climate change, and transport: A review. Nat. Hazards **118**, 1341–1360 (2023). https://doi.org/10.1007/s11069-023-06054-2
8. H. Apel, S. Vorogushyn and B. Merz, Brief communication: Impact forecasting could substantially improve the emergency management of deadly floods: Case study July 2021 floods in Germany. Nat. Hazards Earth Syst. Sci. **22**(9), 3005–3014 (2022). https://doi.org/10.5194/nhess-22-3005-2022
9. Ethnos, The Podoniftis flood: 30 years ago, 17 dead and images from Greece's future [in Greek]. Ethnos (2023). https://www.ethnos.gr/todayinhistory/article/284995/hplhmmyratoypodonifthprinapo30xronia17nekroikaieikonesapotomellonthselladas
10. D. Koutsoyiannis, T. Iliopoulou, A. Koukouvinos and N. Malamos, A stochastic framework for rainfall intensity–time scale–return period relationships. Part I: Theory and estimation strategies. Hydrol. Sci. J. **69**, 1082–1091 (2024). https://doi.org/10.1080/02626667.2024.2345813

11. L. Cea, M. Álvarez and J. Puertas, Estimation of flood-exposed population in data-scarce regions combining satellite imagery and high-resolution hydrological–hydraulic modelling: A case study in the Licungo basin (Mozambique). J. Hydrol.: Reg. Stud. **44**, 101247 (2022). https://doi.org/10.1016/j.ejrh.2022.101247
12. E. Martínez-Gomariz, M. Gómez, B. Russo and S. Djordjević, A new experiments-based methodology to define the stability threshold for any vehicle exposed to flooding. Urban Water J. **14**, 930–939 (2017). https://doi.org/10.1080/1573062X.2017.1301501
13. Hampton Roads Transit, Inclement weather plan. Hampton Roads Transit (2024). https://gohrt.com/customer-service/inclement-weather-plan/
14. Metropolitan Transportation Authority, Weather service guide: Storm, flood and hurricane service. MTA (2024). https://www.mta.info/guides/weather-service-guide/storm-flood-hurricane-service

Impact of resettlement process on spatial and social sustainability in Plosokerep, Yogyakarta, Indonesia

Ardhya Nareswari[1,*] and *Arina* Fatchurrochmah[1]

[1]Department of Architecture and Planning, Universitas Gadjah Mada, Yogyakarta, Indonesia

Abstract. The 2010 eruption of Merapi volcano in Yogyakarta, Indonesia, affected thousands houses within hazard areas and required inhabitant to resettle on safer village owned land. The concept of resettlement extended beyond housing replacement to relocating entire neighbourhoods to preserve their social environment. After years of inhabitation, relocation site of Plosokerep (Huntap Plosokerep) exhibits low occupancy, minimal spatial development, and weakened social cohesionin contrast with other relocation sites. This study examined the relationship between the relocation process and the social and spatial development of Huntap Plosokerep. The research method employed a mixed-method approach by integrating physical observations, in-depth interviews, and thematic analysis. Findings show that gradual relocation, combined with low community participation and cultural gap between rural context and urban layouts, has led to changes in neighbourhood spatial configuration, hindered habitation, and undermined social activities development in the new area, ultimately affecting the quality of housing and infrastructure.

1 Introduction

The 2010 Merapi Eruption claimed 346 lives and temporarily displaced 151,336 people [1]. Aside from destruction, this disaster also altered the hazard risk map due to increased vulnerability of southeastern settlements prior to the collapse of a lava dome on the southern slope of Mount Merapi, where a total of 3.726 houses were identified at risk. By 2012, 2,608 of these houses had been relocated, primarily through the REKOMPAK (Community-Based Settlement Rehabilitation and Reconstruction) program, which collaborated with the community, NGOs, private sectors, academia, the government, and independent actors [2].

The program aimed to relocate any entire *padukuhan* (hamlet) affected by the eruption to communal permanent housing sites (*hunian tetap/huntap*) situated on village-owned land (*Tanah Kas Desa/TKD*) [2]. Each household received an earthquake-resistant core house, supported by infrastructure for social, economic, and utility needs. Most of the houses constructed under the REKOMPAK program were concentrated in 15 *huntap* sites across 5 villages within the Cangkringan Sub-District, Sleman Regency, with 12 of these sites developed collectively in communal settings [2].

* Corresponding author: nares@ugm.ac.id

However, initial observations of *huntap* sites revealed notable differences, particularly between *Huntap* Plosokerep and other sites. While most *huntap* sites are largely occupied, only about half of the houses in Huntap Plosokerep remain inhabited by their original intended residents (see Table 1). Besides being less populated, spatially, the site exhibits minimal physical development, and socially, there is a notable lack of communal activity. A key distinction observed early in the relocation process was the gradual and prolonged construction timeline in Huntap Plosokerep, which remains unfinished (see Fig 1).

Table 1. Collective huntap sites occupation rate.

Village	No.	*Huntap*	Sum of houses	Construction year	Not occupied by original owner (%)
Umbulharjo	1	Karangkendal	81	2011	<10%
	2	Plosokerep	115	2012, 2014, 2015, 2018, 2020, 2023, 2024 (not yet finished)	50%
Kepuharjo	3	Batur	252	2011, 2012, 2013, 2020	6,7%
	4	Pagerjurang	272	2011, 2012, 2013, 2020	4,4%
	5	Bulaksusukan	19	2012	5%
Wukirsari	6	Gondang 2	89	2012	N/A
	7	Gondang 3	36	2012	N/A
	8	Dongkelsari	147	2012	8%
Glagaharjo	9	Gading	62	2013	N/A
	10	Jetisumur	54	N/A	N/A
	11	Banjarsari	177	2013	N/A
Argomulyo	12	Kuwang	151	2012	10%
	13	Randusari	109	2012	N/A

Source: Author based on informants (2025)

Fig 1. Left: development in Huntap Kuwang; Right: abandoned houses in Huntap Plosokerep.
Source: Author (2025)

This indicates a relation between the prolonged relocation process and the social and spatial sustainability of Huntap Plosokerep. Accordingly, this study aims to delve into the connection between the relocation process and the sustainability of the relocation sites, both socially and spatially. It is expected that this research will offer further insights into how resettlement strategies influence the sustainability of the new settlement.

2 Theoretical Background

2.1 Relocation processes and inhabitation in post-disaster resettlement

Several studies have empirically reported similar patterns of weakened social cohesion and even abandonment of homes, which reveals several factors contributing to low occupancy levels, stemming from both the reconstruction process and its output.

- **Relocation timeline and governance gaps**

  A study conducted in Zimbabwe [4] highlighted the impact of delays in the construction process on inhabitation. When a project is not completed within the designated timeframe, it becomes difficult for the community to participate effectively, particularly in terms of inhabiting the housing. The study also identifies the major reason for the failure in the construction timeline as a shortage of adequate resources, which in turn is attributed to poor governance in the disaster management program.

  The United Nations Office for Disaster Risk Reduction (UNDRR) emphasizes the importance of effective implementation and equitable enforcement of reconstruction mechanisms, supported by strong leadership and good governance [5].

- **Community participation**

  Several studies emphasize the significance of low community participation in the reconstruction process as one of key factors affecting inhabitation [4–6–7]. A study from Nepal reveals the lack of community participation in house design [7], while in the post-disaster relocation at Tsholotsho District, Zimbabwe, residents were not consulted on numerous issues, such as their willingness to be relocated, relocation site selection, house design, preferred house structure, and livelihood restoration [4–6], all of which negatively impacted the inhabitation process.

  The Global Facility for Disaster Reduction and Recovery promote the community participation and inclusion as core principles in post-disaster reconstruction [8]. It outlines five core building blocks of participation: inform, consult, involve, collaborate, and empower [8].

- **Livelihood access and economic recovery**

  Several studies also emphasize limited access of livelihoods as a key factor affecting inhabitation [4–6]. Livelihood and economic recovery are among the core principles of settlement recovery, playing a crucial role in building community resilience [9] such as in the post-disaster reconstruction sites.

  The United Nations Development Programme (UNDP) outlines six approaches to support local economic recovery [9]. Given that most livelihoods in the original location fell within the informal sector, the fourth approach, which is supporting the informal economy, should be examined in depth, particularly in terms of how the specific needs of the informal economy were addressed during the recovery process [9].

- **Physical output of reconstruction**

  A few studies also highlighted the physical output of the relocation sites such as unmet housing needs [7] and inadequate infrastructure [6] as key factors affecting inhabitation levels. These shortcomings were linked to limited community participation in planning process, resulting in output that failed to address the housing and livelihood needs of the relocated communities. These physical factors contributed to the low levels of inhabitation observed in both studies.

2.2 Social sustainability

The general overview of social development in Huntap Plosokerep suggests a case of low inhabitation and a lack of communal activity. This reflects a situation in which the use of housing and infrastructure remains minimal, indicating that these assets may not align with the residents' expectations. Such conditions point to a case of social depreciation, which leads to a concurrent shift in attitudes toward the acceptability and value of certain uses within the settlement, which are the result of changes in social values in the community [10].

These dynamics suggest that the current trajectory of social development is insufficient to sustain key social assets, such as community cohesion, shared meaning, and collective use of infrastructure. This form of unsustainable social development undermines what is commonly referred to as social assets, defined as social fabric and network through which people interact, build relationships, and share resources on a daily life [11]. In many cases, relocating community to a new site often results in unsustainable social development. Preserving the original village layout and ensuring the availability of public spaces in the new location are essential to support social sustainability.

Therefore, to further investigate this sustainability of social assets in Huntap Plosokerep, it is necessary to examine the prevailing social values, patterns of housing use (i.e., inhabitation), and infrastructure use, as reflected in social-economic activities.

2.3 Spatial sustainability

The general overview of spatial development in Huntap Plosokerep suggests a case of minimal development and signs of settlement abandonment. These conditions indicate functional depreciation, which refers to the declining capacity of physical assets to serve the needs of the community [10]. Functional depreciation itself is manifested through intolerable structural conditions and under maintenance of the settlement [10].

Physical assets refer to the basic infrastructure and goods needed to support livelihoods, such as housing, infrastructure, or transportation [11]. While this definition emphasizes availability, it often overlooks the adequacy and capacity of such assets. In the context of housing or infrastructure, capacity is critical, as communities tend to develop demographically with evolving needs. Demographic development in the future requires physical assets that offer sufficient space, not only to meet the present needs, but also to accommodate future expansion. Similarly, evolving needs require the physical assets not only to function effectively for current needs, but also to possess the flexibility to be expanded or transformed in response to future challenges. These considerations suggest that space and function, as the core of the spatial aspect of the assets, are essential to ensuring the long-term spatial sustainability.

Therefore, to further investigate the sustainability of the spatial assets in Huntap Plosokerep, it is necessary to assess the spatial sustainability of these assets, both from the negative perspective, such as asset degradation (i.e. housing and infrastructure), and conversely, from the positive perspective, such as spatial capacity enhancement or expansion within the settlement.

2.4 Regulatory framework in the Merapi relocation process

The relocation following the Merapi eruption was governed by a set of regulations issued across multiple administrative levels. At the national level, this included regulations regarding relocation mechanism (National Disaster Management Agency/ *Badan Nasional Penanggulangan Bencana/ BNPB*) as well as building requirements (Ministry of Public Works). At the provincial level, regulation addressed the formation of implementing teams, financial management, and land governance. Meanwhile, at the regency level, policies covered the mechanisms for permanent housing, the selection of relocation sites and permanent housing recipients.

The designation of permanent housing recipients was based on whether an area was located within the designated disaster-prone zone, with some informants stating that all residents of affected hamlets (*padukuhan*) within this zone were automatically deemed eligible. Entire *padukuhan* was relocated to a new site, a process locally referred to as *bedol desa*, which denotes a full-scale community relocation.

At the time, land governance allowed *BPBD* to purchase *TKD* as primary source of land for resettlement [2]. However, a policy shift occurred in 2017 regarding land governance, particularly concerning Sultan Ground – state land under the authority of the Sultanate of Yogyakarta, including *TKD* – which prohibited the transfer of ownership beyond the status of *hak pakai* (right of use) [12].

In this context, the dynamics of policies across administrative levels become crucial in shaping both the relocation process and settlement development in Huntap Plosokerep, particularly by determining how land is acquired and how the settlement is constructed, distributed, and evolves.

3 Material and methods

3.1 Study area

This study focuses on a post-disaster community that became divided into two locations, both situated in rural areas: Padukuhan Pangukrejo, the original hamlet that was designated as a hazard zone after eruption, and Huntap Plosokerep, the intended relocation site established after the 2010 Mount Merapi eruption in a safer area. This division resulted from the reluctance of some residents to move to the relocation area. These two locations are approximately 5 km apart, with Padukuhan Pangukrejo situated closer to Mount Merapi at a higher elevation, as shown in Fig 2. Padukuhan Pangukrejo, consists of three *dusun* (sub-hamlets): Sidorejo, Pangukrejo, and Ngrakah. Examining the social and spatial development in each area is essential for understanding the effects of the relocation process that divides the community into these two locations.

Residents of original hamlets – including Padukuhan Pangukrejo – were predominantly engaged in agrarian and livestock-based livelihoods. In their original hamlet, they typically owned extensive plots of land for housing, with livestock pens located in their backyard, as well as additional land for farming or gardening. In contrast, the land provided in *huntap* – including Huntap Plosokerep – is considerably smaller than their original land and allocated solely for housing, with buildings positioned closer to one another than in their original hamlet, thereby mimicking urban layouts. Meanwhile, land designated for agricultural activities remains in their original location, and livestock pens are now provided as communal facilities.

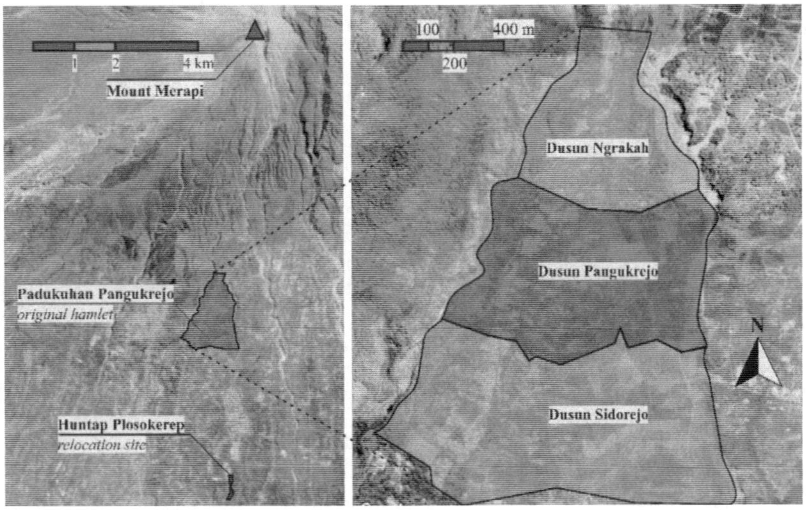

Fig 2. Location map of Padukuhan Pangukrejo (original hamlet) and Huntap Plosokerep (relocation site) on the slopes of Mount Merapi.

Source: Author, modified from Google Maps, *dusun* boundaries were reconstructed based on information provided by the hamlet leader (2025)

3.2 Data collection

This study draws on mixed-method fieldwork conducted in March 2025, 13 years after first resettlement phase, supplemented by key informant interviews carried out through May 2025. The research aimed to explore the relocation process as well as the social and spatial development of Huntap Plosokerep.

In-depth interviews were conducted using a semi-structured method with residents (n=4), a renter (n=1), and both formal and informal local leaders (n=2). Residents were selected from each construction stage, ensuring diversity in gender, life stage, occupation, and house ownership status to obtain comprehensive insights from multiple perspectives. Resident informants were asked a range of questions, primarily focusing on individuals and community social values, socio-economic activities, and the spatial development of their homes and the settlement. Social values were explored through residents' perceptions of their settlement and their sense of community within it. Local leaders were mainly asked about the reconstruction and recovery processes in Huntap Plosokerep, as well as the social values upheld in both the original hamlet and the *huntap*. Interviews on local leader from other huntap were also conducted (n=3) to assess the reconstruction and recovery process in a comparative context. Additionally, some informants (n=4) were interviewed informally in an unstructured manner during the transect walks to complete the observational data within the housing area. All interviews were anonymized and manually transcribed, with attention given to conversational nuances in order to capture implicit meanings and better understand the underlying social values.

3.3 Thematic analysis

The acquired data were organized into three categories: the relocation and recovery process, social development, and spatial development. The analysis began with thematic analysis to identify key patterns within the data [13]. The resulting themes were then examined in relation to one another to explore possible interconnections. This discussion was further enriched by referencing existing literature on similar themes or phenomena, enabling

 https://doi.org/10.1051/e3sconf/202565103004

a deeper contextual understanding and positioning the findings within the broader scholarly discourse.

4 Results and Discussion

4.1 Gradual relocation in Huntap Plosokerep

The construction of Huntap Plosokerep was divided into seven stages and remains incomplete, with stages 4 to 7 located in the northern part of the area. These stages refer to the construction phases, as illustrated in Fig 3, with the legend provided in Table 2. Construction began in 2012, continued in 2014 and 2015, and the final phase commenced after 2015 eruption, around 2018.

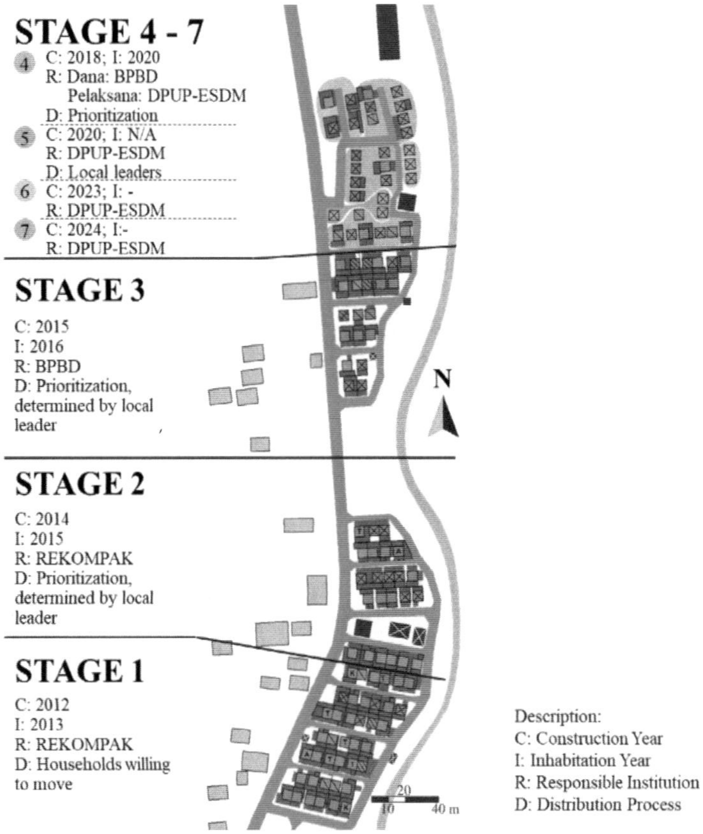

Fig 3. Construction stages of Huntap Plosokerep.
Source: Author, based on informants (2025)

Table 2. Profiles of habitation in Huntap Plosokerep

Symbol	Legends	**Number** (percentage of total number of houses)				
		Total	Stage 1	Stage 2	Stage 3	Stage 4-7
	Total Number of Houses	**115**	**40**	**23**	**21**	**31**
■	Place of Origin: Pangukrejo	**58**	16	16	17	9
■	Place of Origin: Ngrakah	**26**	21	0	0	5
■	Place of Origin: Sidorejo	**24**	3	7	4	10
■	Place of Origin: Manggong	**1**	0	0	0	1
■	The owner has not yet received the land certificate	**6**	0	0	0	6
		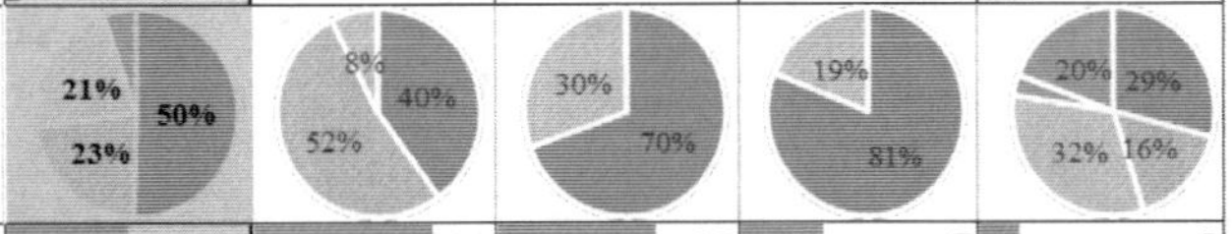				
□	Occupied by the original owner	**57**	30	15	7	5
⧅	Rented out	**11**	4	0	3	4
⧅	Sold out and occupied	**5**	2	0	3	0
⊠	Vacant	**37**	3	8	4	22
⊠	Sold out and vacant	**2**	1	0	1	0
		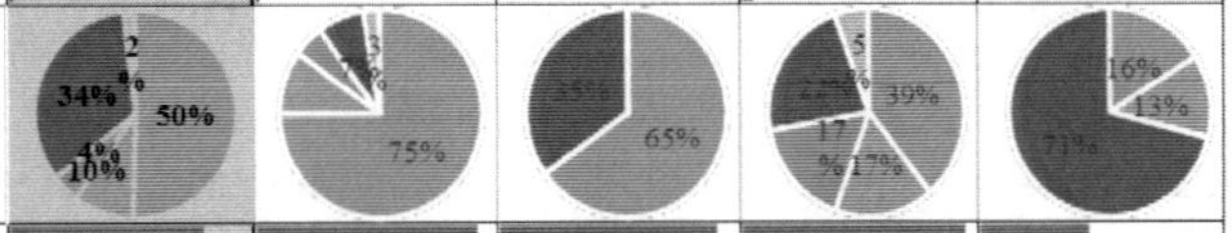				
■	Expansion	**92**	37	21	20	14

Source: Author, based on informants (2025)

A previous study has highlighted the differences between Padukuhan Pangukrejo and other surrounding hamlets, particularly regarding the initial willingness to relocate. Unlike other communities, residents of Padukuhan Pangukrejo were initially reluctant to move [14]. According to the residents, this reluctance stemmed from policy uncertainty. At first, the government informed them that their original land, much of which was relatively large in size, would be exchanged for a *huntap* unit with a land area of approximately 100 m^2, and that their former land would be reclassified as state-owned land (*tanah negara*). However, the policy was later revised, stating that they would retain ownership of their original land under the new designation of disaster-prone area. Due to this miscommunication, many residents feared that accepting a *huntap* unit would mean forfeiting their large plots of land.

As a result, relocation in Huntap Plosokerep was carried out gradually. The first stage of relocation involved resettling families from Ngrakah and Pangukrejo - 35 households who initially agreed to move. This early agreement is likely related to the original locations of their homes, as Ngrakah is the *dusun* closest to Mount Merapi and was among the most severely affected areas, followed by Dusun Pangukrejo, which lies just below Ngrakah in terms of elevation. The extent of destruction reshaped residents' risk perception of the disaster, prompting them to relocate. Another statement indicates that some of the first relocatee households agreed to move because they did not originally own land in their previous hamlet, as they were still living in the same house with their parents. The placement of residents during the first stage was arranged to reflect the spatial layout of their original houses. As a result, rows of houses in the first stages were mostly composed of extended families and neighbours who previously lived within the same RT (neighbourhood unit), preserving pre-existing social bonds.

The second and third construction stages housed the second wave of agreed-upon households (approximately 50 households). The composition of residents in these stages was determined through discussions among those who were planning to relocate. This wave primarily consisted of households from Pangukrejo and Sidorejo, the latter being the lowest lying *dusun* in Padukuhan Pangukrejo.

The last area (upper area of Huntap Plosokerep) was built through another 4 construction stages, 20 houses in 2018, 4 houses in 2020, 3 houses in 2023, and the last 3 houses in 2024.

Observation results in Fig 3 and Table 2 show a high portion of abandoned houses, up to 35% of overall houses in Plosokerep, with less than 50% of them remaining occupied by their original owners, and the rest are rented or sold. This trend is remarkably increasing in the later stages of construction, particularly in the northern area of Huntap Plosokerep. This suggests a relation between gradual relocation and occupancy rate in Huntap Plosokerep.

A previous study highlights a link between relocation timeline and occupancy rate, which stems from the delay of construction [4]. While a construction delay did not happen in Huntap Plosokerep, a gradual and prolonged relocation process did occur. Different from the previous study where the construction delay resulted from a shortage of adequate resources [4], the gradual and prolonged relocation resulted from the reluctance of the residents to move.

Due to initial reluctance toward the relocation process, more than half of the residents of Padukuhan Pangukrejo chose to rebuild their homes in the affected area after the eruption, as shown in Fig 4. Once they were assured that they would not lose ownership of their original land, many of them eventually accepted the *huntap* units, resulting in a situation where most residents came to possess two houses.

The majority of residents were engaged in agrarian and livestock-based livelihoods. Livestock, particularly cattle, were said to thrive better and produce more milk in the cooler temperatures, conditions found in the original hamlet. Additionally, much of the residents' farmland was located in their former settlement area. For practical reasons, many residents are maintaining proximity to their livelihood, where they continued to live and work in their original hamlet, leading to low levels of permanent inhabitation in Huntap Plosokerep.

Fig 4. Left: current housing condition in original hamlet (Padukuhan Pangukrejo). Right: development of original hamlet into jeep tourism.
Source: Author (2025)

This situation reveals a gap in the recovery process, particularly in addressing the specific needs of the informal economy [9], as these economic activities are closely tied to location and are vital for sustaining daily livelihoods. Since these needs were not adequately considered in the relocation strategy, the economic pressures faced by residents contributed to the low occupancy rate observed in the new settlement. This issue was further compounded by construction delays experienced by some beneficiaries due to the gradual and prolonged construction process, all of which collectively led to reduced levels of inhabitation.

Such governance shortcomings resonate with Henri Levebre's concept of *representations of space*, referring to the planned and designed spaces associated with the 'order' (planning rationality) and the production of the space [15]. In this context, the livelihood dimension of relocation planning was conceived in ways that failed to incorporate the economic necessities of the community. As a result, the relocated space was not functionally adequate to support livelihood recovery, and the economic pressure contributed to the low occupancy rate observed in the new settlement.

4.2 Changes in neighbourhood spatial configuration

As explained in Section 2.4, relocations were originally carried out for entire *padukuhan*, with the distribution of huntap units arranged to mimic the spatial layout of the original hamlet. Each house was organized according to its respective dusun, *RW (rukun warga*/ community unit), and *RT (rukun tetangga*/ neighborhood unit) area.

However, based on data from the distribution process across construction stages and the current arrangement of houses on-site as shown in Fig 3, it became evident that this initial distribution mechanism was not implemented in Huntap Plosokerep. The recipients of each construction stage were not selected based on their sub-regional affiliation in the original hamlet, making it impossible to reconstruct the neighbourhood layout accordingly. While an effort was made in the first stage to align the distribution with the original spatial configuration, this approach seems to not be continued in subsequent stages. As a result, the spatial integration of the later stages disrupted the initial pattern. This finding indicates that the overall neighbourhood spatial structure of Huntap Plosokerep has significantly changed from that of the original hamlet.

This change in spatial arrangement is reflected in the residents' social condition, where individuals perceive themselves as coming from different areas, despite having originated from the same *padukuhan*.

This statement implicitly suggests a detachment between individuals and their immediate surroundings, which is spatially manifested in the altered neighbourhood configuration. These conditions point to fragmentation across both social value and spatial organization. Such fragmentation signals emotional detachment and a breakdown of shared meaning. These forms of symbolic disengagement align with Henri Levebre's concept of representational space, which refers to how individuals and communities emotionally and symbolically relate to space, embedding it with meanings, values, and lived experiences [15]. Levebre argues that when these symbolic and affective dimensions are eroded, space ceases to function as a meaningful place, leading to alienation and a breakdown in communal meaning [15]. In this context, the symbolic disengagement emerged in part due to the change in neighbourhood spatial configuration, which disrupted the established community bonds rooted in the original hamlet and leads to fragmented community within Huntap Plosokerep,

4.3 Shifting meaning of the relocation process

Aside from the gradual relocation (see Section 4.1), a critical distinction between Huntap Plosokerep and other *huntap* sites lies in the fact that the hamlet leader, along with several *RT* and *RW* leaders, did not relocate. The hamlet leader's original house, located in the southernmost part of the original hamlet, was not severely affected by the eruption, and he therefore felt no need to move.

In rural communities like Padukuhan Pangukrejo, local leaders are often regarded not only as administrative authorities, but also as role models/unifying figures [3]. During the interviews, several perceptions about the relocation site emerged, including the idea that the *huntap* served as a secondary or emergency shelter. Some even considered it acceptable to rent out the huntap unit to outsiders rather than to live in it themselves. These views were expressed by both residents and local leaders. This indicates a reinterpretation of the relocation program, where it is no longer seen as a site for permanent resettlement.

Given the influential role of local leaders as figures of reference within the rural context, it is likely that this perceptual shift was shaped by their choice and example. The distorted meaning of the relocation sites has eroded the intended symbolic meaning of the settlement as permanent settlements. As discussed in the Section 4.2 this erosion of symbolic meaning causes space to ceases to function as a meaningful place, ultimately leading to alienation and a breakdown in communal meaning [15]. As a result, this evolving interpretation contributed to the low occupancy rates in Huntap Plosokerep, where residents perceived it as acceptable not to inhabit the relocation site themselves.

4.4 Stateless community

In other *huntap*, the administrative territory of the original hamlet was typically divided functionally, where the original hamlet designated as a place for agrarian or other livelihood activities, and the relocation site as place of residence. In such cases, community meetings and hamlet-related activities were held in the relocation site where the residents now lived. However, this was not the case in Huntap Plosokerep, where the administrative territory remained anchored in the original hamlet. Not only the administrative territory, community meetings and other hamlet-related activities continued to be held in the original hamlet.

This condition seems to have relation with earlier findings regarding changes in the neighbourhood spatial configuration in Huntap Plosokerep (see Section 4.2), as well as the fact that several local leaders did not relocate (see Section 4.3). Transferring administrative territory would have been difficult in the absence of spatial continuity and leadership presence in the new site. As a result, Huntap Plosokerep was left without an official administrative territory (A5).

This lack of territorial recognition was reflected residents' expressions about their settlement. This sense of disconnection reflects a deeper issue: although physically resettled, the residents were not institutionally anchored in the new site. The absence of administrative territory transfer contributed to the loss of territorial legitimacy as citizens, where community lacks formal recognition within any official administrative boundary. This disrupted the social bonds between individuals and government, which as explained in Section 4.2, erosion of symbolic meaning causes space to ceases to function as a social space, ultimately leading to alienation and a breakdown in communal meaning [15] that leads to fragmented community within Huntap Plosokerep.

4.5 Selective huntap allocation

Unlike other *huntap* sites where relocation occurred collectively and land was acquired in a single purchase to accommodate all households, in Padukuhan Pangukrejo the relocation was implemented gradually (see Section 4.1), consequently, land acquisition was also carried out in stages depending on the number of households confirmed to relocate.

As the process progressed, a shift in land governance policy as explained in Section 2.4 made further land purchases impossible. This policy change resulted in a critical constraint: the land already acquired was insufficient to accommodate all designated recipients. Although many residents continued to expect that they would eventually receive *huntap* units, especially given their official listing as eligible beneficiaries, the land shortage meant not all could be accommodated.

Faced with this limitation, the allocation process became increasingly selective. Prioritization was conducted by the local leader, who distributed units first to those deemed most in need such as elderly or newly formed families. However, the fifth construction stage was notably allocated to individuals holding local leadership positions, including the hamlet head and RT/RW leaders who had not yet received units. This illustrates how power relation and local authority influenced access to post-disaster resources, resulting in what can be termed a selective *huntap* allocation.

This phase of construction became stigmatized among residents. Testimonies revealed that this area was widely perceived as reserved for the hamlet leader and remaining RT/RW officials, often with negative nuances, and reflect an erosion of communal trust, where unequal access to resources acted as a catalyst that undermined trust toward the relocation process, disrupted the social bonds between individuals and their government. This disrupted the social bonds between individuals and government, which as explained in Section 4.2, led to erosion of symbolic meaning causing space to ceases to functioni as a social space, ultimately leading to alienation and a breakdown in communal meaning [15] that ultimately lead to fragmented community within Huntap Plosokerep.

4.6 Community participation and inhabitation

The gradual relocation construction was carried out by different implementing agencies across each construction stage as shown in Fig 3. These agencies adopted varying approaches to housing development. REKOMPAK, in collaboration with BPBD, involved residents in both the planning and construction processes. Although REKOMPAK program was discontinued after 2014, residents reported that BPBD continued the construction of *huntap* in Stage 3 using the same principles. In Stage 4, BPBD allocated funds for the construction of *huntap*, which was carried out contractually by DPUP-ESDM DIY through a third party. Upon completion, the buildings were handed over to BPBD, who then conducted the key handover to the residents. In the subsequent stages, both the funding and construction were fully managed by DPUP-ESDM DIY. Based on these data, it can be concluded that REKOMPAK and BPBD incorporated community participation in the planning and construction phases, whereas DPUP-ESDM DIY adopted a contractual model that excluded resident participation.

A comparison of stages with and without community participation reveals a notable difference in occupancy rates, as shown in Table 3. Construction stages that incorporated community participation demonstrated significantly higher levels of inhabitation compared to those without such engagement. This pattern suggests that community participation is positively associated with the occupancy rate in relocation process.

Table 3. Program design and inhabitation for each construction stages.

Construction stage (Year)	Implementing agencies	Program design	Number of houses occupied by original owners	
			%	Σ
Stage 1 (2012)	REKOMPAK	involving the community in planning and construction	75%	30/40
Stage 2 (2014)			65%	15/23
Stage 3 (2015)	BPBD		33%	7/21
Stage 4 (2018)	DPUP-ESDM DIY	contractual with a third party, without community involvement	17	4/23
Stage 5 (2020)			25%	1/4
Stage 6 (2023)			0%	0/3
Stage 7 (2024)			0%	0/3

Source: Author (2025)

Community participation has been linked to physical output in other studies [6–7], where poor physical output, due to lack of community participation, contributed to the low levels of housing occupancy. However, this does not appear to be the case in Huntap Plosokerep, where the quality of housing and infrastructure is comparable to that of other *huntap*, built under the same construction standards and with similar infrastructure provisions. In this context, the lack of community participation did not directly affect the physical quality, but instead shaped residents' perceptions, leading some to view the housing as inadequate for living, which in turn negatively affected the occupancy levels.

This situation reveals a gap in the recovery process, particularly in incorporating community participation [8], especially during the later stages of relocation. Community participation is essential to ensure that the planning and construction processes reflect the actual interests and needs of the whole community [8]. Such governance shortcomings resonate with Henri Levebre's concept of *representations of space*, as discussed in Section 4.1, where planned and designed spaces should be conceived according to the 'order' that will shape production of the space [15]. In this context, the absence of participatory strategies in the later relocation stages might result in spatial outcomes that failed to meet residents' preferences and disrupted individuals' sense of ownership toward the place. This, coupled with economic pressures as discussed in Section 4.1, contributed to the low occupancy rate observed in the new settlement. economic pressures faced by residents contributed to the low occupancy rate observed in the new settlement.

4.7 Inhabitation and spatial development

Spatial development was examined from both positive and negative perspectives – positively through housing expansion and negatively through asset degradation. A comparison between occupancy status and house expansion (see Table 4) revealed a clear trend: all houses occupied by owners (either original owners or subsequent buyers) had undergone expansion. On the other side, houses that were rented out or left vacant showed expansion in only 50-72% of cases (see Table 4), where the scale of expansion was generally insignificant. Rented houses were typically expanded to increase rental value and minor expansions observed in some vacant houses may reflect their perceived function as potential future refuge.

Table 4. Comparison of occupancy status and house expansion

Occupancy status	House expansion	
	%	Σ
Occupied by the original owner	100%	57/57
Sold out and occupied	100%	5/5
Rented out	72%	8/11
Sold out and vacant	50%	1/2
Vacant	62%	23/37

Source: Author (2025)

Visual assessment of expanded and occupied houses as shown in Table 5, compared to expanded but unoccupied ones, revealed a clear difference in the level of maintenance. Occupied houses appeared significantly better maintained, while unoccupied units are undermaintained, showed evident signs of neglect, particularly in the plot areas, where overgrowth of weeds has taken over much of the land.

Table 5. Visual assessment on various occupancy status and settlement environment

Construction stage (occupation rate)	Expanded and occupied houses	Expanded and unoccupied houses	Rented houses	Settlement environment
Stage 1 (75%)				

Stage 2 (65%)				
Stage 3 (39%)				
Stage 4-7 (16%)				
Visual assessment	Houses tend to be well-maintained	Houses tend to be undermaintained (overgrowth weeds)	Houses exhibited average level of maintenance	Higher occupancy level areas are better maintained than those with lower occupancy

Source: Author (2025)

Rented houses exhibited an average level of maintenance, where the buildings and land plots were generally kept clean, but they lacked signs of personalization or resident identity. However, maintenance on damaged or broken components was often overlooked. This was evident in the case of one renter. The renter's windows are already damaged, yet neither the renter nor the owner has carried out any repairs. The renter may be reluctant to fix them because it is not their own property, while the owner likely does not perceive the damage as urgent since it does not directly impact their daily life.

A visual assessment on settlement environment (see Table 5), when compared to the occupation rate of each area, revealed a clear distinction: areas with higher occupancy levels appeared significantly better maintained than those with lower occupancy.

The way residents inhabit and use the settlement, resonates with Henri Levebre's concept of *spatial practice*, which concerns the actual pattern of use and inhabitation of space [15]. The low occupancy rate indicate that individual dwellings are materially underutilized. These inhabitation and usage issues of individual dwelling can be traced to earlier governance shortcomings, including the gradual relocation process and inadequate livelihood access (see Section 4.1), both of which failed to align with community needs, thereby disrupting the settlement's intended function, lack of community participation (see Section 4.6) that disrupted the individual's sense of ownership toward the place, and the shifting meaning of the relocation sites as a space not necessarily meant for permanent use (see Section 4.3) that hindered actual use of the houses. Collectively, they contributed to the low inhabitation issues of dwelling in Huntap Plosokerep.

Spatial practice also refers to the competence and performance of the physical manifestation of space [15], which evident in the abandoned housing condition and its surrounding which show clear signs of physical deterioration, rooted from the inhabitation and usage issues of the dwelling and the environment. In contrast, owner-occupied houses exhibit significantly better maintenance and expansion, demonstrating the stability that comes with permanent occupancy in lower-risk location. This contrast illustrates how regular use and long-term commitment to place can reinforce spatial practices that support habitability and social function. Conversely, the deterioration of unoccupied or poorly maintained homes signals a breakdown in the spatial practices necessary to produce a sense of function in the relocated settlements.

4.8 Low occupation in relation with social value

When accumulated, the house occupied by the original owner from Padukuhan Pangukrejo is less than half of the number of houses as calculated in Table 2. The community, which was originally formed based on territorial affiliation, has experienced a fundamental transformation. It no longer consists solely of individuals from the same geographic area, as outsiders have begun occupying the space. Furthermore, the majority of original members are no longer present in the settlement, largely due to their decision not to inhabit the allocated house.

This inhabitation shift, as well as the perception voiced by residents, reflects a weakening of shared locality and the erosion of social cohesion. This disrupted the social bonds between individuals and its community, which as explained in Section 4.2, erosion of symbolic meaning causes space to ceases to function as a social space, ultimately leading to alienation and a breakdown in communal meaning [15] and that leads to fragmented community within Huntap Plosokerep.

4.9 Unsustainable social activity

Padukuhan Pangukrejo is a hamlet known for its rich social and cultural life, including activities such as *kenduri*, *tahlil*, Labuhan Merapi, rotating religious study group (*pengajian gilir*), youth organisations, *arisan* (community savings groups), night patrols (*ronda*), and children's Qur'anic education (*TPA*), and many more. Despite the vibrant social fabric in the original hamlet, there is little to no comparable social activity in Huntap Plosokerep. In the early stages of resettlement, efforts were made to replicate these social activities in Huntap Plosokerep – such as youth organisations, *arisan*, *kenduri*, *tahlil*, and night patrols. However, these initiatives were not sustained and have gradually disappeared from the social life of the resettlement area.

Certain forms of social participation still persist, though in fragmented and limited forms across the *huntap* area. All *huntap* owners, regardless of whether they occupy their units, are included in the residents' group responsible for basic infrastructure maintenance. Among productive-age men, some continue to participate in social activities in their original hamlet, although *kerja bakti* is now mostly limited to their immediate surroundings. Women of productive age are generally not involved anymore in social activities. Meanwhile, elderly residents tend to remain in the huntap and socialize primarily within their peer group. Youth engagement in community organizations has declined, with most no longer active in the original hamlet's youth groups. However, several social routines remain, such as children actively attend religious education (*TPA*) at the local mosque and two housing rows have their own night patrol groups supported by community-built posts.

Despite the presence of some social activities, their fragmented and limited nature reflects the unsustainability of social activity in the relocation site. This condition is likely related to previous findings, where the emergence of fragmented community, the experience of being stateless community, and the rise of social distrust, appear to hinder collective initiative and reduce the likelihood of replicating social activity in the original settlement.

These unsustainable social activities are reflected in the varying levels of infrastructure maintenance. A comparison between the presence of social or economic activities and visual assessments of infrastructure condition (see Table 6) reveals a clear distinction: infrastructure associated with active social or economy use tends to be better maintained, whereas those linked to inactive or discontinued activities exhibit clear signs of under maintenance.

Table 6. Visual assessment on infrastructure usage

	Social/ economy infrastructure	

	Community hall	**Security post**	**Mosque**	**Communal economy infrastructure**	**Visual assessment**
Social and economy activity	Community meetings	Night patrols	TPA/ Friday prayer	Livestock, mushroom farming	
Active					Well-maintained
Potentially disconti-nued					Undermain-tained
Inactive					Undermain-tained

Source: Author (2025)

Similar to the inhabitation of dwelling space discussed in Section 4.7, the unsustainable nature of social activities indicate that communal areas are materially underutilized. Infrastructure underuse, rooted from the shift of social meaning of the settlement as explained in previous Sections (4.2, 4.3, 4.4), that hinder the sustainability of social activities in Huntap Plosokerep.

As discussed in Section 4.7, spatial practice which concerns the actual pattern of use and inhabitation of space also refers to the competence and performance of the physical manifestation of space [15], which is evident in the unused infrastructure condition which show clear signs of physical deterioration, rooted from the usage issues of the infrastructure. Such deterioration signals a breakdown in the spatial practices necessary to produce a sense of function in the relocated settlements.

5 Conclusion

The findings on case of Huntap Plosokerep reveal a connection between relocation process and settlement development, showing that when top-down relocation planning fails to integrate principle of reconstruction and recovery, the result is a settlement that is neither socially nor spatially sustainable.

The governance shortcomings in the relocation process, particularly in the gradual relocation process, inadequate livelihood access, and lack of community participation, as well as the shifting meaning of relocation sites hindered social sustainability in the settlement. Particularly on the use and inhabitation of individual dwelling which manifested spatially in its insignificant expansion and undermaintained condition, showing degradation of spatial sustainability on housing and its surrounding.

Other governance shortcomings in the relocation process, particularly the change in neighbourhood spatial configuration, the absence of administrative territory transfer, and the selective *huntap* allocation, collectively eroded the symbolic meaning of the settlement as social space. These issues, compounded by a leadership gap that distorts the meaning of the relocation sites as permanent settlement and further affected the sustainability of social activities, which in turn lead to underutilization of infrastructure in the settlement which manifested spatially in the under maintenance of infrastructure, showing degradation of spatial sustainability on infrastructure.

Such spatial and social developments are commonly observed in cases of urban blight, where settlements experience degradation in the spatial aspect, particularly from the functional dimensions, which are referred to as functional depreciation, and in the social aspect, referred to as social depreciation [10]. These forms of degradation hinder the sustainability of social and spatial assets, as their development trends are not progressing but instead regressing. This decline threatens the adequacy of these assets not only for future needs but also for meeting present demands.

It is important to acknowledge that these dynamics may be amplified by the specific rural context of the relocated community, where social cohesion has a disproportionate relation with the social sustainability of the settlement.

Based on the case of Huntap Plosokerep, several key policy recommendations can be proposed to support more sustainable post-disaster resettlements:

- Integrate socio-spatial continuity in relocation process.
  Relocation planning should preserve the existing spatial neighbourhood structure to maintain social bonds and symbolic attachments to place. Elements that support symbolic and affective ties to the new place, such as shared spaces or landmarks should be incorporated into the resettlement planning. Legal and administrative recognition of the new settlement, including the transfer of RT/RW structures and integration into local governance systems should be coordinated to sustain the community's sense of belonging and legitimacy.
- Institutionalize leadership inclusion and transfer.
  Relocation of local leadership figures should be made integral to the resettlement process as their presence legitimizes the new settlement as a civic and social entity and strengthens representational space.
- Ensure equitable and transparent allocation.
  Resources allocation must be perceived as fair to avoid social distrust.
- Promote participatory planning and construction.
  Community involvement should not be optional, but a built-in component of post-disaster planning since participatory approaches foster ownership and long-term inhabitation.
- Secure livelihood continuity and accessibility.
  The relocation site must be planned in consideration of the community's economic activities by integrating livelihood assessments in site selection and post-occupancy.

Given the unique traditional and rural context of the original hamlet of Huntap Plosokerep, future studies should include comparative research with other post-disaster resettlement sites in different contexts such as urban or culturally distinct settings, to provide broader insights into the diverse factors influencing successful recovery.

Author Contributions Statement. Ardhya Nareswari contributed to the research design; contacting authorities and stakeholders at the research site; conducting the initial survey; managing the research methods and scope; supervised and coordinate the entire research process; and finalizing the publication article.
Arina Fatchurrochmah contributed to conducting literature study; data collection; data analysis; and writing the initial publication article.

References

1. Badan Penanggulangan Bencana Daerah Kabupaten Sleman, *Laporan Tanggap Darurat Erupsi Merapi 2010* (Sleman, 2011).
2. E. Maly, K. Iuchi, & A. Nareswari, Community-Based Housing Reconstruction and Relocation: REKOMPAK Program after the 2010 Eruption of Mt. Merapi, Indonesia.

地域安全学会論文集, **27** (2015) 205–214. https://doi.org/https://doi.org/10.11314/JISSS.27.205.

3. C. D. A. Depari & M. K. Lindell, "Moving or not?": Factors affecting community responses to environmental disruption. *International Journal of Disaster Risk Reduction*, **95** (2023) 103898. https://doi.org/10.1016/j.ijdrr.2023.103898.
4. N. K. Ngulube, H. Tatano, & S. Samaddar, Factors Impacting Participatory Post-Disaster Relocation and Housing Reconstruction: The Case of Tsholotsho District, Zimbabwe. *International Journal of Disaster Risk Science*, **15** (2024) 58–72. https://doi.org/10.1007/s13753-024-00536-y.
5. United Nations Office for Disaster Risk Reduction (UNDRR), *Words into Action Guidelines: Build Back Better in Recovery, Rehabilitation and Reconstruction* (Geneva, 2017).
6. T. Ncube & U. Murray, Managing two locations: Relocation, resettlement and negotiated immobility of climate-displaced communities in Zimbabwe. *World Development Perspectives*, **38** (2025) 100676. https://doi.org/10.1016/j.wdp.2025.100676.
7. B. Shrestha, S. Uprety, & M. M. Keitsch, A Longitudinal Study of Post-Disaster Resettlement in Nepal: Insights into Building Back Better. *Architecture*, **5** (2025) 17. https://doi.org/10.3390/architecture5010017.
8. Global Facility for Disaster Reduction and Recovery, *Guidelines for Community Participation in Disaster Recovery* (Washington, 2021).
9. United Nations Development Programme (UNDP), *Building resilience through livelihoods and economic recovery* (New York: United Nations Development Programme, 2022).
10. G. E. Breger, The Concept and Causes of Urban Blight. *Land Economics*, **43** (1967) 369–376.
11. A. E. Piggott-McKellar, J. Pearson, K. E. McNamara, & P. D. Nunn, A livelihood analysis of resettlement outcomes: Lessons for climate-induced relocations. *Ambio*, **49** (2020) 1474–1489. https://doi.org/10.1007/s13280-019-01289-5.
12. Provinsi DI Yogyakarta, Peraturan Daerah Istimewa Nomor 1 Tahun 2017 tentang Pengelolaan dan Pemanfaatan Tanah Kasultanan dan Tanah Kadipaten. (2017).
13. V. Braun & V. Clarke, Using thematic analysis in psychology. *Qualitative Research in Psychology*, **3** (2006) 77–101. https://doi.org/10.1191/1478088706qp063oa.
14. B. Ramadhan & G. Bidari Adninda, Community strategies in the development of Merapi permanent settlements. *E3S Web of Conferences*, **340** (2022) 04003. https://doi.org/10.1051/e3sconf/202234004003.
15. H. Lefebvre, *The Production of Space* (Oxford: Blackwell (Wiley-Blackwell), 1991).

Preparedness and strengths in disaster risk management in Lurigancho-Chosica, Lima, Peru: toward resilience of Lima Metropolitan

Jheyder Perez[1,2,*], *Nathaly* Cabrera[2], and *Alejandra* Reza[2]

[1]GERDIS Research Group, Department of Engineering, Pontificia Universidad Católica del Perú, Lima, Peru
[2]Community Resilience Laboratory, Department of Research, Toward Resilience Corporation S.A.C., San Martin de Porres, Peru

Abstract. Globally, disasters have been increasing in frequency, causing significant economic and human losses; however, there remains a clear lack of knowledge focused on disaster preparedness. Lima, the capital of Peru, faces multiple hazards, including debris flows, floods, and earthquakes. Chosica, a peripheral area of the city, is intersected by several streams that are highly prone to flooding and intense flows. Through field visits and in-depth interviews, our objective was to assess the level of disaster risk management (DRM) preparedness in relation to the activation of the Pedregal, Santo Domingo, and Yanacoto streams. The field visits allowed us to identify both the strengths of the human settlements in Chosica and the measures required to address existing hazards. The research included an analysis of the dimensions of community resilience—social, institutional, and physical—along with an evaluation of resilience levels based on community participation and organization. The findings revealed that each ravine presents a distinct situation with respect to hazard level and the implementation of preventive measures. Settlement in the ravines initially arose informally, driven by Lima's housing demand and compounded by unregulated construction. Although earlier studies had already called for preventive measures before occupation, these warnings were disregarded, leaving the communities vulnerable to the imminent threat of debris flows. Over time, certain mitigation and prevention measures were implemented, but they remain insufficient, leaving the population in persistent danger. Among the three sites, the Pedregal ravine has been the most affected historically yet still lacks effective measures to address future events, demonstrating that past experiences have not always translated into learning or action. Overall, this study concludes that effective DRM begins with active community participation and organization, complemented by continuous and collaborative intervention from local government authorities.

* Corresponding author: jheyder.perez@pucp.edu.pe

1 Introduction

Climate conditions are changing and triggering natural phenomena that affect different regions of the world. Natural phenomena that have become disasters are mainly caused by floods, tornadoes and earthquakes, causing economic losses of £2.97 trillion between 2000 and 2019 [1]. Irregularities in the water cycle are increasing over time, manifesting themselves in floods or droughts depending on the pole in which they occur. For example, in 2024, certain parts of South America and southern Africa faced a severe drought season, while Asia, Europe and central Africa experienced abnormal wet conditions. In 2024, the average annual surface temperature reached 1.55°C (±0.13°C), making it the hottest year on record and placing it above the pre-industrial baseline of 1850-1900 [2]. The problem is that climate change is an impoverishing agent for low- and middle-income countries, as the economic losses associated with disasters limit the development of communities. Unfortunately, according to a United Nations report, climate-related disasters increased by more than 80% between 2000 and 2009, compared to events recorded between 1980 and 1999 [1]. In this regard, current needs require exposed communities to develop resilience to disasters.

Research on community resilience at the international level has experienced significant growth in various disciplines and settings. Its objective is to refine its definition, redefine its application, and significantly influence disaster science, establishing itself as a prominent concept. As a result, it seeks to involve stakeholders, both socially and scientifically, at a higher level in discussions on Disaster Risk Management (DRM). Cooperation is particularly important during the preparedness and recovery stages, as it highlights the crucial role played by coordination between governments, communities, and private sector actors in strengthening resilience [3]. However, this interdisciplinary process often reduces direct contact with the most vulnerable among the wide range of stakeholders, particularly residents of communities most exposed to risk. These communities face multiple risks due to their diverse geographical conditions. Therefore, to fully understand the most vulnerable, it is proposed to assess their preparedness and capacities in DRM. This approach is expected to generate a positive response from communities and promote greater resilience after a major disaster, such as an 8.5 Mw earthquake in Lima or an extraordinary coastal El Niño phenomenon.

Lima, the metropolitan capital of Peru, is a city known for its cultural and economic diversity, but it is also vulnerable to various risks. This vulnerability is due to the fact that thousands of residents settle in areas prone to natural hazards, such as ravines susceptible to the geological hazard of mass movements and earthquakes, which increases the overall risk to the city. Water infiltration into the soil as a result of heavy rainfall reduces the stability of slopes and deposits, favouring the occurrence of mass movements such as falls, landslides and debris flows, one of the most destructive natural phenomena for communities [2]. Furthermore, Lima is a highly seismic area due to its location in the 'Pacific Ring of Fire', which, combined with informal construction practices, creates a scenario of constant risk. As a result, Lima is a city in continuous development amid multiple hazards that require multidisciplinary and collaborative intervention. Therefore, the participation of local communities is a crucial element for effective DRM [3,5].

This research is being carried out in the ravines of the Lurigancho-Chosica district through a series of interviews and field visits. These ravines were declared risk areas in 2005 by the Peruvian institution INDECI [6] due to the dangers of mass movements; however, they have been occupied over time despite these designations. It is worth mentioning that most of Lima was occupied illegally, encouraged by migration from other Peruvian provinces. Therefore, the human settlements are informal occupations without title deeds and lacking basic services. Lurigancho-Chosica, like the rest of Lima, was occupied through

 https://doi.org/10.1051/e3sconf/202565103005

invasions and illegal land sales. However, with the creation of the Agency for the Formalisation of Informal Property in 1996, these settlements were formalised. This agency was responsible for the physical and legal reorganisation of informal properties to facilitate the issuance of property titles. Although attention was given to the legality of the occupations, this does not ensure that the risk to which they are exposed has been effectively mitigated.

The methodology used in this study is qualitative and includes in-depth interviews with stakeholders in the Pedregal, Santo Domingo, and Yanacoto ravines. Although the danger of rockfalls, landslides, and debris flows has been evident in the study areas, the latter is of greatest concern due to its impact in past events. In addition, field visits are a key tool for visualising current conditions, with the support of existing literature. In this way, interviews, field visits, and literature review are combined to provide a comprehensive analysis. It is important to note that each of the ravines has unique characteristics, which will be systematically analysed in this research.

2 Method

2.1 Study area: Lurigancho-Chosica district

Lurigancho-Chosica is a district east of Lima, with 310,087 inhabitants and an area of 236.47 square kilometres in the central basin of the Rimac River. Officially, the district includes nineteen streams, which have become active over the years, causing damage to both people and property [7] (Figure 1). These streams can produce water and mud flows, known as debris flows, driven by geological conditions and heavy rainfall. The district experiences heavy rainfall from January to March, which often causes debris flows, rockfalls and flooding. The patterns shown in Figure 1 are the result of a combination of geographical, climatic, and human factors. The figure also highlights the major mudslides between 1980 and 2017 and their relationship to coastal El Niño phenomena, ranging in intensity from weak to extraordinary [8]. Each El Niño coastal phenomenon has unique effects and impacts coastal cities in Peru differently. In particular, the extreme El Niño phenomena of 1983 and 1998 worsened the situation in Lurigancho-Chosica.

The ravines studied are Pedregal, Santo Domingo and Yanacoto, which form part of the urban area, each of which presents notable differences in their organization as human settlements and in DRM. These ravines are the result of informal occupation and are gradually being formalized. Although all three ravines show signs of continued urban expansion, Pedregal is currently one of the largest (see Figure 2). Table 1 indicates that Pedregal has six times more dwellings than Santo Domingo and more than twice as many as Yanacoto. Furthermore, historical records of hazards caused by heavy rainfall highlight the damage suffered in Pedregal. In this context, its current location, the quality of its buildings and its disaster preparedness contribute to its vulnerability to debris flows, rockfalls and earthquakes.

Table 1. Number of dwellings, households, and residents in the neighbourhood's studied. From the 2017 National Census [9]

Item	Pedregal	Santo Domingo	Yanacoto	Total
Dwellings	2099	349	1748	4196
Households	2376	375	1787	4538
Residents	7804	1144	3444	12392

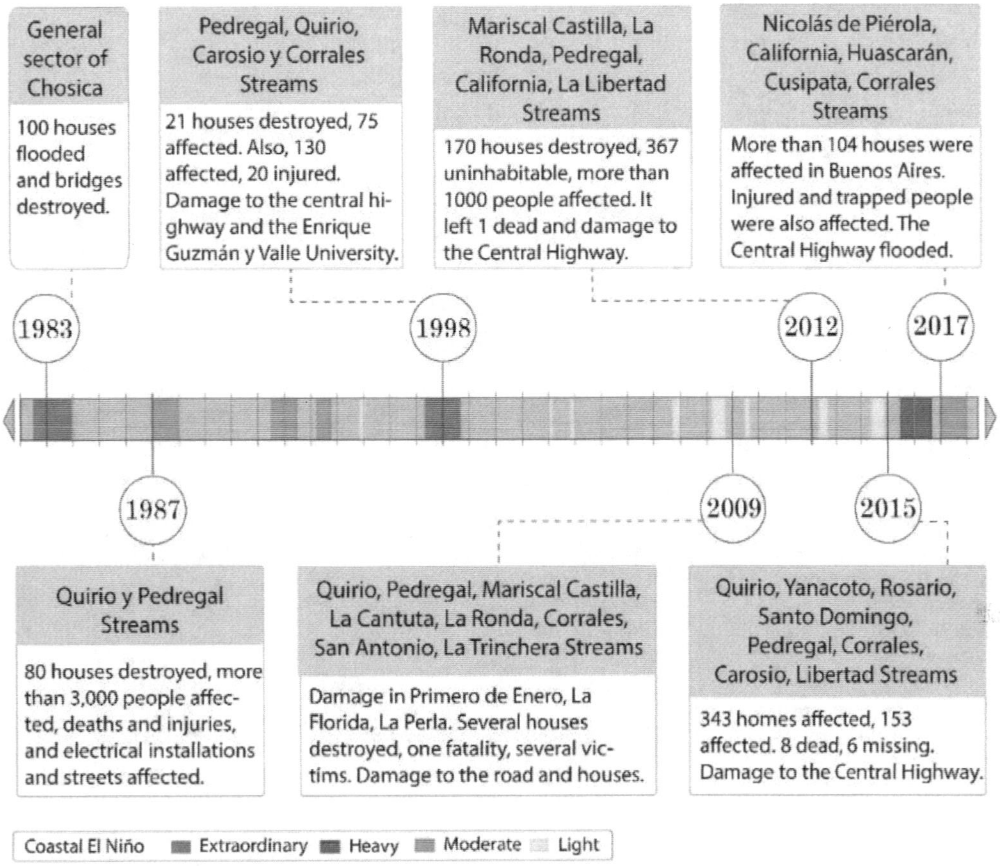

Fig. 1. Relationship between major debris flows and coastal El Niño phenomena. Adapted from ENFEN, 2024, and IGP, 2023.

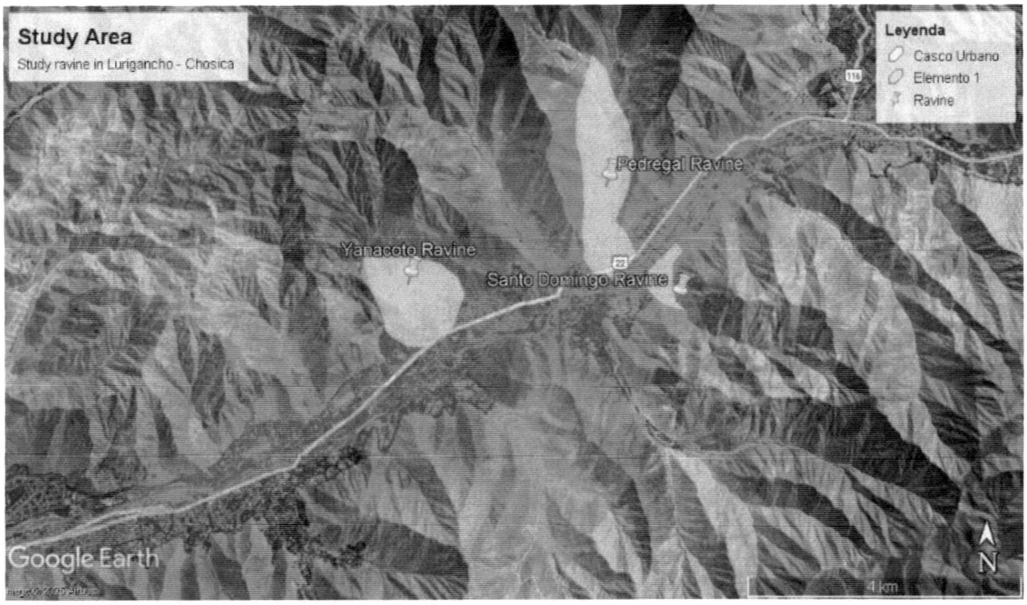

Fig.2. Study ravines in Lurigancho – Chosica.

2.2 In-depth interviews

In-depth interviews were conducted with community leaders (see Table 2). The questions focused on current conditions, hazard perception, DRM activities, and the community organization system. 1-6: San Antonio de Pedregal Alto, 6-12: Santo Domingo, 13-16: La Rinconada de El Bosque de Yanacoto.

Table 2. List of community leaders interviewed. (C) refers to completed studies and (I) refers to incomplete studies

N°	Name and last name	Age	Education level	Years of residence	Occupation	Community role
1	Raul Palacios	56	University Superior (C)	20	Employee	Secretary of Press and Propaganda
2	Carlos Tello	40	Technical Superior (I)	40	Municipal Employee	Secretary
3	Carlos León	68	Secondary (C)	50	Independent Worker	Secretary of the Economy
4	Leoncia Álvarez	72	Secondary (C)	45	Retired	Secretary of Environmental Health
5	Casilda Camargo	71	Technical Superior (I)	52	Retired	Neighbor support member
6	Franco Yuánuco	53	Secondary (C)	53	Independent worker	Citizen Security
7	Claudia Javier	68	Secondary (I)	50	Independent worker	Social Worker on its board of directors
8	Venturo Meza	46	Secondary (C)	45	Independent worker	Records and Archives Secretary
9	Irene Melendez	65	Technical Superior (C)	55	Independent worker	Secretary of the organization
10	Ingrid Oré	58	Technical Superior (I)	52	Independent worker	Secretary of Economy
11	Emma Rubio	60	Technical Superior (C)	45	Independent worker	Neighbor support member
12	Lucia Falcón	62	Secondary (C)	48	Independent worker	Neighbor support member
13	Alfonso Espinoza	84	Primary (C)	23	Retired	Neighbor support member
14	Josefa Camarena	50	Technical Superior (I)	38	Independent Worker	Representative
15	Juvenal Coica	55	Secondary (C)	30	Yanacoto S.A.C	Representative
16	Hector Coica	47	Secondary (C)	30	Independent Worker	Neighbor support member

2.3 Field visits

Field visits (Figure 3) involve tours of the area to verify the testimonies of interviewees and current literature. This allows for a visual inspection of the buildings, distribution of lots, identification of some non-structural measures, and assessment of the state of structural measures in the DRM, among other aspects.

(a) (b) (c)

Fig. 3. Field visits with tours of the area to verify the interviewees' testimonies. (a) AA.HH. San Antonio de Pedregal Alto, (b) AA.HH. Santo Domingo, and (c) AA.HH. La Rinconada de El Bosque de Yanacoto.

3 Results

3.1 Dimensions of community resilience

The community's resilience has been evaluated across social, institutional, and physical aspects. Table 3 provides a summary of the current conditions identified.

Table 3. Current conditions in the study areas.

Current conditions		**Pedregal**	**Santo Domingo**	**Yanacoto**
Access to basic services	Water	Availability on alternate days.	Availability on alternate days.	Availability on alternate days.
	Electricity	Continuous access.	Continuous access.	Continuous access.
	Sanitation	Although most sectors have access to the sewage network, some households continue to use silo-type latrines.	Although most sectors have access to the sewage network, some households continue to use silo-type latrines.	Access to the sewage network.
Type of construction	Structural typology	Upper-middle section: wooden houses with corrugated metal roofs and confined masonry. Lower section: confined masonry houses.	Upper section: wooden houses with corrugated metal roofs. Lower section: confined masonry houses.	Predominance of confined masonry houses.
	Constructio n formality	Self-construction.	Self-construction.	Self-construction.
Land commodification		Informal land occupation for sale.	A single case of land sale in a hazardous area.	Informal land occupation for sale
Local government intervention		Inefficient and slow procedures.	Inefficient and slow procedures.	Inefficient and slow procedures.

3.1.1 Social dimension: Growth of vulnerable human settlements

Among the ravines studied, Pedregal has been the most affected by debris flows, as shown by the condition of its dikes (see Figure 4), compared to Santo Domingo. Despite this, its population has grown significantly over the years, creating differences between the lower, middle, and upper parts. The lower part experienced the events of 1998, when many lost their homes, and afterward, some chose to relocate. Meanwhile, the lateral part faces the most serious problems, with residents invading the streets for their own benefit and leaving neighbors without access. In contrast, the central part demonstrates some urban planning, with wider roads than the lateral part. Finally, the upper part is experiencing land commercialization, so it is likely to become populated in the coming years. Despite the human and material damage caused by debris flows, a lack of awareness leads people to settle in risk areas. Although Santo Domingo and Yanacoto face similar issues, they are not affected to the same extent. Santo Domingo has a different terrain than Pedregal, which limits land commercialization. Conversely, Yanacoto is equally extensive, but organization by sectors and adherence to regulations have helped maintain order.

(a) (b)

Fig. 4. Comparison of the condition of the dams in the Santo Domingo (a) and Pedregal (b) streams.

3.1.2 Institutional dimension: Work of the government

Peru's DRM legal framework is established by Law No. 29664, which created the *Sistema Nacional de Gestión del Riesgo de Desastres* (SINAGERD) in 2011. This law outlines the processes involved in DRM, including risk assessment, prevention, risk reduction, preparedness, response, rehabilitation, and reconstruction. It assigns reactive responsibilities to the *Instituto Nacional de Defensa Civil* (INDECI), while the *Centro Nacional de Estimación, Prevención, Reducción del Riesgo de Desastres* (CENEPRED) handles preventive activities. CENEPRED collaborates with institutions such as the *Instituto Geológico, Minero y Metalúrgico* (INGEMMET), the *Autoridad Nacional del Agua* (ANA), and the *Instituto Geofísico del Perú* (IGP), which provide reports for planning purposes. In Lurigancho-Chosica, even before this law was enacted, INDECI developed a hazard map and land use plan in 2005, identifying the area as a critical sector. According to INDECI, disaster mitigation constituted an essential requirement for habitability; nevertheless, settlement in the area continued unabated. After the law's enactment, INGEMMET and ANA issued reports on geological hazards, vulnerable points, and flood-prone areas due to heavy rains. The IGP has also conducted seismic zoning studies for earthquake risk management. Additionally, the *Autoridad Nacional de Infraestructura* (ANIN) is executing a comprehensive project to mitigate risks from flooding, river erosion, and mass movements in the Rimac River basin. The project aims to protect 23 streams over 63.5 km of the Rimac River, including Lurigancho-Chosica, safeguarding more than 178,000 residents.

At the regional level, the Metropolitan Municipality of Lima has allocated resources for desilting and cleaning parts of the Rimac River basin. Additionally, it has conducted technical inspections to identify homes at risk. It also coordinates with local government authorities to carry out prevention and preparedness activities for natural disasters. Local governments, such as Lurigancho-Chosica, are responsible for DRM activities within their districts. They are required to develop a *Plan de Prevención y Reducción del Riesgo de Desastres* (PPRRD) during their term of office to identify hazards, vulnerabilities, and risks, along with proposed measures for disaster reduction and prevention. This plan is supported by a budget allocated to the local government for its implementation, through Budget Program 0068: Vulnerability Reduction and Disaster Emergency Response. The most recent PPRRD for Lurigancho-Chosica covers 2022-2025 and was developed by the municipality with technical assistance from CENEPRED. Despite this, vulnerable communities in the AA.HH. do not see the results, as they must perform additional tasks to address their needs.

Table 4. Structural and non-structural measures in the study streams

		San Antonio de Pedregal Alto	**Santo Domingo**	**La Rinconada de El Bosque de Yanacoto**
Measures for DRM	**Structural measures**	Dikes are in poor condition, requiring maintenance or reconstruction. Flexible barrier in need of maintenance. Presence of retaining walls.	Retaining wall 2 flexible barriers 1 dike	Non-existent
	Non-structural measures	Presence of sandbags. Lack of cleaning and desilting works. No early warning systems are in place. Lack of community participation in DRM Lack of training on DRM.	Presence of sandbags. Periodic cleaning and desilting works, carried out by central coordinating committee and local government. No early warning systems in place. Lack of training on DRM.	Presence of sandbags. Periodic cleaning and desilting works is the result of community organization efforts. No early warning systems are in place. Lack of training on DRM.

Given the lack of initiative on the part of the local government, communities at risk have taken the initiative to submit project proposals, such as cleaning and dredging the riverbed before heavy rains and building retaining walls. In terms of community development, they have also requested the construction of stairways, paving of streets and a sewerage system. As a result, Yanacoto managed to install its sewerage system and pave its streets. Although this system is operational, it is inefficient and slow, as it is necessary to provide technical documentation, constantly supervise the work and supply labor, which requires time and money. Santo Domingo has a staircase construction project underway, but this is due to the efforts of the previous board of directors. Elections for the central committee of this ravine are held every four years, illustrating how long the process with local government can be. Problems of access to the upper parts of the ravines pose additional challenges, as they require the local government to carry out cleaning and dredging work. As a result, cleaning efforts focus on the lower area, leaving the riskiest and affected area exposed to future events. Table 4 shows the structural and non-structural measures implemented in the Pedregal, Santo Domingo, and Yanacoto streams.

3.1.3 Physical dimension: Basic services and type of construction

The AA.HH. of Pedregal, Santo Domingo and Yanacoto are the result of informal settlements, which explains why the current conditions shown in Table 3 are consistent. Informal occupation affects access to basic services such as water, electricity and sewerage, as well as the existence of roads and public spaces, and increases vulnerability to natural hazards. The dwellings in the ravines studied are examples of informal construction, characterized by confined masonry systems that do not comply with the guidelines established by Peruvian regulations (RNE E.030) and the construction of stone walls without technical advice. In terms of basic services, electricity is the first to be managed, so it has a continuous supply. Water is supplied by pumps, so it is available every two days. For drainage, there is a sewerage system, although in some areas, due to new occupations, homes use silos. Currently, land continues to be commercialized, and in the upper parts of Pedregal and Yanacoto, plots are openly sold despite being in a risk area.

Fig. 5. Sale of plots along Pedregal Stream.

3.2 Level of community resilience

3.2.1 Preparedness: Community organization

Community organizations are typical of emerging communities, whose objective is to achieve common goals. In the ravines of Lurigancho-Chosica, organization is carried out through committees that represent different sectors of the population based on their geographical location. In this way, a central committee oversees all the committees, allowing them to work together as a unified human settlement.

Pedregal is one of the largest ravines in Lurigancho-Chosica, with approximately 41 committees and one central committee. Unfortunately, power struggles between former and current leaders of the central committee have caused divisions within the Pedregal ravine community. On the other hand, the Santo Domingo ravine is smaller, with only six committees. This has allowed it to manage community projects more effectively with the local government. Both ravines operate with similar systems, but organizational conflicts have been an obstacle to the implementation of projects. In contrast, Yanacoto shows clear sectorization, both physical and organizational, and each sector works independently to carry out its projects with the local government.

Table 5. Community Resilience

Community resilience	Pedregal	Santo Domingo	Yanacoto
Organization	41 local committees and one central coordinating committee.	6 local committees and one central coordinating committee	Unknown.
Participation	Poor participation.	Poor participation.	Poor participation.

3.2.2 Strengths: Community participation

The communities in the areas studied do not show active engagement with DRM, despite facing multiple hazards, as evidenced by their low participation in activities and meetings with central committees. It is important to note that this was not always the case. According to residents' testimonies, the community used to be much more involved in the past, but participation has declined since its creation. Community participation declined with the gradual allocation of land titles and the provision of basic services, a process further exacerbated by the COVID-19 pandemic.

A tree-planting project was implemented in Pedregal prior to the pandemic; however, due to insufficient supervision, the seedlings failed to survive. At present, prevention initiatives are overseen by the central committee of each ravine, with community participation reduced to financial contributions from only part of the population. The central committee is primarily responsible for requesting and supervising government projects oriented toward debris-flow prevention and local development.

4 Discussion

Community participation reflects a community's response to everyday needs; therefore, when these needs are met, the value of community work begins to decline. This marks a critical turning point for the communities of San Antonio de Pedregal Alto, Santo Domingo, and Yanacoto. Declining community participation hinders development, but this problem is not unique to these areas; it is observed in several communities in the Lima metropolitan area. One example is Pueblo Joven Casa Huertas, which faced numerous legal challenges related to property titling, and after resolving these issues, community participation declined again. The fundamental problem of declining participation stems from community stagnation and dangerous conditions, which should not be tolerated. It should be noted that debris flows and earthquakes are cyclical hazards that require preventive rather than reactive management strategies. Taking action to strengthen resilience through prevention and preparedness would significantly reduce the effects of disasters [10].

Certain conditions influence DRM, including community organization, the state of structural measures, and the terrain of the ravine. Difficulties in community organization increase with the number of members. The community of San Antonio de Pedregal Alto is much larger than that of Santo Domingo, which leads to greater mistrust of community leaders and more power struggles. In addition, Pedregal has been more severely affected, and its structural measures are in poor condition and in urgent need of repair. In contrast, Santo Domingo has embankments built after a high-intensity debris flow, which has not recurred, so they remain in good condition. The terrain of the Pedregal and Yanacoto ravines allows for continuous expansion, which encourages illegal land sales. Yanacoto does not experience high-intensity debris flows like Pedregal, so although it faces similar problems, the impact is less severe. It is important to note that debris flows vary in intensity, but this cannot be taken for granted in the case of other hazards, such as a possible large-scale earthquake. Lurigancho-Chosica is located in a seismic zone; therefore, preventive measures are equally necessary in these ravines.

A culture of prevention is still developing in Lurigancho-Chosica, where the community and local authorities are aware of the imminent hazards, but there is not yet sufficient awareness to translate this knowledge into concrete action. DRM tasks should not be limited to identifying hazards, as a significant reduction in disasters requires an understanding of vulnerability and the development of proactive strategies [11]. Furthermore, the participation of academia, government agencies at all levels, and the community as a whole is essential. These conclusions are supported by research conducted by Cordova and Perez (2024) [12]. Therefore, technical and scientific studies should be shared with decision-makers in land management so that, together with legal and financial support, they can develop mitigation and prevention strategies. Furthermore, it is essential to involve the community in this process, as their empirical knowledge provides a broad perspective that can serve as a basis for adopting more appropriate measures. It is important to note that leadership in this initiative lies with the local government, specifically the municipalities in each district. This is because municipalities are allocated an annual budget for their administrative functions, which includes a specific fund for DRM. Unfortunately, the lack of technical knowledge and professional guidance has hampered the implementation of disaster prevention projects.

Current measures, both structural and non-structural, are insufficient in the Lurigancho-Chosica ravines. At the structural level, it is essential to manage a maintenance plan for the conservation of existing structural measures, especially in ravines with conditions similar to those in Pedregal. At the non-structural level, "resilience centres" could be implemented. These are innovative concepts that have gained relevance in recent years, as their purpose is to support communities in overcoming various adversities, promoting their well-being in emergency situations [13]. Strengthening the resilience of communities to disasters is essential, as it facilitates their recovery and reconstruction; over time, they also become stronger and better prepared for future adverse events [10]. Although the above measures are necessary to address the current situation, the expansion of the area exposed to danger must be halted. In this regard, it is crucial that the local government asserts its authority and takes concrete measures around DRM, such as conducting on-site inspections to prevent construction on the ravine bed or the illegal sale of land. There are projects proposed by *Ministerio de Vivienda, Construcción y Saneamiento* of Peru that could be carried out in Lurigancho-Chosica, such as the development of linear parks to help prevent settlement in risk-prone areas.

5 Conclusions

Current conditions underscore the urgent need for Peruvian authorities to prioritize disaster risk management in order to reduce the likelihood of future catastrophes. The evidence reflects a persistent lack of preventive action, with responses predominantly oriented towards reactive measures. Consequently, the recurrent activation of streams, heavy rainfall, and uncontrolled urban expansion in alluvial fans perpetuate a state of chronic vulnerability in Lurigancho-Chosica. This cyclical exposure, exacerbated annually by local climatic conditions, is further amplified by climate change, which significantly increases the potential for large-scale disasters if no preventive measures are implemented.

Addressing this challenge requires decisive action at the local government level. Authorities must implement strategies grounded in technical evidence provided by Peruvian scientific and governmental institutions, while simultaneously promoting awareness among residents to discourage reckless practices and to foster community engagement. The role of local governments is particularly critical, as they possess both a dedicated DRM budget and the legal authority to regulate hazardous land use practices, such as the illegal sale of plots and the construction of housing in riverbeds. Moreover, when reactive management is insufficient, local governments retain the power to mobilize support from national institutions, including INDECI and provincial governments, thereby enhancing their response capacity.

Community participation constitutes a cornerstone of resilience. Active involvement in preventive actions and collective organization strengthens local adaptive capacity and reduces long-term vulnerability. However, current dynamics reveal high levels of mistrust toward governmental institutions and social fragmentation within communities, fostering individualism over collective action. This shift has eroded traditions of collaborative work that previously contributed to DRM effectiveness. The lack of knowledge and awareness further perpetuates hazardous behaviours that increase risk, underscoring the necessity of renewed strategies for capacity-building and social cohesion.

Building resilience in Lurigancho-Chosica must transcend survival and embrace transformation. Achieving this goal demands sustained and collaborative participation among government, academia, and local communities. Integrating empirical community knowledge with scientific research creates a foundation for evidence-based decision-making by competent authorities. Importantly, such collaboration must extend beyond short-term projects and encompass both preventive and reactive measures. The enactment of the SINAGERD law represented a milestone in Peru's DRM legal framework; its effective implementation remains critical to realizing its intended purpose of safeguarding communities against future disasters.

References

1. Centre for Research on the Epidemiology of Disasters & UN Office for Disaster Risk Reduction, Human cost of disasters: an overview of the last 20 years (2000–2019). (2020). https://www.undrr.org/quick/50922
2. World Meteorological Organization, State of Global Water Resources report 2024. (2025). https://doi.org/10.59327/WMO/WATER/2024
3. B. Vu, O. Obaitor, L. Grobusch, D. Sett, M. Hagenlocher, U. Schinkel, L. Nguyen, F. Bachofer, S. Ngo, M. Garschagen, Enablers and barriers to implementing effective disaster risk management according to good governance principles: Lessons from Central Vietnam. Int. J. Disaster Risk Reduct. 120, 105344 (2025). https://doi.org/10.1016/j.ijdrr.2025.105344
4. Y. He, M. Ding, Y. Duan, H. Zheng, W. He, J. Liu, Debris flows dynamic risk assessment and interpretable Shapley method-based driving mechanisms exploring – A case study of the upper reach of the Min River. Ecol. Indic. 173, 113400 (2025). https://doi.org/10.1016/j.ecolind.2025.113400
5. V. Acuña, S. Valdivieso, L. Juzam, Dignificando la Gestión de Riesgo de Desastres: liderazgos femeninos y estrategias comunitarias en el Campamento Dignidad, Santiago de Chile. REDER 5, 91–106 (2021). https://doi.org/10.55467/reder.v5i2.79

6. Instituto Nacional de Defensa Civil, Estudio mapa de peligros y plan de usos de suelo y medidas de mitigación ante desastres de la ciudad de Chosica. (2005). https://sigrid.cenepred.gob.pe/sigridv3/documento/123
7. Instituto Geofísico del Perú, Análisis y evaluación histórica de precipitaciones en Chaclacayo, Chosica y áreas aledañas. (2023). https://repositorio.igp.gob.pe/server/api/core/bitstreams/7ab825b4-0aca-4275-95a8-4243998bbe4c/content
8. Estudio Nacional del Fenómeno "El Niño" ENFEN, Eventos El Niño y La Niña Costeros. http://met.igp.gob.pe/elnino/lista_eventos.html
9. Instituto Nacional de Estadística e Informática, Censos Nacionales 2017. (2017). https://censo2017.inei.gob.pe/resultados-definitivos-de-los-censos-nacionales-2017/
10. J. Lin, J. Chao, C. Lin, Evaluation of community resilience to debris-flow disasters: A case study of Nantou, Taiwan. Int. J. Disaster Risk Reduct. 81, 103258 (2022). https://doi.org/10.1016/j.ijdrr.2022.103258
11. O. Dewa, D. Makoka, O. Yusuf, Assessing capacity and implementation status of the Disaster Risk Management Strategy for Health and Community Disaster Resilience in Malawi. Int. J. Disaster Risk Sci. 12, 673–688 (2021). https://doi.org/10.1007/s13753-021-00369-z
12. C. Cordova, J. Perez, Quantitative indicators for multi-hazard community resilience assessment of informal settlements. 18th World Conference on Earthquake Engineering (WCEE2024), Milan, Italy (2024). https://www.researchgate.net/publication/381847001_Quantitative_Indicators_For_Multi-Hazard_Community_Resilience_Assessment_Of_Informal_Settlements
13. T. Ciriaco, S. Wong, Review of resilience hubs and associated transportation needs. Transp. Res. Interdiscip. Perspect. 16, 100697 (2022). https://doi.org/10.1016/j.trip.2022.100697

Critical success factors in humanitarian logistics operations in disaster response

Haliza Mohd Zahari[1,*], *Nurhana* Mohamad Rafiuddin[2], *Sharifah Aisha* Syed Ali[3] and *Ruzanna Mat Jusoh*[3]

[1]HADR Research Centre, Strategic & Security Studies Institute, National Defense University of Malaysia, Kuala Lumpur, Malaysia,
[2]Faculty of Management and Defense Studies, National Defense University of Malaysia, Kuala Lumpur
[3]Faculty of Defense Science and Technology, National Defense University of Malaysia, Kuala Lumpur

Abstract. Inefficiency in disaster response is often reported to occur due to failures in humanitarian logistics operations. This operation involves a long process involving various agencies. The study was conducted to explore the critical success factors that contribute to the effectiveness of humanitarian logistics operations, emphasizing the importance of an efficient logistics system in mitigating operational inefficiencies that could impede relief efforts. This study was conducted using a qualitative methodology with the phenomenological research approach. The sampling strategy implemented is to select research participants who have experience related to this study. Five Focus Group Discussions (FGDs) were conducted with the research participants who are representatives from the agencies involved in Humanitarian Logistics (HL) operations during disasters. Each FGD comprises 6 to 10 research participants. The FGDs were guided by a semi-structured interview protocol, by adopting a logistics process and action. Thematic analysis was employed to systematically analyze the data collected from the FGD. Based on the results, the researchers have analyzed and categorized the HL operations challenges and limitations translated to five key critical success factors (CSFs): Man, Machine, Material, Mechanism, and Monetary (5M). It was found that the inefficiencies of 5M created a big impact on HL operations.

1 Introduction

The frequency of disasters is on the rise, posing significant challenges to disaster response and management efforts. Climate change, for instance, has increased the severity of extreme weather such as floods. To exemplify, in Malaysia, climate change worsens the weather patterns, humidity, and rainfall, which leads to the frequency and severity of floods [1]. When a disaster takes place, immediate and coordinated action is performed to alleviate suffering and economic loss.

* Corresponding author: haliza.mz@upnm.edu.my

Disaster response aims to minimize the impact of a disaster. Disaster response involves activities before, during, and after a disaster, which requires efficient resource management. The disaster response effort faces growing complexities and requires resistance strategies. During Malaysia's flood in December 2021, among the disaster response complexity factors that are being highlighted are the absence of a timely government agency response caused by a lack of coordination and communication. According to Davidson [2] it is the responsibility of the agencies to provide logistics support during disaster response. One of the reasons the agencies failed to respond effectively during Malaysia's flood in December 2021 was due to a lack of preparedness and coordination [3].

The complexity of humanitarian logistics (HL) operations comes from the unpredictable nature of disasters, requiring swift responses while facing limited resources and challenging operational conditions. For instance, the 2004 Aceh Tsunami highlighted challenges such as inadequate logistics information, resulting in duplicated efforts and misallocation of resources, leading to delays and wastage [4]. Similarly, the humanitarian response to the devastating Haiti earthquake faced significant coordination issues among local agencies and international organizations due to poor planning, which contributed to prolonged suffering and delayed recovery efforts [5]. Therefore, disaster response can be significantly improved by enhancing logistics support operations.

Effective logistical support is crucial in alleviating the suffering of affected populations by ensuring the timely and efficient delivery of relief aid. However, failures in humanitarian logistics operations can exacerbate the crisis, leading to secondary disasters such as the inefficient distribution of relief supplies or the overcrowding of temporary shelters due to inadequate coordination among the agencies.

Successful humanitarian logistics operations rely significantly on resource availability and efficient management, as well as seamless coordination among stakeholders [6]. In this context, factors such as human resources, financial resources, assets, equipment, and materials are critical to effectively delivering aid. The ability to manage these resources efficiently has a direct impact on the overall success of logistics operations. Achieving this requires well-coordinated efforts in accordance with the principles of the right place, right time, and right location across various domains to ensure timely and effective disaster response.

2 Literature review

Humanitarian logistics (HL) has emerged as a critical discipline within disaster management and operations management. HL aims to alleviate suffering and reduce the number of victims. It encompasses the planning, implementation, and cost control of the efficient and cost-effective flow and storage of goods, materials, and information from the point of origin to the point of consumption to meet the needs of affected populations during disasters. According to Yusoff & Yusoff, [1] HL includes coordination of food, water, supplies, and other relief aid, as well as transporting and distributing them effectively. In order to deliver the relief aid at the right time, right location, right quantity and quality, it requires a quick and accurate assessment of the number of affected people and their basic needs.

Humanitarian logistics plays a pivotal role in disaster response and relief operations, bridging the gap between aid availability and accessibility. According to Kembro [7], logistics and supply chain management emphasize the preparation for, response to, and recovery from a humanitarian disaster, to preserve lives and mitigate the suffering of those affected by it. The success of HL operations is contingent upon the availability and effective management of critical resources, including human resources, financial assets, equipment, and materials [8]. Without a robust logistics framework, disaster response efforts may become fragmented and inefficient, leading to resource misallocation and prolonged human suffering. In large-scale disasters, resource shortages are a significant concern, and the complexity of relief operations often exacerbates logistical challenges. Additionally, effective coordination among agencies is paramount to ensuring that aid reaches affected populations in a timely manner. Since not all agencies are equally equipped with the necessary resources, coordination in asset sharing and logistics planning is critical to optimize relief efforts and minimize response delays.

The concept of Critical Success Factors (CSFs) in humanitarian logistics has gained considerable attention as these factors are fundamental in ensuring the effectiveness and efficiency of disaster response and recovery efforts. CSFs refer to the key operational elements that must be achieved to optimize HL performance [9]. Mohd Zahari [10] has investigated the optimization of humanitarian logistics operations against supply chain disruptions. Their findings indicate inefficiencies in procurement, transportation, and coordination systems. On the other hand, several frameworks in disaster logistics highlight coordination, stakeholder networks, and information systems as fundamental components. What has been insufficiently highlighted, however, is an integrated framework that explains why and how specific characteristics emerge as CSF. This study employs a 5M concept: Man (people), Machine (equipment, vehicles, technology), Material (aid relief), Mechanism (governance, coordination, and information sharing and Monetary (financial resources). This adaptation emphasizes findings that the critical component, when effectively managed, will significantly enhance performance. This concept is grounded in previous disaster logistics, such as Kovács [11].

To achieve efficient and successful humanitarian logistics operations, it is important to prioritize key operational goals, including speed, coordination, adaptability, and accountability. Speed is essential to ensure the timely delivery of aid to affected populations, minimizing the impact of the disaster. Coordination facilitates synergy among various agencies, preventing duplication of efforts and enhancing resource utilization. Adaptability is crucial to responding effectively to the dynamic and often unpredictable nature of disaster environments, ensuring flexibility in logistics planning and execution. Lastly, accountability is vital to maintain transparency and foster trust among stakeholders, ensuring that aid reaches the intended beneficiaries without mismanagement or misuse of resources.

This study aims to explore the critical success factors that contribute to the effectiveness of humanitarian logistics operations, emphasizing the importance of an efficient logistics system in mitigating operational inefficiencies that could impede relief efforts.

3 Methods

Many studies have explored the humanitarian logistics operation; however, very limited studies have explored the CSF of humanitarian logistics in disaster response. This study employed a qualitative research approach to explore the challenges and limitations faced by the agencies involved in HL operations during disasters and to identify the CSFs that contribute to effective operations.

A phenomenological research approach was used in this study to understand the research participants' experiences. The sampling strategy is by selecting research participants who can provide rich, detailed descriptions of the experience being investigated [12]. Using this sampling helps in describing the nature of the research participants' experiences. In order to get the right research participants, the researchers' selection is based on the research participants who have had experience in managing and overseeing HL during the recent disasters. This will assist in obtaining experiences of broad phenomena and the richness of data [12]. The next process was conducting a Focus Group Discussion (FGD) to capture views from practitioners involved in HL operations. FGD is chosen as the primary method of data collection because it allows for dynamic discussion, interaction, and consensus-building among the research participants. A total of five FGDs were conducted, where each FGD comprises 6 to 10 research participants who were chosen through purposive sampling to ensure relevance, focusing on the practitioners with substantial experience in HL operation, disaster management, disaster response coordination, and logistics. Snowball sampling was then employed, wherein the earlier participants suggested further stakeholders in order to expand viewpoints and mitigate bias. The number of research participants involved was recommended by Weller [13] in ensuring diversity of expertise and roles while maintaining a manageable group size for discussion. All research participants in the FGD were informed of their right to confidentiality, and they requested that their names and identities not be disclosed. In accordance with their preferences, general descriptions such as Agency A, B, C, D, and E were used throughout the FGD, and no personally identifiable information has been disclosed. This approach produced a varied assembly comprising government officials who were directly involved in HL operations.

The discussion was guided by a semi-structured protocol by adopting the logistics process and action by Bastos [14] based on their review of various logistics operational guides used in disaster response operations. The areas include: mobilization, acquisition, warehousing and inventory management, distribution, transportation, quality control and monitoring, and coordination between agencies

Based on these areas, the researchers have prepared questions for the FGD, which are attached in Appendix A. The questions are designed to provoke thoughtful discussion, uncover gaps, and identify actionable insights for improving the HL operations in disaster by identifying challenges, barriers to coordination, resource allocation, and potential CSF for speed, coordination, adaptability, and accountability in HL operations. Ethical clearance was obtained prior to data collection. Research participants were informed of the purpose and assurance of confidentiality and anonymity. Written consent was obtained, and research participants were given the freedom to withdraw at any stage without repercussions. The protocol used is aligned with the research objectives and encourages the research participants to share detailed insights into the operational challenges and perceptions of success factors. This is to ensure that the discussion is grounded in practical, real-world challenges and solutions. Thematic analysis was employed to systematically analyze the data collected from the FGD by following Braun & Clarke [15] six-phase framework. The researchers have successfully captured an in-depth understanding of the collective experience and opinions of the agencies and enabled the identification of shared experiences, challenges, and perspectives, which have assisted the researchers in exploring nuanced insights. Finally, the key insights from the analysis were presented in the findings section [13].

4 Results

This section presents key findings from the study focusing on challenges and limitations in HL operations. These factors align with CSF, which is essential for effective disaster response and logistics operations. CSF in HL operations diagram is illustrated in Figure 1.

4.1 Mobilization

Mobilization emerged as a crucial process in humanitarian logistics, involving Agencies engaged in humanitarian logistics operations. They are professional entities doing disaster-related tasks. Their involvement in the disaster is a secondary task, as they have other primary responsibilities. Nevertheless, their organization is highly structured, and they are directly involved in the event of a disaster. This emphasizes the significance of preparedness and rapid personnel deployment.

During the disaster response, Agency A and Agency B were informed that the number of personnel they have was limited, and the remaining personnel are reliant on volunteers who have registered with their agencies. These volunteers are trained to assist in humanitarian logistics response and can be quickly mobilized.

> *"We keep a database of qualified responders ready to be called and used within hours of an incident. Regular drills and training courses enable our staff members to be ready for an emergency and to act fast." (Agency B)*

This reflects not only the reliance on pre-identified human resources but also the role of continuous capacity-building to ensure operational readiness. It highlights how the *Man* dimension of the 5M framework is strengthened when institutional systems and training protocols are embedded in organizational practice.

Similarly, Agency A described how rapid reaction teams serve as the backbone of initial response

> *"Rapid reaction teams are crucial; depending on the degree of the emergency, they can be mobilized right away from standby. Effective response depends on strong coordination between local emergency services and staff mobilization, streamlining processes. We used a mass notification system to alert staff quickly. Effective communication is essential in ensuring that everyone knows where to go and how to help." (Agency A)*

This demonstrates that mobilization is not merely about the number of responders but about the effectiveness of communication and coordination systems. Connecting trained personnel with technology-driven alert systems, mobilization becomes streamlined, reducing delays and maximizing impact.

Taking these into account, it shows that mobilization relies on the synergy between human preparedness (Man), man and technology enabler (Machine), and coordination (Mechanism). The findings indicate that mobilization should not be perceived as a separate stage, but rather as an integrative process in which agencies' preparedness, communication systems, and inter-agency coordination converge to influence the speed and efficiency of the humanitarian response.

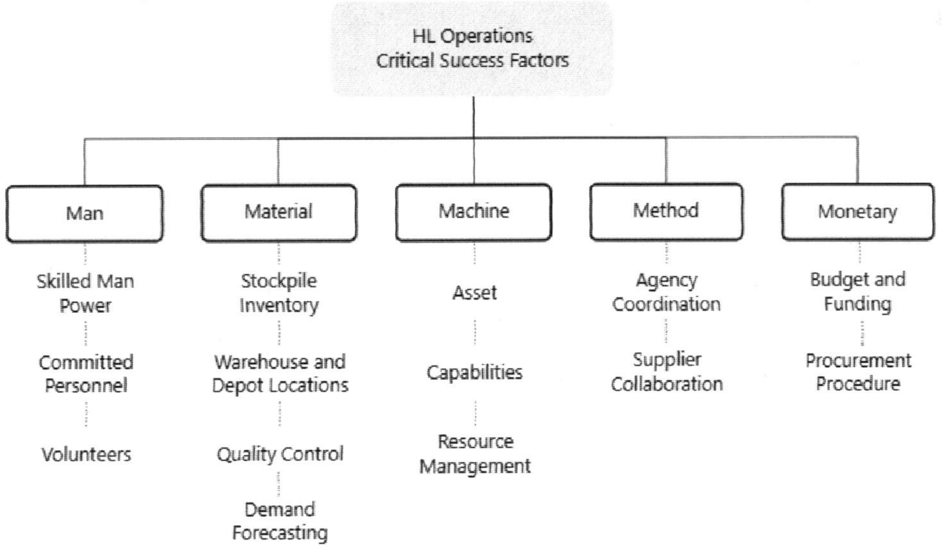

Fig. 1. CSF in HL Operations.

4.2 Acquisition

Not all five agencies engage in acquisition activities. Only Agency C was involved in the acquisition. The agency procures assistance relief, including food, clothing, hygiene kits, and specialized provisions for children, seniors, and those with disabilities.

Agency C outlines the procedure for completing a needs assessment to ascertain the resources necessary during a disaster based on the scale of the disaster. They possess a comprehensive schedule for procurement planning. The most crucial period is from October to March during the monsoon season. The acquisition process will take place annually from April to August.

Agency C usually has the identified products they need to purchase, but at times, they will survey the quality of the product and make adjustments for the new product. A good relationship with suppliers will guarantee the delivery of high-quality supplies in adequate quantities, at the appropriate time and location.

Acquisition was recognized as a persistent impediment in humanitarian logistics, where the conflict between formal procurement frameworks and the pressing demand for adaptability frequently compromised swift response. Agency C articulated this challenge:

> *"We begin with a needs analysis to determine the most important supplies. This is derived from ground-based comments. Our procurement is very structured, which is not flexible enough to let us buy the required goods quickly. We need to get approval from a higher level. This is our biggest challenge when there are many locations flooded, where there is insufficient aid relief, and the worst part is when the flood takes longer than we expected." (Agency C)*

This highlights a significant inconsistency in humanitarian logistics. Although structured procurement processes ensure accountability, they can also slow down response when emergencies demand immediate action. From a 5M perspective, the issue illustrates the interdependence of Monetary (financial approvals) and Mechanism (decision-making and governance). The bureaucratic inflexibility constrains the responsiveness of acquisition processes, leaving affected communities vulnerable when disasters exceed expected duration.

Simultaneously, Agency C emphasized the importance of supplier dependability and quality assurance:

> *"We have a list of reliable suppliers; however, we also routinely evaluate their ability to deliver. Essential is quality control before they are sent to the field; all supplies satisfy specific requirements." (Agency C)*

Quality control is essential before deployment, ensuring that all supplies meet particular criteria. This represents an effort to reconcile expediency with integrity, guaranteeing that relief supplies adhere to basic requirements before distribution. It illustrates how Material (the supply) is influenced by Mechanism and Monetary (financial allocations). Although quality assurance is crucial for sustaining trust and effectiveness in relief operations, the processes indicate a compromise: more regulations on supplier selection and quality may prolong the procurement process.

These findings indicate that acquisition in humanitarian logistics is not simply about procuring items, but also managing the conflict between speed and accountability. These findings indicate the need for hybrid procurement mechanisms systems that uphold integrity, maintain openness and quality while permitting flexibility for rapid decision-making in emergencies. This provides insight into how acquisition processes reflect both structural strengths and operational vulnerabilities in disaster response.

4.3 Warehousing and inventory management

Warehousing and inventory management emerged as essential and complex components of humanitarian logistics, balancing the necessity for preparedness with the unpredictability of disasters. As discussed in the acquisition section above, Agency C performed the acquisition, and from there, they will keep the inventory in the warehouse. This warehouse and inventory are crucial during humanitarian logistics operations because they directly impact the agency's performance, speed, efficiency, and effectiveness of delivering aid to affected populations. Therefore, the warehouse functions to store critical items such as food, water, hygiene kits, shelter kits, and clothing. Proper inventory management will ensure supplies are ready when needed, which will reduce delay, prevent shortages, overstocking, and avoid wastage. Agency C explains:

> *"We hold inventory based on assessment, water, and food are crucial. In terms of how much we need to keep, we calculate based on our system that analyzes historical data, and we also perform a risk assessment on whether to keep more or less inventory". (Agency C)*

This highlights the growing dependence of organizations on data-driven instruments—historical demand analysis and risk assessments—to optimize inventory levels. These approaches reflect an attempt to professionalize humanitarian logistics by integrating material management with mechanisms, including systems and risk protocols. Nevertheless, while these models offer a framework, they also pose a risk of underestimating requirements when disasters exceed previous trends or occur in unexpected patterns. This conflict between predictive planning and real-time demand highlights the vulnerability of inventory systems in volatile crisis contexts.

In addition to stock level estimation, physical warehousing faced significant challenges. As indicated by Agency C:

> *"We faced many challenges, such as the warehouse was initially built in safe areas and proximity to disaster-prone areas; however, as time goes by, some of the warehouses are in the disaster areas. On the other hand, when a disaster takes place, we have difficulty distributing supplies because of the damaged infrastructure." (Agency C)*

This statement illustrates how evolving risk landscapes could harm the facilities intended for resilience. Warehouses that were formerly strategically positioned have become progressively susceptible due to growing cities and changing risk areas. Furthermore, compromised infrastructure during disasters affects the transportation of supplies, reducing the strategic objective of prepositioning. This directly links to Machine (assets) and Mechanism (strategic planning), illustrating how misalignment between risk planning and physical realities compromises efficiency.

Overall, these results indicate that warehousing and inventory management exceed stockpiling goods; they embody a continuous negotiation among prediction models, geographical realities, and infrastructure limitations. The findings indicate that although systems-based assessments enhance inventory planning, they require supplementary adaptive strategies such as flexible warehouse relocation, mobile storage units, or regional stock-sharing agreements to remain effective in facing evolving risks.

4.4 Distribution

It is noteworthy that due to the annual occurrence of disasters such as floods, Agencies A, B, and C mentioned that their preparedness regarding personnel, resources, and flood relief supplies commences in September each year. Agencies A and B are tasked with the evacuation of flood victims and assisting Agency C in distributing aid; thus, resources and staff are crucial for these two agencies. Nonetheless, the limitations encountered by these two agencies include the inadequacy of their assets and a scarcity of personnel available for operational assignments. Despite the limited workforce of Agencies A and B, both agencies possess a substantial number of volunteers available for mobilization during operations.

Agency C's strategy is quite different. This agency plans to procure relief aid from the beginning of the year and distribute it to all depots throughout Malaysia. When there is a flood, these stocks will be released for distribution to disaster-affected areas. Since Agency C has constraints in terms of assets and personnel, this agency will coordinate distribution together with Agencies A and B. Agency C explained the structural constraints concisely:

> *"One of the challenges we have in distribution is the constraint of resources, including assets and personnel. Not all agencies have complete assets. Consequently, inter-agency collaboration is essential." (Agency C)*

Distribution emerged as a highly resource-dependent process, where inter-agency collaboration often determined whether aid reached affected communities in a timely manner. The statement above indicates that distribution capacity is unevenly allocated among humanitarian actors. Agency C faces limitations in both manpower (Man) and assets (Machine), which makes collaboration with more resourceful agencies not only beneficial but essential. This highlights the importance of the Mechanism, which involves the governance and coordination arrangements that enable resource pooling and joint distribution efforts. Agency B provided an actual illustration of how such coordination emerges in practice.

> *"Agency B has numerous assets and volunteers available for the provision of aid relief. However, there are instances when our resources are unable to access the flood-affected areas due to their insufficient elevation to traverse the deep floodwaters. We will disseminate information regarding the areas/zones that are inaccessible to us and collaborate with Agency A for the distribution of aid relief, as Agency A possesses a large and high-capacity 5-tonne trucks". (Agency B)*

This statement illustrates how resource asymmetries across agencies are strategically managed through communication and coordination. Even Agency B, which has sufficient volunteers and assets, is constrained by the physical limits of its assets, showing how challenges, for instance, deep floodwaters reveal limitations in capacity. By sharing real-time information on inaccessible areas and utilizing Agency A's higher-capacity assets, Agency B builds an alternate distribution strategy.

These findings indicate that distribution is not just a logistical "last mile" task but a relational process influenced by resource limitations, environmental challenges, and inter-agency governance. The evidence confirms that no singular agency holds the complete range of assets necessary for efficient distribution in disaster situations. Effectiveness depends on a collaborative approach in which the strengths of one agency balance the weaknesses of another. In the 5M, distribution represents the relationship of Man, Machine, and Material, facilitated by Mechanism (coordination) and limited by Monetary resources. This relationship illustrates that the distribution process is a logistical and collaborative issue, requiring strong inter-organizational trust and proactive collaboration.

4.5 Transportation

Transportation is essential for delivering supplies, personnel, and equipment to affected areas. It connects all other logistics functions, such as warehousing, procurement, and distribution. Nevertheless, it was discovered that not all agencies have sufficient assets, such as vehicles. Consequently, it was frequently highlighted as a critical and vulnerable element of humanitarian logistics, where the suitability of vehicles directly impacted the efficiency and extent of relief distribution. Agency A recognized its limitations while emphasizing the significance of collaboration.

> *"We have a limited number of vehicles, but we can assist Agency C in the distribution of relief aids." (Agency A)*

This highlights how agencies with limited transport fleets address operational gaps for others, showcasing the relationship of Machine (vehicles) and Mechanism (inter-agency collaboration) for enabling shared responsibility. Although capacity may be limited, being prepared to collaborate turns transportation from a single organizational function into a shared resource. Agency C provided a more detailed report on reliance and vulnerability:

> *"We are heavily reliant on Agency A and Agency B. We have only two trucks, and their utilization is somewhat restricted. Additionally, there is an issue wherein supplies must occasionally be made to impacted locations. However, the agency's trucks were unable to withstand the elevated flood levels. In my assessment, Agency A's vehicle is quite effective at traversing this area. Furthermore, should the disaster grow more severe, Agency D will take control. Collaboration with Agency E is made easier by their extensive assets and capabilities." (Agency C)*

These statements highlight that challenges with transportation are not solely technical, as discussed in many previous studies, such as limited assets and poor road conditions, but also relational and organizational. Asset limitations (Machine) are made worse by environmental limits, such as heightened flood levels, showing gaps in fleet resilience among agencies. Inter-agency collaboration is both essential and strategic for overcoming these limitations, as bigger organizations offer support to less capable ones. This illustrates transportation as a networked function reliant on resource complementarity, trust, and structured escalation mechanisms. Within the 5M strategy, transportation illustrates the interdependent relationship of Machine, Material, and Man; however, it is ultimately regulated by Mechanism, which includes the organizational structures that govern who transports what, when, and how.

4.6 Quality control and monitoring

Ensuring that aid items are delivered in good condition, meeting the standards, and reaching the right people at the right time is the most crucial in humanitarian logistics operations. Besides that, monitoring the supply chain to minimize wastage, delays, and errors while maintaining accountability is also important. Agency C described how food provision strategies changed before and after the COVID-19 pandemic:

> *"As we are responsible for food preparation and basic necessities, we have used various techniques to efficiently fulfill requirements and reduce waste. For instance, prior to COVID-19, during a flood event, we purchased fresh supplies and delivered them to Temporary Shelters. The victims will work together and prepare three meals daily. It saves huge amounts of government money. Since the onset of COVID-19, victims are isolated to prevent the spread of the disease; hence, we give packaged food. However, not many people like pre-packaged food, as it is prepared prior to consumption and sometimes becomes rotten and stale by the time they wish to consume it. We conducted a study with the victim, who expressed a preference for simple meals that may be consumed at any time. Consequently, we offer retort meals." (Agency C)*

This statement illustrates that quality control is not just technical; it requires an understanding of the cultural preferences and life experiences of victims of disasters. Pre-packaged meals provided safety during COVID-19 but were rejected for being tasteless or unpleasant, underscoring the conflict between efficiency and acceptance. The transition to retort meals exemplifies adaptive quality control, integrating Material (food supply), Mechanism (decision-making rules), and Man (beneficiary feedback) collaboratively. It also demonstrates that quality monitoring is continuous: victim feedback serves as an input for enhancing humanitarian supply strategies.

Agency C also highlighted proactive measures to guarantee supply availability and quality:

> *"Apart from dispatching essential supplies to depots in flood-prone areas prior to the monsoon season, we also provide them to forward bases. Recently, the basic supplies provided have become insufficient. Our new strategy is to engage in contracts with many supermarkets to address potential shortages. A credit note is used for the payment process. We collaborate with suppliers to ensure timely delivery and high-quality packaging." (Agency C)*

This demonstrates how supplier dependability, timeliness, and packaging are all aspects of quality control that go beyond the actual product. By bridging the gaps between Monetary (financing flexibility), Mechanism (supplier agreements), and Material (quality of supplies), the technique of entering into contracts with supermarkets and utilizing credit notes demonstrates an inventive mechanism that guarantees both speed and quality.

These findings demonstrate that quality control and monitoring are not static checkpoints but rather dynamic, context-sensitive procedures. They have to adjust to fluctuating supplier landscapes, victim expectations, and disaster conditions (such as pandemic limits). Within the 5M, quality emerges as an interdependent function that requires financial flexibility (Monetary), reliable suppliers and systems (Mechanism), culturally appropriate goods (Material), and active feedback from beneficiaries. This integrated viewpoint emphasizes that sustaining quality in disaster response involves more than just compliance; it also involves maintaining operational effectiveness, integrity, and trust.

4.7 Coordination between agencies

Integrating related agencies and all levels of government in the joint implementation of mobilization preparations is a crucial element. Coordination between agencies was consistently identified as both the most critical enabler and the most persistent challenge in humanitarian logistics. Agency A, which often assumes the central coordinating role, explained:

> *"We serve as the secretariat for disaster management. Coordination poses significant challenges in management. It is challenging to instruct agencies to comply with directives and provide resources. Nevertheless, due to our quarterly meetings, we were able to address difficulties and challenges prior to the onset of the monsoon season, with several agencies attempting to assist us in managing logistical requirements." (Agency A)*

Two aspects of coordination are highlighted in this statement: the relational problem of maintaining compliance among independent actors and the structural power to gather and provide instruction. The quarterly meetings serve as a mechanism to anticipate bottlenecks, but this statement emphasizes how challenging it may be to turn directives into operational collaboration during a disaster.

Additionally, the need for civil-military cooperation was highlighted. According to Agency E:

> *"Civil-military coordination can only be done if the resource mobilization on the effectiveness of logistics response operations can be managed thoroughly." (Agency E)*

Military involvement was seen as strengthening overall logistics performance due to its extensive resources, disciplined command structure, and established logistics infrastructure. This result is consistent with past research on humanitarian logistics, which portrays the military as both important partners and sometimes dominant players. The efficiency of military coordination relies on the proper integration of resource mobilization into the civilian-led response.

Agency D reinforced the importance of preparedness through simulation, noting:

> *"Coordination requires enhancement in simulation drills; however, all agencies lack a realistic understanding of relief capabilities." (Agency D)*

This observation points to a gap between formal coordination structures and practical readiness. Without joint exercises that reflect real operational constraints, agencies may overestimate their own capacities and underestimate the interdependencies needed for effective response.

These findings demonstrate that coordination is a multi-layered process including authority, resources, and practice rather than a single mechanism. Coordination is the 5 M's most direct reflection of Mechanism, but it also mediates all other elements, including funding (Monetary), supply (Material), vehicles and assets (Machine), and human resources (Man). Based on these analyses, effective coordination necessitates relational practices that include trust, civil-military cooperation, simulation drills, as well as structural mechanisms such as secretariat duties and regular meetings.

5 Discussion

This study's findings indicate that humanitarian logistics concerns in Malaysia should not be perceived as isolated issues, but as interconnected processes influenced by human resources, assets, supplies, governance, and financial systems. This study employs the 5M: Man, Machine, Material, Mechanism, Monetary, and the logistics process model; mobilization, acquisition, warehousing, distribution, transportation, quality control, and coordination to explain the interaction of critical success factors (CSFs) and their impact on disaster response outcomes.

A critical finding is that weaknesses in one dimension cascade into others, creating systemic vulnerabilities. For instance, delays in Monetary approvals slow procurement (Acquisition), which reduces stock availability (Material) and increases the pressure on distribution networks (Machine and Man). Similarly, poor Mechanism (coordination) weakens mobilization and transportation, even when resources are available. The findings confirm these observations but extend them by illustrating interdependencies, specifically the broader effects that arise when one CSF fails, amplifying weaknesses across the humanitarian logistics chain.

The results both validate and contest the previous study by Bastos, Campos, and Bandeira [14], which delineated seven fundamental logistical processes, highlighting structural obstacles in procurement, storage, and transportation. However, this research corroborates their framework while introducing complexity by demonstrating that these processes are both operational and relational, necessitating inter-agency coordination and adaptive mechanisms. Likewise, previous studies frequently characterize civil-military collaboration as a coordinating asset. This study's findings confirm that the military provides essential resources and logistical support; however, this collaboration is effective only when civilian agencies incorporate mobilization and simulation exercises that accurately represent actual capabilities. The Malaysian case enhances the literature by demonstrating that coordination cannot depend exclusively on formal authority or resource dominance, but must be rooted in trust, practice, and adaptive governance.

Although this study is focusing on Malaysia, the findings are applicable to a broader context in Southeast Asian disasters. Countries, for instance, Indonesia and the Philippines, encounter similar constraints of decentralized governance, limited transport infrastructure, and insufficient funding mechanisms. This research suggests that the interdependencies of CSFs, specifically the link between financial agility, supplier reliability, and transport resilience, are transferable lessons for other disaster-prone areas. At the same time, the Malaysian experience highlights the importance of customizing strategies to local cultural and institutional realities, such as bureaucratic approval systems and community preferences in food provision. These insights contribute to the discussion on how humanitarian logistics operations can become more agile and resilient in complex, multi-stakeholder environments.

Strengthening humanitarian logistics necessitates concrete actions across the 5M dimensions, which requires investing in pre-disaster training and simulation exercises (Man), sustaining adaptable and resilient assets (Machine), pre-positioning critical supplies through flexible warehousing solutions (Material), creating interoperable information systems and conducting multi-agency drills (Mechanism), and preparing contingency funding mechanisms for rapid resource allocation (Monetary). In addition to broader concerns for coordination and finance, these initiatives represent practical actions that agencies might take to enhance organizational resilience.

6 Conclusion

This study investigated the challenges faced by humanitarian logistics operations in Malaysia by using the 5M approach within a logistics process-oriented perspective encompassing mobilization, acquisition, warehousing, distribution, transportation, quality control, and coordination. The findings reveal that these dimensions are significantly interconnected, with deficiencies in one area extending throughout the entire logistical chain. The study validates existing knowledge on the significance of coordination and preparedness by referencing previous work, while also enhancing this understanding through evidence of systematic interdependencies and contextual factors, including bureaucratic approval processes and evolving risk environments.

Even though the study focuses on Malaysia, it provides broader lessons applicable to Southeast Asia and other disaster-prone areas with similar governance and resource limitations. Significantly, it highlights that enhancing humanitarian logistics necessitates not only better coordination and funding but also actionable strategies across the 5M dimensions, which consist of skilled personnel, robust fleets, adaptable stockpiling, interoperable coordination structure, and responsive financial systems. These findings enhance both academic study findings and practical strategies for developing more efficient and effective humanitarian logistics systems.

Acknowledgements. This research is fully supported by the FRGS grant R0163-FRGS/1/2024/SS02/UPNM/02/2. The authors fully acknowledge the Ministry of Higher Education Malaysia, Yayasan TM, and UPNM for the approved funding, which makes this important research viable and effective.

References

1. S. Yusoff, N.H. Yusoff, Disaster risks management through adaptive actions from human-based perspective: Case study of 2014 flood disaster. Sustain. 14, 7405 (2022). https://doi.org/10.3390/su14127405

2. L. Davidson, Key Performance Indicators in Humanitarian Logistics. MIT Center for Transportation and Logistics (2006).
3. S. Rahman, Malaysia's floods of December 2021: Can future disasters be avoided? Perspective: ISEAS Yusof Ishak Institute 26, 1–15 (2022).
4. R. Arcala Hall, Civil-military cooperation in international disaster response: the Japanese Self-Defense Forces' deployment in Aceh, Indonesia. Korean J. Defense Anal. 20, 4, 383–400 (2008). https://doi.org/10.1080/10163270802507310
5. M.A. Salam, S.A. Khan, Lessons from the humanitarian disaster logistics management: A case study of the earthquake in Haiti. Benchmarking 27, 4, 1455–1473 (2020). https://doi.org/10.1108/BIJ-04-2019-0165
6. R.M. Zain, H. Mohd Zahari, Inter-agency information sharing coordination on humanitarian logistics support for urban disaster management in Kuala Lumpur. Front. Sustain. Cities 5, 1149454 (2023).
7. J. Kembro, N. Kunz, L. Frennesson, D. Vega, Revisiting the definition of humanitarian logistics. J. Bus. Logist. 45, 2, 1–18 (2024). https://doi.org/10.1111/jbl.12376
8. M.A.G. Bastos, V.B.G. Campos, R.A.M. Bandeira, Logistic processes in a post-disaster relief operation. Procedia Soc. Behav. Sci. 111, 1175–1184 (2014). https://doi.org/10.1016/j.sbspro.2014.01.152
9. M. Azmat, S. Kummer, Importance of key success factors for local and international NGOs in humanitarian supply chain. Logforum 15, 4, 545–555 (2019). https://doi.org/10.17270/J.LOG.2019.372
10. H.M. Zahari, R. Zain, A. Ismail, N.A.M. Zainol, S. Yaacob, N.I.R.C. Ali, A framework for humanitarian logistics coordination in disaster relief. IOP Conf. Ser. Earth Environ. Sci. 1479, 012061 (2025). https://doi.org/10.1088/1755-1315/1479/1/012061
11. G. Kovács, I. Falagara Sigala, Lessons learned from humanitarian logistics to manage supply chain disruptions. J. Supply Chain Manag. 57, 1, 41–49 (2021). https://doi.org/10.1111/jscm.12253
12. D.E. Polkinghorne, Phenomenological research methods. In: Existential-Phenomenological Perspectives in Psychology, 41–60 (Plenum Press, New York, 2011). https://doi.org/10.4135/9781412995658
13. S.C. Weller, et al., Open-ended interview questions and saturation. PLoS One 13, 6, e0198606 (2018). https://doi.org/10.1371/journal.pone.0198606
14. M.A.G. Bastos, V.B.G. Campos, R.A.M. Bandeira, Logistic processes in a post-disaster relief operation. Procedia Soc. Behav. Sci. 111, 1175–1184 (2014). https://doi.org/10.1016/j.sbspro.2014.01.152
15. V. Braun, V. Clarke, Thematic analysis. In: APA Handbook of Research Methods in Psychology 2, 57–71 (2012). https://doi.org/10.1037/13620-004

Modeling of Spatial Utilization Based on Tsunami Disaster Mitigation Areas Affected by Seram 1899, Maluku

Ferad Puturuhu[1,*], *Evert* R. Sopacua[2], *Anawat* Suppasri[3], *Teuku* Andri Renaldi[4], *Hertine* M Kesaulya[5]

[1]Soil Science Study Program, Faculty of Agriculture, Pattimura University, Indonesia
[2]Urban and Regional Planning Study Program, Faculty of Engineering, Pattimura University
[3]International Research Institute of Disaster Science (IRIDeS), Tohoku University, Sendai, Miyagi Prefecture, 9800845, Japan
[4]Civil Engineering Department, Faculty of Engineering, Universitas Syiah Kuala, Banda Aceh, 23111, Indonesia
[5]Urban and Regional Planning Study Program, Faculty of Engineering, Pattimura University, Indonesia

Abstract. Teluk Elpaputih District, located in Central Maluku, faces high tsunami risk due to its coastal geography and historical events, including the 1899 Seram tsunami. This study aims to (1) analyze tsunami vulnerability levels and (2) propose spatial utilization models for mitigation. Using GIS-based overlay analysis, parameters such as distance from earthquake source, land elevation, slope, coastal morphology, and presence of barrier islands were evaluated. Results show three vulnerability classes, with high-risk zones concentrated in the southern coastal areas and low-risk zones in the northern mountainous region. Based on these findings, spatial utilization was divided into four zones: protection, restricted, conditional, and development. The proposed model provides guidance for safer, productive, and sustainable spatial planning, supporting disaster risk reduction and regional resilience in Central Maluku.

1 Introduction

Geographically, Indonesia is an archipelagic nation located between tectonic plates, along with active volcanic mountains and earthquake faults. This geographical location makes Indonesia vulnerable to geological disasters such as earthquakes and tsunamis [1] [2]. Tsunamis are natural disasters that cause both material and human losses. According to the Directorate of Volcanology and Geological Disaster Mitigation (DVMBG), 28 regions in Indonesia are prone to earthquakes and tsunamis, including Maluku Province [3].

Based on the Arica tsunami incident in Aotearoa, New Zealand, from 1868 to 2018, ceremonies were still held to commemorate the disaster 150 years ago [4]. Commemorative events held in the weeks, months, or years after a disaster may play an important role in

* Corresponding author: feradputuruhu@gmail.com

community recovery and foster community resilience. In addition to the Tsunami Event in New Zealand there is also the 2011 Great East Earthquake and Tsunami Event in Japan 2011, Japan is one of the countries that is believed to be the most prepared country in the world to face the tsunami disaster, recognizing that the variability in the size interval and periodization of large earthquakes in the subduction zone cannot be resolved well only by seismological studies, especially for the occurrence of extreme events with a periodization interval of 500 to 1000 years [5]. From the Arica Tsunami tsunami in Aotearoa-New Zealand and the 2011 Great East Earthquake and Tsunami Event in Japan 2011 it can be concluded that the tsunami event that occurred in the past and has a long periodization also deserves to be watched out for because the tsunami is a passive disaster that occurs but if it occurs it has a very large impact of damage. On August 19, 2024, the Maluku Meteorology, Climatology, and Geophysics Agency (BMKG) issued a high-risk warning for the Maluku region, which is at risk of a megathrust earthquake. This megathrust earthquake is located in the Banda Sea and has the potential to trigger a large earthquake and tsunami. Areas in Maluku potentially affected include Ambon, Seram, Ternate, and the Banda Islands [6].

Based on the Central Maluku Regency Spatial Plan for 2011 – 2031, Central Maluku Regency, which is located in Maluku Province, is an area with a high level of seismicity and is an area that is flanked by the meeting of 3 continental and oceanic plates that collide with each other, thus having the potential for earthquakes and tsunamis. The tsunami that occurred in Central Maluku Regency, namely in Teluk Elpaputih District in 1899, is also known as the "Seram Hazard." The tsunami that occurred in Teluk Elpaputih District resulted in the submersion of most of the Paulohy-Samasuru area, claiming 2,400 lives and impacting several surrounding villages [7].

Teluk Elpaputih District is located in the southern coastal area of Central Maluku Regency, where most of the people live and work in this coastal area. A tsunami disaster that can occur at any time will cause negative impacts such as damage to physical infrastructure supporting social and economic activities, as well as environmental damage that disrupts social and economic activities of the community and threatens the lives of people living in tsunami-prone areas [8].

Based on historical records of disasters and the potential for tsunamis in the Teluk Elpaputih District, disaster management efforts are needed. Disaster management is a dynamic, integrated, and sustainable process to improve the quality of steps related to a series of activities, including prevention, mitigation, preparedness, emergency response, evacuation, rehabilitation, and reconstruction [9]. Tsunami Disaster Mitigation Efforts are needed to reduce the negative impacts of an impending Tsunami [10]. Based on the results of initial interviews, disaster mitigation efforts in Teluk Elpaputih District have not been carried out optimally due to the lack of adequate disaster mitigation support infrastructure, which can create enormous impacts and losses if a Tsunami occurs at any time. To minimize the impact and losses of the Tsunami disaster that occurred in Teluk Elpaputih District in the future, disaster mitigation efforts are needed through better spatial planning.

The goal of disaster mitigation-based spatial planning is to create regional resilience against disasters [11] [12]. One aspect of disaster mitigation-based spatial planning is directing spatial use in tsunami-prone areas. This spatial use guidance is necessary to anticipate the threat of tsunamis in the Teluk Elpaputih District. Therefore, the aim of this study is to determine a model for spatial utilization guidelines based on disaster mitigation in Teluk Elpaputih District, Central Maluku Regency.

2 Methodology

The location used as the object of research is Teluk Elpaputih District. Teluk Elpaputih District has five villages, namely Elpaputih, Sahulau, Liang, Tananahu, and Waraka. The research location can be seen in Figure 1, a map of the research location.

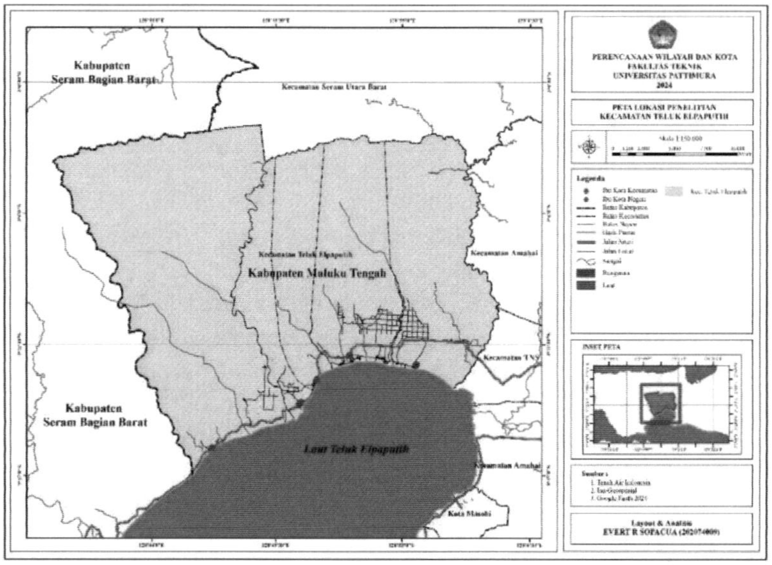

Fig. 1. Research Location Map.

The relationship between research objectives, variables and research indicators is presented in Table 1 below.

Table 1. Research Variables and Indicators.

No	Research purposes	Variables	Indicators	Parameters	Sources/ References
1	Analyzing the level of tsunami disaster vulnerability in Teluk Elpaputih District, Central Maluku Regency	Tsunami Disaster Vulnerability	Tsunami Disaster Vulnerabi lity Level	• Distance from Earthquake Source (km) • Land Elevation (m) • Slope (%) • Distance from River (m) • Land Protection • Presence of Barrier Islands • Coastline Morphology • Distance From Shoreline (m)	(Sidik et al., 2022)
2	Proposing guidelines for disaster mitigation-	Disaster Mitigation-Based Spatial	Tsunami Disaster Vulnerability	Tsunami Disaster Vulnerability	

No	Research purposes	Variables	Indicators	Parameters	Sources/ References
	based spatial utilization based on tsunami vulnerability levels in Teluk Elpaputih District, Maluku Regency.	Utilization Guidelines	Tsunami History Notes	Tsunami Reach	Ihsan, F. & Pramukanto, Q. (2017)
			Existing Condition	Land Use Spatial Pattern Plan	Yuniartanti R. (2021).

The results of the Tsunami Vulnerability Level in Teluk Elpaputih District, Central Maluku Regency, were used [13]. Obtained through the overlay of various existing landslide parameters. Tsunami vulnerability is an integral part of this research; based on the research title, we focus more on spatial planning modeling as a basis for creating a model for spatial utilization guidelines based on disaster mitigation. After that, designating disaster-prone zones as protected or restricted development areas is a strategic step in mitigation-based spatial planning. This method aims to protect communities and the environment from the impacts of disasters with a risk-based approach. The following explains the disaster mitigation-based Spatial Planning Modeling method as carried out for the research location with the following provisions.

2.1 Utilization of Tsunami Vulnerability Levels

a) High-risk zones are prohibited for development; b) Protective functions are enhanced through vegetation conservation, green belts, and controlled spatial use; c) Development activities are restricted or prohibited in vulnerable zones; d) Spatial use is directed towards non-permanent functions such as agriculture, conservation, or green open spaces; e) Reinforced by regulations such as PUPR Ministerial Regulation No. 22 of 2007 and Presidential Decree No. 32 of 1990; f) The results of the disaster vulnerability analysis are used as the basis for determining safe zones for development.

2.2 Spatial Planning

a) Settlements are directed to low-risk zones, away from active faults, river banks, and steep slopes; b) Vital infrastructure (hospitals, schools, disaster posts) is placed in strategic locations with high accessibility and minimal risk; c) Spatial planning is based on the principles of reducing vulnerability and increasing regional capacity.

2.3 Zoning Regulations

a) Establish geological protection zones and development-restricted zones in disaster-prone areas; b) Regulate the types of activities permitted in each zone, for example: High-risk zones: only for conservation or green open spaces, Medium-risk zones: conditional development with engineering, Safe zones: permitted for residential areas and public facilities; c) Zoning regulations serve as an instrument for controlling spatial use.

The spatial planning direction zone division map was created through GIS analysis using ArcGIS software, and the modeling using SketchUp is a 3D modeling software [14].

 https://doi.org/10.1051/e3sconf/202565103007

3 Results and Discussion

The level of tsunami disaster vulnerability in Teluk Elpaputih District [13], consists of 3 classes, namely low vulnerability level, medium vulnerability level and high vulnerability level. The tsunami vulnerability level of Teluk Elpaputih District. The high tsunami vulnerability area is located in the southern part of Teluk Elpaputih District, which is on the coast with an area of 3429.01 ha, while the low tsunami vulnerability area is in the northern part of Teluk Elpaputih District, which is in a mountainous area with an area of 40513.38 ha (Figure 2).

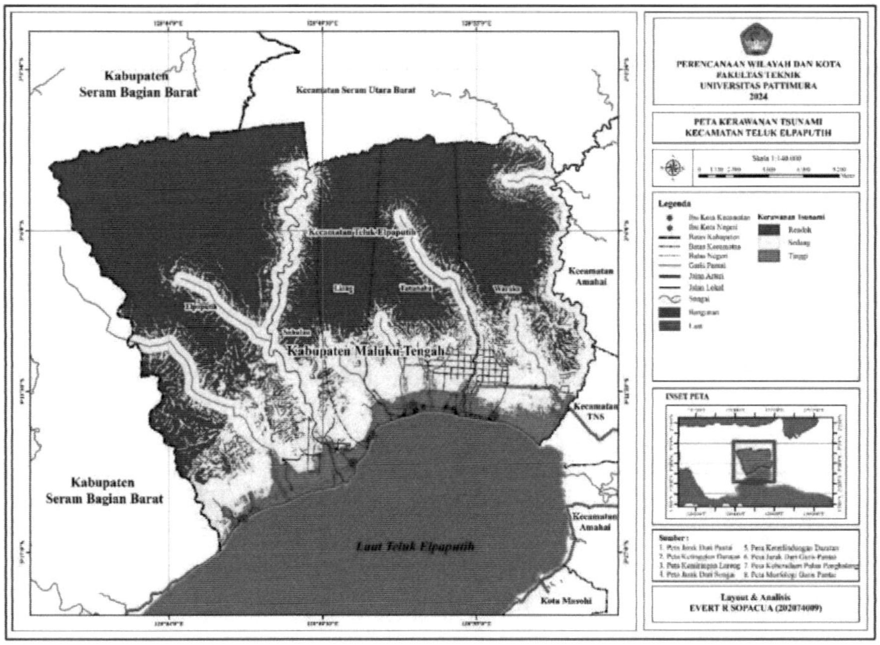

Fig. 2. Tsunami hazard Map. Source: Sopacua (2024) [13].

The Spatial Utilization Guidelines for Teluk Elpaputih District were developed based on Disaster Mitigation to create a safe, comfortable, productive, and sustainable space. The spatial utilization guidelines for Teluk Elpaputih District were implemented by considering the history of the Teluk Elpaputih District tsunami (Seram hazard 1899), with the highest tsunami reach of 270 m from the coastline. Existing conditions and the results of the analysis of the level of vulnerability to tsunami disasters. Based on the results of consideration of the direction of spatial use, Teluk Elpaputih District was developed into 4 zones, namely the protection zone which is at the high KRB level with a distance from the coast of 0 - 100 meters, the limited zone which is at the High KRB level with a distance from the coast >100 meters, the conditional zone which is at the Medium KRB level with a distance from the coast of 500 - 1000 meters, and the development zone which is at the Medium KRB level with a distance from the coast >1000m and the Low KRB level with a distance from the coast >1000 meters. The direction of space use in the 4 zones is explained as follows:

3.1 Protection Zone

Planning for the protection zone is optimized as a protected area integrated with coastal tourism activities. This zone serves as a coastal boundary that functions as a protected area, providing coastal conservation to reduce tsunami wave energy reaching land during a tsunami disaster. The spatial use guidelines for the Protection Zone include coastal embankments, tourism areas, coastal conservation, and evacuation routes.Coastal conservation as a barrier has a vegetation density of between 40-60 trees per 100 m2, depending on the shape and characteristics of the trees planted. It is recommended that coastal border areas that have been designated as protected areas be equipped with structural and non-structural mitigation facilities, and that no further development activities be allowed in the protected zone.

The use of residential space within existing protection zones is not permitted for the addition of new settlements. Existing settlements must be equipped with evacuation facilities and structural reinforcements to withstand earthquakes and tsunamis. Based on the analysis of residential areas within protection zones, they are highly vulnerable. To minimize the impact on these areas, settlements located in high tsunami hazard zones are being directed to be relocated. Further studies are needed regarding relocation locations and special policies for settlements in areas prone to tsunamis.

3.2 Restricted Zone

A restricted zone is a zone designated as an area permitted for limited space utilization activities. Directions for space utilization in restricted zones include plantations, agricultural land, open spaces, low-density settlements, evacuation routes, and temporary evacuation sites (TES). Evacuation facilities in restricted use zones include evacuation routes and temporary evacuation sites (TES). TES utilize existing tall buildings that meet the criteria. Permitted open spaces include public cemeteries, sports fields, and parks. The permitted use of space in restricted zones must take into account river boundary areas in accordance with PUPR Regulation No. 28 of 2015 concerning the Determination of River and Lake Boundary Lines.

3.3 Conditional Zone

Conditional zones are zones designated as areas permitted for conditional spatial utilization activities. Directions for spatial utilization in conditional zones include plantations, agricultural land, open spaces, medium-density settlements, evacuation routes, and temporary evacuation sites (TES). Evacuation facilities in the conditional zone are in the form of Evacuation Routes and Temporary Evacuation Sites (TES). TES utilizes existing tall buildings that meet the criteria. Allowed open spaces include public cemeteries, sports fields, and parks. Directions for the use of space permitted in the conditional zone must pay attention to river boundary areas in accordance with PUPR Ministerial Regulation No. 28 of 2015 concerning the Determination of River and Lake Boundary Lines.

3.4 Development Zone

The Development Zone is planned as a development area for Teluk Elpaputih District with a spatial utilization arrangement as a medium-high density residential area, public services, government, industry, education, commercial, production forest, protected forest, evacuation routes and final evacuation site (TEA). Final evacuation sites (TEA) utilize green open spaces with trees and public facilities, such as places of worship, sports fields, and parks, that are deemed safe. Facilities built in low-risk zones (Development) must meet earthquake and tsunami safety standards, particularly for facilities intended to serve as evacuation sites. Therefore, the feasibility of facilities must be maintained even after being hit by a large magnitude earthquake. In addition, the evacuation sites provided must be easily recognized by the public with evacuation site signs. Directions for permitted spatial use in the development zone must pay attention to river boundary areas in accordance with PERMEN PUPR No. 28 of 2015 concerning the Determination of River and Lake Boundary Lines. Specifically for protected forests in this zone, development activities are prohibited (in accordance with the Central Maluku Regency Spatial Plan 2011-2031).

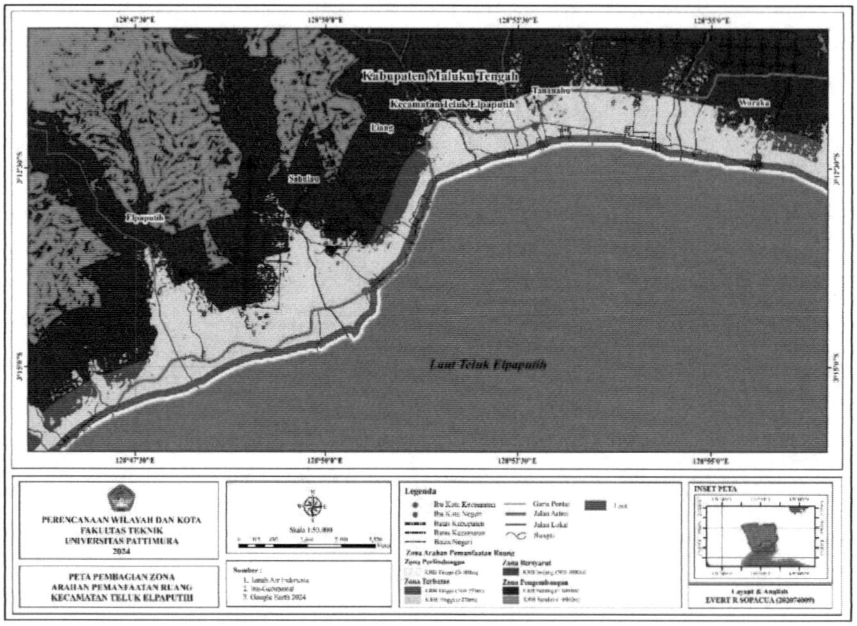

Fig. 3. Map of Zone Division for Spatial Utilization Directions.

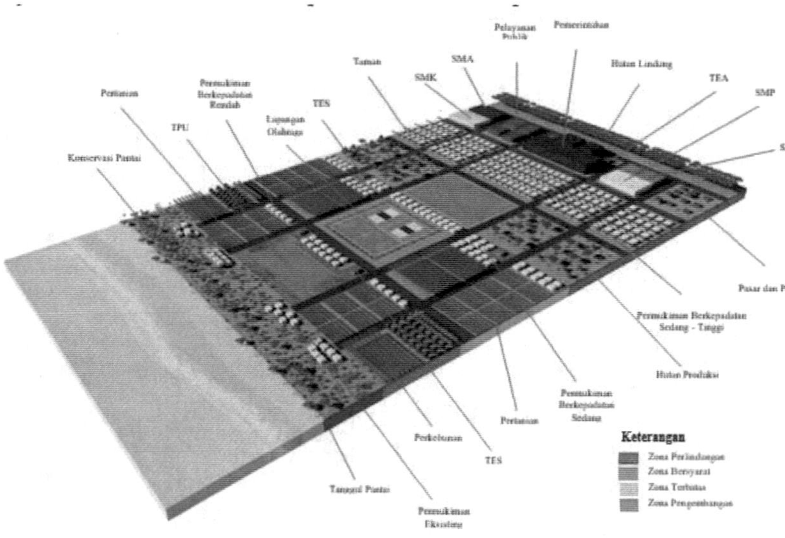

Fig. 4. Modeling of Directions for Space Utilization Based on Existing Conditions in Teluk Elpaputih District. Source: (SketchUp Analysis Results, 2024).

Compared to the division of zones according to the Sendai Framework for Disaster Risk Reduction 2015–2030 issued by UNDRR (United Nations Office for Disaster Risk Reduction), and emphasizing the importance of area-based risk management and safe and resilient use of space, there are differences. The Sendai Framework's zoning principles are: Risk Zoning: Identification and classification of areas based on risk levels; Protection Zoning: Establishment of protected areas for conservation and mitigation; Safe Use Zoning: Placement of economic and social activities in low-risk zones; and Evacuation and Response Zoning: Arrangement of strategic evacuation routes and locations [15]. The restricted zone in this study has the same function as the restricted zone in the Sendai Framework, and the development zone also has the same function as the safe utilization zone in Sendai.

4 Conclusion

The spatial utilization guidelines for Teluk Elpaputih District take into account the history of the Teluk Elpaputih District tsunami (Seram hazard 1899) with the highest tsunami reach of 270 m from the coastline, existing conditions, and the results of the tsunami disaster vulnerability analysis. Based on the results of the considerations, the spatial utilization guidelines for Teluk Elpaputih District were developed into 4 spatial utilization guidelines zones, namely the protection zone, the limited zone, the conditional zone, and the development zone. The division of the spatial utilization zones for the Teluk Elpaputih District is as follows:

1. Protection zones at the High KRB level with a distance from the coast of 0–100 meters.
2. Restricted zones at the High KRB level with a distance from the coast of >100 meters.
3. Conditional zones at the Medium KRB level with a distance from the coast of 500–1000 meters.
4. Development zones at the Medium KRB level with a distance from the coast of >1000 meters and Low KRB levels with a distance from the coast of >1000 meters.

Based on the Disaster Mitigation-Based Spatial Utilization Directions, it is recommended that: 1) The results of this study can be input and consideration in revising the Regional Regulation of Central Maluku Regency No. 1 of 2012 concerning the Regional Spatial Planning (RTRW) of Central Maluku Regency in the future; 2) The results of this study are expected to be a reference in compiling the Detailed Spatial Planning (RDTR) of Teluk Elpaputih District.

References

1. Schlehe, J. (2010). Anthropology of religion: Disasters and the representations of tradition and modernity. *Religion,* **40**(2), 112–120. https://doi.org/10.1016/j.religion.2009.12.004
2. Ahmad Pratama Putra, *Penataan Ruang Berbasis Mitigasi Bencana Kabupaten Kepulauan Mentawai*, *Jurnal Dialog Penanggulangan Bencana (JDPB),* **2**(1) (2011).
3. Pusat Vulkanologi dan Mitigasi Bencana Geologi, *Laporan khusus analisis geologi gempa bumi di wilayah Kabupaten Nabire, Provinsi Papua Tengah* (Kementerian Energi dan Sumber Daya Mineral Republik Indonesia, 2025).
4. Thomas, K. L., Kaiser, L., Campbell, E., Johnston, D., Campbell, H., Solomon, R., … & Northern, A., Disaster memorial events for increasing awareness and preparedness: 150 years since the Arica tsunami in Aotearoa-New Zealand, *Aust. J. Emerg. Manag.* **35**(3), 71–78 (2020).
5. Satake, K. (2015). Geological and historical evidence of irregular recurrent earthquakes in Japan. Philosophical Transactions of the Royal Society A: Mathematical, Physical and Engineering Sciences, 373(2053), 20140375. https://doi.org/10.1098/rsta.2014.0375
6. BMKG, Buletin Stasiun Badan Meteorologi, Klimatologi, dan Geofisika (BMKG, 2024).https://stamar-ambon.bmkg.go.id/wp content/uploads/pdf%20edisi/12.BULETIN_DESEMBER%20Print.pdf
7. Martin, S. S., Cummins, P. R., & Meltzner, A. J. (2022). Gempa Nusantara: A Database Of 7380 Macroseismic Observations For 1200 Historical Earthquakes In Indonesia From 1546 To 1950. *Bulletin of the Seismological Society of America,* **112**(6), 2958–2980. https://doi.org/10.1785/0120220047

8. Jokowinarno, D., Mitigasi Bencana Tsunami Di Wilayah Pesisir Lampung, *Jurnal Rekayasa* **15**(1), (2011).
9. Leelawat, N., Laosunthara, A., Tang, J., Suppasri, A., Ruangrassamee, A., Akkharaprathompong, P., & Imamura, F., The 2011 Great East Japan Earthquake and Tsunami: A message from Japan to Thailand, *J. Disaster Res.* **16**(6), 908–913 (2021). https://doi.org/10.20965/jdr.2021.p0908
10. Suppasri, A., Shuto, N., Imamura, F., et al., Lessons learned from the 2011 Great East Japan tsunami: Performance of tsunami countermeasures, coastal buildings, and tsunami evacuation in Japan, *Pure Appl. Geophys.* **170**, 993–1018 (2013). https://doi.org/10.1007/s00024-012-0511-7
11. Wijaya, N., Bisri, M., Aritenang, A., & Mariany, A., Spatial planning, disaster risk reduction, and climate change adaptation integration in Indonesia: Progress, challenges, and approach, in *Disaster Risk Reduction in Indonesia: Progress, Challenges, and Issues*, pp. 235–252 (Springer, Cham, 2017). https://doi.org/10.1007/978-3-319-54466-3_9
12. Novi Rahmawati, *Penataan Ruang dan Pertanahan: Penataan Ruang Berbasis Bencana* (Universitas Gadjah Mada Fakultas Geografi, 7 Juni 2010). https://geo.ugm.ac.id/2010/06/07/penataan-ruang-dan-pertanahan-penataan-ruang-berbasis-bencana/.
13. Sopacua, E., *Mitigasi bencana tsunami di Teluk Elpaputih Kabupaten Maluku Tengah* (Skripsi, Prodi Perencanaan Wilayah Kota, Universitas Pattimura, 2024).
14. Ihsan, F., & Pramukanto, Q., *Pemodelan pemanfaatan ruang berbasis mitigasi bencana* (Pusat Penelitian dan Pengembangan Perumahan dan Permukiman, Kementerian Pekerjaan Umum dan Perumahan Rakyat, Jakarta, 2017).
15. UNDRR, *Sendai Framework for Disaster Risk Reduction 2015-2030* (United Nations Office for Disaster Risk Reduction, 2015). https://www.undrr.org/publication/sendai-framework-disaster-risk-reduction-2015-2030.

Women's Leadership, Local Wisdom and Household Resilience during Multiple Crisis (Tsunami, Conflict, Pandemic) in Ambon, Indonesia

Ismiatul Ramadhian Nur[1], *Rizka* Puspitasari[2], *M.* Arifin[2], *Bokiraiya* Latuamury[3], and *Rina Suryani* Oktari[4,*]

[1]Tsunami and Disaster Mitigation Research Center (TDMRC) Universitas Syiah Kuala and Department of Islamic Community Development, Faculty of Da'wah and Communication, UIN Ar-Raniry, 23111, Banda Aceh, Indonesia
[2]Tsunami and Disaster Mitigation Research Center (TDMRC) Universitas Syiah Kuala, 23111, Banda Aceh, Indonesia
[3]Tsunami and Environmental Science Study Program, Faculty of Agriculture, Universitas Pattimura, 97223, Ambon, Indonesia
[4]Tsunami and Disaster Mitigation Research Center (TDMRC); Department of Family Medicine, Faculty of Medicine, Universitas Syiah Kuala, 23111, Banda Aceh, Indonesia

Abstract. Women's leadership is critical to disaster resilience, yet its role in *compound crises* remains underexplored. This study examines how women in Ambon, Indonesia, strengthened community resilience through three consecutive crises: the 1999 social conflict, the COVID-19 pandemic, and the continuing tsunami threat. Using a descriptive qualitative design, we conducted in-depth interviews with 19 women leaders and applied thematic analysis informed by social capital (bonding, bridging, linking) and resilience capacities (absorptive, adaptive, transformative). Findings show that women safeguarded families and mediated peace during the conflict, sustained household economies and psychosocial support during the pandemic, and led evacuation and low-tech early-warning initiatives in response to tsunami risk. By activating social capital, drawing on local wisdom such as *pela gandong*, and fostering adaptive innovations, Ambonese women built multi-layered resilience that bridges domestic and public spheres. These results underscore the importance of integrating women's leadership and intersectional perspectives into disaster-risk governance to foster long-term, transformative capacity in island regions facing recurrent, overlapping hazards.

1 Introduction

Community resilience depends not only on infrastructure and policy but also on women's agency as local social actors. Classical disaster literature often frames women as a vulnerable group exposed to multiple risks because of biological factors, social roles, and unequal access

* Corresponding author: okta@usk.ac.id

to resources. More recent scholarship, however, highlights women as agents of change who build community capacity, foster peace, secure household economies, and maintain social cohesion [1][2]. Yet little is known about how women exercise such leadership when multiple crises, such as social conflict, pandemic, and natural hazards occur sequentially and interact. This study aims to address that gap.

Three theories inform the analysis. Social capital contains three elements: bonding (close in-group relations, e.g., church congregations or families), bridging (connections between social groups, e.g., religion or ethnicity), and linking (vertical relations to powerful actors or institutions). Social capital enables women to mobilize resources, build trust, and develop good communication channels in the process of and following a disaster [3].

Second, gender-sensitive resilience highlights that gender norms, household and public role divisions, and unequal access to resources define women's experiences and capacities in times of crisis. Intersectional characteristics such as religion, class, and local traditions also define how women both face risk and organize responses. From this perspective, women are not passive recipients of impacts, but active agents who initiate adaptive processes, ranging from household food management to policy advocacy and cross-community facilitation [4].

Third, resilience capacity consists of three dimensions: (1) absorptive capacity, the ability to withstand initial shocks through family networks or informal coping systems; (2) adaptive capacity, the ability to adjust to new conditions through livelihood innovation, health-protocol adaptation, or online communication; and (3) transformative capacity, the ability to enact long-term structural change such as strengthening women's positions in decision-making forums or embedding local wisdom in formal policies [5].

Ambon has a long history of multidimensional crises. The 1999 social conflict produced religion-based spatial segregation, eroded social cohesion, and impoverished vulnerable groups such as women and children. In the post-conflict period, women from Ambon were more involved in the economy, typically on insecure conditions under which they had to adjust rapidly [1]. COVID-19 laid down mobility restrictions, loss of earnings, more domestic workload, and heightened gender-based violence risks. Being an archipelagic region subject to a high risk of earthquakes and tsunamis, Ambon faces interrelated hazards requiring locally contextualized, intersectional, and gendered resilience responses [2]. In this capacity, women do several roles as peace activists, community leaders, interfaith facilitators, and household stewards. These roles are particularly critical for reconciliation, economic recovery, and disaster risk reduction.

Local wisdom, such as *pela gandong*, economic solidarity, and cross-community forums, forms part of the social capital that strengthens family and community resilience [6][4]. Guided by these frameworks, this study responds to four research questions: (1) What are the women leaders in Ambon's roles in building community resilience at the phases of social conflict, COVID-19, and natural hazard caused disasters? (2) What leadership strategies and initiatives strengthen family functioning and social cohesion? (3) How do women utilize bonding, bridging, and linking social capital, as well as local wisdom like pela gandong, in mitigation, recovery, and reconciliation after a crisis, and (4) What are the enabling and constraining factors shaping their double roles as community leaders and household managers? These perspectives are conceptual background and analytical map to account for empirical stories and constructing a gendered resilience model specific to Ambon.

2 Methods

2.1 Study Design

This research employed a descriptive qualitative design in exploring the experiences and coping strategies of women in contributing to community resilience amidst three types of crises in Ambon, namely social conflict, COVID-19, and the imminent tsunami threat. This research design was employed to offer rich contextual descriptions and tease out emergent themes.

The research was conducted in Maluku Province, Ambon city, an area with a long history of social conflict, high vulnerability to natural hazard caused disasters, particularly tsunamis due to its archipelagic location, and also hit by the COVID-19 pandemic in 2019–2021. Its other unique characteristic is the intense social capital based on local wisdom such as *pela gandong*. These characteristics provide a unique and relevant context for investigating women's leadership in times of crisis.

2.2 Data Collection

Informants participated in this study were 19 prominent women (coded P1–P19), representing diverse roles such as women leaders, female academics, and women activists. Data were collected through in-depth interviews conducted by enumerators over a period of two months. Three of the 19 participants were also involved in a workshop on a similar theme as part of a data triangulation.

Eligibility criteria were residence in Ambon for at least twenty years and demonstrated community leadership experience through one or more crises. Informant selection was done in consultation with local community networks to ascertain that the informant had relevant experience and knowledge. All participants gave informed written consent before the interviews, and a multi-layered anonymization strategy was employed by replacing names with participant codes (P1–P19) and removing identifying details from transcripts and reports.

Table 1. Characteristics of Informants.

Code	Name/Initials	Background/Role
P1	FS	Former Ambon Official and Women's Rights Activist
P2	FM	Academic and Women's Rights Activist
P3	NS	Academic, Gender Studies Researcher in the Nusantara Context
P4	YT	Academic
P5	RFF	Maluku Government Official
P6	WP	Ambon Government Official
P7	YT	Civil Servant
P8	BT	Community Leader
P9	HP	Community Leader

Code	Name/Initials	Background/Role
P10	AH	Member of Parliament
P11	SM	Journalist
P12	ECM	Religious Leader and Academic
P13	LU	Government Official
P14	AL	Member of Parliament
P15	DL	Women's Rights Activist
P16	JH	Women's Rights Activist
P17	IL	Businesswoman
P18	OL	Religious Leader
P19	SMR	Community Leader

In-depth interviews were conducted using a semi-structured guide with open-ended questions to enable participants to provide rich and contextually grounded narrative responses. These questions were designed to explore the experiences, strategies, and reflections of women leaders regarding their roles during the three crisis phases. The interview guide consisted of 17 core questions, as presented in Table 2.

Table 2. List of Key Interview Questions.

No	Interview Question	Research Question	Relevant Theoretical Basis
1	Experiences during the 1999 social conflict, women's roles	RQ1	Gender-based resilience; absorptive capacity
2	Greatest challenges faced by women during the conflict	RQ4	Gender-based resilience; constraining factors
3	How women-maintained family and child safety during the conflict	RQ1	Bonding social capital; absorptive capacity
4	Women's roles as mediators or connectors between parties	RQ1 & RQ3	Bridging social capital; adaptive capacity
5	Women's roles in post-conflict social reconciliation	RQ1 & RQ3	Bridging & linking social capital; transformative capacity
6	Forms of essential support after the conflict	RQ4	Bonding & linking social capital; enabling factors
7	Role in BPBD during the COVID-19 pandemic	RQ1	Linking social capital; adaptive capacity

8	Strategies to sustain household economy during the pandemic	RQ2	Adaptive capacity; gender-based resilience
9	Balancing BPBD work and household responsibilities	RQ2 & RQ4	Adaptive capacity; gender-based resilience
10	Division of household tasks during the pandemic	RQ2	Bonding social capital; absorptive capacity
11	Role of local wisdom *pela gandong* in disaster situations	RQ3	Bridging social capital; absorptive capacity
12	Mechanisms of aid coordination based on *pela gandong*	RQ3	Bridging & linking social capital; adaptive capacity
13	Factors that make *pela gandong* effective	RQ3	Bridging social capital; transformative capacity
14	Enabling factors for women's dual roles	RQ4	Bonding & linking social capital; enabling factors
15	Constraining factors for women's dual roles	RQ4	Gender norms; constraining factors; transformative capacity
16	Public perceptions of women active in the field	RQ4	Gender-based resilience; transformative capacity
17	Aspects that need strengthening for future women's empowerment	RQ4 & RQ2	Transformative capacity; gender-based resilience

2.3 Data Analysis

Data analysis was conducted in several stages to ensure depth and accuracy of interpretation. All interview recordings were transcribed verbatim to preserve participants' original wording and meaning. The transcripts were read repeatedly to gain a full understanding of the narrative context. Using an inductive approach, meaning units were first identified and labeled through open coding, allowing themes to emerge from the data without predetermined categories. These codes were then organized into broader thematic categories through axial coding.

At this stage, the researchers applied the established theoretical framework, social capital (bonding, bridging, linking), gender-based resilience [3], and resilience capacities (absorptive, adaptive, transformative) [4] as analytical lenses to interpret relationships among categories and to align them with the study's four research questions (RQ1–RQ4). Finally, data condensation was carried out to summarize narratives into a concise form while retaining the essence of participants' experiences.

3 Results

The analysis of the workshop transcripts reveals that women in Ambon play simultaneous roles in both the public sphere (leadership, advocacy, facilitation) and the domestic sphere (household management). They draw upon social capital and local wisdom to strengthen community resilience across three crisis phases: the post-conflict period following the 1999 conflict, the COVID-19 pandemic, and the tsunami threat.

Table 3 presents a general summary of the analytical results, illustrating how the interviews generated relevant information that addresses the research questions and aligns with the theoretical framework adopted in this study. Further explanations are elaborated in subsequent sub-sections, each organized according to the respective crisis phase.

Table 3. Summary of Findings.

Crisis Phase	Research Question	Thematic Category	Theoretical Framework	Informants	Example Verbatim
Social Conflict 1999	RQ1 – Role of female leaders	Family protection & peaceful mediation	Bonding & Bridging – Absorptive Capacity	P1, P2, P6, P9, P14	"We set up a communal kitchen for all displaced people regardless of religion, because hunger knows no differences. Peace is the only way forward."
Post-Conflict	RQ2 – Leadership strategies	Interfaith reconciliation & economic recovery	Bridging & Linking – Adaptive & Transformative Capacity	P2, P5, P7, P12	"In our women's forum, we invited members from different faiths while also offering sewing training so we could engage in positive activities together."
COVID-19 Pandemic	RQ2 & RQ3 – Strategies & social capital	Household economic management & health education	Bonding & Linking – Adaptive Capacity	P4, P8, P10, P16	"I started a home-based cake business while encouraging community members to follow health protocols."
Natural hazard-caused disaster	RQ3 – Utilization of social capital & local wisdom	Activation of *pela gandong* for rapid response	Bridging & Linking – Absorptive & Adaptive Capacity	P3, P6, P11, P15, P18	"As soon as news of the earthquake came, our *pela* relatives immediately sent boats to evacuate coastal residents."
Cross-Phase	RQ4 – Supporting factors	Family support & social networks	Bonding & Linking – Adaptive Capacity	P1, P10, P13, P19	"My husband said: do what you can for the community, and we'll share the household responsibilities."
Cross-Phase	RQ4 – Constraining factors	Mobility restrictions & gender norms	–	P5, P9, P15, P17	"It's hard to attend evening meetings because it's considered inappropriate for women, especially when transportation is limited during heavy rain."

3.1 Social Conflict Phase

The analysis indicates that during the period of social conflict in Ambon, women played a pivotal role in sustaining family life and rebuilding fractured social relations. They mobilized various forms of social capital, ranging from bonding (strong intra-group ties) and bridging (connections across different groups) to linking (relationships with authorities), to coordinate assistance, create a sense of safety, and initiate reconciliation processes.

Women were able to perform dual roles, both within the household and in the public sphere, despite facing constraints such as threats of violence, limited mobility, and heavy domestic burdens. Support from family networks, solidarity among women, and social legitimacy as messengers of peace served as strong enabling factors. Overall, this experience demonstrates that post-conflict community resilience is shaped not only by formal policies or external aid but also by grassroots women's initiatives and leadership that safeguard essential functions, adapt to change, and spearhead social recovery efforts.

3.1.1 Role of Women Leaders in Building Community Resilience During Conflict

The social conflict that erupted in Ambon in 1999 led to the creation of religiously based spatial segregation, which physically divided communities and disrupted the trust networks that had previously connected residents across different identities. This situation triggered uncertainty, violence, and limited access to essential resources, including food, clean water, and healthcare services. With the collapse of governmental structures and formal institutions, women assumed strategic roles to ensure the survival of their families while sustaining their communities.

Women's initiatives included evacuating vulnerable family members, especially children and the elderly, to safer locations, organizing communal kitchens that served residents' food needs regardless of religion, and coordinating the fair distribution of logistics.

The strength of social capital underpinned these efforts. Emotional bonds and trust within homogeneous groups, whether formed through kinship ties, shared religious affiliation, or common experiences, enabled rapid and effective coordination in aid distribution. This bonding capital reinforced logistical capacity and fostered the community's psychological resilience.

As P1 stated:

"We, the women in the village, had to make sure all the children ate; no one could go hungry. Sometimes we had to walk long distances to get rice or vegetables, even when tensions were still high. Many men didn't dare go out for fear of being attacked, so it was us women who took action."

This quotation shows women's willingness to take physical risks to meet community needs. Beyond addressing food needs, women also played strategic roles in community security management, as described by P2:

"I remember clearly, at that time we had to keep watch at night, taking turns with our neighbors. If there were suspicious noises, we immediately informed each other. It was the women who organized the shifts and made sure every household had someone on guard."

3.1.2 Leadership Strategies and Contributions to Family Function and Social Cohesion During Conflict

Based on the analysis of the 1999 Ambon social conflict, women's leadership strategies pursued two primary objectives: safeguarding family well-being and restoring social cohesion that had been damaged by religious segregation. These strategies combined direct field actions with tactical decisions to achieve long-term benefits.

At the family level, women prioritized providing food, protecting their children, and maintaining emotional stability within the household. They organized the distribution of food from communal kitchens, managed household food supplies, and ensured that children continued informal learning despite school closures.

As P1 remarked:

"If the family is not at ease, we cannot work outside. So we made sure the children were safe first, and only then could we help the community."

At the community level, women-initiated communication forums and cross-group collaboration. Even in a tense atmosphere, they utilized informal networks, such as religious study groups, rotating savings gatherings (*arisan*), and extended family meetings, to discuss security, logistics, and psychosocial support.

P2 shared:

"*We shared ideas on how this situation must improve quickly. Nobody enjoys living in conflict... Living in peace is the best.*"

These actions demonstrate how women leveraged both informal and formal community ties to safeguard families and repair relationships during the conflict.

3.1.3 Utilization of Social Capital and Local Wisdom

The 1999 social conflict in Ambon severely disrupted the foundations of social cohesion. Yet local wisdom such as *Pela Gandong*, a traditional pact of brotherhood between villages, remained a moral reference, although its application in the early stages of the conflict was limited. As spatial segregation deepened, the practice of *pela gandong* required systematic efforts from local leaders, including women.

Bonding social capital provided the initial foundation for community survival, while bridging social capital began to emerge through women's efforts to initiate cross-group communication.

P8 described this process:

"*We actually have ancestral ties... Pela Gandong does not recognize religious differences. But during the conflict, it was hard to put into practice. So it was the women who started reaching out to relatives in other villages, asking them to look out for each other.*"

This social capital extended beyond personal relationships to include informal mechanisms for exchanging assistance.

P15 explained:

"During those hard times, we sent each other food supplies through safe routes. Sometimes they went by sea at night so as not to be seen. All of this was done through family and pela networks."

Women also used religious forums to rebuild inter-community communication.

P12, a female religious leader, recounted:

"We used religious study groups and church gatherings to pass on messages of peace. Sometimes it was through joint prayers attended by leaders from both sides. Not many, but enough to remind us that we are all family."

Despite these efforts, formal support channels remained weak in the early phase of the conflict.

P1 acknowledged:

"At that time, there was no official channel directly involving women. So we just moved on our own in whatever way we could."

3.1.4 Enabling and Constraining Factors for Women's Dual Roles

During the 1999 Ambon social conflict, women who carried out dual roles as community leaders and family managers encountered a mix of enabling and constraining factors. Enabling factors arose from strong social networks and deeply rooted cultural values. Support from extended families, solidarity among women, and local wisdom such as P*ela Gandong* [7] and the practice of *gotong royong* (mutual cooperation) served as crucial assets.

P5 noted:

"We strengthened each other, especially among women. If someone was in trouble, we would immediately bring food or help take care of her children. It has always been our habit."

Community trust in women's ability to maintain confidentiality, avoid provocation, and convey messages of peace also provided important support.

P9 explained:

"People trust that when women speak, the atmosphere becomes calmer. We are seen as able to cool down heated emotions."

Constraining factors were equally significant. Fear of violence, limited access to transportation and logistics, and the lack of formal support from government or security institutions posed major challenges.

P17, a female entrepreneur, shared:

"Even going out to deliver aid supplies was difficult, we were afraid of being attacked. Sometimes we had to wait until night or take a detour route."

Heavy domestic workloads also restricted participation.

P14 emphasized:

"If a child is sick or the house is not safe, we simply cannot go out to take care of others. We have to be smart in dividing time and energy."

3.2 Pandemic Phase

The COVID-19 pandemic has posed significant challenges for the people of Ambon, particularly in the economic sector and mental health. Women leaders have played a central role in initiating economic activities, maintaining social cohesion, and providing psychological support amid activity restrictions.

3.2.1 The Role of Women Leaders in Building Community Resilience

The COVID-19 pandemic introduced new challenges to Ambon society, including mobility restrictions, declining household income, and a heightened risk of gender-based violence. In this context, women community leaders again became the front line in sustaining community resilience. They were actively involved in disseminating health protocols, providing food assistance, and establishing solidarity networks for residents affected by the virus.

P16 recounted:

"We set up a communal kitchen again, but this time to help those in isolation. It was women who cooked, delivered the food, and made sure the sick did not feel alone."

This testimony highlights women's direct engagement in meeting urgent community needs during the pandemic.

3.2.2 Leadership Strategies and Contributions to Family Functions and Social Cohesion

Women exercised adaptive leadership by combining technology, local social networks, and time-tested values of mutual cooperation (gotong royong) developed in earlier crises. Simple innovations, such as creating WhatsApp groups, were used to coordinate aid distribution, identify residents in need, and monitor their health conditions. Social media platforms like Facebook and Instagram served as educational channels for disseminating information about health protocols, vaccination, and immunity-maintenance tips.

At the neighborhood level, women revitalized RT/RW-based mutual aid systems, which had weakened because of social restrictions, ensuring that aid coordination could continue without violating health protocols.

P13 illustrated this strategy:

"If we can't meet in person, we use our phones. So even from home, we can still coordinate aid and monitor who needs help."

In addition to their public leadership roles, women played a crucial role within families. They reorganized domestic responsibilities to ensure children's online learning could run effectively, while also safeguarding the emotional stability of household members.

P10 emphasized this challenge:

"This pandemic makes people more easily stressed. So apart from taking care of meals and the children's schooling, we also have to make sure the house stays calm."

3.2.3 Utilization of Social Capital and Local Wisdom

The social capital formed during the post-conflict period became a strategic resource when the COVID-19 pandemic struck. Social networks built through shared experiences in previous crises, such as the 1999 conflict, had created trust and solidarity that facilitated the mobilization of aid.

Women utilized bonding capital through close ties with family, neighbors, and local communities; bridging capital by fostering cooperation across groups, including interdenominational religious organizations; and linking capital by connecting communities to formal institutions such as the Regional Disaster Management Agency (BPBD), the Social Affairs Office, and donor organizations.

P18, a female religious leader, described this role:

"We used church and mosque networks to distribute aid. We didn't look at religion—what mattered was that everyone got it."

The strength of this social capital was also enriched by local wisdom, which is part of the Ambonese cultural identity. The tradition of exchanging food (bakudapa makanan) was used as a means to strengthen social bonds while also providing emotional support for affected residents.

P19 added:

"If someone is sick, we usually send vegetables or cooked fish. That's our way of saying 'we care.'"

In addition, women made use of religious forums, arisan groups, and social media as channels to identify community needs and deliver assistance quickly. This social capital functioned not only as a social safety net but also as an efficient information-distribution mechanism amid mobility restrictions.

3.2.4 Supporting and Inhibiting Factors for Women's Multiple Roles

The supporting factors for women's multiple roles during the COVID-19 pandemic in Ambon included social solidarity built since the post-conflict period, advancements in communication technology that facilitated remote coordination, and strong support from the nuclear family. Solidarity tested in previous crises provided a foundation of trust that made resource mobilization easier. Family support enabled women to remain active in the public sphere without completely neglecting domestic responsibilities.

P4 described this situation:

"The children understand when their mother is busy with aid work, so they can study independently."

However, there were also significant obstacles. Limited internet access posed challenges for coordination and for children's online learning. Domestic workloads increased due to homeschooling, which meant women had to divide their time between public duties and household matters. In addition, fear of virus transmission often became a dilemma when they needed to assist affected residents directly.

P6 expressed this dilemma:

"I'm actually scared, especially if I have to go to a sick person's house. But if we don't go, then who will?"

3.3 Tsunami Disaster

As an archipelagic region located within the Pacific Ring of Fire, Ambon is highly vulnerable to natural hazard caused disasters, particularly earthquakes and tsunamis. The history of major earthquakes that have struck the Maluku region, coupled with early warnings from the Meteorology, Climatology, and Geophysics Agency (BMKG) regarding tsunami potential, makes this threat a tangible reality for the community. Unlike crises caused by social conflict and the pandemic, natural hazard-caused disasters are sudden, high-risk, and require a swift, coordinated response.

In this context, women play a strategic role, not only as recipients of information but also as agents in disseminating preparedness messages at the family and community levels. The social capital developed in the post-conflict period and strengthened during the pandemic serves as a crucial foundation for organizing evacuation drills, disseminating safe route information, and ensuring that vulnerable groups such as children, the elderly, and persons with disabilities receive special attention in emergency response plans.

In line with the concepts of absorptive, adaptive, and transformative capacity [5], the experiences of Ambonese women in facing previous crises have provided them with valuable assets to enhance preparedness for natural hazard caused disasters. However, preparedness for tsunami threats requires a more systematic approach, given the nature of the disaster that can destroy physical infrastructure while also undermining the community's psychological resilience.

3.3.1 The Role of Women Leaders in Building Community Resilience

The tsunami threat in Ambon's coastal areas demands preparedness that integrates both technical capacity and social cohesion. Women leaders serve as key actors in disseminating information, guiding evacuation drills, and ensuring the participation of all groups—including children, the elderly, and men—in the process. The social trust they possess makes their messages more readily accepted across diverse groups.

P3 emphasized the urgency of emergency communication skills:

"We need training to be able to use technology for emergency communication. Sometimes we are unsure how to send messages when the signal is weak. If the power goes out, activities stop, so we need alternatives."

3.3.2 Leadership Strategies and Contributions to Family Function and Social Cohesion

Women's leadership during this phase focuses on mobilizing residents and ensuring household preparedness. They divide evacuation roles among family members, prepare emergency supplies, and ensure that everyone understands evacuation routes. Infrastructure barriers are often addressed through the use of informal forums, such as *arisan* or religious gatherings, to insert preparedness messages.

P7 revealed structural challenges:

"Sometimes we want to act quickly, but there is no internet access. Government assistance also sometimes arrives late, forcing residents to wait a long time."

3.3.3 Utilization of Social Capital and Local Wisdom

The lack of formal infrastructure has driven women to create local innovations for early warning, such as using mosque bells, wooden gongs, or quick messages delivered by motorcycle riders. This strategy leverages bonding and bridging social capital to accelerate the dissemination of information.

One female leader explained:

"If there is no siren, we use the mosque bell or wooden gong. Young people are also trained to quickly inform others by riding motorcycles if they see danger signals."

3.3.4 Factors of Women's Double Roles

Supporting factors include social trust in women's leadership, cross-gender community networks, and the ability to utilize routine activities as disaster education platforms. Moral support from the nuclear family also strengthens their public roles. However, significant obstacles include the limited availability of early warning tools, power outages, and the lack of women's involvement in formal disaster planning.

The tsunami threat phase in Ambon demonstrates that women are not only drivers of participation but also local innovators filling the gaps in formal systems. The use of bonding and bridging social capital enables the creation of effective early warning mechanisms, although limited linking capital and infrastructure constrain long-term transformation. Integrating women's leadership into formal disaster management systems is key to strengthening the sustainable resilience of coastal communities against tsunamis.

4 Discussion

4.1 Women's Role in the 1999 Ambon Conflict

Women took on the core role of community survival in the 1999 Ambon conflict, working in an unstructured and dangerous environment in which state government institutions were either inoperative or abolished outright. Without state authority, women drew on bonding social capital, the networks of trust and reciprocity among their close families and dense communities. Through the establishment of public kitchens, women ensured that households would be fed in even times of resource shortages and destructive insurgencies. Similarly, they instituted night vigils to help provide security, which again indicates their ability to mobilize social capital because of a crisis. These reactions are the finest instances of absorptive capacity, or the ability of communities or people to absorb a shock and get on with vital operations regardless of disruptions. Here, women's work to maintain food, security, and minimal community services functioning despite the breakdown of normal governance infrastructure testifies to their role in sustaining core operations under shock [3,5,7].

In addition to these survival strategies in the here-and-now, women also spearheaded efforts that promoted enhanced community cohesion. They held interfaith dialogue forums, where members of different religious groups would sit together to foster cohesion and reduce sectarian violence. Interfaith forums played a significant role during the war in keeping society intact by filling gaps between antagonistic factions. Besides, women built rotating savings associations, a type of collective economic buffer that allowed women access to finance during moments of need and provided a cushion against financial risk generated by war. Additionally, women persisted in pela gandong, a customary kinship system under which members of different communities or villages assist one another during moments of need, thus fostering bridging social capital. These initiatives indicate a shift from merely maintaining livelihood for family and community (absorptive capacity) to disseminating the social network through inter-group relations, increasing the adaptive capacity of the community when faced with crisis [3,5,6,7].

Even if grassroots movements were effective, research indicates that bonding capital—the link between local communities and formal institutions such as the government—remained poor. Women's work to build community networks was not institutionally acknowledged in post-conflict decision-making. Not being institutionally acknowledged, their contribution never reached its potential for transformative capacity, or the ability to change systems and structures after the crisis. Women's leadership, significant as it was, was largely still in the domestic or informal sphere, reinforcing gendered power structures [1,2,6]. Therefore, even if women had an important role to play in bridging gaps and adapting to new circumstances, their ability to contribute towards structural change was limited by institutional barriers.

4.2 Women's Role during COVID-19 Pandemic

When the COVID-19 pandemic hit, the social networks established in the 1999 Ambon conflict were utilized as a source of trust. The women eventually mobilized bonding social capital once more, distributing resources in neighborhoods and households. But even that aside, they utilized bridging and linking social capital to extend beyond their local communities and connect local need with broader institutional support. As an example, females used platforms like WhatsApp, Facebook, and Instagram to disseminate health information, offer support, and interact with local institutions like the Regional Disaster Management Agency (BPBD) and the Social Affairs Office. The use of these technology advances, along with traditional methods of cooperation like gotong royong (mutual support) and *bakudapa makanan* (food sharing programs), is indicative of a striking continuity in strategies of resilience during the conflict decades. This testifies to women's resilience, wherein they can divert their social capital and resources to address new and emergent issues brought about by the pandemic [3,5,6,7].

In addition to their public roles, household women were again on the front line of the crisis, rearranging home chores to adapt to the emergent contexts of the pandemic. They took on the double responsibility of their families' physical and emotional well-being. Women facilitated children's online learning, planned food preparation, and maintained mental health in the family home, thus leveraging their traditional domestic roles as coping strengths. As one other member demonstrated, "This pandemic causes people more easily stressed. So aside from taking care of meals and the schooling of the children, we also have to ensure the house remains calm." This is how gender roles traditionally considered domestic can be reframed as valuable resources toward keeping the household stable in a time of crisis [2,3]. Pandemic forced migration to online communication revealed underlying limits. Women were disproportionately disadvantaged by limited internet access, heightened domestic burdens, and psychological pressure from juggling multiple demands. These factors limited full transformation to transformative capacity, encompassing deeper system change. All this aside, these structural and infrastructural loopholes hindered women's capacity to fully realize their leadership potential and contribute to an increased transformation in society. According to the research, increased linking capital, e.g., increased connectivity to digital infrastructure and official recognition of the women's leadership, is key to lasting resilience in the next crisis [5].

4.3 Women's Roles in Tsunami Preparedness

Women's leadership in tsunami preparedness was another domain where they demonstrated their flexibility and leadership. Understanding the possibility of a tsunami occurring, women once more emerged as key stakeholders in community preparedness. They established national early warning systems in neighborhoods and utilized WhatsApp for quick alerts so that the community could be evacuated in case of an emergency. Females were also entrusted with spreading evacuation plans and maintaining communication channels when there was a risk of a tsunami, without any official infrastructure. Both absorptive capacity (the ability to maintain essential communication when interruption happens) and adaptive capacity (utilizing low-tech and locally appropriate tools such as WhatsApp combined with traditional community practices such as alarm bells or neighborhood watches) [5] are evidenced through these activities. Their imaginative use of available resources, even despite technological shortcomings, shows how social networks and local knowledge can overcome gaps in formal infrastructure. While these responses were significant, they also indicated some serious weaknesses in the community's ability to cope with disaster. Poor infrastructure (e.g., limited access to early warning systems and power outages) hindered the success of women's leadership [9,10]. Additionally, the invisibility of the substantial involvement of women in official disaster planning reinforced gendered exclusion in the official disaster management system. The lack of bridging funds among community programs and official disaster management chains limits the potential for transformative capacity [11]. Women's leadership at the community level, as important in ensuring basic disaster preparedness, was not effectively facilitated or mainstreamed into overall governance or disaster management schemes.

5 Conclusion

This study underscores the multidimensional nature of women's leadership in Ambon in strengthening community resilience to three distinct crises: social conflict, the COVID-19 pandemic, and tsunami threats. The findings reveal that resilience in Ambon is shaped not merely by formal policies or external assistance, but critically by women-led social capital, local knowledge, and adaptive innovation. Embedding women's roles into disaster risk governance is therefore an urgent imperative. Strengthening linking capital through strategic partnerships among communities, governmental bodies, and humanitarian agencies will be pivotal for enhancing transformative capacity and securing sustainable resilience in island contexts vulnerable to compound and recurring crises.

Acknowledgements. This research was supported by the Ministry of Education, Culture, Research, and Technology under the Penelitian Kerja Sama Dalam Negeri (PKDN) scheme. The submission fee for this article was funded by the Dana PTNBH UPT Mitigasi Bencana USK for the 2025 fiscal year.

References

1. Udasmoro, W. Women's labor force participation in insurgency and ethno-religious conflict: The cases of Aceh and Ambon. *Int. Feminist J. Polit.*, **24**(3), 395–414 (2022). https://doi.org/10.1080/14616742.2022.2084138
2. Hasanah, U. Women and the social resilience in post-conflict areas of Lampung. *Martabat: Jurnal Perempuan dan Anak*, **7**(2), 221–240 (2024). https://doi.org/10.21274/martabat.2024.7.2.221-240
3. Shirleyana, S., et al. Community resilience and women's social capital in disaster risk reduction. *Int. J. Disaster Risk Reduction*, **98**, 104090 (2023). https://doi.org/10.1016/j.ijdrr.2023.104090
4. UN Women & UN OCHA. Call for an integrated approach to strengthen Indonesia's resilience against disaster, social conflict, and radicalization. *United Nations Entity for Gender Equality and the Empowerment of Women* (2024). https://asiapacific.unwomen.org/en/stories/press-release/2024/10/call-for-an-integrated-approach-to-strengthen-indonesias-resilience-against-disaster
5. Folke, C., Carpenter, S. R., Walker, B., Scheffer, M., Chapin, T., & Rockström, J. Resilience thinking: Integrating resilience, adaptability, and transformability. *Ecol. Soc.*, **15**(4), 20 (2010). https://doi.org/10.5751/ES-03610-150420
6. Widjojo, M. S., & Nasution, M. Pela Gandong: Solidaritas sosial dan rekonsiliasi di Maluku. LIPI Press (2004).

7. Pattinama, J. Mengidentifikasi dan menyelesaikan konflik menggunakan kearifan lokal: Sebuah studi kualitatif di Maluku. *J. Int. Soc. Sci. Res.*, **12**(4), 221–235 (2024). https://doi.org/10.5296/ijssr.v12i4.33145
8. Aini, F., Maulidya, R., & Wijaya, T. Literasi digital untuk mitigasi bencana di pedesaan Indonesia. *J. Inform. Commun. Dev.*, **14**(2), 88–102 (2023). https://doi.org/10.24156/icdj.2023.14207
9. Lestari, P., & Nugraha, H. Tantangan infrastruktur di wilayah pesisir rawan bencana di Indonesia. *J. Stud. Disaster*, **11**(4), 399–412 (2024). https://doi.org/10.25012/jds.2024.11409
10. Putra, B., Handayani, S., & Rauf, A. Sistem peringatan dini berteknologi rendah untuk komunitas terpencil: Efektivitas dan keterbatasannya. *J. Int. Dis. Manag*., **6**(2), 77–91 (2023). https://doi.org/10.1080/ijdm.2023.60207
11. Syafruddin, D. Gender dan kesiapsiagaan bencana: Studi kasus dari Indonesia Timur. *J. Disast. Risk Reduct.*, **78,** 103–115 (2022). https://doi.org/10.1016/j.jdrr.2022.103115

AUTHOR INDEX